Synthetic Glycomes

Chemical Biology

Titles in the series:

How to obtain future titles on publication:
A standing order plan is available for this series. A standing order will bring delivery of each new volume immediately on publication.

For further information please contact:
Book Sales Department, Royal Society of Chemistry, Thomas Graham House, Science Park, Milton Road, Cambridge, CB4 0WF, UK
Telephone: +44 (0)1223 420066, Fax: +44 (0)1223 420247,
Email: booksales@rsc.org
Visit our website at www.rsc.org/books

Synthetic Glycomes

Edited by

Wanyi Guan
Hebei Normal University, China
Email: guanwanyi@hebtu.edu.cn

Lei Li
Georgia State University, USA
Email: lli22@gsu.edu

and

Peng George Wang
Georgia State University, USA
Email: pwang11@gsu.edu

ROYAL SOCIETY OF CHEMISTRY

Chemical Biology No. 11

Print ISBN: 978-1-78801-164-8
PDF ISBN: 978-1-78801-657-5
EPUB ISBN: 978-1-78801-777-0
Print ISSN: 2055-1975
Electronic ISSN: 2055-1983

A catalogue record for this book is available from the British Library

The Royal Society of Chemistry is a charity, registered in England and Wales, Number 207890, and a company incorporated in England by Royal Charter (Registered No. RC000524), registered office: Burlington House, Piccadilly, London W1J 0BA, UK, Telephone: +44 (0) 20 7437 8656.

For further information see our web site at www.rsc.org

Printed in the United Kingdom by CPI Group (UK) Ltd, Croydon, CR0 4YY, UK

Preface

Carbohydrates (or glycans), one of the most complex and diversified components of all living organisms, were once thought to be more for structures than for function. Tremendous advances made during the last several decades have now made it clear that glycans play critical roles in virtually every aspect of biological processes. In addition, many glycans or glycoconjugates have proved to have promising diagnostic and therapeutic potential. Functional studies, practical applications, and even analysis of these extremely complex molecules rely on the access of structurally well-defined glycoforms.

The development of routine procedures for automated chemical synthesis of oligonucleotides and peptides has revolutionized modern biology. Unfortunately, no general methods are available for the preparation of complex carbohydrates. This is majorly due to the non-template driven biosynthesis, and the multiple connectivities and branching nature of glycans. Regardless of tremendous efforts and advances made over the past three decades in chemical synthesis, enzymatic assembly, and fermentations, synthetic glycans remain relegated to small quantities and specialized laboratories. Most problematically, the synthesis of even a single complex glycan can take a well-trained researcher months to years to accomplish. Transforming strategies and approaches that streamline the synthesis of complex glycans and are sufficiently robust to permit the preparation of large collections of various types of glycans (what we name "synthetic glycomes") are in urgent demand for glycoscience to move ahead.

The book *Synthetic Glycomes* contains descriptions of the most recent advances in accessing complex glycans and glycoconjugates, including state-of-art approaches such as streamlined chemoenzymatic synthesis, reverse-synthesis (separation) and carbohydrate amplification. It also

---

Chemical Biology No. 11
Synthetic Glycomes
Edited by Wanyi Guan, Lei Li and Peng George Wang

Published by the Royal Society of Chemistry, www.rsc.org

summarizes chemical methodologies for generating glycosidic bonds, useful synthetic enzymes, and commercial availabilities of glycans as well as glycan microarrays.

The book begins with an introduction to the glycomics and developed glyco-toolboxes (Chapter 1), as well as an overall summary of chemical methodologies for generating various glycosidic bonds (Chapter 2) and a summary of most synthetic-useful glycosyltransferases applied in the preparation of mammalian glycans (Chapter 3). The following eight chapters describe the most advanced strategies and technologies in streamlining the preparation of various mammalian glycans and glycoconjugates, including *N*-glycans (Chapters 4 and 5), *O*-mannosyl glycans (Chapter 6), glycopeptides and glycoproteins (Chapter 7), glycosaminoglycans (Chapters 8 and 9), glycosphingolipids (Chapter 10), and human milk oligosaccharides (Chapter 11). Chapter 12 describes the synthesis of marine polysaccharides/oligosaccharides and their derivatives. The next four chapters present some of the most advanced methodologies in preparing glycans other than traditional chemical and enzymatic synthesis, including bacterial fermentation (Chapter 13), solid-phase synthesis (Chapter 14), reverse synthesis (Chapter 15) and *O*-glycomics amplification (Chapter 16). The last section of the book contains chapters summarizing commercially available complex carbohydrates (Chapter 17) and glycan microarrays (Chapter 19), as well as the application of glycan microarrays in unmasking glycan functions (Chapter 18).

We would like to sincerely thank all authors who contributed to individual chapters. Without their significant efforts, the book would not have been possible. We also acknowledge all chapter reviewers, and all staff members of the Royal Society of Chemistry who helped us with the publication of this book.

Wanyi Guan
Lei Li
Peng George Wang

Contents

Chemical Biology No. 11
Synthetic Glycomes
Edited by Wanyi Guan, Lei Li and Peng George Wang

Published by the Royal Society of Chemistry, www.rsc.org

CHAPTER 1

Introduction: Glycome and the Glyco-toolbox

SHUAISHUAI WANG[a], GARRETT A. EDMUNDS[a], LEI LI[a], CONGCONG CHEN[a,b,c] AND PENG G. WANG*[a]

[a]Department of Chemistry, Georgia State University, Atlanta, GA 30303, USA; [b]National Glycoengineering Research Center, Shandong University, Jinan, Shandong 250012, China; [c]School of Pharmaceutical Science, Shandong University, Jinan, Shandong 250012, China
*E-mail: pwang11@gsu.edu

1.1 Introduction

Biologists have broken down the different aspects in the study of life into various "-omics". While the exact boundaries are increasingly ill-defined and nebulous, an "-omic" can generally be defined as the study of the structures, functions, and – most importantly – the roles of the biomolecules that make up the "-ome" of interest. Genomics is the study of the complete genome of a living thing, encompassing DNA, its structure, and how the information contained within the genetic code is encoded, modified, and passed on. Similarly, proteomics is the study of the entirety of an organism's proteome and their interactions.

Building off these definitions, glycomics is the study of the glycome – and its interactions – present in an organism. It is a study not just of individual components, but of the entirety of glycans and how they interact. Glycomics is an underdeveloped field compared with the sister fields of genomics and proteomics. This discrepancy is largely due to the time these

Chemical Biology No. 11
Synthetic Glycomes
Edited by Wanyi Guan, Lei Li and Peng George Wang

Published by the Royal Society of Chemistry, www.rsc.org

fields have had to mature, the general interest by researchers and funding agencies, and the ease of manipulating the biomolecules of interest. More recently, as the roles of glycans have been more understood in biological systems, developing the field of glycomics has been increasingly prioritized by researchers. Key to this progress is the development of the synthetic glycome.

With the development of new technologies, the biological functions of carbohydrates were found to span the spectrum from development, growth, maintenance, and survival of the organism. For example, studies into clinical applicability arose early with the discovery of the human blood groups, with studies showing evidence that some blood antigens were glycans, such as ABO blood type and P1Pk type. The ABO blood group is determined by the glycan information on the surface of a red blood cell. Different modifications of the core glycan structure, the H antigen, are responsible for the different ABO blood groups. When the H antigen is present on a red blood cell's surface, anti-A and anti-B antibodies will be generated, meaning only type O can be accepted when transferring the blood (Table 1.1). On the other hand, a person that displays A and B antigens – blood type AB – without the anti-A and anti-B antibodies, can accept blood from the other three blood types. Of course, considerations should be made for other kinds of blood group systems, such as the Rh system.

As each "-omic" field matures and progresses, success and progress in research have largely depended on three ideas: access to the desired biomolecules, accurate means for analysis of those biomolecules, and the ability to manipulate and direct these structures for desirable applications. As more tools for various -omic toolboxes are developed for access, analysis, and application, foundational understandings give way to more complete and developed concepts and designs. In order to better understand these concepts, we must first understand the large diversity within the glycome and how it impacts synthetic efforts.

Table 1.1 ABO blood types and carbohydrate antigens. Symbols and abbreviations: ■ *N*-acetylgalactosamine (GalNAc), ● galactose (Gal), ■ *N*-acetylglucosamine (GlcNAc), ▲ fucose (Fuc).

Blood type	Type O	Type A	Type B	Type AB
Antigen	H antigen 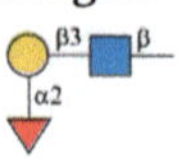	A antigen 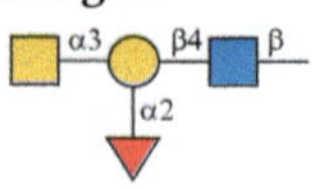	B antigen 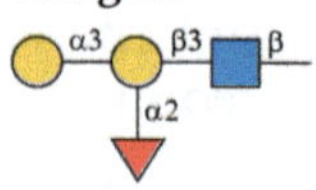	A antigen and B antigen
Antibodies in plasma	Anti-A and Anti-B	Anti-B	Anti-A	None

1.2 Diversity of Glycans

One of the major stumbling blocks in the development of glycomics is the sheer size of the theoretical glycome. This is in large part due to the diversity seen within the glycome. Across all walks of life, the diversity of monosaccharides used as building blocks for glycan sequences is significantly larger than that for the residues in proteins or nucleic acids. Nucleic acids are composed of only four bases, A, T, G, and C for DNA and A, U, G, and C for RNA. Proteins are composed of just 20 amino acids. Both systems have one point for binding to an acceptor and another for receiving the next unit, making these molecules linear in nature. For glycans, there are more than 30 monosaccharides that exist in mammals with a multitude of potential branching points. The nine most common are displayed in Figure 1.1.

Many bacterial glycans show a wide diversity in structures and often feature sugars not found elsewhere in nature. It is widely known that the diversity of carbohydrates is much greater in bacteria than in mammals. Compared with mammalian glycans, the bacterial glycans have a more than ten-fold greater diversity at the monosaccharide level.[1]

The diversity of biopolymers was not only influenced by building blocks, but also the linkages between them. Even we only count nine common monosaccharides in mammals, due to numerous linkages, glycans have a more vast diversity than nucleic acids and proteins. Nucleic acids and proteins are both linear in style. Nucleotides were linked by a phosphodiester

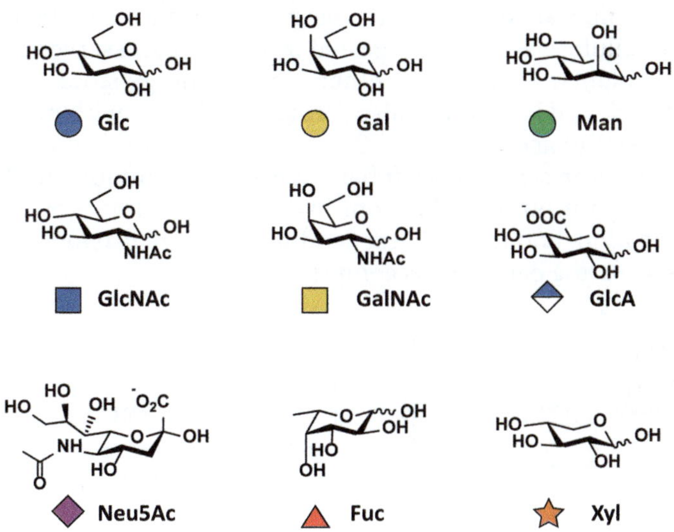

Figure 1.1 Nine common monosaccharides in mammalian cells.

bond that was formed by the 3′ carbon atom of one sugar molecule of nucleotide and the 5′ carbon atom of another nucleotide; proteins were linked with a peptide bond and the bond was condensed by the carboxyl group of one amino acid molecule with the amino group of another amino acid molecule.

Comparing with nucleic acids and proteins, glycans have flexible linkages and branch structures because of the diverse glycosidic bonds. The glycosidic bond is formed between the anomeric carbon of one monosaccharide and a hydroxyl group of another. Firstly, based on the relative stereochemistry of the anomeric carbon and the stereo-center furthest from C1 carbon, the glycosidic bond has α/β distinction. An α-glycosidic bond is formed when both carbons have the same stereochemistry, whereas a β-glycosidic bond occurs when the two carbons have different stereochemistry. Secondly, unlike nucleotides and amino acids, which only have one possible linkage between two building blocks, monosaccharides have various forms of linkage.

For example, if only one kind of nucleotide or one amino acid is used to construct nucleic acid and protein, only one possible nucleic acid or protein can be formed. However, for glycans, due to the presence of multiple hydroxyl groups, there are numerous possible structures. For instance, both starch and cellulose are glucose polymers, with the major difference between them being the glucose linkage (Figure 1.2). Most linkages in starch are Glcα1–4Glc, while cellulose is only defined by the Glcβ1–4Glc linkage. This subtle difference of α- and β- linkage between starch and cellulose gives them disparate characteristics. Starch can be digested by humans into glucose to serve as an energy source and as metabolic precursors, while cellulose cannot be digested by humans and serves as structural support for plants. Glycogen, another kind of energy storage in humans, animals, fungi, and bacteria, is also formed by glucose with multibranched α1–4 and α1–6 linkages.

Rarely are glycan structures left bare. It is not uncommon for oligosaccharides to display various modifications in nature, including acetylation, sulfation, methylation, and more. These modifications are usually included after the oligosaccharide backbone is completed.

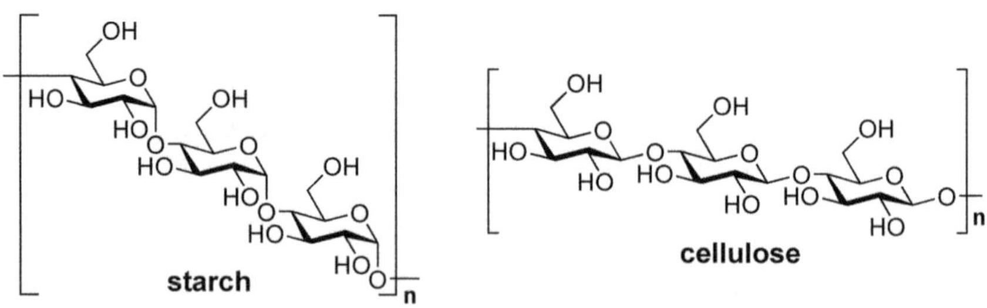

Figure 1.2 Structure of starch and cellulose. Starch was constituted by Glcα1–4Glc while cellulose was formed by Glcβ1–4Glc.

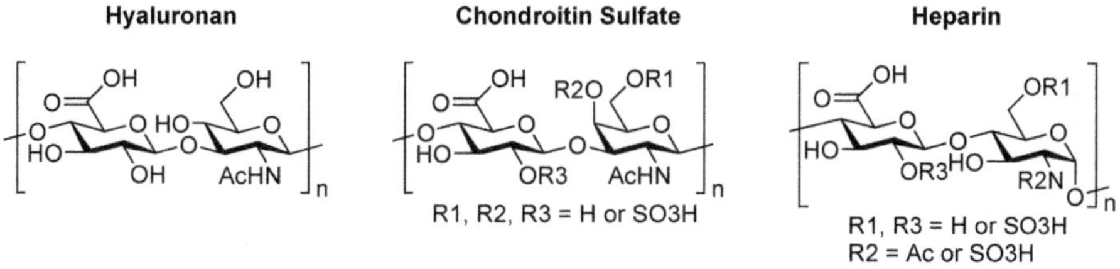

Figure 1.3 Structure of glycosaminoglycans.

As an example, glycosaminoglycans (GAGs) are a class of polysaccharides known for their diversity in size and modifications (Figure 1.3). They are often highly sulfated with sulfate groups in various positions and configurations. The exact locations of these sulfations vary from one GAG family to the next and are not consistently repeated throughout an individual biomolecule. The hexosamine unit of a GAG can be either acetylated or non-acetylated, thus the variety in a single biomolecule grows immensely.

1.3 Limited Glycan Backbone Structures

Unlike nucleotides and peptides, glycans are not derived from a template. Instead, glycosyltransferases and other enzymes are the mediators that determine order and connectivity of glycans. The most predominant class of glycosyltransferases are known as Leloir glycosyltransferases, named after the Argentinian Scientist who received the 1970 Nobel Prize in Chemistry for studying their biochemical pathway. These glycosyltransferases use sugar-nucleotides as donors to transfer a monosaccharide onto an acceptor and are precise in their regio- and stereo-selectivity. Normally, each enzyme is responsible for one specific glycosylation with little promiscuity, meaning a monosaccharide is bound to a specific acceptor at a specific location with a specific regiochemistry. This has led to the common adage of "One enzyme-one linkage". A one-enzyme-class/one-saccharide-linkage paradigm applies for almost all individual steps of glycan biosynthesis.[2] The possible glycan linkages were shown in Table 1.2.

With the large number of sugars and the variety of possible connections between them, the theoretical glycan structures in mammalian glycome are nearly unlimited in size. However, the natural glycome is limited by the number of glycosyltransferases that nature has made available. There are 236 glycosyltransferases that humans use to create the entire glycome in the human body. This leads to a much more manageable 10 000–20 000 glycans that make up the natural glycome.[35] To make matters simpler, researchers can instead focus on the ~7000 glycans that comprise 90% of the meaningful human glycome.[36] Some signature glycan epitopes are shown in Figures 1.4–1.11.

Table 1.2 Glycan linkages in mammalian glycome. Adapted from ref. 2 with permission from Elsevier, Copyright 2006.

	Glc	Gal	Man	GlcNAc	GalNAc	Fuc	GlcA	Neu5Ac	Xyl
UDP-Glc	α1–2 [3] α1–3 [4] α1–4 [5] α1–6 [6]	α1–2 [7]	α1–3 [4]	—	—	β1–3 [8]	—	—	—
UDP-Gal	β1–4 [9]	α1–3 [10] α1–4 [11] β1–3 [12]	—	β1–3 [13] β1–4 [14]	β1–3 [15]	—	—	—	β1–4 [12]
GDP-Man	—	—	α1–2 [16] α1–3 [16] α1–6 [16]	α1–4 [17] β1–4 [16]	—	—	—	—	—
UDP-GlcNAc	—	β1–3 [9] β1–6 [18]	β1–2 [19] β1–4 [19] β1–6 [19]	β1–4 [20]	β1–3 [15] β1–6 [15]	β1–3 [21]	α1–4 [22] β1–4 [23]	—	—
UDP-GalNAc	—	α1–3 [24] β1–3 [25] β1–4 [26]	—	β1–3 [27] β1–4 [28]	α1–3 [15] α1–6 [15]	—	β1–4 [22]	—	—
GDP-Fuc	—	α1–2 [29]	—	α1–3 [30] α1–4 [29] α1–6 [31]	—	—	—	—	—
UDP-GlcA	—	β1–3 [12] β1–4 [19]	—	β1–3 [23] β1–4 [22]	β1–3 [22]	—	—	—	β1–3 [19] β1–4 [19]
CMP-Neu5Ac	—	α2–3 [32] α2–6 [32]	—	—	α2–6 [15]	—	—	α2–8 [33]	—
UDP-Xyl	α1–3 [34]	—	—	—	—	—	α1–3 [19]	—	α1–3 [34]

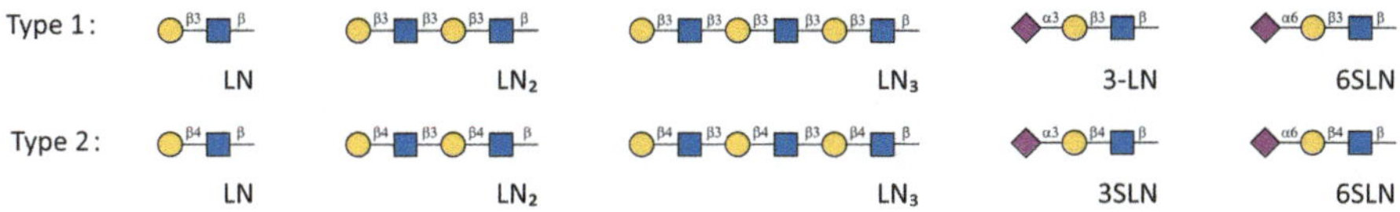

Figure 1.4 Prominent glycan epitopes in mammalian glycans: LacNAc series.

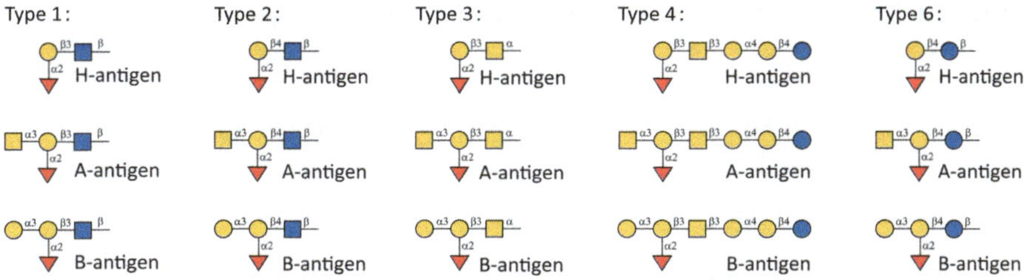

Figure 1.5 Prominent glycan epitopes in mammalian glycans: ABO blood type series.

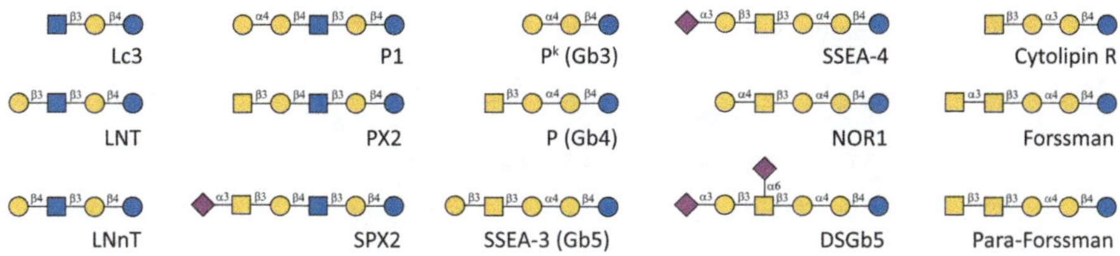

Figure 1.6 Prominent glycan epitopes in mammalian glycans: Globo series.

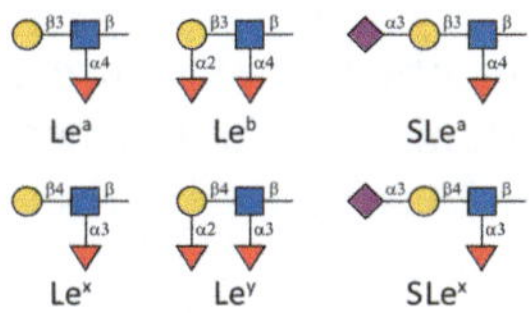

Figure 1.7 Prominent glycan epitopes in mammalian glycans: Lewis series.

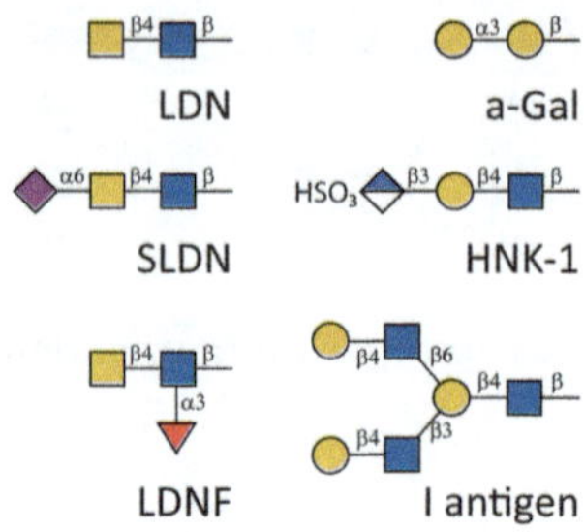

Figure 1.8 Prominent glycan epitopes in mammalian glycans.

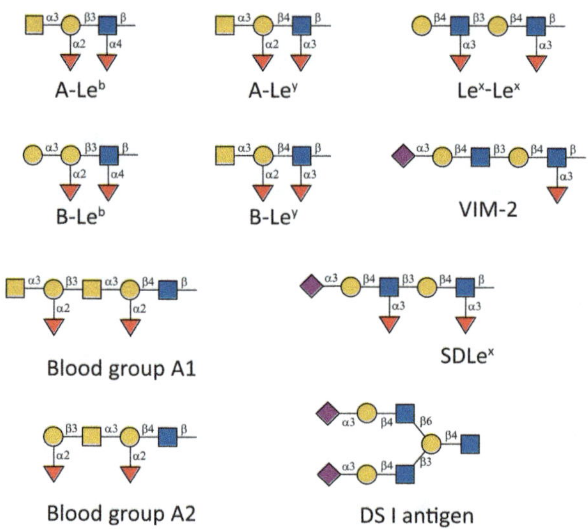

Figure 1.9 Prominent glycan epitopes in mammalian glycans: Tandem Epitope series.

1.4 Access

The tools within the "glyco-toolbox" are still underdeveloped when compared with other "-omics". In particular, tools to access large numbers of complex glycans of interest are largely lacking. In contrast, virtually all known peptide sequences can be synthesized *via* solid-phase peptide synthesis. This technology was pioneered by Merrifield in 1963 and has since developed into a mature field, such that researchers can access a specific peptide by pushing a button on a table-top peptide synthesizer.[37] For the synthesis of nucleotide, as the advance of genomics, the price for gene synthesis is as low as 0.07 US dollars per base pair, which makes it feasible to access any genes of interest.

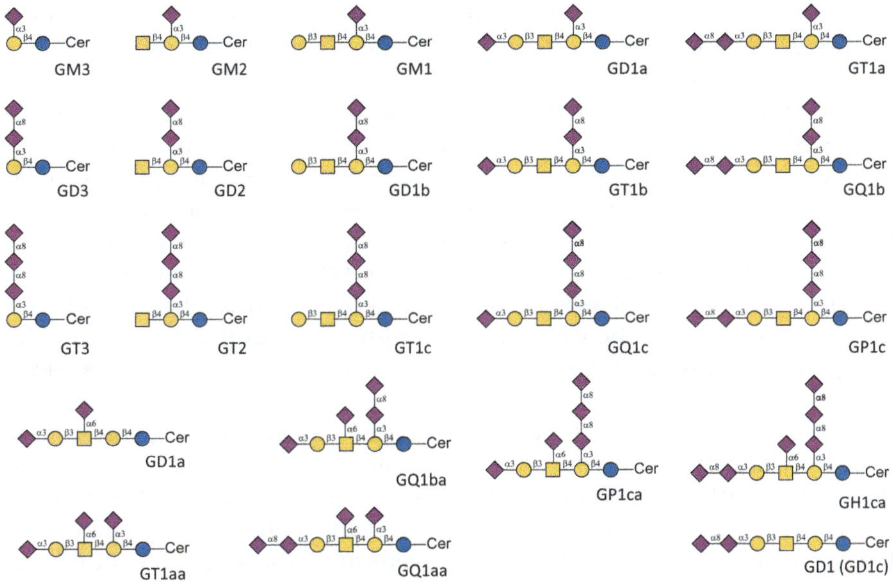

Figure 1.10 Prominent glycan epitopes in mammalian glycans: Gangliosides.

Figure 1.11 Prominent glycan epitopes in mammalian glycans: *O*-GalNAc Glycan.

Furthermore, isolating and purifying glycans from natural sources is prohibitively inefficient, due to their inherent natural complexity and diversity. To access complex glycans and glycoconjugates, numbers of chemical methodologies and enzymatic strategies have been developed, which are usually time-consuming and require years of training and experience in the specific field.

Recently, a number of chemoenzymatic approaches have been reported to access large numbers of complex glycans.[38–41] For example, a systematic and efficient strategy of producing diverse asymmetrically branched *N*-glycans was developed by Boons' group.[40] In addition, an efficient chemoenzymatic strategy, Core Synthesis/Enzymatic Extension (CSEE), was developed by the Wang Lab for rapid production of various asymmetric *N*-glycan isomers,[41] in which eight core structures were firstly synthesized chemically then enzymatically extended to 73 complex *N*-glycans, including 63 isomers. Additionally, this CSEE strategy was used for the assembly of various classes of mammalian glycans, such as human milk oligosaccharides,[42] and tandem epitope *N*-glycans.[43] Most recently, an *O*-mannosyl glycans library was prepared by adapting this strategy.[44]

Programmed and/or automatic chemical synthesis of glycans were also proposed and developed to solve the unavailability of glycans, especially for non-specialists. Two decades ago, Wong developed a computer program OptiMer, a database search tool and guide for the selection of building blocks for the one-pot assembly of a desired oligosaccharide or a library of individual oligosaccharides. OptiMer evaluates a glycan target and suggests a method for a one-pot synthesis based on relative reactivity values.[45] In addition, Seeberger has developed the Glyconeer, a commercially available machine that can perform automatic chemical synthesis of glycans from a large set of chemical building blocks.[46]

Alternatively, enzymes have been introduced into automatic glycan synthesis, which can efficiently control the reactions in a stereo- and regio-selective manner. In 2010, Nishimura and coworkers developed an artificial Golgi apparatus, which contains several recombinant mammalian glycosyltransferases in solutions. It enables the synthesis of simple oligosaccharide (*e.g.*, sialyl Lewis X) in a fully automatable manner.[47] Most recently, we have developed a machine-driven fully-automated system for oligosaccharides synthesis through enzymatic glycosylation.[48] By employing a thermosensitive polymer and a commercially available peptide synthesizer, the glycosylation reactions take place in an aqueous environment controlled by the peptide synthesizer. The growing oligosaccharide chain was purified by the precipitation of the covalently attached polymer, and finally released *via* a specific chemistry to afford final products.

1.5 Application

It is important to note that as glycan synthetic methodologies are further developed, new inroads towards applications will open, in much the same way that therapies targeting genes and peptides have matured over the past few decades. Because of the roadblocks and hindrances naturally present in the study of glycomics, their applications are necessarily delayed. However, that has not stopped researchers from developing novel diagnostic and therapeutic methodologies.

Glycans play important roles in mature protein expression, adhesion, and cell signaling, and can even be indicators of diseases. For example, certain cancers express modified oligosaccharides on the cell surface.[49] It has not escaped the attention of researchers that these near ubiquitous structures could be key markers and handles for diagnostic and therapeutic purposes. For example, P-selectin Glycoprotein Ligand 1 (PSGL-1) had been implicated in the recruitment of leukocytes to areas of inflammation and may aid cancer cells metastasize.[50] A therapeutic drug based on this glycan structure may aid in preventing cancer metastasis, and as such has been the subject of intense research.[51–53] As another example, Globo-H, an oligosaccharide that is overexpressed in many cancer lines, has been suggested to be a potential cancer vaccine.[54,55]

To illustrate the interaction between glycans and glycan binding proteins (GBPs), glycan microarrays were developed.[56,57] Glycans are immobilized or "printed" onto a surface in a regular array. These arrays can then be exposed to various GBPs that may have affinities for the epitopes that these glycans display. Most recently, a new multiplex glycan bead array (MGBA), which allows simultaneous analyses of 384 samples and up to 500 glycans in a single assay, was developed.[58] This technology makes high-throughput screening for binding affinity of glycans with GBPs easier.

1.6 Conclusion

Glycomics, relatively speaking, is late to the game. Genomics and proteomics have the benefit of decades of intense research and extensive funding. On the other hand, glycomics does have a few advantages. The groundwork that other biologists have made can be readily applied to the study of the glycome. New research into the structure and behavior of the cells reveals the significances of glycans in biological systems, triggering greater interest in this field from industry and academia alike. The National Institute of Health, USA, has set-up a Glycoscience Common Fund to facilitate the development of new glycomics tools. The field of glycomics is rapidly catching up with its sister fields, and it is only a matter of time before non-specialists will have the same access to the Synthetic Glycome as they do to the synthetic genome and proteome.

References

1. G. Daniels, *Human Blood Groups*, John Wiley & Sons, NJ, 2013.
2. S. K. Ohtsubo and J. D. Marth, *Cell*, 2006, **126**, 855.
3. E. Alvarado, L. Ballou, L. M. Hernandez and C. E. Ballou, *Biochemistry*, 1990, **29**, 2471.
4. C. M. De Praeter, G. J. Gerwig, E. Bause, L. K. Nuytinck, J. F. Vliegenthart, W. Breuer, J. P. Kamerling, M. F. Espeel, J. J. Martin, A. M. De Paepe, N. W. Chan, G. A. Dacremont and R. N. Van Coster, *Am. J. Hum. Genet.*, 2000, **66**, 1744.
5. L. Leickt, A. Grubb and S. Ohlson, *J. Mol. Recognit.*, 1998, **11**, 114.
6. B. L. Slomiany, A. Slomiany and G. B. Glass, *Eur. J. Biochem.*, 1978, **84**, 53.
7. E. F. Hounsell, *Nucl. Magn. Reson.*, 2009, **38**, 346.
8. K. Kozma, J. J. Keusch, B. Hegemann, K. B. Luther, D. Klein, D. Hess, R. S. Haltiwanger and J. Hofsteenge, *J. Biol. Chem.*, 2006, **281**, 36742.

9. H. C. Krivan, L. D. Olson, M. F. Barile, V. Ginsburg and D. D. Roberts, *J. Biol. Chem.*, 1989, **264**, 9283.
10. H. Smit, W. Gaastra, J. P. Kamerling, J. F. Vliegenthart and F. K. de Graaf, *Infect. Immun.*, 1984, **46**, 578.
11. A. Karlsson, M. Markfjall, H. Lundqvist, N. Stromberg and C. Dahlgren, *Anal. Biochem.*, 1995, **224**, 390.
12. F. Barry, *Biochem. Soc. Trans.*, 1990, **18**, 197.
13. C. Bavington and C. Page, *Respiration*, 2005, **72**, 335.
14. B. L. Gillece-Castro, A. Prakobphol, A. L. Burlingame, H. Leffler and S. J. Fisher, *J. Biol. Chem.*, 1991, **266**, 17358.
15. J. Chia, G. Goh and F. Bard, *Biochim. Biophys. Acta*, 2016, **1860**, 1623.
16. R. M. Schmaltz, S. R. Hanson and C. H. Wong, *Chem. Rev.*, 2011, **111**, 4259.
17. N. Firon, I. Ofek and N. Sharon, *Carbohydr. Res.*, 1983, **120**, 235.
18. Y. Suzuki, *Prog. Lipid Res.*, 1994, **33**, 429.
19. H. Manya and T. Endo, *Biochim. Biophys. Acta, Gen. Subj.*, 2017, **1861**, 2462.
20. P. A. Grange, M. A. Mouricout, S. B. Levery, D. H. Francis and A. K. Erickson, *Infect. Immun.*, 2002, **70**, 2336.
21. D. J. Moloney, A. I. Lin and R. S. Haltiwanger, *J. Biol. Chem.*, 1997, **272**, 19046.
22. S. Yamada, K. Sugahara and S. Özbek, *Commun. Integr. Biol.*, 2011, **4**, 150.
23. S. Li, S. Wang, X. Fu, X. W. Liu, P. G. Wang and J. Fang, *Carbohydr. Polym.*, 2017, **178**, 221.
24. A. Coddens, E. Valis, J. Benktander, J. Angstrom, M. E. Breimer, E. Cox and S. Teneberg, *PLoS One*, 2011, **6**, e23309.
25. L. L. Cooling, T. A. Koerner and S. J. Naides, *J. Infect. Dis.*, 1995, **172**, 1198.
26. A. H. Gori, K. Ahmed, G. Martinez, H. Masaki, K. Watanabe and T. Nagatake, *Am. J. Trop. Med. Hyg.*, 1999, **61**, 473.
27. M. O. Sheikh, S. M. Halmo and L. Wells, *Glycobiology*, 2017, **27**, 806.
28. K. Y. Do, S. I. Do and R. D. Cummings, *Glycobiology*, 1997, **7**, 183.
29. M. Aspholm-Hurtig, G. Dailide, M. Lahmann, A. Kalia, D. Ilver, N. Roche, S. Vikstrom, R. Sjostrom, S. Linden, A. Backstrom, C. Lundberg, A. Arnqvist, J. Mahdavi, U. J. Nilsson, B. Velapatino, R. H. Gilman, M. Gerhard, T. Alarcon, M. Lopez-Brea, T. Nakazawa, J. G. Fox, P. Correa, M. G. Dominguez-Bello, G. I. Perez-Perez, M. J. Blaser, S. Normark, I. Carlstedt, S. Oscarson, S. Teneberg, D. E. Berg and T. Boren, *Science*, 2004, **305**, 519.
30. P. Huang, T. Farkas, W. Zhong, M. Tan, S. Thornton, A. L. Morrow and X. Jiang, *J. Virol.*, 2005, **79**, 6714.
31. A. D. Calderon, Y. Liu, X. Li, X. Wang, X. Chen, L. Li and P. G. Wang, *Org. Biomol. Chem.*, 2016, **14**, 4027.
32. J. A. Garnett, Y. Liu, E. Leon, S. A. Allman, N. Friedrich, S. Saouros, S. Curry, D. Soldati-Favre, B. G. Davis, T. Feizi and S. Matthews, *Protein Sci.*, 2009, **18**, 1935.

33. S. Jayaraman, S. Eswaramoorthy, D. Kumaran and S. Swaminathan, *Proteins*, 2005, **61**, 288.
34. G. E. Whitworth, W. F. Zandberg, T. Clark and D. J. Vocadlo, *Glycobiology*, 2010, **20**, 287.
35. N. R. Council, *Transforming Glycoscience: a Roadmap for the Future*, National Academies Press, 2012.
36. R. D. Cummings, *Mol. BioSyst.*, 2009, **5**, 1087.
37. R. B. Merrifield, *J. Am. Chem. Soc.*, 1963, **85**, 2149.
38. O. Blixt and N. Razi, *Methods Enzymol.*, 2006, **415**, 137–153.
39. B. S. Hamilton, J. D. Wilson, M. A. Shumakovich, A. C. Fisher, J. C. Brooks, A. Pontes, R. Naran, C. Heiss, C. Gao, R. Kardish, J. Heimburg-Molinaro, P. Azadi, R. D. Cummings, J. H. Merritt and M. P. DeLisa, *Sci. Rep.*, 2017, **7**, 15907.
40. Z. Wang, Z. S. Chinoy, S. G. Ambre, W. Peng, R. McBride, R. P. de Vries, J. Glushka, J. C. Paulson and G. J. Boons, *Science*, 2013, **341**, 379.
41. L. Li, Y. Liu, C. Ma, J. Qu, A. D. Calderon, B. Wu, N. Wei, X. Wang, Y. Guo, Z. Xiao, J. Song, G. Sugiarto, Y. Li, H. Yu, X. Chen and P. G. Wang, *Chem. Sci.*, 2015, **6**, 5652.
42. Z. Xiao, Y. Guo, Y. Liu, L. Li, Q. Zhang, L. Wen, X. Wang, S. M. Kondengaden, Z. Wu, J. Zhou, X. Cao, X. Li, C. Ma and P. G. Wang, *J. Org. Chem.*, 2016, **81**, 5851.
43. Z. Wu, Y. Liu, C. Ma, L. Li, J. Bai, L. Byrd-Leotis, Y. Lasanajak, Y. Guo, L. Wen, H. Zhu, J. Song, Y. Li, D. A. Steinhauer, D. F. Smith, B. Zhao, X. Chen, W. Guan and P. G. Wang, *Org. Biomol. Chem.*, 2016, **14**, 11106.
44. S. Wang, Q. Zhang, C. Chen, Y. Guo, M. R. Gadi, J. Yu, U. Westerlind, Y. Liu, X. Cao, P. G. Wang and L. Li, *Angew. Chem., Int. Ed. Engl.*, 2018, **57**, 9268.
45. Z. Zhang, I. R. Ollmann, X.-S. Ye, R. Wischnat, T. Baasov and C.-H. Wong, *J. Am. Chem. Soc.*, 1999, **121**, 734.
46. H. S. Hahm, M. K. Schlegel, M. Hurevich, S. Eller, F. Schuhmacher, J. Hofmann, K. Pagel and P. H. Seeberger, *Proc. Natl. Acad. Sci. U. S. A.*, 2017, **114**, E3385.
47. T. Matsushita, I. Nagashima, M. Fumoto, T. Ohta, K. Yamada, H. Shimizu, H. Hinou, K. Naruchi, T. Ito, H. Kondo and S.-I. Nishimura, *J. Am. Chem. Soc.*, 2010, **132**, 16651.
48. J. Zhang, C. Chen, M. R. Gadi, C. Gibbons, Y. Guo, X. Cao, G. Edmunds, S. Wang, D. Liu, J. Yu, L. Wen and P. G. Wang, *Angew. Chem., Int. Ed.*, 2018, **57**, 16638–16642.
49. D. H. Dube and C. R. Bertozzi, *Nat. Rev. Drug Discovery*, 2005, **4**, 477.
50. R. P. McEver, *Glycoconjugate J.*, 1997, **14**, 585.
51. V. R. Krishnamurthy, A. Dougherty, M. Kamat, X. Z. Song, R. D. Cummings and E. L. Chaikof, *Carbohydr. Res.*, 2010, **345**, 1541.
52. M. Pudelko, J. Bull and H. Kunz, *ChemBioChem*, 2010, **11**, 904.
53. V. R. Krishnamurthy, M. Y. R. Sardar, Y. Ying, X. Z. Song, C. Haller, E. B. Dai, X. C. Wang, D. Hanjaya-Putra, L. J. Sun, V. Morikis, S. I. Simon, R. J. Woods, R. D. Cummings and E. L. Chaikof, *Nat. Commun.*, 2015, **6**, 6387.

54. S. F. Slovin, G. Ragupathi, S. Adluri, G. Ungers, K. Terry, S. Kim, M. Spassova, W. G. Bornmann, M. Fazzari, L. Dantis, K. Olkiewicz, K. O. Lloyd, P. O. Livingston, S. J. Danishefsky and H. I. Scher, *Proc. Natl. Acad. Sci. U. S. A.*, 1999, **96**, 5710.
55. S. J. Danishefsky and J. R. Allen, *Angew. Chem., Int. Ed.*, 2000, **39**, 836.
56. S. Angeloni, J. L. Ridet, N. Kusy, H. Gao, F. Crevoisier, S. Guinchard, S. Kochhar, H. Sigrist and N. Sprenger, *Glycobiology*, 2005, **15**, 31.
57. C. Y. Huang, D. A. Thayer, A. Y. Chang, M. D. Best, J. Hoffmann, S. Head and C. H. Wong, *Proc. Natl. Acad. Sci. U. S. A.*, 2006, **103**, 15.
58. S. Purohit, T. Li, W. Guan, X. Song, J. Song, Y. Tian, L. Li, A. Sharma, B. Dun, D. Mysona, S. Ghamande, B. Rungruang, R. D. Cummings, P. G. Wang and J. X. She, *Nat. Commun.*, 2018, **9**, 258.

CHAPTER 2

Methodologies in Chemical Syntheses of Carbohydrates

WEI-LIN LENG[†a], JING-XI HE[†a,b], HUI YAO[†a] AND XUE-WEI LIU[*a]

[a]Division of Chemistry and Biological Chemistry Nanyang Technological University, School of Physical & Mathematical Sciences, 21 Nanyang Link, Singapore 637371; [b]School of Chemical and Biomedical Engineering, Nanyang Technological University, 62 Nanyang Drive, Singapore 637459
*E-mail: xuewei@ntu.edu.sg

2.1 Introduction

Carbohydrates are ubiquitous in nature and have important roles in numerous biological pathways, ranging from simple glucose as vital energy source, to complex polysaccharides like a 92-mer arabinogalactan.[1–3] Although the study of the structure–activity relationship of these glycans and their conjugates is desirable, nature often presents them in heterogeneous forms. Hence, chemical syntheses of carbohydrates is required to provide access to homogeneous forms of these glycans. However, the chemical assembly of these complex molecules requires deliberate design with several considerations. This chapter describes the factors for such considerations starting from the availability and protecting groups of building blocks, to reactivity and stereoselectivity during glycosylation. The final assembly of these building blocks into complex molecules will also be covered briefly.

[†]These authors contributed equally to this work.

Chemical Biology No. 11
Synthetic Glycomes
Edited by Wanyi Guan, Lei Li and Peng George Wang

Published by the Royal Society of Chemistry, www.rsc.org

2.2 Before Glycosylation

Prior to the glycosylation step, preparative steps to access the carbohydrate starting materials should be carefully devised. From a practical viewpoint, the preference of chiral monosaccharides as building blocks for complex sugar is apparent because of their abundance in nature, such as glucose, mannose, and galactose. Due to the presence of multiple hydroxyl groups of relatively similar reactivity, judicious design of orthogonal protecting groups for the monosaccharides has to be first considered to allow selective deprotection and formation of specific glycosidic bonds in the construction of complex sugars. Besides the readily available monosaccharides, *de novo* syntheses from non-sugar starting materials also allows access to rare sugars. Such considerations are important as they will result in downstream impact on the reactivity and stereoselectivity during glycosylation.

2.2.1 Protecting Groups

As unprotected carbohydrates are insoluble in many common organic solvents, carbohydrate chemistry typically involves the protected forms during functionalisations before the final deprotection step.[4] In addition, the polyhydroxyl groups lead to the challenge in selective chemical derivatization. Hence, a typical protocol for oligosaccharide synthesis usually involves multi-step protection and deprotection, through acylation, alkylation and silylation, blocking the functionalities and only releasing them when required. Key advances in protecting strategies over the last decade, in terms of regioselectivity, orthogonality, conciseness and tracelessness, will be discussed here.

Since oligosaccharide synthesis usually involves regioselective glycosylations at varied positions, protecting groups installed have to be orthogonal to each other, to achieve selective deprotection and functionalisations later on. Unfortunately, increased types of protecting groups also mean increased efforts in their installation and purification. This major limitation was overcome by Hung and Ye in 2007.[5] In their work on one-pot protection of carbohydrates, installation of protecting groups proved to be highly regioselective and multi-step transformations could be conducted sequentially without isolation to yield 2-, 3-, 4- and 6-mono-unprotected carbohydrates (Scheme 2.1). It is worth noting that reaction conditions and sequences of installation were carefully designed to prevent cleavage of protecting groups in subsequent steps and avoid retarding the reactivity of other hydroxyl groups.

Other groups also studied the regioselective mono-protection of carbohydrates extensively. In an attempt to eliminate the toxic stannylated species commonly used to achieve regioselectivity, Taylor reported a milder organoboron-catalyzed monoalkylation of carbohydrates.[6] Conceptually similar to stannylene acetal and stannyl ether, with utilization of the tetracoordinating ability of boron, borinate complex was utilized to achieve regioselective installation of a series of alkyl groups, such as benzyl, naphthylmethyl,

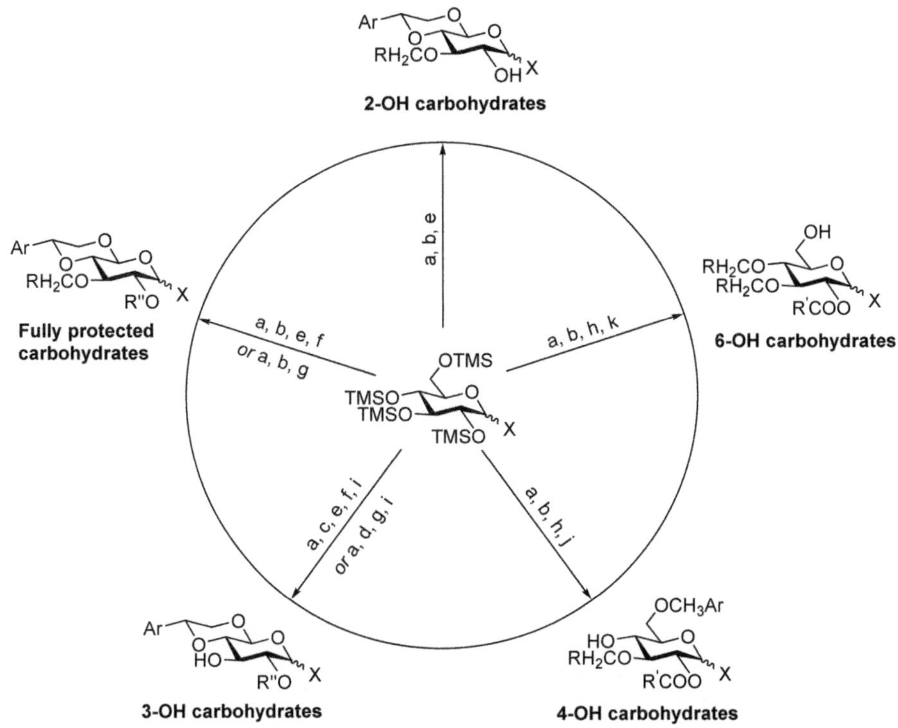

Scheme 2.1 Regioselective one-pot protection of carbohydrates: (a) cat. TMSOTf, ArCHO; (b) RCHO, Et_3SiH; (c) 4-OMePhCHO, Et_3SiH; (d) 2-$C_{10}H_7CHO$, Et_3SiH; (e) TBAF; (f) base, electrophile; (g) acid anhydride; (h) $(R'CO)_2O$; (i) DDQ; (j) HCl, $NaCNBH_3$; (k) BH_3/THF.

4-bromobenzyl and benzyloxymethyl groups. While the methodology is general for hydroxyl groups at various positions, it does not apply to sugars without *cis*-diols. In this regard, Dong reported the use of a $[Fe(dibm)_3]$ catalyst to conduct regioselective mono-protection with a wider scope of sugar substrates, both *cis*- and *trans*-diols of carbohydrates through a proposed cyclic dioxolane/dioxane-type intermediate (Scheme 2.2).[7] In addition, catalytic site-selective acylation was recently reported by Liu and Tang, in which (*R*)- and (*S*)-benzotetramisole could be applied to *trans*-1,2-diols to obtain C3 and C2-acylated products regioselectively, driven by stabilizing cation-*n* interaction (Scheme 2.3).[8]

The abovementioned approaches can be utilized in typical glycosylation protocol, with pre-formed protected sugars as the substrates which can be deprotected later for further derivatizations. While excellent regioselectivity can be achieved, it is often a strenuous and mundane process, particularly when repeating it for oligosaccharide synthesis. Hence, an ideal solution would be to conduct it under protecting-group(s)-free conditions. Most of these reactions are only able to be demonstrated on *N*-glycosides,[9] such as Fairbanks's protecting-group-free glycoconjugate synthesis using

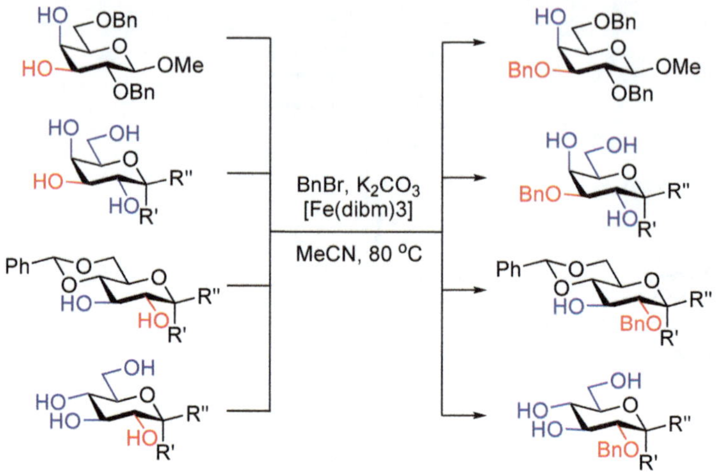

Scheme 2.2 [Fe(dibm)$_3$]-catalyzed regioselective alkylation of carbohydrates.

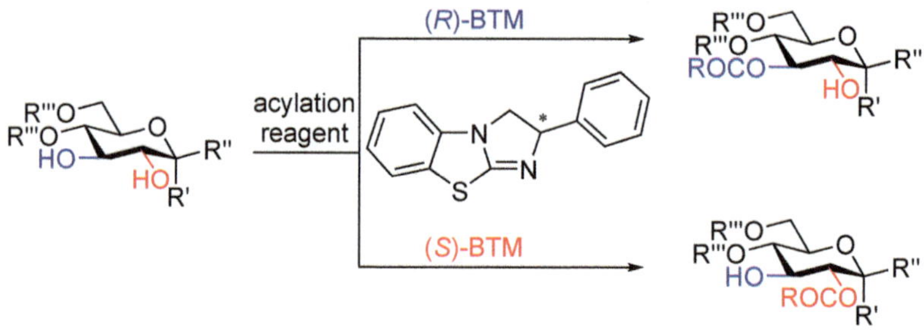

Scheme 2.3 Catalytic regioselective acylation of carbohydrates by chiral catalysts.

Cu-catalyzed Huisgen cycloaddition, in which conjugation of alkynes and unprotected reducing sugars was conducted in aqueous solution.[10] However, such a concept cannot be easily applied in oligosaccharide synthesis as donors and acceptors contain many hydroxyl groups, posing a great challenge to the achievement of regioselectivity. An alternative approach is to install transient protecting groups during glycosylation, which can be easily removed without additional steps. Here, the key goal is to find a sufficiently stable, yet easily deprotected group and compatible glycosylation conditions. In this regard, Liu's group developed a temporal boron-masking strategy in oligosaccharide synthesis (Scheme 2.4).[11] The boron-masking group not only directs both regioselectivity and stereoselectivity, it also allows flexibility in positions and numbers of hydroxyl groups on a variety of carbohydrates, ease of swapping roles between glycosyl donor and acceptor without additional steps as well as facile removal through simple workup, which is especially convenient for syntheses of complex branched oligosaccharides and polysaccharides.

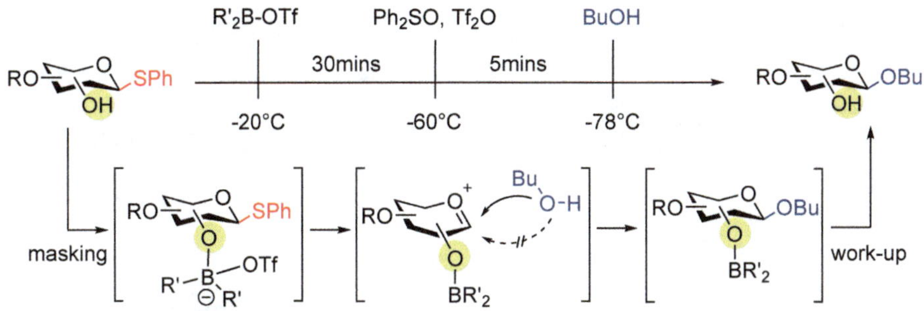

Scheme 2.4 Boron-mediated transient protection in glycosylation with partially protected substrates.

2.2.2 *De Novo* Syntheses of Carbohydrates

Another important part before the assembly of oligosaccharides is *de novo* syntheses of the building blocks. While most common monosaccharide building blocks are commercially available, even with customizable protection patterns, the access to uncommon sugars or special substituents on the sugar backbone remains a great challenge due to their low abundance in nature. With extraction from natural sources out of consideration, chemical syntheses of these sugars have been adopted to circumvent this problem. *De novo* syntheses generally start from non-sugar substrates to offer an alternative access to sugar building blocks. Furthermore, diverse protecting groups may be pre-installed on the fragments to allow for easier access to complex protection patterns on the sugar starting material. Herein we will discuss the syntheses of (1) uncommon hexoses, (2) hexoses with special protecting groups, (3) deoxysugars and (4) unsaturated sugars.

Besides the abundant hexoses glucose, galactose and mannose, other carbohydrates are rare and cannot be easily extracted from natural sources. While many efforts have been devoted to this area, the first to apply such a synthetic method to all eight hexoses were Masamune and Sharpless.[12,13] (*E*)-4-Diphenylmethoxybut-2-en-1-ol was employed as the starting material for Sharpless asymmetric epoxidation with chiral ligand (+)-DIPT. After two Payne rearrangements, the key tetrose fragment diol was obtained, which was later transformed into the respective hexoses. Later on, Danishefsky demonstrated the use of 1,3-dioxanone in Diels–Alder cycloaddition to furnish uncommon sugar allose, altrose, talose and galactose (Scheme 2.5a).[14,15] A hetero-Diels–Alder cycloaddition/elimination was first conducted with (*E*)-((4-methoxybuta-1,3-dien-2-yl)oxy)trimethylsilane and aldehyde using chiral activator (+)-Eu(hfc)$_3$ to generate *trans*-dihydropyran, which was subjected to manganese(III) acetate-promoted acetoxylation. A subsequent Luche reduction gave the glycal, which can be dihydroxylated to access various hexoses.

An alternative approach to obtain asymmetric hexoses is by aldol reaction using organocatalyst. Macmillan employed L-Proline to perform the first

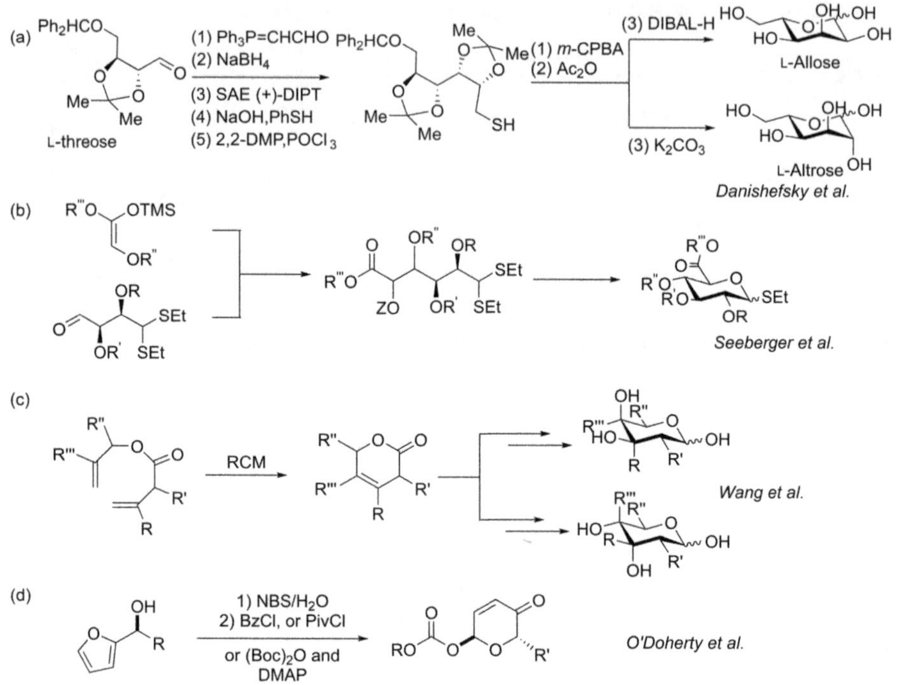

Scheme 2.5 Typical *de novo* synthesis strategies towards carbohydrates: (a) uncommon sugars, (b) C6 oxidized uronic acids, (c) 2,6-dideoxysugars and (d) glycal derivatives.

aldol reaction, following which a Lewis acid-catalyzed aldol reaction and subsequent cyclisation occurred to give the different hexoses.[16,17] Alternatively, thiazole could be utilized as the organocatalyst in Dondoni's approach.[18]

Inclusion of certain functionalities on the hexoses, such as the carbonyl group at C6 position, is also a challenge in the synthesis of derivatives. Seeberger's group synthesized uronic acid thioglycosides from protected threose, using Evans aldol reaction as the key step (Scheme 2.5b)[19] while Reissig's group obtained L-cymarose from L-lactic acid in ten steps.[20] Chirality-bearing compounds offer the advantage of pre-existing chiral centers, which make them convenient substrates for monosaccharides with multiple stereogenic centers. Recently, Legionaminic acid as an analog of sialic acid was first chemically synthesized from D-threonine by Seeberger.[21]

Of the hexoses, deoxysugar is a common scaffold in natural products as well and plays an important role in drug discovery. It is somewhat surprising that mono-deoxysugar is relatively scarce in nature, hence accounting for the need for synthetic approaches to gain access to this starting material. In 2002, Wang developed a *de novo* synthesis of 2,6-dideoxysugars *via* a ring-closing metathesis reaction (Scheme 2.5c).[22] Various β,γ-unsaturated δ-lactones were synthesized from terminal dienes in the presence of Grubbs' catalyst, which then underwent reduction to yield 3,4-*cis*-dihydroxy-2,6-dideoxypyranoses.

Glycals were usually synthesized from the common hexoses as an alternative to access glycosides which could not be readily prepared from saturated sugars, but the reactivity of these common glycals were not sufficient on some occasions. For example, Feringa's and O'Doherty's groups concurrently used the glycal pyranones to increase the potential of transition metal catalyzed decarboxylative allylation. The glycal pyranone starting materials were gained through *de novo* syntheses as shown in Scheme 2.5d.[23,24] O'Doherty prepared the glycal pyranones starting from furan alcohols through Achmatowicz reaction and subsequent hemiacetal protection. While anomeric hydroxyl groups could be acylated regioselectively (>20 : 1) at low temperature (−78 °C), a 1 : 1 mixture of anomeric epimers were obtained at room temperature.[24]

2.3 Amidst Glycosylation

Glycosylation is perhaps the most important reaction in carbohydrate chemistry. Unfortunately, the construction of glycosidic bonds often poses great challenges.[25] Hence, assembling strategies from monosaccharide building blocks have to be carefully designed to ensure high efficiency and practicability, with two major considerations: reactivity and stereoselectivity, during the glycosylation step. In the fine-tuning of reactivities of each building block, sequential glycosylations can be done in one-pot without isolation, saving cost and effort. In addition, absolute stereocontrol precludes the need for tedious anomer separation. Through enhancing the yields for glycosylations and using concise synthetic schemes, oligosaccharide syntheses has become increasingly convenient.

2.3.1 Reactivity

With the aforementioned tools at hand to manipulate protection patterns on various monosaccharides, it will be desirable to understand their effects on glycosylation reactions. Specific protecting groups can enhance reactivity while the others impede glycosylation rate, termed as 'armed' or 'disarmed' effects by Fraser-Reid.[26] Generally, an armed donor consists of predominantly weakly electron-withdrawing ether protecting groups such as benzyl groups, while a disarmed donor contains strongly electron-withdrawing ester functionalities such as benzoyl or acetyl groups. Under mild iodonium dicollidine perchlorate (IDCP) activating conditions, the disarmed donor serves only as a glycosyl acceptor, leaving the anomeric functionality unreacted as it requires stronger activating conditions like a combination of *N*-iodosuccinimide (NIS) and triflic acid (Scheme 2.6).

Following Fraser-Reid's discovery of armed/disarmed glycosyl donors, Bols and Madsen worked on expanding the scope of protecting groups to 'superarmed' or 'superdisarmed' (Scheme 2.7).[27,28] The former mainly constitutes bulky silyl ether protections which strongly repel each other, forcing the sugar ring into fixed conformation with axial groups which stabilizes

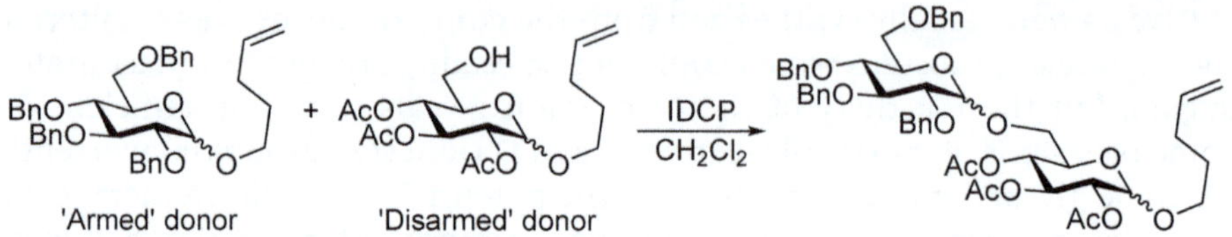

Scheme 2.6 Selective activation of an armed donor through reactivity discrimination.

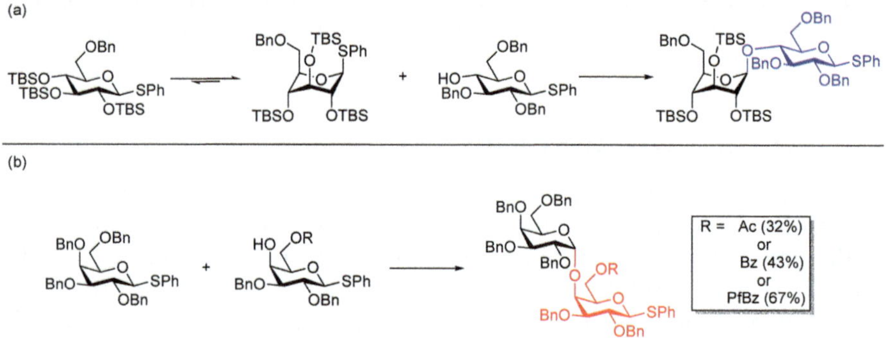

Scheme 2.7 Illustrations of using (a) TBS protections to superarm a donor or (b) a pentafluorobenzoyl protecting group to superdisarm a donor.

the cationic intermediate formed during glycosylation.[29] On the other hand, the latter demands installation of fluorinated alkyl or benzoyl groups which are extremely electron-withdrawing and significantly lower the reactivity of the protecting groups. Notably, detailed study into this effect has been conducted by Demchenko *et al.* in recent years.[30]

However, it is not as straightforward to categorize the reactivities of different combinations of protecting groups into the four effects as described above, especially without any measurable value for comparison. As such, Wong and Hung came up with an HPLC-based approach to quantify relative reactivity values (RRVs) for a library of protected thioglycoside donors.[31,32] Variations of monosaccharides as well as chemical properties and positions of protecting groups were recorded as the authors closely monitored the consumption progress of each donor for plotting of their respective reaction kinetics, measured against the least reactive peracetylated thiomannoside, which was normalized to have RRV = 1 (Figure 2.1). These quantifiable values significantly contributed to the prediction of reactivities and choices of protecting groups.

Besides the manipulation on protecting groups, careful designs of the anomeric leaving group have also been reported for tuning the donor's reactivity and expediting syntheses of complex carbohydrates.[25] Pioneering works by Koenigs-Knorr and Fischer on glycosyl halides and thioglycosides in early 1900s often demanded stoichiometric amounts of expensive

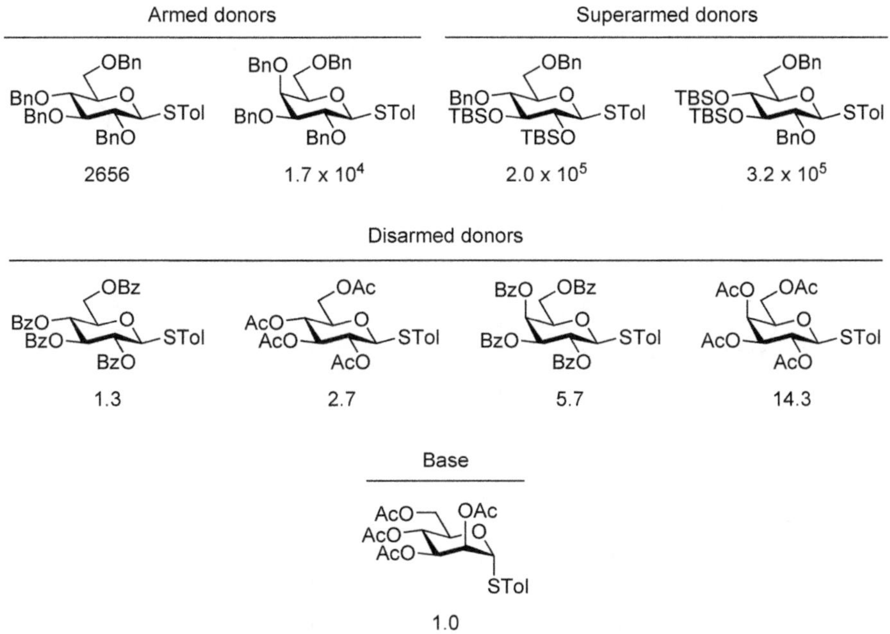

Figure 2.1 RRVs of selected examples of thioglycosides.

or toxic promoters. This was later circumvented by the discovery of glycosyl trichloroacetimidate (TCAI) by Schmidt and Michel in 1980, which could be readily activated by Lewis acid catalyst and led to better reactivity control (Scheme 2.8a).[33] In their subsequent studies, the authors noted that trifluoroacetimidate (TFAI) offered higher stability in reaction with a significantly lower reactivity as trade-off. Yu later reported an improved design, *N*-phenyltrifluoroacetimidates (PTFAI), which was able to construct glycosidic bonds in situations where rearrangement or decomposition of TCAI was likely to occur.[34] However, the improved reactivity of this leaving group was at the cost of shelf-life, which was a common problem of armed donors. In a later report, the same author came up with another design, *ortho*-alkynylbenzoate, with longer shelf-life and excellent reactivity during glycosylation upon addition of catalytic amount of promoter. The applicability of this leaving group was also demonstrated in the synthesis of cyclic triterpene saponin.[35]

Alternatively, reactivity can also be tuned after installation of the anomeric leaving group, through converting a 'latent' group to its 'active' form (Scheme 2.8b). The utility is best demonstrated in one-pot oligosaccharide synthesis by its repeating glycosylation-activation sequences. Roy and Boons reported their initial studies on the latent leaving group through the use of 4-nitrophenol and 3-butene-2-ol respectively.[36,37] The former is activated by reduction and subsequent acetylation of nitro group to form acetamide, while the latter requires a rearrangement from allyl

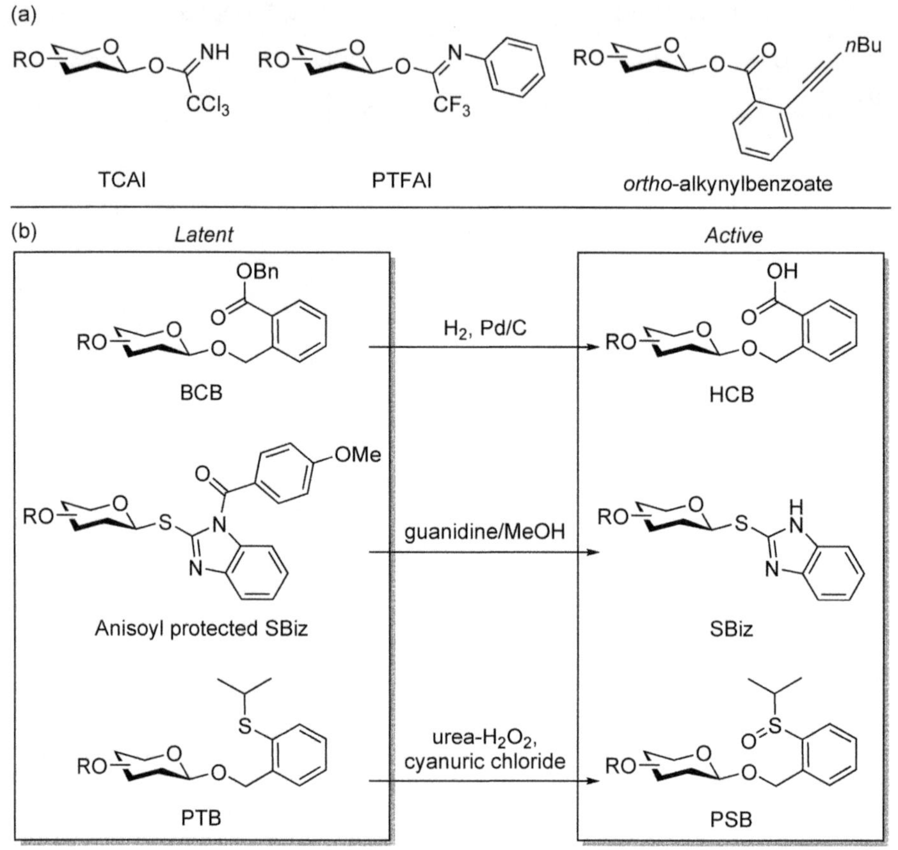

Scheme 2.8 Design of anomeric leaving group for improved efficiency in practice of oligosaccharide synthesis based on: (a) reactivity and shelf-life improvement or (b) latent/active donor strategy.

group to vinyl group. Earlier this century, Kim also reported his use of a 2-(benzyloxycarbonyl)benzyl (BCB) latent group, which was highly effective for β-mannosylation.[38] When utilized in a glycosyl acceptor, it remained inert but after its debenzylation into a 2-(hydroxycarbonyl)benzyl (HCB) group, the carbohydrate became a highly reactive glycosyl donor. Similarly, anisoyl protected *S*-benzimidazolyl (SBiz) may also be activated after deprotection.[39] In a different approach, Wan adapted an interrupted Pummerer reaction into his design of latent/active donor with emphasis on its regenerability.[40] It was worth noting that the conversion from a 2-(2-propylthiol) benzyl (PTB) latent form to a 2-(2-propylsulfinyl)benzyl (PSB) active form could be reversed after glycosylation, by recovering the free PSB-OH and reducing it back to PTB-OH.

Apart from the efforts made to tune glycosyl donors' reactivity, it is worth noting that glycosyl acceptors may also exhibit different reactivities. The prevalent use of such difference is seen in complexed carbohydrate synthesis, when sugars often play the role of both donor and acceptor. A carefully designed protecting pattern for the acceptor may allow for some hydroxyls to be exposed without interference to glycosylation. These can be applied to sequential glycosylation reactions with certain protection and deprotection steps skipped.[41,42]

2.3.2 Stereochemistry

After addressing the issue of reactivity during glycosylation, the coupling of the glycosyl donor and glycosyl acceptor presents another challenge: the stereochemistry about the anomeric position. A series of factors could affect the anomeric selectivity: (1) properties of the glycosyl donor; (2) properties of the glycosyl acceptor; (3) reaction conditions. With the vast library of available stereoselective strategies, the coverage here would be based on key stereochemistry-directing approaches with the focus being dual-directing approaches, whereby both α- and β-anomers could be obtained, through slight modifications.

To date, one of the most popular and well-studied methods towards selective glycosylations is through modulating the glycosyl donor. Through changing the stereochemical and electronic properties of the glycosyl carbons as well as functionalities on them, the desired stereoselective outcome could be achieved. Before looking into specific dual-directing strategies using a glycosyl donor, it is instructive to describe the classifications of the common strategies: anomeric effect, anomeric leaving group, neighboring group participation, remote directing group, conformational constraint (such as ring locking) as well as intramolecular aglycon delivery (Scheme 2.9).

Starting from the anomeric position, which is the crucial reaction site for glycosylation, the stabilization of an intermediate, such as oxocarbenium ion, can result in preferred formation of anomer. In the case of anomeric effect, it stems from the repulsive interactions between the lone-pair electrons on anomeric substituent and that of ring oxygen as well as the stabilization of electron-withdrawing axial substituent *via* hyperconjugation of O5 non-bonding orbital and C1 antibonding orbital in periplanar orientation. This accounts for the preference of α-glycosidic bond formation, particularly for *O*-glycosides. Notably, Lemieux *et al.* reported halide ion-catalyzed glycosylations, or *in situ* isomerization, where a rapid equilibrium exists between the stable α-halide and reactive β-intermediate upon adding Bu_4NBr, resulting in S_N2-type attack of reactive intermediate and achieving α-selectivity.[43]

Besides modulating the reactivity, the anomeric leaving group also affects the stereochemical outcome in some cases of glycosylations as well. While it holds trivial meaning for S_N1-type glycosylation, the α- or β-orientation of the leaving group has a significant effect on S_N2-type glycosylation. As such, 1,2-*cis* glycosides can be obtained from 1,2-*trans* glycosyl donors and *vice versa*.[44]

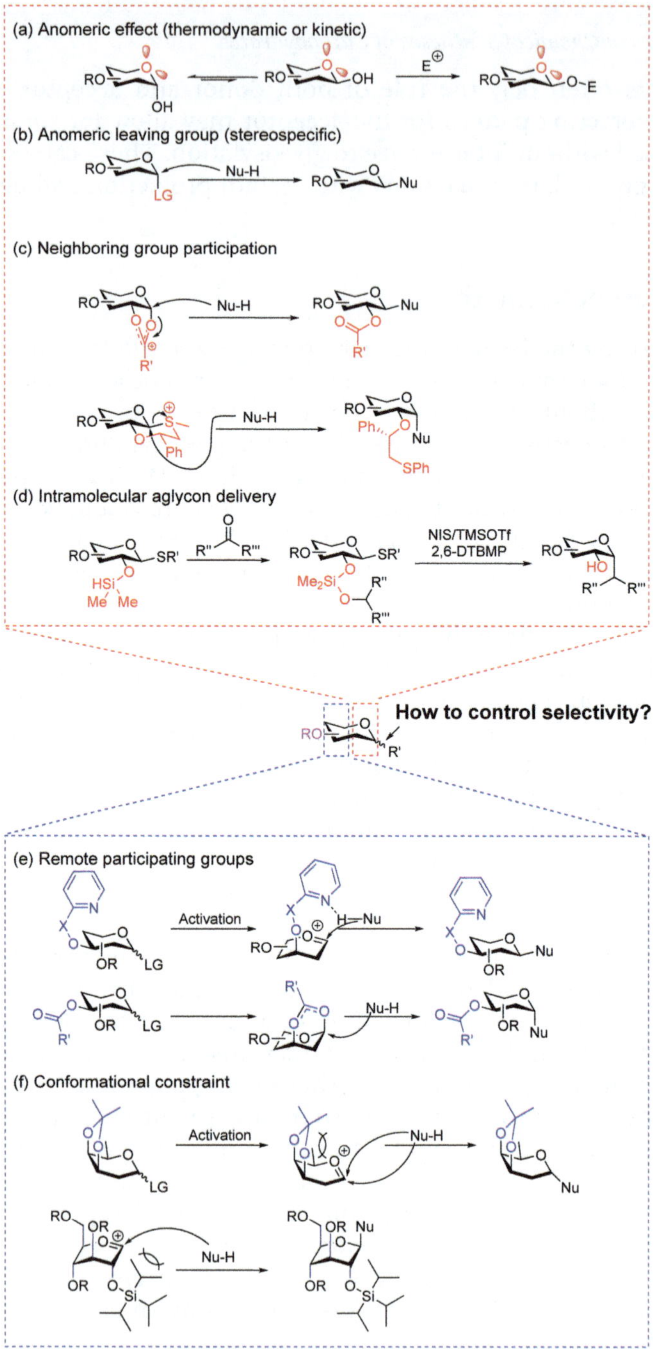

Scheme 2.9 Typical glycosyl donor-controlled stereoselective glycosylations. The control could be realized through manipulation of the vicinity of anomeric centre in terms of (a) anomeric effect, (b) positioning of anomeric leaving group, (c) blocking of one side by a neighboring group or (d) delivery of aglycon from a neighboring group. Alternatively, protections in the far end might also exert influence on the stereo outcome *via* (e) hydrogen bonding or covalently blocking or (f) re-orienting the monosaccharide by conformational constraint.

Moving one carbon center away from the anomeric position, stereochemistry can still be directed using neighboring group participation (NGP) or anchimeric assistance. The classical NGP uses 2-*O*-acyl protected glycosyl donor and proceeds *via* 5-membered ring intermediate to form 1,2-*trans* glycosides in S_N2-type attack. Interestingly, redesigning of NGPs to form 6-membered ring intermediates by Boons,[45] Turnbull[46] and Fairbanks[47] allows construction of 1,2-*cis* glycosidic linkages.

Another neighboring effect drives stereoselective glycosylation through direct delivery instead of docking effect. Intramolecular aglycon delivery (IAD) exclusively targets 1,2-*cis* glycosides through C2 tethering and subsequent intramolecular glycosylation. Various functionalities, such as ketal, silyl acetal, 2-iodoalkylidene acetal, benzylidene acetal and hemiaminal ether, have been utilized for this reliable glycosylation approach.[48] Recent IAD works include usage of silyl acetal by Montgomery *et al.*[49] and napthylidene acetal by Ishiwata and Ito *et al.*[50]

Advancing further from the anomeric center, the remote directing group can still target anomeric selectivity directly or indirectly. The straightforward approach occurs *via* direct intramolecular attack of anomeric position by remote functionalities, such as 3-, 4- and 6-*O*-acyl groups. Formation of dioxocarbenium ion was proven by Kim[51] and Yu.[52] Demchenko *et al.* demonstrated this approach indirectly with remote *O*-picolyl and picolinyl groups through a hydrogen bond-mediated aglycon delivery.[53]

The conformational assistance of glycosyl donor is another useful strategy. This method ultimately affects anomeric effect, sterics and transition state (TS) energies. An investigation into silyl protected $^1C_4$ ring-flipped xylopyranosyl donors led to finding the preferred formation of axial anomer due to anomeric effect.[54] Yamada *et al.* later reported that glucosyl donor gave rise to β-selectivity due to the sterics arising from near $^1C_4$ skew-boat conformation.[55] Similarly, Toshima *et al.* adopted conformationally rigid 3,4-*O*-isopropylidene group, which could favor α-anomer through strong 1,3-diaxial interaction between glycosyl acceptor and C3 substituent.[56] Whitfield *et al.* explored the TS energy of α- and β-face attack and its preference for the anomer with lower-energy TS.[57]

The understanding of the above donor-controlled approaches serves as a foundation for dual-directing approaches through modulating the properties of glycosyl donor (Figure 2.2). While the IAD approach could effectively yield α- and β-glycosides from D-mannose and D-glucose respectively, such examples would not be covered here.

Matsuda and Shuto examined the conformational constraint through various xylose derivatives, whereby the ring conformations are locked in $^4C_1$ (2,3-*O*-cyclic-diketal and 3,4-*O*-cyclic-diketal substrates) and $^1C_4$ (2,4-*O*-cyclic-phenylboronate and 2,3,4-tri-*O*-TIPS substrates) orientations by different protecting groups.[58] Consequently, the anomeric effect was altered, leading to reversed stereochemical outcomes.

Walczak[59] and O'Doherty[24] demonstrated the effect of the anomeric leaving group on α-/β-selectivity through palladium (Pd) catalyzed cross couplings.

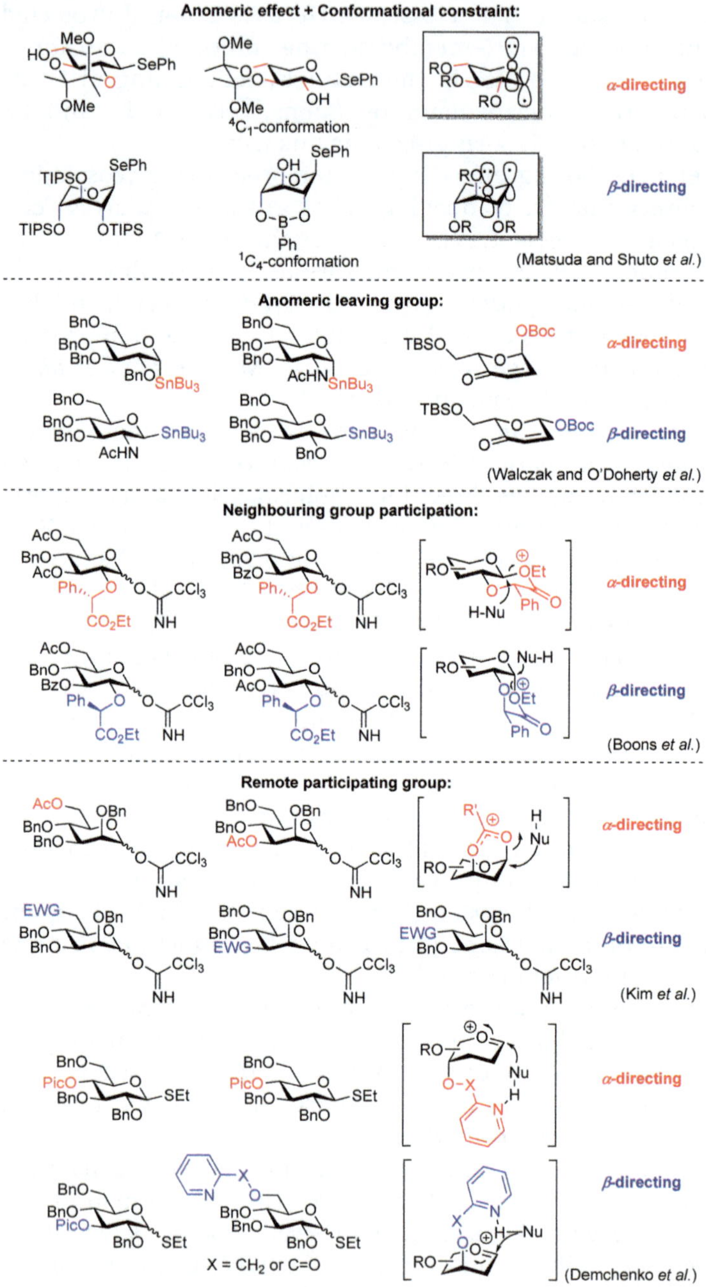

Figure 2.2 Glycosyl donor-controlled dual-directing glycosylations.

In the Stille type coupling, the stereochemistry of the stannylated substrate was conserved through transmetallation and subsequent reductive elimination. On the other hand, double inversion occurred in the Tsuji–Trost type reaction, whereby the Pd complex attacked from opposite face to the leaving group and nucleophile then attacked from the opposite face to the π-allyl Pd complex, leading to retained stereochemistry.

An interesting finding by Boons during his study on 6-membered ring intermediate for neighboring group participation was the installation of chiral auxiliaries at the C2 position of glycopyranosyl donors.[60] (*R*)- and (*S*)-(ethoxycarbonyl)benzyl groups were used to obtain *cis*-fused and *trans*-fused dioxolenium ion intermediate respectively, resulting in 1,2-*trans*- and 1,2-*cis*-glycosides. Such moiety could be easily removed through Birch reduction.

As described earlier, works by Kim[51] and Demchenko[53] utilized the remote participating groups of *O*-acyl as well as *O*-picolyl and *O*-picolinyl groups to effect stereochemistry. In the presence of remote acyl coordinating groups, a dioxolenium ion was formed, forcing the attack of nucleophile to be from the opposite face. The electron-withdrawing groups have strong β-directing effect for mannosyl oxocarbenium ion. On the other hand, the glycosyl donor and acceptor could be involved in a H-bond tethering through remote *O*-picolinyl and *O*-picoloyl groups, giving rise to 1,2-*cis*- or 1,2-*trans*-glycosides.

Comparatively, little attention is given to utilizing the glycosyl acceptor in the control of stereoselective glycosylation. This approach is advantageous the lengthy and time-consuming preparative route of glycosyl donor can be avoided. Reversal of anomeric selectivity can also be achieved through modifying the properties of the glycosyl acceptor, so long as a glycosyl donor with sufficient reactivity is employed (Scheme 2.10).

Early works by Woerpel *et al.* explored the correlations between the nucleophilicities of glycosyl acceptors and stereoselectivities in glycosylations. In the employment of weak C-nucleophiles, they proposed that α-stereochemical outcome was based on S_N1 reaction.[61,62] On the other hand, strong C-nucleophiles depended on the reaction conditions, whereby erosion of selectivities occurred without coordinating counterions while β-glycosides were favored in the presence of triflate counterions *via* S_N2 pathway.[61] Later, O-nucleophiles were applied as well.[62] Similarly, weak O-nucleophiles could yield α-glycosides while the stereoselectivity was eroded for strong O-nucleophiles, owing to the rate of the stereoselectivity-determining step being close to the diffusion limit.

In a significant study by Liu, breakthrough was achieved for this acceptor-controlled glycosylation by employing a carefully selected donor, bearing 2-cyanobenzyl ether auxiliary.[63] This synergistic approach combined the donor-controlled strategies such as arming-disarming strategy for reactivity, solvent effect, neighboring group participation, hydrogen bonding and intramolecular aglycon delivery with nucleophile-derived selectivity to yield

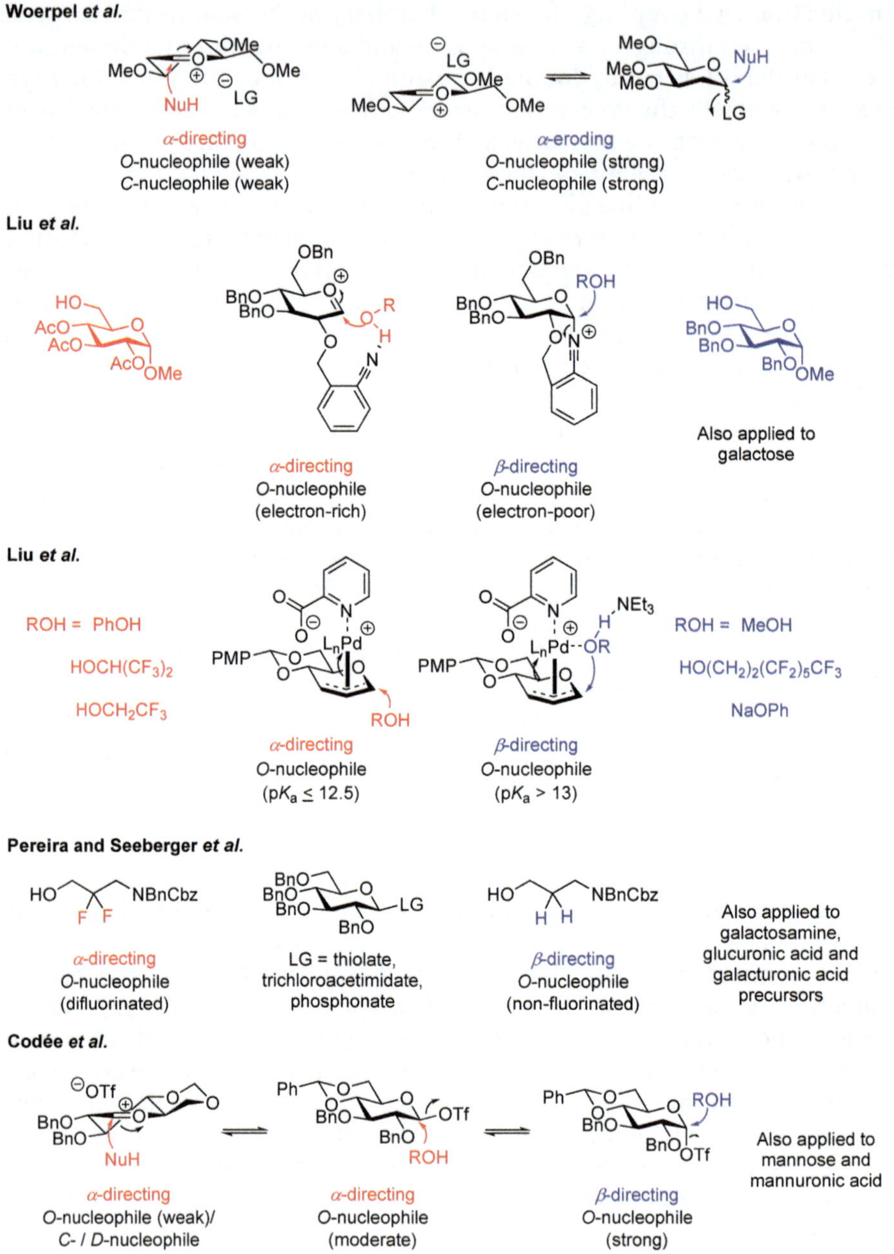

Scheme 2.10 Examples of glycosyl acceptor-controlled dual-directing glycosylation.

glycosides with exclusive stereochemistry. The activated oxocarbenium ion can be attacked by the nitrile group to form the nitrile ion and the equilibrium exists between two species. For electron-rich O-nucleophiles, the displacement of nitrilium ion occurs in S_N2 fashion. In contrast, electron-poor O-nucleophiles attack the oxocarbenium ion directly and the α-selectivity is accounted by hydrogen bond-mediated aglycon type delivery. Formations of 1–2, 1–3, 1–4, 1–6 glycosidic linkages were demonstrated. In addition, galactose type donor was also screened.

In another study, Liu *et al.* tapped on the inner-sphere/outer-sphere mechanism raised by Trost,[64] whereby pK_a values indicative of hardness and softness of the glycosyl acceptors could be used to predict the α-/β-selectivities.[65] Hard O-nucleophiles with pK_a values more than 13 undergo inner-sphere attack through coordinating to Pd cation while soft O-nucleophiles with pK_a value less than 12.5 proceed by outer-sphere mechanism to yield α-glycosides.

Seeberger and Pereira adopted another approach by fusing acceptor-controlled strategy with solvent effect.[66] Fluorinated amino alcohol resulted in the preferred formation of β-glycosides with diminished selectivity upon increasing the temperature to beyond −40 °C. On the other hand, non-fluorinated amino alcohol favored α-glycoside formation but required temperatures higher than −40 °C to drive its reactivity. Ethereal and acetonitrile solvents were shown to further enhance the selectivities. While the detailed mechanism remained to be studied, a variety of substrates as well as leaving groups were explored. Their work also exhibited the feasibility of employing it in oligosaccharide syntheses using automated glycan assembly as the fluorinated linker did not have an adverse effect on glycan immobilization and lectin binding.

Building on Seeberger and Pereira's works, Codée *et al.* recently subjected fluorinated, partially fluorinated and non-fluorinated nucleophiles to three glycosylation systems: glucose, mannose and mannuronic acid donors.[67] While the glucosyl system could be easily modulated with acceptor nucleophilicity, the two other systems were more robust. They proposed that switchable stereoselectivity from β- to α-selectivity arose from the shift in reaction mechanism from S_N2 to S_N1 as the nucleophilicity of glycosyl acceptor decreased.

Reaction conditions often play an important role in facilitating the departure of leaving group on glycosyl donor, facial-selective nucleophilic attack of glycosyl acceptor due to steric and electronic properties of the intermediate formed as well as stabilization of reaction intermediates. In this context, the effects of solvent, concentration, activator and catalyst on anomeric stereoselectivity are being examined (Scheme 2.11).

The use of solvent to direct stereoselectivity is attractive as a simple workup could remove the stereo-directing factor. Choice of non- or weakly dissociating (diethyl ether and dioxane) and coordinating solvents (nitrile) could lead to reversed anomeric stereochemistry.[68] The α-directing effect of ethereal solvent results from the formation of β-onium adduct, while β-directing effect arises from the generation of α-nitrilium ion by nitrile attacking to the oxocarbenium ion. An interesting study by Mong *et al.* reported the use of DMF solvent additive to modulate stereoselectivity. In the presence of DMF, excellent α-selectivity was achieved due to entrapment of oxocarbenium ion as glycosyl imidate, while glycosylation occurred to give α-/β-mixture without DMF.[69]

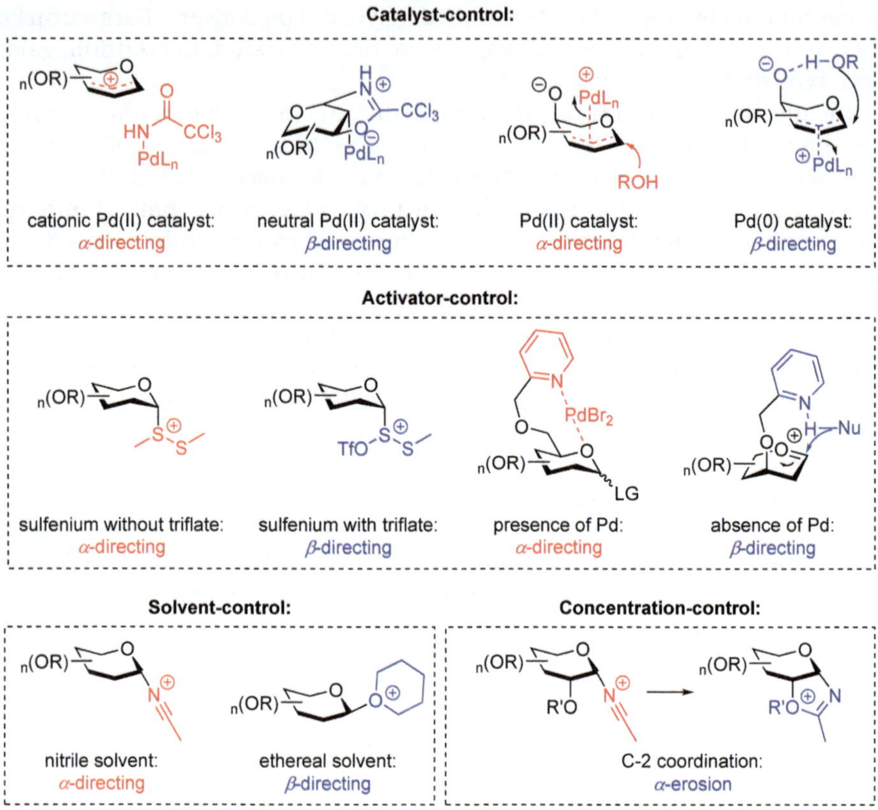

Scheme 2.11 Reaction condition-controlled dual-directing stereoselective glycosylation.

There are very rare cases of concentration-dependent stereoselectivity. Contrary to the solvent effect mentioned earlier, Mong *et al.* observed a shift to α-selectivity in lowered glycosyl donor and acceptor concentrations in nitrile system. The authors later proposed that the reaction proceeded *via* a glycosyl oxazolinium intermediate through C2 participating ether.[70] Yu also found that glycosyl alkynylbenzoates could undergo remote participation in low substrate concentrations.[71] This accounted for the preference of intramolecular attack of oxocarbenium ion to form dioxalenium intermediate over intermolecular reaction. Toshima also demonstrated this control in organo photoacid catalyzed photoinduced glycosylation, yielding β-glycosides at high concentrations and α-glycosides at low concentrations.[72]

Activators could affect stereochemical changes through modification of intermediates. In particular, Bennett explored the direct glycosylation with a series of deoxy sugar donors, using different promoter systems, such as potassium hexamethyldisilazane/*p*-toluenesulfonic anhydride and oxalyl bromide.[73,74] Davis *et al.* utilized two different activating conditions, dimethyl disulfide-triflic anhydride and dimethyl(methylthio) sulfonium triflate conditions, to generate sulfenium ions with differing nature *in situ*, giving rise to α- and β-glycosides respectively.[75] The opposing deactivating effect was investigated by Demchenko using 6-*O*-picolinyl and picoloyl groups. Addition of a metal (Pd) temporarily blocks nitrogen coordination through metal complexation and H-bond-assisted delivery cannot proceed, yielding α-glycosides as opposed to β-glycosides mentioned earlier.[76]

Control of stereoselectivity using catalyst appears to be a daunting task when comparing the catalyst to reactants ratio. Remarkably, Ye and Jacobsen successfully proved this possibility by exploiting catalyst-directed approach in stereoselective glycosylations.[77,78] Lee, Nguyen and Liu used the different nature of Pd-ligand complexes to target both α- and β-glycosides selectively.[79,80] A switch from $P(OMe)_3$ to DTBBP ligand could result in a shift from α- to β-glycoside.[79] When using cationic Pd species, the coordination of imidate nitrogen led to ionization to allylic cation and subsequent attack from α-face while neutral Pd caused coordination to double bond and cyclisation-induced rearrangement occurred to give β-*N*-glycosides.[80] In a similar manner, Pd(II) could coordinate to galactal carbonate oxygen from the β-face to yield α-*O*-glycosides while Pd(0) coordinated to double bond from opposite of the leaving group and H-bond-mediated delivery led to β-anomers when phenols were employed.[81] Tang demonstrated the dynamic kinetic diastereoselective acylation of pseudo-glycals and subsequent Pd-catalyzed glycosylation, whereby anomeric selectivity could be reversed by changing the chirality of benzotetramisole catalyst from (*R*)- to (*S*)-form.[82]

Interestingly, other factors such as the addition rate of substrates,[83] reaction time[84] and temperature[85,86] could also affect the stereochemical outcome.

2.4 Applications Beyond Glycosylation

The aforementioned preparations of building blocks and numerous considerations for glycosylations are crucial for the downstream assembly into more complex molecules, such as oligosaccharides and polysaccharides, glycoconjugates as well as natural products. Besides repeating glycosylation for further elongations, other moieties can also be incorporated through diverse transformations, furnishing natural forms and mimetics of the bioactive components for biological studies. The strategic applications to gain access to these complex carbohydrates and carbohydrate-derived compounds are discussed in this section.

2.4.1 Oligosaccharides and Polysaccharides

Glycosylation serves as a powerful tool for assembly of glycans. Using monosaccharide building blocks, oligosaccharides and polysaccharides can be synthesized through mainly *O*-glycosidic linkages. In particular, bacterial glycan has been of great interest to researchers due to its prevalence and significant role in biological processes.[87] Recent designs of new donors and acceptors help to expedite syntheses of these bacterial glycans.

A noteworthy example of such a design is Kim's phthalic anhydride-mediated direct glycosylation using hydroxyl glycosides as the glycosyl donors.[88] The hydroxyl group was first converted to phthalate which then formed the oxocarbenium ion with the addition and elimination of phthalic anhydride (Scheme 2.12a). This method was later applied to the synthesis of a tetradecasaccharide arabinan motif found on the cell wall of mycobacteria for in-depth study of *Mycobacterium tuberculosis* (Scheme 2.12b).[89] Repetitions of the phthalate-mediated step could afford the dodecasaccharide key intermediate.

The challenge of synthesizing these complex carbohydrates can be addressed by strategic selection of building blocks and concise design of a synthetic route. Core fragments are prepared first and used as modules for subsequent assemblies.[90,91] This, in turn, necessitates tools with sufficient capability to overcome reactivity hurdles when multiple modules are present in its total synthesis as the identification of reactivity gradient of donors and acceptors can be a daunting task. In particular, preactivation based glycosylation strategy was shown to be a competent alternative in this situation to calculate the design of donor's and acceptor's reactivities. Generally, the glycosyl donor has to be pre-activated before the addition of the acceptor. In this manner, the acceptor can be left unactivated as the activator or promoter has completely reacted with glycosyl donor before its introduction, eliminating the chance of side reactions from competitive activation of acceptor. An earlier work adopting this strategy was Huang's synthesis of dimeric Lewis^{X} octasaccharide. Recently, Ye *et al.* reported a total synthesis of 92-mer arabinogalactan with pre-activation based glycosylation, breaking the record of the largest glycan ever chemically synthesized (Figure 2.3).[3,92] In the latter case, arabinogalactan was dissected into two arabinan and one galactan modules which are further fragmented for ease of synthesis and this pre-activation strategy was proven to be an indispensable help in this assembly of large numbers of fragments.

2.4.2 Glycoconjugates

The synthetic oligo- and polysaccharides obtained can then be covalently linked with other chemical molecules, like lipids, peptides and proteins, yielding glycoconjugates. Quick and concise chemical approaches to obtain glycoconjugates include: (1) olefin metathesis; (2) Click reaction; (3) multicomponent reactions; (4) native chemical ligation (NCL); (5) Staudinger ligation; (6) sugar-assisted ligation. A brief discussion of these key synthetic methods will hopefully serve as a prelude for subsequent chapters (Scheme 2.13).

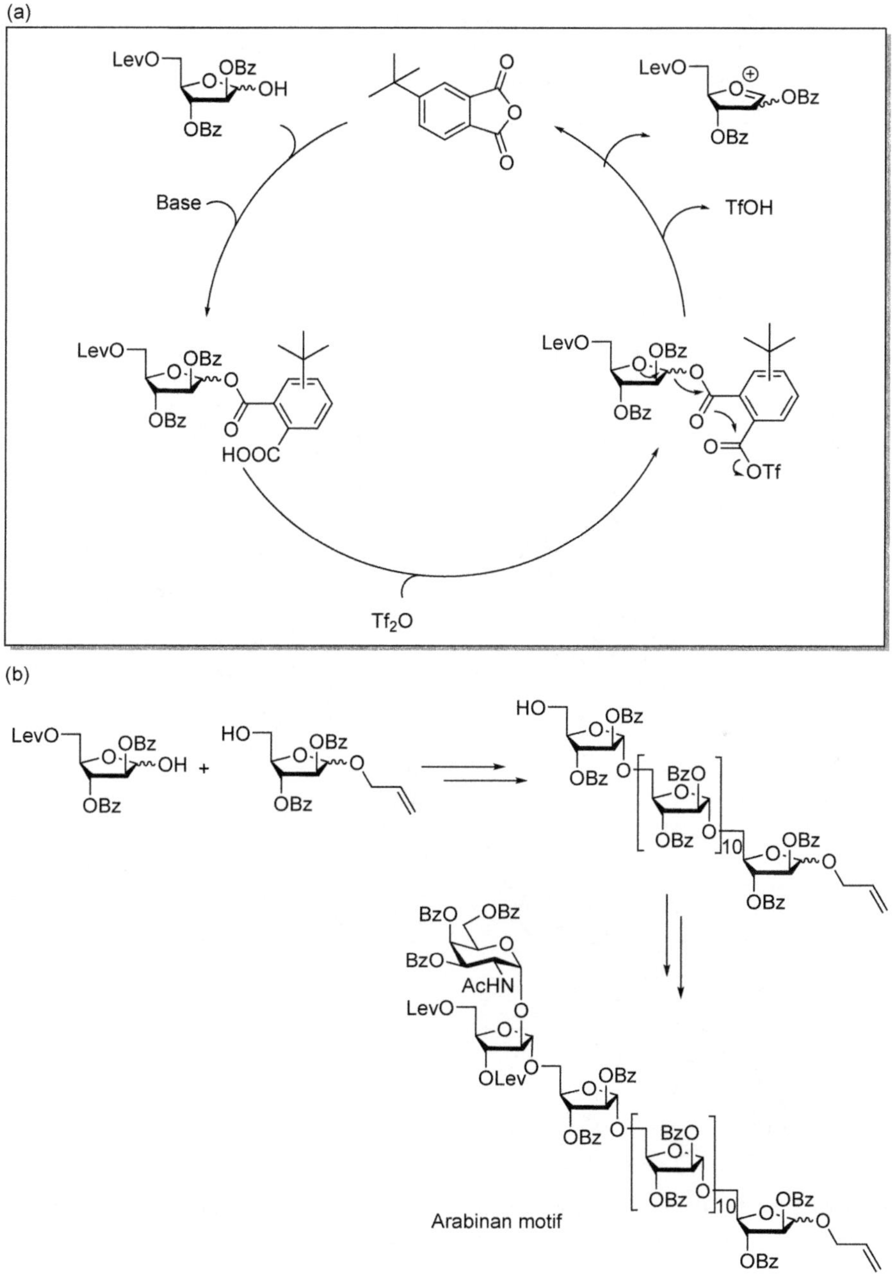

Scheme 2.12 (a) Mechanism and (b) application of a phthalic anhydride mediated direct glycosylation.

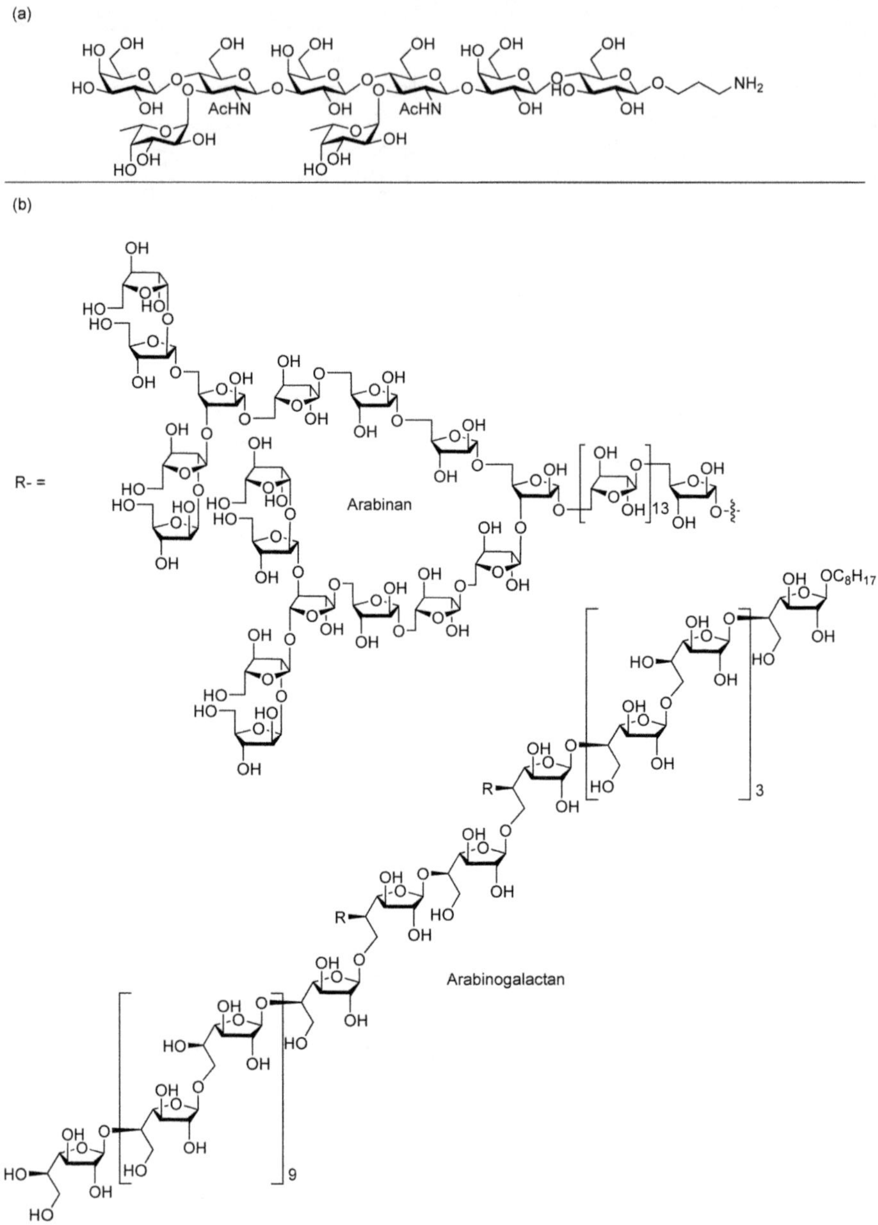

Figure 2.3 (a) Dimeric Lewis^{X} octasaccharide and (b) arabinogalactan which have been synthesized by preactivation based glycosylation strategy.

Metal catalyzed olefin metathesis approach can be adopted to link the alkylene glycosyl fragment with another alkene-bearing conjugate, connecting them by a new carbon–carbon double bond.[2] Danishefsky successfully conjugated oligosaccharide and polypeptide using this approach, forming Gb3 glycosylamino acid in the presence of a Grubbs–Hoveyda

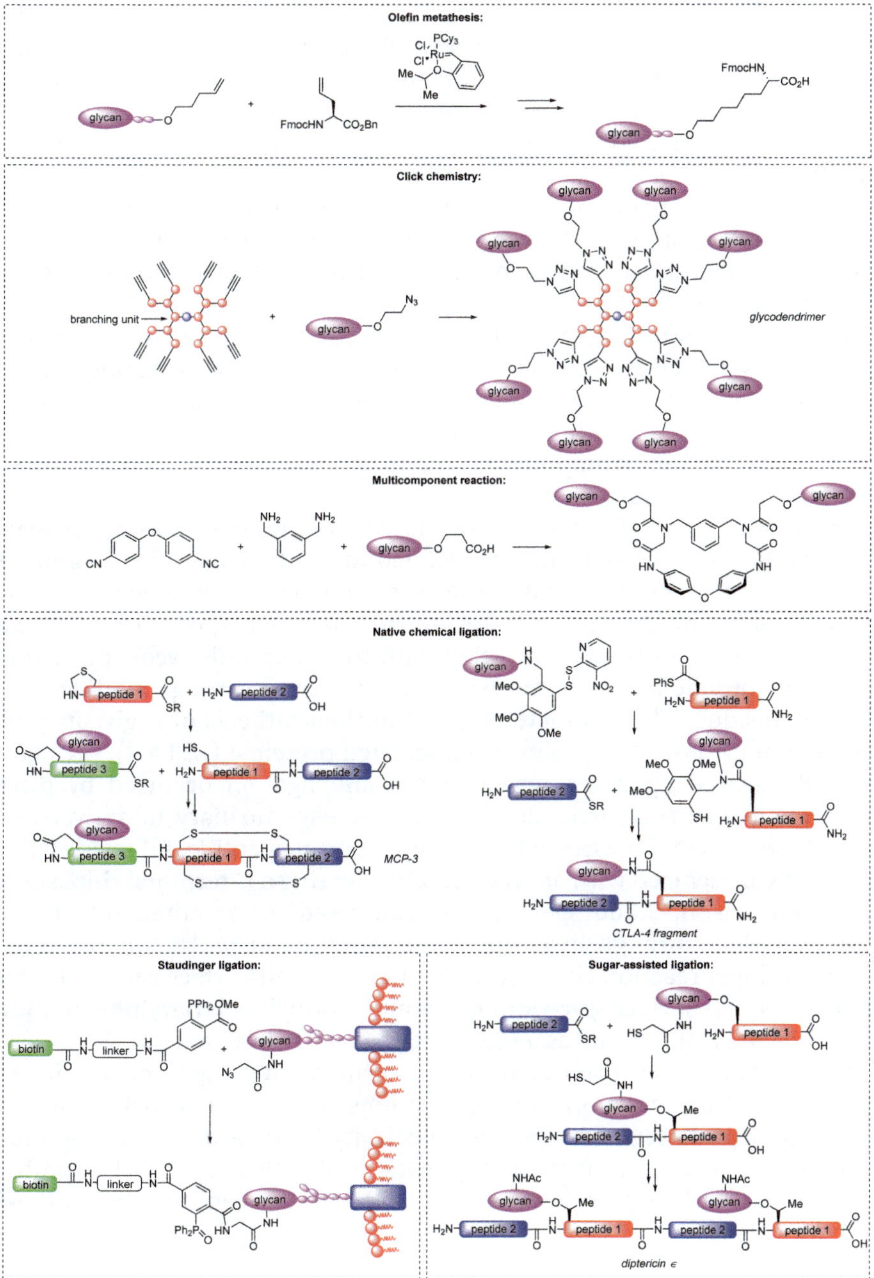

Scheme 2.13 Commonly adopted synthetic routes to access glycoconjugates.

catalyst.[93] The utility of this method lies in its specificity and minimal side product generated.

In another transition metal-catalyzed approach by Sharpless, high yields were observed for a wide variety of substrates using click chemistry. The 1,2,3-triazole formation traditionally requires copper(I) catalyst and is obtained from 1,3-dipolar cycloaddition of azides to terminal alkynes, whereby the glycosyl group could have either of the functionalities and the coupling partner possesses the other moiety. Later, thio-ene coupling and copper-free variants of click reaction were developed. An example was the synthesis of glycolipid clusters or glycodendrimers by clicking the azido sugars with alkynes substituents, and this was later applied to binding interactions with lectins.[94,95]

Alternatively, multicomponent reactions can be used to obtain convenient access to a diverse library of compounds through varying one or more component(s) of the reaction.[96] This combinatorial approach demonstrated its utility for high-throughput screening systems in Ugi and Passerini reactions.[97] Ugi-type multicomponent reaction was also used for peptide macrocyclisations with sugar moieties.[98]

Undoubtedly, one of the most powerful techniques in glycoconjugate synthesis is NCL, which links unprotected peptide and glycopeptide fragments. Dawson and Kajihara reported synthesis of monocyte chemotactic protein-3 (MCP-3) using this strategy with two peptide fragments, one bearing N-terminal cysteine residue and the other with thioester and glycopeptide thioester.[99] Requirement of cysteine residue was overcome through modifications of other residues[100,101] and later applied to the synthesis of *N*-glycopeptide segment of cytotoxic T-lymphocyte-associated protein 4 (CTLA-4).[102] In Liu's carefully-devised approach, glycosylation and ligation occurred by three-component dual-NCL, whereby *N*-linked glycosyl auxiliary underwent the first NCL to join to thioester of N-terminal aspartate peptide, following which second NCL occurred with another peptide bearing a C-terminal thioester.

In comparison, Staudinger ligation can avoid the requirement of thiol group for NCL, through installing azido group on sugar.[103] Notably, modified Staudinger reaction could be applied to label cell-surface carbohydrates, using azido cell-surface glycoconjugates and biotinylated triarylphosphine *in vivo*, showing its potential as a tool for intracellular interactions.[104]

An alternative approach is through sugar-assisted ligation, which proceeds by careful installation of functionalities on the sugar, which could help bring the peptide segments in close proximity for peptide bond formation. Wong and co-workers probed the applications of such an approach to both *N*-linked and *O*-linked glycopeptides[105,106] and later applied it to the synthesis of diptericin ϵ.[107]

2.4.3 Natural Products

Apart from the application of carbohydrates as macromolecules, they can be utilized as affordable biomass materials and natural chiral pools for natural product and drug syntheses.[108] Many efforts have been devoted to this area and numerous complicated functional molecules could be conveniently synthesized from carbohydrates.[109] Herein, recent representative examples are chosen based on their usage of *O*-, *N*- and *C*-glycosides as the starting materials.

Besides being the core units of oligo- and polysaccharides, *O*-glycosides are also widely applied in natural product syntheses. For example, C2-bromoglucal was successfully applied to synthesize the highly substituted chromans and isochromans with *O*-glycosylation being the key step by Werz in 2010 (Scheme 2.14).[110] By tapping the rich stereochemistry on carbohydrate starting materials and using the Pd-catalyzed domino approach, the complex oligocyclic molecules could be generated with minimal effort.

N-Glycosides are indispensable components of DNA, RNA, cofactors and various antibacterial, anti-inflammatory, anti-viral and anti-neoplastic agents. Of them, 2-amino-2-deoxyglycopyranosides are commonly found in prokaryotic and eukaryotic glycoconjugates. Using a sequential rhodium-catalyzed aziridination and Barbier allylation of D-glycal, the *N*-glycoside intermediate can be converted to a highly-substituted [1,2,3]-oxathiazocan e-2,2-dioxide and can be later applied to the total synthesis of sialic acid *N*-acetylneuraminic acid (Neu5Ac) (Scheme 2.15).[111]

While *N*-glycosides are mostly found in the biological components, *C*-glycosides occurred more broadly in natural products, particularly for 2,3-unsaturated glycosides. One of their applications is the rational design of intramolecular decarboxylative allylation to form the β-*C*-glycoside which was applied as the key step of the formal synthesis of aspergillide A (Scheme 2.16).[112]

As described above, *O*-, *N*- and *C*-glycosides play important roles in drug discovery. Vankar further demonstrated this in his report on bicyclic hybrid sugar molecules comprising of oxa-aza, oxa-oxa, and oxa-carbasugar-fused

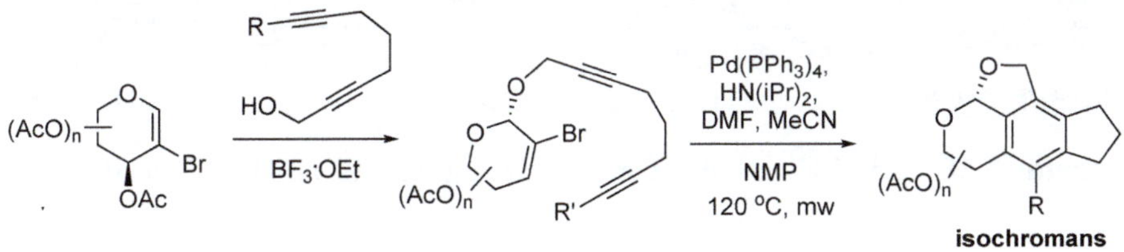

Scheme 2.14 Synthesis of isochromans from glycals.

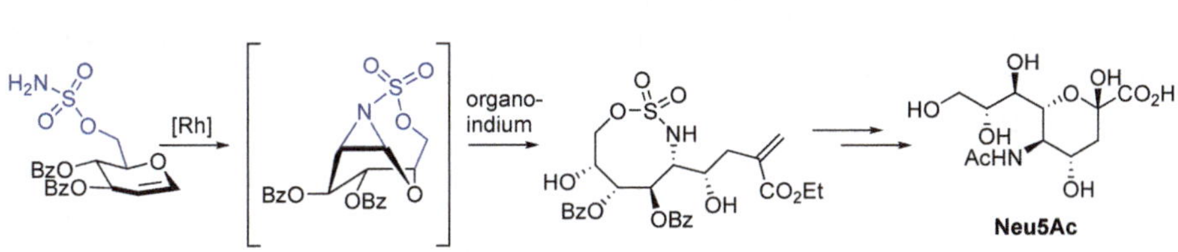

Scheme 2.15 Total synthesis of sialic acid by aziridination and allylation of D-glycal.

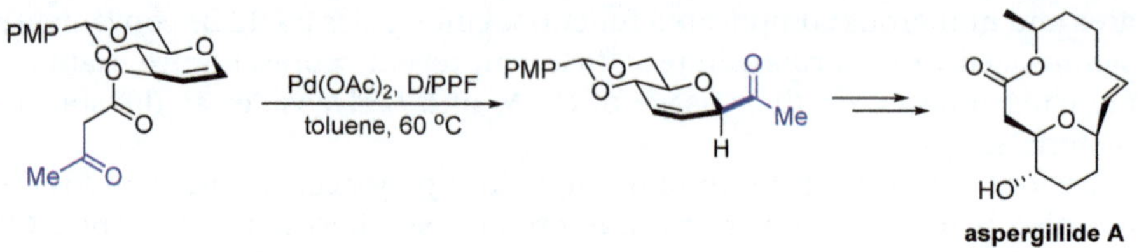

Scheme 2.16 Intramolecular decarboxylative allylation for *C*-glycosylation.

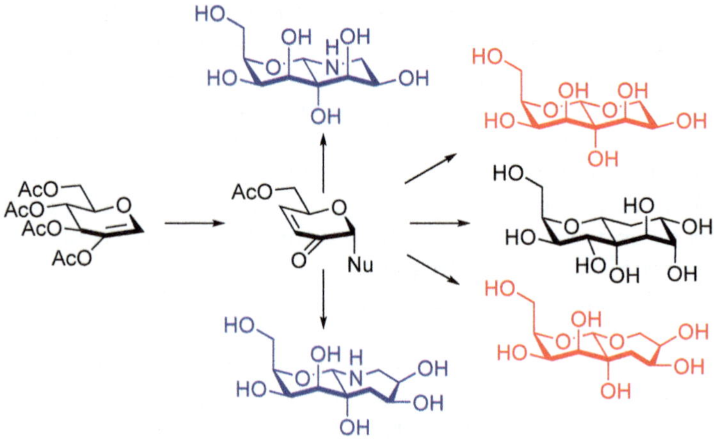

Scheme 2.17 Synthesizing glycosidase inhibitors from *C*-, *O*- and *N*-glycosides.

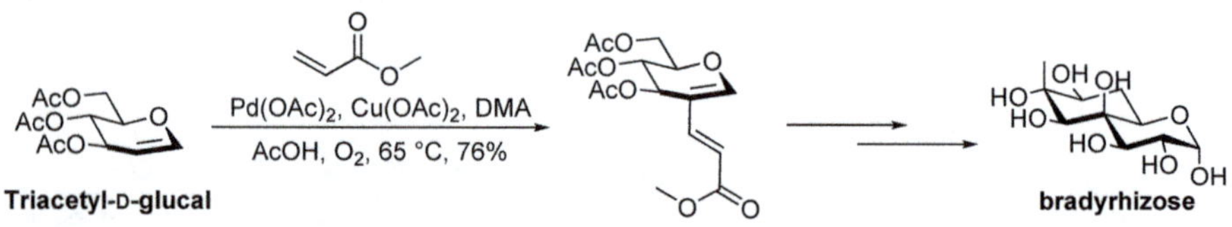

Scheme 2.18 Pd-catalyzed direct cross-coupling reaction and its application.

skeletons, synthesizing from *O*-, *N*- and *C*-glycosides originating from C2-acetoxyglucal (Scheme 2.17).[110] These hybrid sugar molecules were found to exhibit good activity as glycosidase inhibitors.

Meanwhile most reactions on glycosylation discussed above are centered about the anomeric carbon of glycals, functionalization of other positions can also provide access to natural products and carbohydrate derivatives. In an example of functionalizing C2 position of glycal with activated alkenes under mild conditions using Pd-catalyzed direct cross-coupling reaction of glycals, it was later applied to the synthesis of bradyrhizose, a unique C10 monosaccharide which is the unit of O-antigen polysaccharide related to a nitrogen fixation independent of Nod-factor (Scheme 2.18).[113,114]

2.5 Conclusion

A plethora of state-of-the-art strategies are now available to synthesize carbohydrates chemically, starting from non-sugar, pseudo-sugar and sugar materials. Considerations for orthogonal, regioselective and one-pot protecting groups on the sugar hydroxyl groups are examined prior to glycosylation steps. Factors affecting reactivity and stereochemical outcome for the key glycosylation step are also discussed. In particular, dual-directing approaches to obtain both α- and β-anomers with minimal changes to glycosyl donors, acceptors or reaction conditions are highlighted. The resultant glycosides have demonstrated their utility in applications such as the syntheses of oligosaccharides, polysaccharides, glycoconjugates, as well as natural products and their mimetics. Given the complexity of these molecules, the search for a universal strategy to gain access to them will continue to drive the field of carbohydrate research.

References

1. L. L. Kiessling and R. A. Splain, *Annu. Rev. Biochem.*, 2010, **79**, 619.
2. B. Lepenies, J. Yin and P. H. Seeberger, *Curr. Opin. Chem. Biol.*, 2010, **14**, 404.
3. Y. Wu, D.-C. Xiong, S.-C. Chen, Y.-S. Wang and X.-S. Ye, *Nat. Commun.*, 2017, **8**, 14851.
4. M. W. Giuliano and S. J. Miller, in *Topics in Current Chemistry*, ed. T. Kawabata, Springer International Publishing AG, 2016, vol. 372, p. 157.
5. (a) C. C. Wang, J. C. Lee, S. Y. Luo, S. S. Kulkarni, Y. W. Huang, C. C. Lee, K. L. Chang and S. C. Hung, *Nature*, 2007, **446**, 896; (b) Y. Wang, X.-S. Ye and L.-H. Zhang, *Org. Biomol. Chem.*, 2007, **5**, 2189.
6. L. N. Chan and M. S. Taylor, *Org. Lett.*, 2011, **13**, 3090.
7. B. Ren, O. Ramstrom, Q. Zhang, J. T. Ge and H. Dong, *Chem. - Eur. J.*, 2016, **22**, 2481.
8. G. Xiao, G. A. Cintron-Rosado, D. A. Glazier, B. M. Xi, C. Liu, P. Liu and W. Tang, *J. Am. Chem. Soc.*, 2017, **139**, 4346.
9. Y. A. Kwase, M. Cochran and M. Nitz, in *Modern Synthetic Methods in Carbohydrate Chemistry*, Wiley-VCH Verlag GmbH & Co. KGaA, 2013, pp. 67–96.
10. D. Lim, M. A. Brimble, R. Kowalczyk, A. J. A. Watson and A. J. Fairbanks, *Angew. Chem., Int. Ed.*, 2014, **53**, 11907.
11. K. Le Mai Hoang, J.-X. He, G. Bati, M. B. Chan-Park and X.-W. Liu, *Nat. Commun.*, 2017, **8**, 1146.
12. S. Y. Ko, A. W. M. Lee, S. Masamune, L. A. Reed, K. B. Sharpless and F. J. Walker, *Science*, 1983, **220**, 949.
13. S. Y. Ko, A. W. M. Lee, S. Masamune, L. A. Reed, K. B. Sharpless and F. J. Walker, *Tetrahedron*, 1990, **46**, 245.
14. S. J. Danishefsky, *Chemtracts: Org. Chem.*, 1989, **2**, 273.
15. S. J. Danishefsky and M. P. Deninno, *Angew. Chem., Int. Ed.*, 1987, **26**, 15.
16. A. B. Northrup and D. W. C. MacMillan, *Science*, 2004, **305**, 1752.

17. A. B. Northrup, I. K. Mangion, F. Hettche and D. W. C. MacMillan, *Angew. Chem., Int. Ed.*, 2004, **43**, 2152.
18. A. Dondoni and A. Marra, *Chem. Rev.*, 2004, **104**, 2557.
19. M. S. M. Timmer, A. Adibekian and P. H. Seeberger, *Angew. Chem., Int. Ed.*, 2005, **44**, 7605.
20. M. Brasholz and H. U. Reissig, *Angew. Chem., Int. Ed.*, 2007, **46**, 1634.
21. S. Matthies, P. Stallforth and P. H. Seeberger, *J. Am. Chem. Soc.*, 2015, **137**, 2848.
22. P. R. Andreana, J. S. McLellan, Y. C. Chen and P. G. Wang, *Org. Lett.*, 2002, **4**, 3875.
23. A. C. Comely, R. Eelkema, A. J. Minnaard and B. L. Feringa, *J. Am. Chem. Soc.*, 2003, **125**, 8714.
24. R. S. Babu and G. A. O'Doherty, *J. Am. Chem. Soc.*, 2003, **125**, 12406.
25. X. Zhu and R. R. Schmidt, *Angew. Chem., Int. Ed.*, 2009, **48**, 1900.
26. D. R. Mootoo, P. Konradsson, U. Udodong and B. Fraser-Reid, *J. Am. Chem. Soc.*, 1988, **110**, 5583.
27. C. M. Pedersen, L. U. Nordstrøm and M. Bols, *J. Am. Chem. Soc.*, 2007, **129**, 9222.
28. T. H. Schmidt and R. Madsen, *Eur. J. Org. Chem.*, 2007, **2007**, 3935.
29. H. Satoh and S. Manabe, *Chem. Soc. Rev.*, 2013, **42**, 4297.
30. M. D. Bandara, J. P. Yasomanee and A. V. Demchenko, in *Selective Glycosylations: Synthetic Methods and Catalysts*, Wiley-VCH Verlag GmbH & Co. KGaA, 2017, pp. 29–58.
31. Z. Zhang, I. R. Ollmann, X.-S. Ye, R. Wischnat, T. Baasov and C.-H. Wong, *J. Am. Chem. Soc.*, 1999, **121**, 734.
32. Y. Hsu, X.-A. Lu, M. M. L. Zulueta, C.-M. Tsai, K.-I. Lin, S.-C. Hung and C.-H. Wong, *J. Am. Chem. Soc.*, 2012, **134**, 4549.
33. R. R. Schmidt and J. Michel, *Angew. Chem., Int. Ed.*, 1980, **19**, 731.
34. B. Yu and J. Sun, *Chem. Commun.*, 2010, **46**, 4668.
35. Y. Li, X. Yang, Y. Liu, C. Zhu, Y. Yang and B. Yu, *Chem. - Eur. J.*, 2010, **16**, 1871.
36. R. Roy, F. O. Andersson and M. Letellier, *Tetrahedron Lett.*, 1992, **33**, 6053.
37. G.-J. Boons, B. Heskamp and F. Hout, *Angew. Chem., Int. Ed.*, 1996, **35**, 2845.
38. K. S. Kim, J. H. Kim, Y. J. Lee, Y. J. Lee and J. Park, *J. Am. Chem. Soc.*, 2001, **123**, 8477.
39. S. J. Hasty, M. A. Kleine and A. V. Demchenko, *Angew. Chem., Int. Ed.*, 2011, **50**, 4197.
40. P. Shu, X. Xiao, Y. Zhao, Y. Xu, W. Yao, J. Tao, H. Wang, G. Yao, Z. Lu, J. Zeng and Q. Wan, *Angew. Chem., Int. Ed.*, 2015, **54**, 14432.
41. D. Crich and V. Dudkin, *J. Am. Chem. Soc.*, 2001, **123**, 6819.
42. L. Bohé and D. Crich, *Trends Glycosci. Glycotechnol.*, 2010, **22**, 1.
43. R. Lemieux, K. Hendriks, R. Stick and K. James, *J. Am. Chem. Soc.*, 1975, **97**, 4056.
44. A. V. Demchenko, *Curr. Org. Chem.*, 2003, **7**, 35.

45. T. J. Boltje, J.-H. Kim, J. Park and G.-J. Boons, *Nat. Chem.*, 2010, **2**, 552.
46. M. A. Fascione, S. J. Adshead, S. A. Stalford, C. A. Kilner, A. G. Leach and W. B. Turnbull, *Chem. Commun.*, 2009, **0**, 5841.
47. G. P. Singh, A. J. Watson and A. J. Fairbanks, *Org. Lett.*, 2015, **17**, 4376.
48. A. Ishiwata, Y. J. Lee and Y. Ito, *Org. Biomol. Chem.*, 2010, **8**, 3596.
49. J. T. Walk, Z. A. Buchan and J. Montgomery, *Chem. Sci.*, 2015, **6**, 3448.
50. A. Ishiwata, A. Sakurai, Y. Nishimiya, S. Tsuda and Y. Ito, *J. Am. Chem. Soc.*, 2011, **133**, 19524.
51. J. Y. Baek, B.-Y. Lee, M. G. Jo and K. S. Kim, *J. Am. Chem. Soc.*, 2009, **131**, 17705.
52. Y. Ma, G. Lian, Y. Li and B. Yu, *Chem. Commun.*, 2011, **47**, 7515.
53. J. P. Yasomanee and A. V. Demchenko, *J. Am. Chem. Soc.*, 2012, **134**, 20097.
54. S. Shuto, Y. Yahiro, S. Ichikawa and A. Matsuda, *J. Org. Chem.*, 2000, **65**, 5547.
55. Y. Okada, O. Nagata, M. Taira and H. Yamada, *Org. Lett.*, 2007, **9**, 2755.
56. K. Toshima, Y. Nozaki and K. Tatsuta, *Tetrahedron Lett.*, 1991, **32**, 6887.
57. T. Nukada, A. Bérces and D. M. Whitfield, *Carbohydr. Res.*, 2002, **337**, 765.
58. H. Abe, S. Shuto and A. Matsuda, *J. Am. Chem. Soc.*, 2001, **123**, 11870.
59. F. Zhu, M. J. Rourke, T. Yang, J. Rodriguez and M. A. Walczak, *J. Am. Chem. Soc.*, 2016, **138**, 12049.
60. J. H. Kim, H. Yang, V. Khot, D. Whitfield and G. J. Boons, *Eur. J. Org. Chem.*, 2006, **2006**, 5007.
61. J. R. Krumper, W. A. Salamant and K. Woerpel, *J. Org. Chem.*, 2009, **74**, 8039.
62. M. G. Beaver and K. A. Woerpel, *J. Org. Chem.*, 2010, **75**, 1107.
63. K. Le Mai Hoang and X.-W. Liu, *Nat. Commun.*, 2014, **5**, 5051.
64. B. M. Trost and D. L. Van Vranken, *Chem. Rev.*, 1996, **96**, 395.
65. S. Xiang, K. Le Mai Hoang, J.-X. He, Y.-J. Tan and X.-W. Liu, *Angew. Chem., Int. Ed.*, 2015, **54**, 604.
66. B. Schumann, S. G. Parameswarappa, M. P. Lisboa, N. Kottari, F. Guidetti, C. L. Pereira and P. H. Seeberger, *Angew. Chem., Int. Ed.*, 2016, **55**, 14431.
67. S. van der Vorm, T. Hansen, H. S. Overkleeft, G. A. van der Marel and J. D. C. Codée, *Chem. Sci.*, 2017, **8**, 1867.
68. K.-K. T. Mong, T. Nokami, N. T. T. Tran and P. B. Nhi, in *Selective Glycosylations: Synthetic Methods and Catalysts*, Wiley-VCH Verlag GmbH & Co. KGaA, 2017, pp. 59–77.
69. S.-R. Lu, Y.-H. Lai, J.-H. Chen, C.-Y. Liu and K.-K. T. Mong, *Angew. Chem., Int. Ed.*, 2011, **50**, 7315.
70. C.-S. Chao, C.-Y. Lin, S. Mulani, W.-C. Hung and K.-K. T. Mong, *Chem. - Eur. J.*, 2011, **17**, 12193.
71. F. Yang, Y. Zhu and B. Yu, *Chem. Commun.*, 2012, **48**, 7097.
72. T. Kimura, T. Eto, D. Takahashi and K. Toshima, *Org. Lett.*, 2016, **18**, 3190.

73. J. M. Nogueira, M. Bylsma, D. K. Bright and C. S. Bennett, *Angew. Chem., Int. Ed.*, 2016, **55**, 10088.
74. J. P. Issa and C. S. Bennett, *J. Am. Chem. Soc.*, 2014, **136**, 5740.
75. K. J. Doores and B. G. Davis, *Org. Biomol. Chem.*, 2008, **6**, 2692.
76. A. K. Kayastha, X. G. Jia, J. P. Yasomanee and A. V. Demchenko, *Org. Lett.*, 2015, **17**, 4448.
77. L. Sun, X. Wu, D. C. Xiong and X. S. Ye, *Angew. Chem., Int. Ed.*, 2016, **55**, 8041.
78. Y. Park, K. C. Harper, N. Kuhl, E. E. Kwan, R. Y. Liu and E. N. Jacobsen, *Science*, 2017, **355**, 162.
79. H. Kim, H. Men and C. Lee, *J. Am. Chem. Soc.*, 2004, **126**, 1336.
80. J. Yang, G. J. Mercer and H. M. Nguyen, *Org. Lett.*, 2007, **9**, 4231.
81. H. Yao, S. Zhang, W.-L. Leng, M.-L. Leow, S. Xiang, J.-X. He, H. Liao, K. Le Mai Hoang and X.-W. Liu, *ACS Catal.*, 2017, **7**, 5456.
82. H.-Y. Wang, K. Yang, D. Yin, C. Liu, D. A. Glazier and W. Tang, *Org. Lett.*, 2015, **17**, 5272.
83. S.-S. Chang, C.-H. Shih, K.-C. Lai and K.-K. T. Mong, *Chem. - Asian J.*, 2010, **5**, 1152.
84. L. O. Kononov, N. N. Malysheva and A. V. Orlova, *Eur. J. Org. Chem.*, 2009, **2009**, 611.
85. J. Kalikanda and Z. Li, *J. Org. Chem.*, 2011, **76**, 5207.
86. K. N. Baryal, D. Zhu, X. Li and J. Zhu, *Angew. Chem., Int. Ed.*, 2013, **52**, 8012.
87. A. Ishiwata and Y. Ito, *Trends Glycosci. Glycotechnol.*, 2009, **21**, 266.
88. K. S. Kim, D. B. Fulse, J. Y. Baek, B.-Y. Lee and H. B. Jeon, *J. Am. Chem. Soc.*, 2008, **130**, 8537.
89. B.-Y. Lee, J. W. Oh, J. Y. Baek, H. B. Jeon and K. S. Kim, *J. Org. Chem.*, 2016, **81**, 11372.
90. Z. Wang, Z. S. Chinoy, S. G. Ambre, W. Peng, R. McBride, R. P. de Vries, J. Glushka, J. C. Paulson and G.-J. Boons, *Science*, 2013, **341**, 379.
91. S. S. Shivatare, S.-H. Chang, T.-I. Tsai, S. Y. Tseng, V. S. Shivatare, Y.-S. Lin, Y.-Y. Cheng, C.-T. Ren, C.-C. D. Lee, S. Pawar, C.-S. Tsai, H.-W. Shih, Y.-F. Zeng, C.-H. Liang, P. D. Kwong, D. R. Burton, C.-Y. Wu and C.-H. Wong, *Nat. Chem.*, 2016, **8**, 338.
92. A. Miermont, Y. Zeng, Y. Jing, X.-S. Ye and X. Huang, *J. Org. Chem.*, 2007, **72**, 8958.
93. J. Zhu, Q. Wan, G. Ragupathi, C. M. George, P. O. Livingston and S. J. Danishefsky, *J. Am. Chem. Soc.*, 2009, **131**, 4151.
94. S.-K. Wang, P.-H. Liang, R. D. Astronomo, T.-L. Hsu, S.-L. Hsieh, D. R. Burton and C.-H. Wong, *Proc. Natl. Acad. Sci. U. S. A.*, 2008, **105**, 3690.
95. P. Wu, X. Chen, N. Hu, U. C. Tam, O. Blixt, A. Zettl and C. R. Bertozzi, *Angew. Chem., Int. Ed.*, 2008, **47**, 5022.
96. A. Dondoni and A. Massi, *Acc. Chem. Res.*, 2006, **39**, 451.
97. O. Lockhoff, *Angew. Chem., Int. Ed.*, 1998, **37**, 3436.
98. D. G. Rivera, O. E. Vercillo and L. A. Wessjohann, *Org. Biomol. Chem.*, 2008, **6**, 1787.

99. N. Yamamoto, Y. Tanabe, R. Okamoto, P. E. Dawson and Y. Kajihara, *J. Am. Chem. Soc.*, 2008, **130**, 501.
100. R. Yang, K. K. Pasunooti, F. Li, X.-W. Liu and C.-F. Liu, *J. Am. Chem. Soc.*, 2009, **131**, 13592.
101. P. E. Dawson and S. B. Kent, *Annu. Rev. Biochem.*, 2000, **69**, 923.
102. H. Chai, K. Le Mai Hoang, M. D. Vu, K. Pasunooti, C. F. Liu and X.-W. Liu, *Angew. Chem., Int. Ed.*, 2016, **55**, 10363.
103. C. I. Schilling, N. Jung, M. Biskup, U. Schepers and S. Brase, *Chem. Soc. Rev.*, 2011, **40**, 4840.
104. E. Saxon and C. R. Bertozzi, *Science*, 2000, **287**, 2007.
105. A. Brik, S. Ficht, Y.-Y. Yang and C.-H. Wong, *J. Am. Chem. Soc.*, 2006, **128**, 15026.
106. R. J. Payne, S. Ficht, S. Tang, A. Brik, Y.-Y. Yang, D. A. Case and C.-H. Wong, *J. Am. Chem. Soc.*, 2007, **129**, 13527.
107. Y.-Y. Yang, S. Ficht, A. Brik and C.-H. Wong, *J. Am. Chem. Soc.*, 2007, **129**, 7690.
108. Y. Yang, X. H. Zhang and B. Yu, *Nat. Prod. Rep.*, 2015, **32**, 1331.
109. P. C. Gao, S. Y. Zhu, H. Cao and J. S. Yang, *J. Am. Chem. Soc.*, 2016, **138**, 1684.
110. A. A. Ansari, P. Rajasekaran, M. M. Khan and Y. D. Vankar, *J. Org. Chem.*, 2014, **79**, 1690.
111. R. Lorpitthaya, Z. Z. Xie, J. L. Kuo and X.-W. Liu, *Chem. - Eur. J.*, 2008, **14**, 1561.
112. J. Zeng, J. Ma, S. Xiang, S.-T. Cai and X.-W. Liu, *Angew. Chem., Int. Ed.*, 2013, **52**, 5134.
113. Y. Bai, J. Zeng, S. Cai and X.-W. Liu, *Org. Lett.*, 2011, **13**, 4394.
114. W. Li, A. Silipo, A. Molinaro and B. Yu, *Chem. Commun.*, 2015, **51**, 6964.

CHAPTER 3

Synthetically Useful Glycosyltransferases for the Access of Mammalian Glycomes

WANYI GUAN*[a], GARRETT A. EDMUNDS[b], ZHIGANG WU[c] AND LEI LI[b]

[a]College of Life Science, Hebei Normal University, Shijiazhuang, Hebei 050024, China; [b]Department of Chemistry, Georgia State University, Atlanta, GA 30303, USA; [c]College of Bioscience and Bioengineering, Hebei University of Science and Technology, Shijiazhuang, Hebei 050018, China
*E-mail: guanwanyi@hebtu.edu.cn

3.1 Introduction

Carbohydrates are one of the major classes of biomacromolecules and are ubiquitous throughout nature, playing roles in numerous biological events, including cell recognition and adhesion, molecular trafficking, signal transduction, tumor progression, immune modulation, *etc.*[1,2] As glycoscience enters the glycomics era, the availability of glycans (oligosaccharide chains)/glycoconjugate can not only facilitate the studies of unveiling their biological and biomedical functions, but also assist in glycomics analysis (by serving as standards) and in exploiting glycan-related therapeutics.[3] However, no universal technologies are available to provide the desired complex glycans/glycoconjugates.[4] Isolation of glycans from natural sources is complicated due to the relatively low abundance and the heterogeneity of glycans. Therefore,

Chemical Biology No. 11
Synthetic Glycomes
Edited by Wanyi Guan, Lei Li and Peng George Wang

Published by the Royal Society of Chemistry, www.rsc.org

syntheses has become the main approach for obtaining structurally defined glycans in sufficient amounts. Structurally, the components of glycans (*i.e.* monosaccharides or their derivatives) contain several hydroxyl groups with similar reactivities, leading to certain limitations associated with the chemical syntheses of glycans. One major drawback is the immense time and effort required to make a single complex oligosaccharide. A good rule of thumb is that each chemical glycosylation requires an average of seven reactions. Additionally, chemical syntheses are rarely quantitative in yield and frequently require purification. A proposed solution to these drawbacks is the use of enzymes for glycosylation.

Enzymes are exceptional in the preparation of molecules with specific regio- and stereo-selectivity, including glycans. Their mild reaction conditions (at room temperature to 37 °C, in aqueous solutions with neutral or slightly acidic or slightly alkaline pH) and high efficiency are very attractive for syntheses. Two major groups of enzymes are employed for the enzymatic or chemoenzymatic synthesis of glycans.[4] One group is glycosyltransferases (GTs, EC 2.4), which are the natural approach of glycan biosynthesis in organisms. GTs are a group of enzymes that can transfer a sugar moiety from its activated donor to the acceptor (either a sugar or a non-sugar biomolecule) by forming a regio- and stereo-specific glycosidic bond. GT-catalyzed reactions are optimal for glycan synthesis, as they afford defined glycan products in a single step under mild conditions. Certain considerations must be taken into account, such as the availability of acceptors and relatively expensive donors, access to appropriate glycosyltransferases, and the limitations on structures imposed by the use of enzymes. Another group is glycosidases (EC 3.2) and their engineered mutants, including glycosynthases and glycoligases. Glycosidases are intrinsically responsible for the hydrolysis of glycans. Under certain conditions, their reverse reactions are favorable and hence are utilized for glycan synthesis. For example, the sialidase BfGH33C from *Bacteroides fragilis* NCTC9343 has been explored to synthesize 6′-sialyllactose (6′-SL) for its strict α2–6-regioselectivity in transglycosylation to lactose.[5] However, the yields of wild-type glycosidase-catalyzed synthesis are relatively low because of its hydrolysis activity, making it less attractive for synthetic scope. To decrease the hydrolysis activity, based on rational design, researchers created glycosynthases by replacing the key nucleophilic residue that is responsible for hydrolysis and keeping the transglycosylation catalytic residue intact.[6–9] One successful application is the mutation of EndoM, an *endo*-β-*N*-acetylglucosaminidase (ENGase, belongs to *endo*-glycosidases) from *Mucor hiemalis* which naturally hydrolyzes the glycosidic bond within the *N,N*′-diacetylchitobiose in the core pentasaccharides of mammalian *N*-glycans. In this example, an N175Q mutation was introduced into EndoM, endowing the mutant excellent capability for transferring *N*-glycans or oxazoline *N*-glycans to peptides/proteins bearing a GlcNAc residue and forming the authentic GlcNAcβ1–4GlcNAc linkage, generating a valuable tool for the preparation of homogeneous glycopeptides/glycoproteins.[10–12] Furthermore, in combination with efficient high-throughput screening methods, directed evolution serves as a powerful tool to improve the synthetic activity and expand substrate specificity of existing glycosynthases.[13–17] Another engineering approach is to create glycoligases from glycosidases, by mutating the general acid/base residue of glycosidases to eliminate the hydrolysis activity while retaining their synthetic capabilities.[17–19] With this principle, Wang and co-workers converted the α1–6-fucosidase from *Lactobacillus casei* (AlfC) into a glycoligase, and efficiently core-fucosylated a variety of *N*-glycopeptides and glycoproteins with full-size *N*-glycans, both high-mannose and complex type included.[20] Although glycosidases and their mutants use inexpensive starting materials, regio-selectivity and conversion ratio are usually low, which require further engineering to improve yields. These concepts are beyond the scope of this review, and excellent reviews have been given elsewhere.[6,17] In this chapter, we will focus on the more specific and efficient GT-catalyzed syntheses of glycans within the scope of mammalian glycome. For enzyme-involved synthesis of glycoconjugates, especially glycoproteins, focused reviews are available elsewhere.[4]

3.2 Glycosyltransferases (GTs) and Their Usage in Mammalian Glycan Preparation

Glycosyltransferases (GTs) are responsible for the formation of specific glycosidic linkages. Based on sequence similarity, GTs have been classified into over 100 families in the Carbohydrate Active Enzymes (CAZy) database (http://www.cazy.org/). According to the stereochemistry of the reactions, GTs can be categorized into retaining GTs or inverting GTs. From a synthetic point of view, they can be grouped from the following aspects.

3.2.1 Leloir GTs and Non-Leloir GTs

GTs can be divided into Leloir GTs and non-Leloir GTs, depending on the type of donors they use.

Leloir GTs, named after the discoverer of the first sugar nucleotide Luis Federico Leloir, utilize sugar nucleotides (also known as nucleotide sugars) as donors for glycosylation. Leloir GTs are responsible for the syntheses of most glycoconjugates in mammals, and have been commonly applied for the syntheses of glycans and glycoconjugates.[21] Despite the tremendous diversity of structures, nine main components exist in mammalian glycans, and the corresponding sugar nucleotides are uridine 5′-diphosphate-D-glucose (UDP-Glc), uridine 5′-diphosphate-D-galactose (UDP-Gal), uridine 5′-diphosphate-*N*-acetyl-D-glucosamine (UDP-GlcNAc), uridine 5′-diphosphate-*N*-acetyl-D-galactosamine (UDP-GalNAc), guanidine 5′-diphosphate-D-mannose (GDP-Man), guanidine 5′-diphosphate-L-fucose (GDP-Fuc), uridine 5′-diphosphate-D-glucuronic acid (UDP-GlcA), uridine 5′-diphosphate-D-xylose (UDP-Xyl), and cytidine monophosphate-sialic

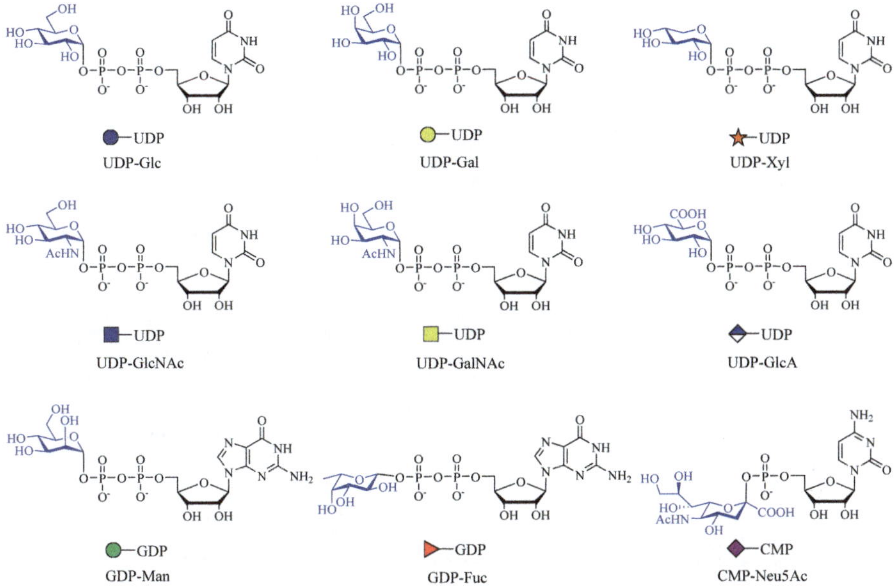

Figure 3.1 Nine common sugar nucleotides in mammalian glycans.

acid (CMP-Sia, including CMP-*N*-acetylneuraminic acid (CMP-Neu5Ac) and CMP-*N*-glycolylneuraminic acid (CMP-Neu5Gc)) (Figure 3.1).[22] These sugar nucleotides, while commercially available, can be prohibitively expensive to acquire. Tremendous efforts have been put into enzymatic preparation of these sugar donors. Taking advantage of salvage biosynthetic pathways of sugar nucleotides, so far, seven out of the nine compounds (excluding UDP-Glc and UDP-Xyl) can be enzymatically prepared through single- or two-step enzymatic reactions in gram scales (Scheme 3.1).[23–32]

Non-Leloir GTs, on the other hand, use non-nucleotide sugar donors, including sugar-1-phosphates, sugar-1-pyrophosphates, lipid phospho-sugar, lipid pyrophospho-sugar, or preexisting oligosaccharides.[33] For instance, in endoplasmic reticulum (ER), the enzyme initiating the *N*-glycan synthesis is a non-Leloir GT, which requires dolicol-pyrophospho-*N*-acetylglucosamine (Dol-P-P-GlcNAc) as sugar donor.[34] The difficulties in the preparation of lipid-diphosphate-sugar limited the application of the corresponding non-Leloir GTs in glycan syntheses. Exceptionally, the non-Leloir D-galactosyl-β1–3-*N*-acetyl-D-hexosamine phosphorylase from *Bifidobacterium infantis* (BiGalHexNAcP) which uses galactose-1-P (Gal-1-P) as sugar donor served as an excellent catalyst for Galβ1–3GlcNAc/GalNAc preparation.[35,36]

3.2.2 Mammalian GTs and Bacterial GTs

In the synthetic effort of mammalian glycans, researchers first turn towards mammalian GTs, the natural catalysts for their biosynthesis. They exhibit exquisite substrate specificity toward both donors and acceptors, reflecting

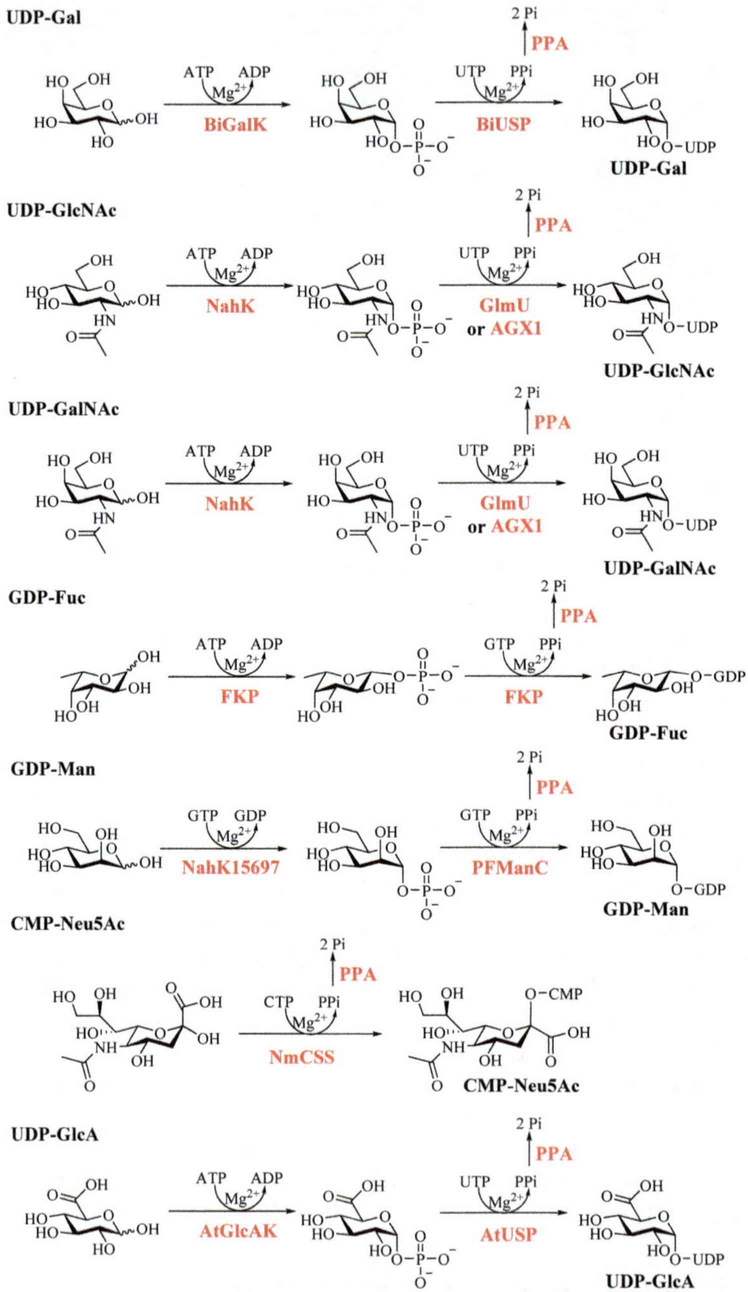

Scheme 3.1 Enzymatic preparation of sugar nucleotides common in mammals.

the stringency of glycan processing in mammals. Hence, they are ideal for the synthesis of mammalian glycans with complex structures. However, to obtain these GTs in an active form is relatively challenging. In early application of mammalian GTs, they were obtained *via* purification from natural sources. As gene technology has advanced, many mammalian GTs have been cloned, expressed as recombinant proteins and characterized.[37] Despite endeavors putting the functional expression of mammalian GTs in recombinant bacterial systems, only a small number of mammalian GTs have been functionally expressed in bacterial system as soluble active truncated catalytic domains,[38–40] or through *in vitro* refolding.[41] Eukaryotic expression systems, which contain post-translational modifications that are normally required for correct folding of proteins, are optimal for recombinant expression of mammalian GTs. Using mammalian cell and baculovirus-insect cell expression systems, the Moremen and Jarvis groups have expressed all human glycan-processing enzymes, including GTs.[3] In their efforts, mainly the catalytic domains were expressed as *N*- or *C*-terminal fusion proteins. The expression levels of GTs varied, but over 92% GTs can be expressed at >10 mg L^{-1} using mammalian hosts, and can be purified at multimilligram level.[3] However, the handling of mammalian cell and insect cell expression systems requires extensive expertise and the expression cost is relatively high compared to the *Escherichia coli* expression system.

Although using a bacterial expression system is not ideal for producing functional eukaryotic GTs, numbers of equivalent GTs have been identified from bacteria and applied in the syntheses of mammalian glycans (see Section 3.3 and Table 3.1). These GTs are most often discovered in pathogenic bacteria, which can produce oligosaccharides that mimic mammalian glycans, including but not limited to T antigen, sialyl-T antigen, type I/II chains, Lewis antigens, histo-blood group antigens, and polysialic acids, and assist in their invasion to hosts.[42] For example, α2–6-sialyltransferase from *Photobacterium damselae* (Pd2,6ST) is able to extend glycans with terminal and internal α2–6-linked sialic acid, mimicking the mammalian glycans.[43,44] α1–3-Fucosyltransferase from *Helicobacter pylori* has been frequently used to prepare glycans containing $Lewis^X$ (Le^X) epitope.[45,46] A bifunctional glycosyltransferase, hyaluronan (hyaluronic acid, HA) synthase from *Pasteurella multocida*, has been applied for the synthesis of HA.[47] Bacterial GTs are usually well-expressed in the *E. coli* expression system as recombinant enzymes and can often be easily purified for synthetic use. Typically, their substrate specificities are broad, *i.e.* a single GT can catalyze the transfer of the corresponding sugar residue to a variety of acceptors. Therefore, they offer an alternative source of GTs for mammalian glycan preparation. However, when opting for bacterial GTs, researchers must bear in mind that the substrate specificity (mostly acceptor specificity) of bacterial GTs is sometimes broader than their mammalian counterparts, potentially leading to the generation of undesired byproducts.

Table 3.1 Synthetically useful glycosyltransferases in mammalian glycan syntheses. Adapted from ref. 4 with permission from American Chemical Society, Copyright 2011.

Enzyme (donor)	Entry	Glycosyl accepter(s)	Known acceptor specificity	Linkage(s) synthesized	Eyzyme name	Source	Reference
GlcNAcT (UDP-GlcNAc)	1	GlcA	Multiple	α1–4-	α4GlcNAcT	Human	62–64
	2	Man	$Man_5GlcNAc_2$	β1–2-	GlcNAcT	Rabbit	65
	3	Gal	LacNAc	β1–3-	B3GNT1	Human	66
	4	Gal	LacNAc	β1–3-	B3GNT2	Human	68,69
	5	Gal	Lac/LacNAc/others	β1–3-	NmLgtA	*N. meningitidis*	70,71
	6	Gal	Multiple	β1–3-	Hpβ3GlcNAcT	*H. pylori*	71,72
	7	GlcNAc	Multiple	β1–4-	β4GlcNAcT	*L. stagnalis*, *A. caulinodans*	74,75
	8	GlcA	Multiple	β1–4-	PmHAS	*P. multocida*	47,76,77
	9	Gal	Internal Gal residue in polyLacNAc	β1–6-	CIGnT6 (GCNT2)	Human	69,78,79
	10	GalNAc	Galβ1–3GalNAcα–	β1–6-	C2GnT	Murine	80,81
	11	GalNAc	Galβ1–3GalNAcα–/ GlcNAcβ1–3GalNAcα–	β1–6-	C2GnT2 (GCNT3)	Human	3,82
	12	Man	Multiple, especially α-linked Man in GlcNAcβ1–2Manα– sequence on α1–6Man branch of *N*-glycans	β1–6-	GnT-V (MGAT5)	Human	3,84–87
	13	Man	*O*-Man glycans, α-linked Man in GlcNAcβ1–2Manα– sequence in *N*-glycans	β1–6-	GnT-Vb (MGAT5b, GnT-IX)	Human	3,83,84,88

GalT (UDP-Gal)	14	Gal	Fucα1–2Gal	α1–3-	GTB	Human	89
	15	Gal	Lac/LacNAc	α1–3-	α3GalT	Bovine	46,90–93
	16	Gal	Fucα1–2Gal	α1–3-	WbnI	*E. coli*	94
	17	Gal	Lac	α1–4-	NmLgtC	*N. meningitidis*	93,95–97
	18	Gal	Lac	α1–4-	α4GalT	*N. gonorrheae*	98
	19	Gal	Galβ–/Galβ1–4Xylβ–	β1–3-	B3GALT6	Human	99,100
	20	GlcNAc	GlcNAcβ1–3Galβ1–4Glcβ–	β1–3-	B3GALT1,2	Human	101,102
	21	GlcNAc	Multiple	β1–3-	B3GALT5	Human	69
	22	GlcNAc	GlcNAcβ1–3Galβ1–4Glcβ–	β1–3-	WbgO	*E. coli*	103
	23	GlcNAc/ GalNAc	Multiple	β1–3-	BiGalHexNAcP[a]	*B. infantis*	35
	24	GalNAc	GalNAcα–	β1–3-	C1GALT (with COSMC)	Human	80,82
	25	GalNAc	GalNAcα–	β1–3-	WbiP	*E. coli*	104
	26	GalNAc	Globotetraose	β1–3-	HiLgtD	*H. influenzae*	105
	27	GalNAc	GalNAcβ1–4(Neu5Acα2–3) Galβ1–4Glc	β1–3-	CjCgtB	*C. jejuni*	106,107
	28	GlcNAc/Glc	Multiple	β1–4-	B4GalT	Bovine	45,68,69,71
	29	Man/GlcNAc/ Glc	Multiple	β1–4-	NmLgtB	*N. meningitidis*	44,71,72,91, 108–110
	30	GlcNAc	Multiple	β1–4-	Hpβ4GalT	*H. pylori*	71,108
	31	Xyl	Xylβ–	β1–4-	B4GALT7	Human	67,111

(*continued*)

Table 3.1 *(continued)*

Enzyme (donor)	Entry	Glycosyl accepter(s)	Known acceptor specificity	Linkage(s) synthesized	Eyzyme name	Source	Reference
FucT (GDP-Fuc)	32	Gal	Terminal Gal in type I (Galβ1–3GlcNAc) or type II (Galβ1–4GlcNAc) glycans	α1–2-	FUT1	Human	69
	33	Gal	Terminal Gal in Galβ1–3GalNAc–/globopentaose	α1–2-	WbnK	*E. coli*	90,94,112
	34	Gal	Terminal Gal in Lac/Galβ1–4Fru/Galβ1–4Man/Galβ1–3GalNAcα–/globopentaose	α1–2-	WbsJ	*E. coli*	95,113
	35	Gal	Lac/LacNAc	α1–2-	WbgL	*E. coli*	89,114
	36	Gal	Galβ1–3GalNAc–	α1–2-	WbiQ	*E. coli*	115
	37	Gal	Prefers Galβ1–3GlcNAc to Galβ1–4GlcNAc	α1–2-	FutC	*H. pylori*	89,116
	38	Gal	Galβ1–3GlcNAc–/Galβ1–3GalNAc–	α1–2-	Teα2FucT (Te2FT)	*T. elongatus*	117
	39	GlcNAc	LacNAc/NeuAcα2–3Galβ1–4GlcNAc/GalNAcβ1–4GlcNAc	α1–3-	FUT6	Human	118,119
	40	GlcNAc	NeuAcα2–3Galβ1–4GlcNAc	α1–3-	FUT7	Human	120
	41	GlcNAc	Distal GlcNAc in polyLacNAc chain	α1–3-	FUT9	Human	69,121
	42	GlcNAc	M3/M3F6/LDNT/LN-pNP/LNnT	α1–3-	CEFT(1-4)	*C. elegans*	122
	43	GlcNAc	LacNAc/NeuAcα2–3Galβ1–4GlcNAc	α1–3-	Hpα3FucT (Hp3FT)	*H. pylori*	53,123–126
	44	GlcNAc	LacNAc	α1–3-	FutA	*H. pylori*	73,127
	45	GlcNAc	LacNAc/NeuAcα2–3Galβ1–4GlcNAc	α1–3-	HhFT1	*H. hepaticus*	128
	46	GlcNAc	LacNAc/NeuAcα2–3Galβ1–4GlcNAc	Predominantly α1–3-	FUT5	Human	69,129–133
	47	GlcNAc	LacNAc/Galβ1–3GlcNAc	α1–3/4-	FUT3	Human	69,134,135
	48	GlcNAc	LacNAc/Galβ1–3GlcNAc	α1–3/4-	Hpα3/4FucT (Hp3/4FT)	*H. pylori*	136
	49	GlcNAc	Galβ1–3GlcNAcβ–	α1–4-	α4FucT	*H. pylori*	137
	50	GlcNAc	Reducing end GlcNAc in *N*-glycan core pentasaccharide, with a free α1–2-GlcNAc on α1–3-Man branch	α1–6-	FUT8	Human	87,138,139

SiaT (CMP-Neu5Ac)	52	Gal	Terminal Gal	α2–3-	ST3GalIV	Human	68,133,141
	53	Gal	LacNAc	α2–3-	α3SiaT	*S. frugiperda*	142
	54	Gal	Fucosylated acceptors (Lewis A and Lewis X)	α2–3-	v-ST3Gal I	Myxoma virus	143
	55	Gal	Multiple	α2–3-	JTFAJ-16	*Vibrio* sp.	126
	56	Gal	Lac/LacNAc/α and β galactosides/multiple	α2–3-	α3SiaT	*N. meningitidis*	106,144–148
	57	Gal	Multiple	α2–3-	α3SiaT	*N. gonorrheae*	149
	58	Gal	Terminal Gal	α2–3-	PmST1	*P. multocida*	124,125, 150,151
	59	Gal	Terminal Gal	α2–6-	ST6GAL1	Human	69,133,141
	60	GalNAc	GalNAc in sialyl-T antigen	α2–6-	ST6GalNAc-IV	Human	152
	61	Gal	LacNAc	α2–6-	α6SiaT	*S. frugiperda*	142
	62	Gal	Lac/LacNAc	α2–6-	α6SiaT	*Photobacterium* sp.	126,153
	63	Gal/GalNAc	Terminal and internal Gal; GalNAc	α2–6-	Pd2,6ST	*P. damselae*	43,56,125
	64	Gal	Terminal Gal	α2–6-	Pd2,6ST mutant	*P. damselae*	56
	65	Gal	Terminal Gal	α2–6-	Pd2,3ST mutant	*P. dagmatis*	54
	66	Gal	Terminal Gal	α2–6-	PmST1 mutant	*P. multocida*	55
	67	Gal/Sia	Gal for α2–3-sialylation; Sia for α2–8-sialylation	α2–3/8-	CjCstII	*C. jejuni*	48
	68	Sia	Multiple	α2–8/9-	α8/9SiaT	*E. coli*	154

(*continued*)

Table 3.1 (*continued*)

Enzyme (donor)	Entry	Glycosyl accepter(s)	Known acceptor specificity	Linkage(s) synthesized	Eyzyme name	Source	Reference
GalNAcT (UDP-GalNAc)	69	Gal	Fucα1–2Gal	α1–3-	GTA	Human	155,156
	70	Gal	Fucα1–2Gal	α1–3-	BgtA	*H. mustelae*	89,157
	71	GalNAc	GalNAcα–PP-alkyl	α1–3-	WbnH	*E. coli*	90,158
	72	GalNAc	Globotetrose	α1–3-	Pm1138	*P. multocida*	163
	73	Gal	Globotriose	β1–3-	CgtE	*C. jejuni*	163
	74	Gal	Globotriose	β1–3-	HiLgtD	*H. influenzae*	105,160
	75	Gal	NeuAcα2–3Galβ1–4Glc	β1–4-	CjCgtA	*C. jejuni*	106
	76	GlcNAc	GlcNAc	β1–4-	β4GalNAcT	Human	159
	77	GlcA	GlcAβ1–3Galβ–	β1–4-	β4GalNAcT	Human	62,162
	78	GalNAc	Multiple	β1–4-	β4GalNAcT	*T. ni*	161
	79	GlcNAc	*N,N′*-diacetylchitobiose	β1–4-	β4GalNAcT	*C. elegans*	118
GlcAT (UDP-GlcA)	80	Gal	Galβ1–3Galβ1–4Xyl–	β1–3-	GlcAT-I	Human	164–166
	81	Gal	Multiple	β1–3-	GlcAT-P, GlcAT-S	Rat	167,168
	82	Gal	Galβ1–3Galβ1–4Xyl–	β1–3-	DmGlcAT-I	*D. melanogaster*	169
	83	Gal	Multiple	β1–3-	DmGlcAT-BS	*D. melanogaster*	169
	84	GalNAc	Chondroitin	β1–3-	K4CP	*E. coli*	170
	85	GlcNAc	Hyaluronan	β1–3-	PmHAS	*P. multocida*	76
	86	GalNAc	Chondroitin	β1–3-	PmCS	*P. multocida*	171
ManT (GDP-Man)	87	Man	Polymannose	α1–2-	α2ManT	*C. albicans*	172,173
	88	Man	Multiple	α1–6-	α6ManT	*M. smegmatis*	174
	89	GlcNAc	*N,N′*-diacetylchitobiose	β1–4-	β4ManT	*S. cerevisiae*	175,176

[a]Using Gal-1-P as sugar donor.

3.2.3 Wild-type GTs and Engineered GTs

Wild-type GTs are commonly used in the syntheses of mammalian glycans. However, certain characteristics and activity of wild-type GTs are not ideal for their synthetic application. Hence, efforts have been put into engineering wild-type GTs to gain the desired characteristics and activity. For example, many GTs, both from mammals and from bacteria, are used in their truncated version to gain increased solubility.[3,47–51,53] To reduce donor hydrolysis and product hydrolysis, Chen and co-workers performed site-directed mutagenesis to *P. multocida* sialyltransferase 1 (PmST1). The resulting mutant PmST1 M144D not only exhibited highly improved efficiency for α2–3-sialylation, but also gained broadened acceptor specificity (toward fucosylated acceptor Le^X).[52] In another case, several groups have attempted to generate a bacterial α2–6-sialyltransferase that can specifically sialylate terminal Gal even with the existence of internal Gal/GalNAc. To this end, the key amino acids in the active sites of two bacterial α2–3-sialyltransferases, Pd2,3ST[54] and PmST1,[55] were redesigned to switch the linkage selectivity (from α2–3- to α2–6-linkage by tuning the orientation of acceptor) but keep their regio-selectivity (terminal sialylation) intact. Conversely, the amino acids in acceptor binding pocket of Pd2,6ST, which naturally lacks regio-selectivity, were engineered to accommodate only terminal Gal/GalNAc residue, hence gaining terminal regio-selectivity.[56] Due to the limited understanding on the structure–activity relationship of GTs, directed evolution coupled with high-throughput screening offers an additional useful approach for engineering GTs.[57] By this approach, α2–3/8-sialyltransferase CstII from *Campylobacter jejuni* (CjCstII) was engineered to exhibit improved ability to transfer a negatively charged Neu5Ac to a fluorescence-labelled neutral glycan.[58] By using an ultra-high-throughput screening approach, a variant of *C. jejuni* β1–3-galactosyltransferase CgtB (CjCgtB) that has broader substrate tolerance and 300-fold higher activity compared with the wild-type enzyme was obtained from a library of $>10^7$ CgtB mutants. Using this variant, two important oligosaccharides, GM1a and asialo GM1 were synthesized.[59]

3.3 Synthetically Useful Glycosyltransferases

A large number of GTs have had their glycosylation activities characterized, yet only a small portion of them have been applied in mammalian glycan preparation. Wong and co-workers reviewed synthetically relevant GTs almost a decade ago.[4] In this chapter, we will include published applications of GTs in the syntheses of mammalian glycans (Table 3.1). Since Glc residues are rarely observed in mature structures of mammalian glycans, with the exception of in the reducing end of lactose, no glucosyltransferases (GlcTs) have been reported to date for mature mammalian glycan preparation. In mammals, Xyl residues are the key residues that link glycosaminoglycan (GAG) chains to core proteins. Therefore, xylosyltransferases (XylTs) are responsible for protein xylosylation[60,61] rather than glycan synthesis, which is out of the scope of this review. Hence, neither GlcTs nor XylTs will be discussed here.

3.3.1 *N*-Acetylglucosaminyltransferases (GlcNAcTs)

GlcNAc is a common sugar residue that is usually located in the core structure or backbone of glycans, *e.g.* core pentasaccharide of *N*-glycans, type I glycan (Galβ1–3GlcNAcβ–R), type II glycan (Galβ1–4GlcNAcβ–R) including poly-*N*-acetyllactosamine (polyLacNAc), I antigen Galβ1–4GlcNAcβ1–6(Galβ1–4GlcNAcβ1–3)Gal, hyaluronic acid (hyaluronan, HA) [(–4GlcAβ1–3GlcNAcβ1–)$_n$], *etc.* Among these structures, the *N*-glycan core pentasaccharide is biosynthesized within ER by a series of membrane integrated GTs that are integrated membrane proteins, therefore, the enzymatic syntheses of these structures is very challenging and is thus usually achieved by chemical preparation.[45,68,126,141] Enzymatic syntheses of GlcNAc-containing glycans usually focus on the preparation of the peripheral portion of glycans. Out of these structures, the ones most adjacent to core structures usually need to be synthesized by mammalian GTs, while the ones close to termini frequently have bacterial mimicry and can be prepared by bacterial GTs (Table 3.1). For example, the branches of complex *N*-glycans start with a GlcNAc attaching to the branching Man within the core pentasaccharide. These GlcNAc residues can be enzymatically assembled using mammalian GlcNAcTs and no bacterial equivalents have been identified to date. For instance, human MGAT5 (also known as GnT-V) can transfer GlcNAc to the α1–6-Man branch, forming a GlcNAcβ1–6 Man linkage. MGAT5 requires a free GlcNAc residue that is α1–2-linked to the α1–6-Man for glycosylation (Figure 3.2A).[177,178] With the aid of recombinant human MGAT5 expressed in a baculovirus-insect expression system, Li and co-workers determined the structure of galactosidase-hydrolyzed glycans and achieved the syntheses of asymmetrically branched *N*-glycans from isolated *N*-glycans.[87] The β1–6GlcNAcT that is used for branching of I antigen is also a mammalian enzyme (Figure 3.2B). With this human β6GlcNAcT (GCNT2), Boons and co-workers prepared various complex human milk oligosaccharides (HMOs) with branches.[69] Another human β6GlcNAcT (GCNT3, also named C2GnT2) was used by Inka and co-workers for the synthesis of core 2 and core 4 *O*-GalNAc core structures (Figure 3.2C).[82] In type I/II glycans, the β1–3GlcNAc is commonly transferred by bacterial β3GlcNAcTs, such as NmLgtA[71] and *H. pylori* β3GlcNAcT (Hpβ3GlcNAcT) (Figure 3.2D).[71,72] The oligomers[77] and polymers[47] of glycosaminoglycan HA can also be prepared by bacterial enzymes, PmHAS, for example. This bifunctional GT has two distinct catalytic domains, one of which is specific for β1–4GlcNAc transfer (Figure 3.2E).[76]

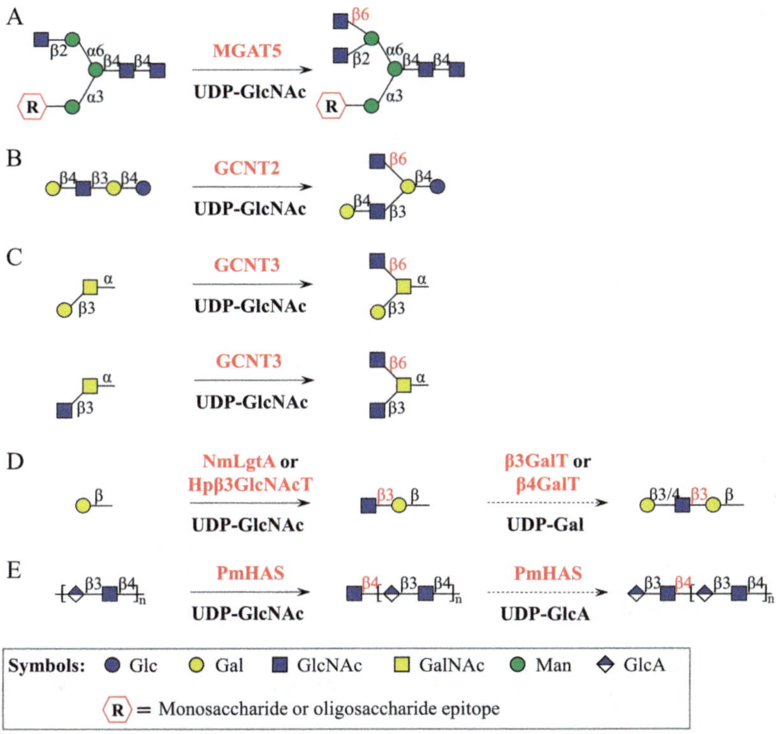

Figure 3.2 Glycan syntheses catalyzed by GlcNAcTs.

3.3.2 Galactosyltransferases (GalTs)

Gal is a hexose that is almost ubiquitous in all types of glycans. It can appear concomitantly with GlcNAc to form type I/II glycans. Both Leloir and non-Leloir GalTs have been used in glycan syntheses (Table 3.1). The Gal directly linked to *N*-glycan core pentasaccharide cannot be easily added by bacterial GalTs. In this case, bovine β4GalT (B4GALT1) is an excellent catalyst (Figure 3.3A).[45] But for peripheral type II glycan (Galβ1–4GlcNAc, also known as *N*-acetyllactosamine, LacNAc) syntheses, bacterial β4GalTs, such as NmLgtB and *H. pylori* β4GalT are robust (Figure 3.3B),[108] and can easily form polyLacNAc.[72] For the syntheses of type I glycans, WbgO from *E. coli* O55 serves as a nice catalyst (Figure 3.3C), and was used in the preparation of lacto-*N*-tetraose (LNT).[103] In the Gal-containing blood B antigen, Gal can be transferred by human GalT (GTB)[89] or bacterial equivalent WbnI from *E. coli* O86 (Figure 3.3D).[94] The xenotransplantation antigen α-Gal, isoglobotriose (with the same sequence Galα1–3Galβ1–4Glc) or linear blood group B (Galα1–3Gal–)[179] can be synthesized *via* bovine α3GalT-catalyzed reaction

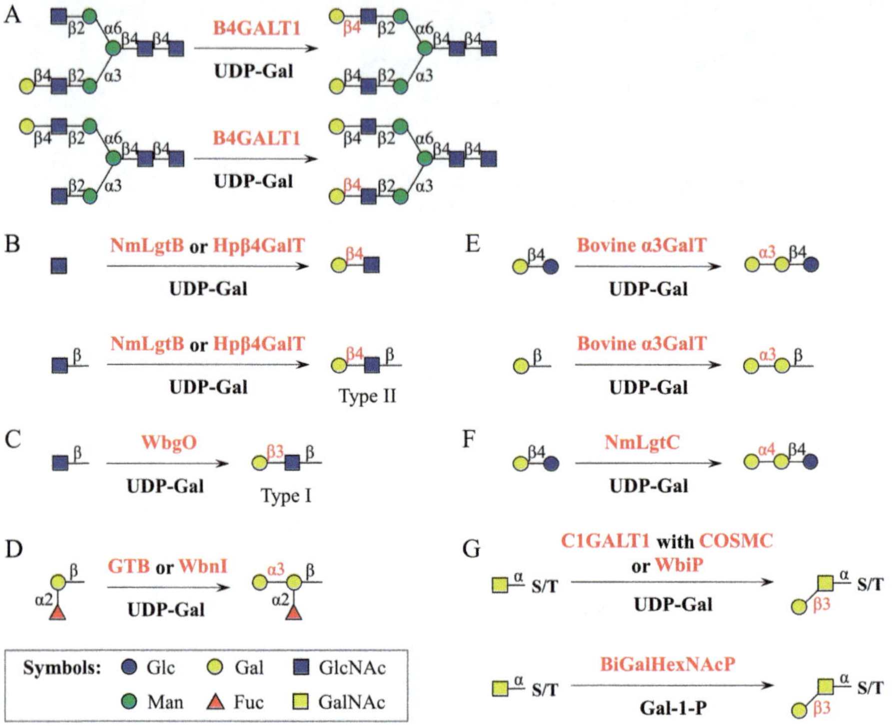

Figure 3.3 Glycan syntheses catalyzed by GalTs.

(Figure 3.3E). Bovine α3GalT is a mammalian enzyme, but can be effectively expressed in *E. coli* in an active form.[40] In contrast, the tumor associated antigen globotriose (Gb3) and P^k blood group antigen (with the same sequence Galα1–4Galβ1–4Glc) has bacterial mimicry in *Neisseria meningitidis*, and can be prepared by the corresponding α4GalT NmLgtC (Figure 3.3F).[96] In the synthesis of another tumor antigen T-antigen, the enzymatic addition of Gal has more options. The authentic mammalian GalT (C1GALT1, also called T-synthase) needs a molecular chaperon (COSMC) to act correctly. The co-expression of T-synthase with COSMC in insect cells was able to provide T-antigen (Figure 3.3G),[82] but the expertise required has limited its application. Alternatively, BiGalHexNAcP, a non-Leloir GT that can transfer Gal from Gal-1-P to GlcNAc or GalNAc, offered an alternative method for the preparation of T-antigen[35] (Figure 3.3G) or its mimics.[36,180] In addition, a bacterial β3GalT (WbiP) from *E. coli* O127 was also reported to efficiently synthesize T-antigen (Figure 3.3G).[104]

3.3.3 Fucosyltransferases (FucTs)

FucTs are the GTs responsible for the transfer of Fuc residues from GDP-Fuc to acceptors (Table 3.1). As a C-6 deoxy L-hexose, Fuc is usually considered as a modification to the glycan backbones and endows the modified glycans a distinctive binding profile to glycan binding proteins. This gives rise to the prominence of various Fuc-containing glycan epitopes, such as ABO histo-blood groups, Lewis antigens and Globo H antigen. In the glycan epitopes, Fuc usually modifies terminal glycan structures by forming α1–2-, α1–3-, or α1–4-linkages. These Fuc-containing glycan structures are commonly present on bacteria, therefore, bacterial FucTs are extensively used in the preparation of mammalian glycans. For example, α2FucT from *E. coli* O126 (WbgL) was utilized for the addition of α1–2-Fuc onto Gal residues of β1–4-linked galactosides,[89] whereas α2FucT from *E. coli* O127 (WbiQ),[115] *H. pylori* (FutC),[89] and *Thermosynechococcus elongatus* (Te2FT)[117] were employed with a preference to add α1–2-Fuc onto Gal residues of β1–3-linked galactosides (Figure 3.4A). In particular, α2FucTs from *E. coli* O86 (WbnK),[94,112] O128 (WbsJ)[95] and *H. pylori* (FutC)[181] had been used for the preparation of tumor-associated carbohydrate antigen Globo H (Figure 3.4B). Hp3/4FT can transfer a Fuc residue onto GlcNAc in type II glycan to form an α1–3-linkage or onto GlcNAc in type I glycan to form an α1–4-linkage, therefore was widely used in the syntheses of Lewis antigens such as Lewis X, Y, A, and B, sialyl Lewis X and sialyl Lewis A (Figure 3.4C), as well as HMOs such as 3-fucosyllactose (3-FL), lacto-*N*-fucopentaose (LNFP) III, and lacto-*N*-difuco-hexaose (LNDFH) II and III.[182] In contrast, Hp3/4FT catalyzes only α1–3-fucosylation, and its truncated version with increased solubility is often used for glycan synthesis.[45,46,53] However, for specific addition of Fuc to GlcNAc in the terminal LacNAc in multi-LacNAc-containing glycans, Hp3FT cannot be used due to its ability of fucosylating both terminal and internal GlcNAc residues. In this case, FUT9, the mammalian FucT specific for α1–3-Fuc-addition to GlcNAc in distal LacNAc (Figure 3.4D),[121] becomes an optimal choice, as illustrated in the syntheses of HMOs.[69] In addition to terminal modifications, "core-fucose" modification, the α1–6-Fuc attachment existing in the innermost GlcNAc residue in mammalian *N*-glycans, is also very common. Core fucosylation is required for antigen recognition of IgG B cell receptors,[183] but decreases the antibody-dependent cellular cytotoxicity (ADCC) of therapeutic monoclonal

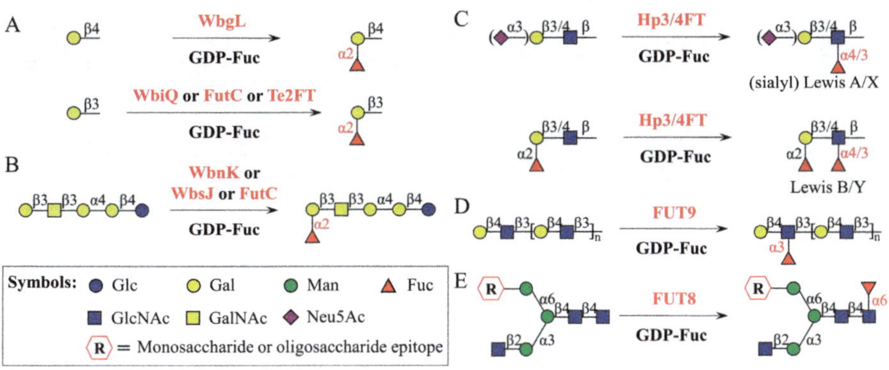

Figure 3.4 Glycan syntheses catalyzed by FucTs.

antibodies.[184] Unfortunately, no bacterial FucTs identified so far has the ability to add this "core-fucose" to *N*-glycan core pentasaccharide or more complex *N*-glycans, hence the authentic mammalian FucT – FUT8 – has to be used for GT-catalyzed preparation of *N*-glycans with core-fucosylation. When preparing glycans with FUT8, acceptor *N*-glycans need to bear a free β1–2-linked GlcNAc residue on the α1–3 Man branch (Figure 3.4E),[87,138,139] but this structural requirement is not necessary if the acceptor glycans are linked to polypeptides.[139] The glycoligase derived from *L. casei* α1–6-fucosidase (AlfC) can also catalyze core-fucosylation of diverse *N*-glycans within *N*-glycopeptides and glycoproteins (see Section 3.1),[20] but does not belong to FucT.

3.3.4 Sialyltransferases (SiaTs)

Like Fuc, Sia is another common modification to glycans that endows distinct binding properties to the glycans. This nine-carbon acidic sugar has four main types, Neu5Ac, Neu5Gc, Kdn (2-keto-deoxynonulosonic acid), and Neu (neuraminic acid, no sugar nucleotide in nature), and they may carry additional modifications, such as *O*-acetylation, *O*-methylation, *O*-sulfation or *O*-lactylation.[185] SiaTs can be used to catalyze the addition of Sia onto various glycans (Table 3.1). Neu5Ac is the most common type in humans, while Neu5Gc is seen more in other mammals. Some mammalian SiaTs are capable of transferring both Neu5Ac and Kdn, whereas the rest can transfer only one form.[185] Four types of linkage are found in nature, *i.e.* α2–3/6/8/9-linkages. α2–3-Linked Sia residue attaches to non-reducing terminal Gal residue, which is often observed in *N*- or *O*-glycans, glycosphingolipid glycans, HMOs, *etc.* For enzymatic extension of terminal Gal with α2–3-linked Sia, both mammalian α3SiaT (ST3Gal-IV)[141] and bacterial PmST1[150] (or its mutant[45,46]) have been used (Figure 3.5A). A PmST1 M144D mutant can even directly sialylate Le^X (Figure 3.5B).[52] In mammals, α2–6-linked Sia is found on Gal (common

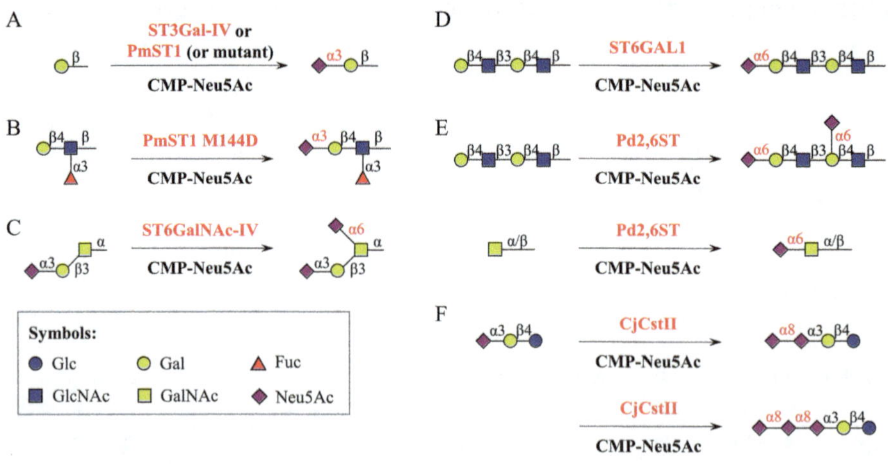

Figure 3.5 Glycan syntheses catalyzed by SiaTs.

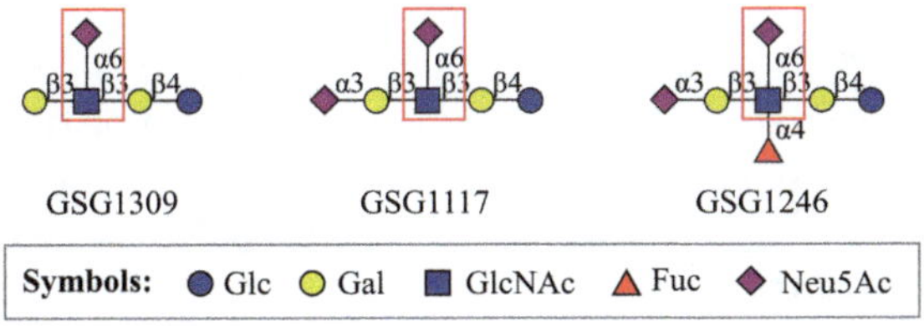

Figure 3.6 Glycans containing Neu5Acα2–6GlcNAc moieties (shown in red boxes). Structures are from Glycosphingolipid database (http://www.lipidbank.jp/), with lipidbank ID GSG1309, GSG1117, and GSG1246.

in *N*-glycans, *O*-glycans, glycolipid glycans and HMOs, *etc*), GalNAc (in *O*-GalNAc glycans), and even GlcNAc (in glycolipid glycans) (Figure 3.6). The specific addition of α2–6-Sia to GalNAc or terminal Gal can be achieved by using mammalian α6SiaTs ST6GalNAc-IV (Figure 3.5C)[152] and ST6GAL1 (Figure 3.5D),[69,141] respectively. Regarding its bacterial counterpart Pd2,6ST, both terminal and internal Gal/GalNAc residues are potential sialylation sites (Figure 3.5E).[43,44] Therefore, unless demanded,[44] attention needs to be paid as the bacterial enzyme can generate undesired byproducts with multiple sialylations when more than one Gal or GalNAc residue exist in the acceptor. To overcome this limitation, rational design was performed to create a bacterial α6SiaT mutant specific for terminal Gal on acceptors (see Section 3.2.3). α2–8/9-Sia can only attach to another Sia, in glycosphingolipid, *e.g.* GD3 and GT3, or in poly-sialic acid (poly-SA) structures.[186,187] To synthesize Siaα2–8Sia linkages in glycosphingolipid glycans, a bifunctional α2–3/8SiaT from *C. jejuni* (CjCstII) is commonly used (Figure 3.5F), but the molar ratio of CMP-Neu5Ac to acceptor glycans needs to be carefully controlled to obtain the desired products, as polysialylation can occur.[188]

3.3.5 *N*-Acetylgalactosaminyltransferases (GalNAcTs)

GalNAcTs can be used to transfer GalNAc residue to diverse glycans (Table 3.1). GalNAc is an acetylated hexosamine that can be found in both core (*O*-GalNAc glycans) or backbone structures (GAGs) and terminal structures (such as Globo series glycolipid glycans, ganglioside glycans, blood group A antigens, and CAD (Sd^a) antigen). The formation of blood group A antigens relies on the addition of the outermost GalNAc. This GalNAc can be added by mammalian GT – GTA,[89] or bacterial homolog BgtA from *Helicobacter mustelae* (Figure 3.7A).[157] For Globo series or ganglioside series glycolipid glycans, bacterial mimicry was identified, hence corresponding GTs are available for synthetic purposes. The bifunctional LgtD from *Haemophilus influenzae* (HiLgtD) was used for GalNAc extension to generate Globo series glycans (Figure 3.7B),[181] whereas β4GalNAcT from *C. jejuni* (CjCgtA) was used for GalNAc addition to generate ganglioside glycans (Figure 3.7C),[189] as well as enzymatic preparation of the CAD antigen (unpublished result) (Figure 3.7C).

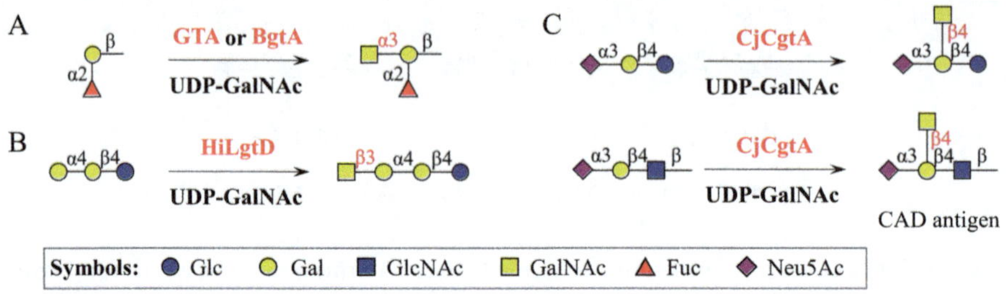

Figure 3.7 Glycan syntheses catalyzed by GalNAcTs.

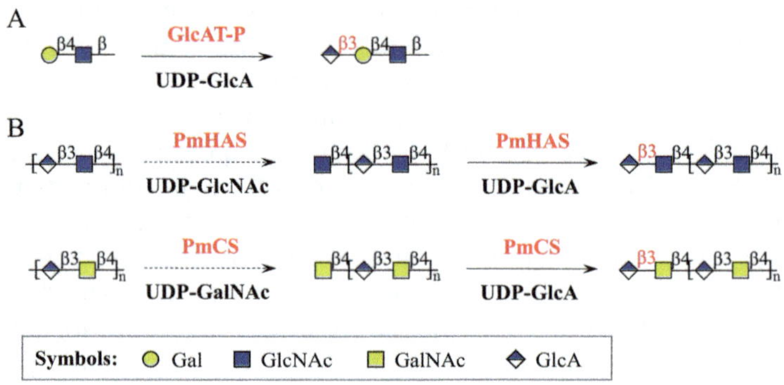

Figure 3.8 Glycan syntheses catalyzed by GlcATs.

3.3.6 Glucuronosyltransferases (GlcATs)

Glucuronic acid (GlcA) is the C6 oxidized glucose, and exists in many mammalian glycan structures, including human natural killer-1 (HNK-1, 3-sulfo-GlcAβ1–3Galβ1–4GlcNAcβ–R) antigen[190] and the GAGs, such as HA, chondroitin, and heparin sulfate.[191] GlcATs can be used to transfer GlcA with UDP-GlcA as donor for glycan syntheses (Table 3.1). The synthesis of an unsulfated HNK-1 precursor has been recently achieved in the preparation of *O*-mannose (*O*-Man) glycans, with the application of a human GlcAT-P heterologously expressed by *E. coli* in cell lysate (Figure 3.8A).[125] For GAGs syntheses, the bifunctional PmHAS was able to extend HA chains by adding β1–3-linked GlcA as well as β1–4-linked GlcNAc to the non-reducing end (Figure 3.8B),[47,77] whereas the bifunctional PmCS (chondroitin synthase from *P. multocida*) formed unsulfated chondroitin by alternating addition of β1–3-linked GlcA and β1–4-linked GalNAc (Figure 3.8B).[171] By swapping the natural acceptors for PmHAS and PmCS, chimeric GAGs (partial HA and partial chondroitin) were prepared.[171]

3.3.7 Mannosyltransferases (ManTs)

ManTs are capable of transferring Man from GDP-Man to acceptors (Table 3.1). Man is a conserved component in *N*-glycan core pentasaccharide, and is also found in the core of *O*-Man glycans and in core glycans of glycosylphosphatidylinositol (GPI) anchors. Comparing with GTs for transferring other common sugar residues, the application of ManTs in mammalian glycan preparation is limited. A recombinant yeast β1–4-ManT (Alg1) was adopted for preparation of trisaccharide Manβ1–4GlcNAcβ1–4GlcNAc (part of *N*-glycan core pentasaccharide).[175,176] By using *E. coli* expressed catalytic domains of yeast ManTs Alg1 and Alg11, together with mammalian cell expressed full length yeast ManT Alg2, the lipid-linked $GlcNAc_2Man_5$ oligosaccharides were chemoenzymatically synthesized from lipid-linked chitobiose.[192] However, Alg2, the ManT responsible for adding α1–3-Man and α1–6-Man to β1–4-Man,[193] is a non-Leloir GT which requires a long lipid tail in acceptors for efficiency,[192] limiting its application in the preparation of *N*-glycans. Instead, the preparation of *N*-glycan core structures had been recently achieved by chemical synthesis.[45,141] Due to enzyme limitations, the core structures for *O*-Man glycans were also prepared chemically.[124,125] There is no report on enzymatic synthesis of core glycan of GPI anchors, either.

3.4 Strategies in GT-catalyzed Glycan Syntheses

3.4.1 Direct GT-catalyzed Reactions

Direct GT-catalyzed reactions are practical applications for enzymatic glycan syntheses. By providing appropriate sugar donors and acceptors, GTs can catalyze glycosylation reactions with the existence of appropriate metal ion and buffering system. Both GTs and the donor sugar nucleotides can be acquired commercially or through enzymatic synthesis. The only issue is the acquisition of acceptors. The acceptors are generally obtained using three approaches – (1) commercial sources, (2) isolation, and (3) enzymatic or chemical syntheses. Normally, only simple glycans are commercially available in large amounts, *e.g.*, lactose. With carefully designed synthetic routes, complex glycans can be prepared by stepwise enzymatic extension, as exemplified by the Boons group for the syntheses of 60 asymmetrical multi-antennary HMOs.[69] Unfortunately, commercially available acceptors are limited and can be costly when they are available. Therefore, alternative methods are required. In some cases, the glycans from natural sources are structurally homogeneous and relatively abundant, and can be isolated to be acceptors. The asialo-glycopeptide prepared from sialylglycopeptide (SGP) that was isolated from chicken egg yolk was used as starting material by the Paulson group, and a series of *N*-glycans containing polyLacNAc[72] or multiple sialylated polyLacNAc[44] extensions were synthesized from there. Still from isolated SGP, after enzymatically trimming the peptide portion and the terminal sialic acids, the Wang group obtained large amounts of a bi-antennary *N*-glycan and prepared additional 34 *N*-glycans with tandem glycan epitopes.[46] By enzymatically removing the terminal sialic acids, they attained the same bi-antennary *N*-glycan and another tri-antennary *N*-glycan from fetuin, and synthesized a variety of asymmetrically branched *N*-glycans from them.[87] Nevertheless, in most cases, acceptor glycans can be neither purchased nor isolated. Therefore, syntheses are the ultimate solution. For acceptors with simple structures, such as LacNAc, one-step enzymatic galactosylation can afford the target glycans. Regarding acceptor glycans with complex structures, chemical preparation has been frequently utilized recently to provide acceptors for GT-catalyzed reactions, especially for the generation of asymmetrically branched glycans. For example, the Boons group was the first to report that asymmetrically branched *N*-glycans can be chemoenzymatically produced by using different GTs with sugar nucleotides and chemically synthesized key acceptors.[141] The Wong group synthesized homo- and mixed *N*-glycans (including high-mannose, hybrid and complex types) by modular chemoenzymatic methods and studied the binding behaviors of hetero-glycan.[126] Wang, Li and co-workers synthesized 73 *N*-glycans (63 isomers), around 30 HMOs and 45 mammalian *O*-mannosyl glycans by a Core Synthesis/Enzymatic Extension (CSEE) strategy.[45,125,194] When several GTs are used to synthesize complex glycans, the sequences in which the different GTs are used need to be carefully designed. Taking VIM-2 as an example (Figure 3.9),[46] GlcNAcylation

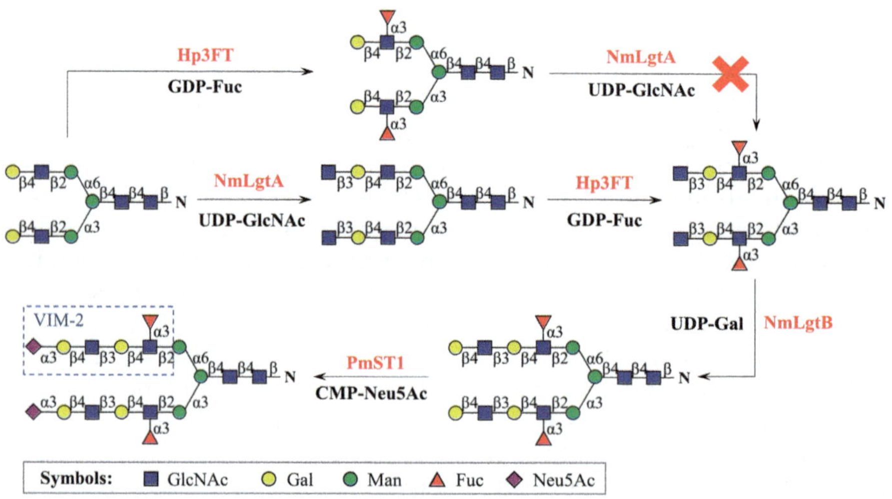

Figure 3.9 Direct GT-catalyzed reaction. Sequences in which the different GTs are used need to be carefully designed. The glycan sequence in the dashed box is VIM-2 epitope. *N* at the reducing end of glycans is asparagine. Reproduced from ref. 46 with permission from The Royal Socitey of Chemistry.

of LacNAc catalyzed by NmLgtA needs to happen before fucosylation catalyzed by Hp3FT, and the resulting glycan can be further galactosylated and sialylated to give the glycan containing VIM-2 epitope. If Hp3FT was used before NmLgtA, the fucosylated glycan cannot be recognized by NmLgtA. These direct GT-catalyzed glycan syntheses are extremely suitable for the generation of a variety of structurally defined homogeneous glycans, with sufficient amounts obtained for each glycan for applications such as glycan microarray, allowing the investigation of their binding profile to glycan-binding proteins (GBPs) and/or viruses.[46,125,126,141]

3.4.2 GT-catalyzed Reaction with Sugar Nucleotide *In Situ* Regeneration

To synthesize oligosaccharides enzymatically on large scales, large quantities of sugar nucleotides are needed. Since the nucleotides required for sugar nucleotide generation were expensive decades ago, a sugar nucleotide regeneration strategy was developed to circumvent direct supply of this expensive starting material.[195] The Wong group was the first to report the enzymatic synthesis of 10 gram scale of LacNAc with sugar nucleotide regeneration strategy (Figure 3.10).[195] By employing the nucleotide-regenerating enzymes, expensive sugar nucleotides can be regenerated *in situ*, thus reducing the cost of synthesis. Three decades later, with the regeneration of sugar nucleotides, including UDP-Gal, UDP-GalNAc, GDP-Fuc,

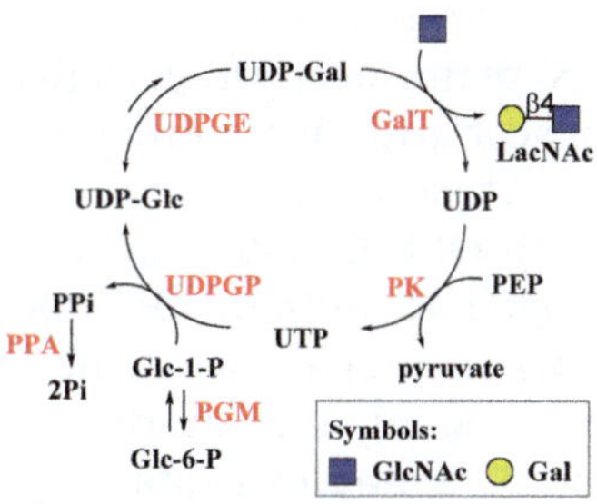

Figure 3.10 Synthesis of *N*-acetyllactosamine (LacNAc) by sugar nucleotide *in situ* regeneration.[195] Abbreviations: UDP, uridine 5′-diphosphate; UTP, uridine 5′-triphosphate; PEP, phosphoenolpyruvate; UDPGP, UDP-glucose pyrophosphorylase; UDPGE, UDP-galactose 4′-epimerase; GalT, galactosyltransferase; PK, pyruvate kinase; PGM, phosphoglucomutase; PPA, inorganic pyrophosphorylase; LacNAc, *N*-acetyllactosamine. Adapted from ref. 195 with permission from American Chemical Society, Copyright 1982.

and CMP-Neu5Ac, they achieved multi-gram synthesis of cancer-associated carbohydrate antigens Globo H and sialyl-globopentaose (SSEA4).[181] To better reuse the enzymes involved in sugar nucleotide regeneration, the Wang group developed superbeads by co-immobilizing the four His_6-tagged recombinant enzymes required for UDP-Gal regeneration onto nickel-nitrilotriacetate (NTA) beads.[196] When combining these superbeads with GalT(s), they successfully synthesized a variety of galactosides at gram-scale or 100 milligram-scale. Another advantage of sugar nucleotide regeneration is the elimination of feedback inhibition caused by the nucleotides – the byproduct of glycosylation. However, as nucleotides become more affordable, the regeneration system that requires several enzymes for a single sugar extension of a glycan appears a time- and labor-consuming process.

3.4.3 One-pot Multienzyme (OPME) Syntheses and Enzymatic Modular Assembly (EMA)

As more enzymes involved in the salvage pathway of sugar nucleotide syntheses have been identified, as well as the cost of nucleotides decreasing, one-pot multienzyme (OPME) systems have been adopted for glycan syntheses by coupling the sugar nucleotide syntheses with glycosylation, mostly catalyzed by Leloir GTs.[197] Additionally, an inorganic pyrophosphatase (PPA) is usually added to digest a byproduct pyrophosphate, which could inhibit the generation of sugar nucleotides. The Chen and Wang groups have established a set of standard systems to generate the nine common sugar nucleotides found in mammalian glycomes which can be readily combined with appropriate GTs to perform the desired glycosylation (Figure 3.11A).[197–199] These standardized OPME systems can also be sequentially used to generate complex oligosaccharides. To demonstrate this technology, Lewis X pentasaccharide, sialyl LNnT pentasaccharide, and sialyl Lewis X hexasaccharide were efficiently synthesized from propyl azide-functionalized lactose using appropriate OPME systems (Figure 3.11B).[197] Considering each OPME system as an enzyme module, the Cao group efficiently synthesized 15 human ABO blood group antigens from five disaccharide acceptors in preparative scale by sequential application of several OPME systems, namely enzymatic modular assembly (EMA).[89] Using the same approach, 58 complex *O*-Man glycans were prepared from five chemically synthesized core structures.[124] OPME syntheses are especially useful for the preparation of glycans containing post-glycosylational modifications, as the substrate promiscuity of enzymes for the sugar nucleotide syntheses and transferring allows the incorporation of monosaccharides bearing modifications into the target glycans.[197] However, production and purification of a large number of enzymes required by OPME syntheses are likely to hinder large-scale (over 100 gram-scale) production of oligosaccharides.

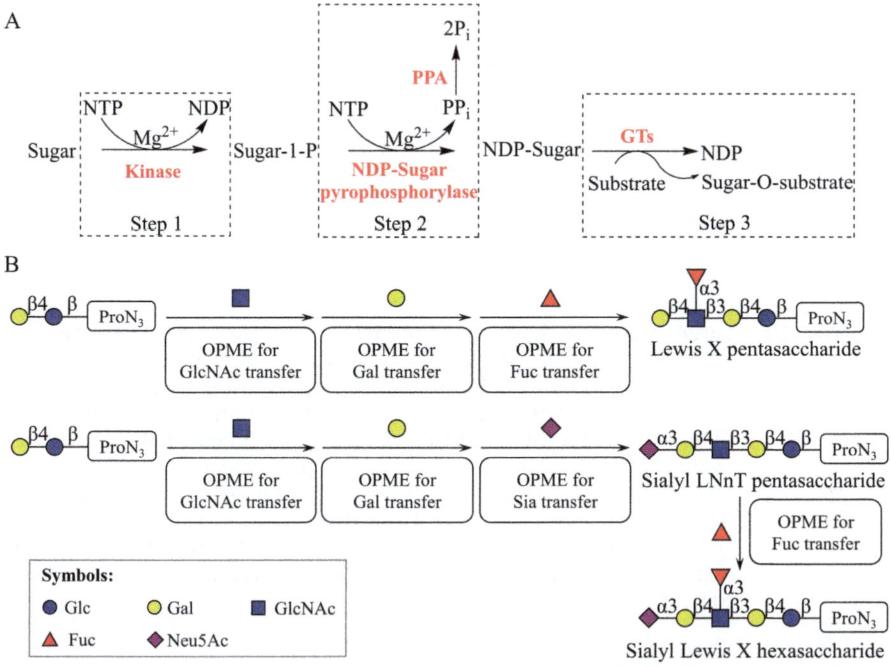

Figure 3.11 Glycan synthesis by OPME strategy. (A) General scheme for OPME strategy. (B) Synthesis of Lewis X pentasaccharide, sialyl LNnT pentasaccharide and sialyl Lewis X hexasaccharide by OPME strategy. Reproduced from ref. 197 with permission from The Royal Society of Chemistry.

3.4.4 Whole-cell Catalysis

In large-scale syntheses of glycans, bypassing enzyme purification would significantly reduce the time, labor and cost. Therefore, whole-cell catalysis was applied by freezing and thawing cells containing the required enzymes, usually recombinant enzymes. A 10–100 gram-scale oligosaccharide production has been developed by Kyowa Hakko Kogyo Co., by combining several metabolically engineered bacteria and using low-cost starting materials (Figure 3.12).[98,200,201] The Wang group developed multi-gram scale production of lactose-based trisaccharides using "superbugs" by sugar nucleotide regeneration or generation in one *E. coli* strain.[202] But the limitation of whole-cell strategies is that the oligosaccharides that are synthesized have relatively simple structures rather other complicated glycans. This strategy has potential to be generalized to the production of glycans with relatively complex structures. The enhanced permeability of cell membrane of freeze–thawed bacteria should allow the entrance of relatively large oligosaccharides, thereby realizing the syntheses of complex glycans by sequential application of freeze–thawed bacteria containing the appropriate GTs.

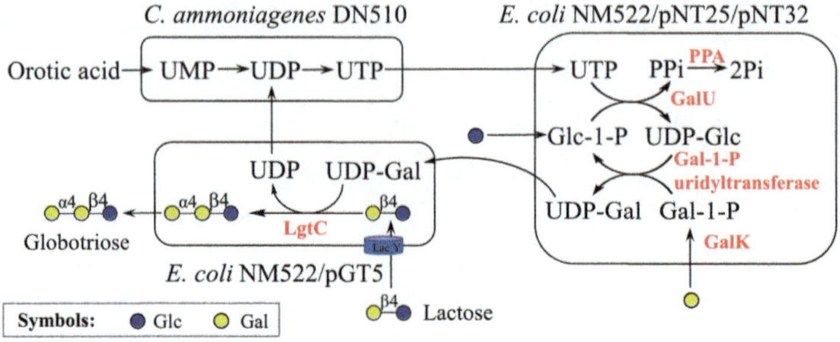

Figure 3.12 Large scale synthesis of globotriose by whole-cell catalysis.[98] *E. coli* NM522/pNT25/pNT32 cells expressing the enzymes glucose-1-phosphate uridyltransferase (GalU), galactose-1-phosphate uridyltransferase, galactokinase (GalK), and inorganic pyrophosphorylase (PPA) are used to produce UDP-Gal. The produced UDP-Gal was used by *E. coli* NM522/pGT5 to synthesize globotriose from lactose. The UTP, required for UDP-Gal synthesis, can be produced by *Corynebacterium ammoniagenes* DN510 from orotic acid or the byproduct UDP generated with the synthesis of globotriose. Abbreviation: LacY, lactose permease. Adapted from ref. 98 with permission from Springer Nature, Copyright 1998.

3.4.5 Living Cell Factory

GTs in cells can also be employed for glycan syntheses while cells are alive. This is one of the best approaches of large-scale synthesis of oligosaccharides. In this "living cell factory" strategy, GTs required for target glycan syntheses are introduced into bacterial cells to perform glycan syntheses. With regard to the sugar nucleotides needed, the endogenous sugar nucleotides produced by the cells can be borrowed, but they may not be sufficient. Therefore, enzymes needed for sugar nucleotide syntheses are also introduced to the cell factory. Meanwhile, natural glycosidases that can degrade the acceptor or target glycans are removed from the cells by gene knockout. In addition, the transporters of acceptor glycans, if there are any, may be incorporated or overexpressed to enhance the uptake of acceptors.[203,204] Then, by providing the cells with the necessary conditions for cell proliferation and protein expression, the target glycans can be produced within cells or secreted into medium as the cells grow when appropriate acceptors are fed. Based on this principle, various attempts of oligosaccharide production have emerged. For simple trisaccharide 2′-fucosyllactose (2′-FL) production from lactose (Figure 3.13), by adopting the *de novo* pathway or salvage pathway for GDP-Fuc synthesis, the Feng group,[205] the Albermann group[206] and the Seo group[207] achieved a yield of 2′-FL at 9.12 g L^{-1}, 20.28 g L^{-1} and 23.1 g L^{-1}, respectively. The Samain group was able to produce lacto-*N*-triose II (LNT-II) (6 g L^{-1}), GD3 (0.83 g L^{-1}) and lacto-*N*-neofucopentaose (LNnFP)-I (1.5 g L^{-1}) from the mineral salt medium supplemented with lactose.[204,208,209] The yields of the structurally more complex

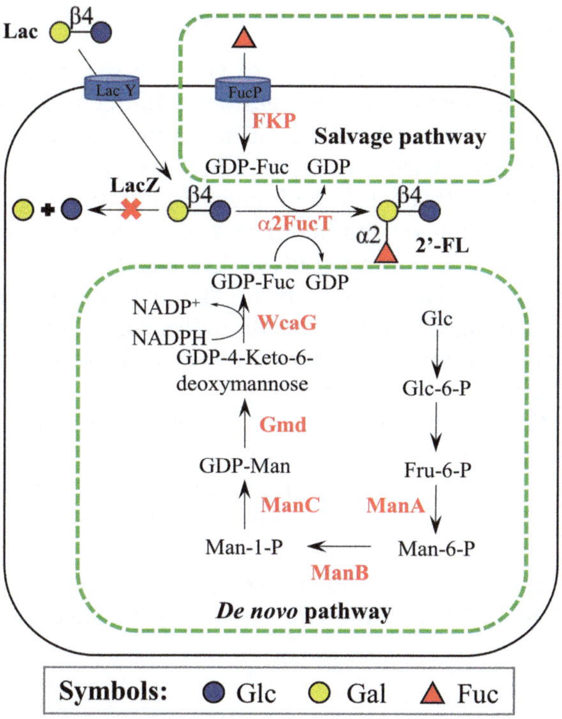

Figure 3.13 2′-Fucosyllactose (2′-FL) synthesis by *de novo* pathway or salvage pathway of GDP-Fuc synthesis in engineered *E. coli*. The cells are engineered to contain enzymes in *de novo* pathway (ManABC, Gmd and WcaG) or in salvage pathway (FKP) of GDP-Fuc biosynthesis, and the produced GDP-Fuc can be used for 2′-FL synthesis *via* α2FucT-catalyzed fucosylation in the living cells. Abbreviations: LacY, lactose permease; LacZ, β-galactosidase; FucP, fucose permease; Lac, lactose.

glycans (tetrasaccharide and pentasaccharide) are relatively low. This is at least partly caused by the loss of control over the sequences of the multiple GTs and over the quality of each glycosylation step when several GTs coexist, and thus the concomitant production of byproducts. For instance, during the production of LNnFP-I, in addition to extending LNnT to provide LNnFP-I, *H. pylori* α2FucT (Hp2FT) can also act on lactose to generate 2′-FL as a main byproduct.[209] Therefore, the living cell factory is a very promising strategy for the production of glycans that need single-step or two-step extension from the acceptors. Considering the difficulty in transmembrane transport of most glycans, this strategy could be most easily applied to the single-step or two-step extension of lactose. For glycans that need multiple extensions, conditions such as the acceptor concentration fed into the medium need to be carefully controlled over the course of fermentation to decrease the generation of undesired intermediate(s).

3.5 Prospective

The high regio- and stereo-specificity of GTs are the basis for their application in glycan syntheses. The progression of biotechnology has advanced the way in which GTs are utilized. Right now, efforts regarding the development of GT-catalyzed glycan syntheses have two main directions: (1) discovery of novel GTs with desired activities; (2) engineering the known GTs to gain desired activities or characteristics. Genetic engineering has facilitated the availability of recombinant GTs. As the genomic sequences of more organisms become available, more useful GTs will be discovered, especially from bacteria. Although bioinformatical analysis of sequences can propose putative GTs, experimental evidence is needed to prove their actual functions. Such identification and characterization of novel GTs are difficult and tedious. High-throughput screening techniques, such as the SAMDI technique,[210] can benefit enzyme discovery. In addition to discovering new GTs, efforts have also been put into engineering currently known GTs to gain the desired activity by rational design or directed evolution, especially with the assistance of structural information of enzymes or with the aid of high-throughput screening. As more desired GTs become available, more glycans will be accessible and affordable and eventually advance the functional glycomics studies.

References

1. K. Ohtsubo and J. D. Marth, *Cell*, 2006, **126**, 855.
2. S. S. Pinho and C. A. Reis, *Nat. Rev. Cancer*, 2015, **15**, 540.
3. K. W. Moremen, A. Ramiah, M. Stuart, J. Steel, L. Meng, F. Forouhar, H. A. Moniz, G. Gahlay, Z. Gao, D. Chapla, S. Wang, J. Y. Yang, P. K. Prabhakar, R. Johnson, M. D. Rosa, C. Geisler, A. V. Nairn, J. Seetharaman, S. C. Wu, L. Tong, H. J. Gilbert, J. LaBaer and D. L. Jarvis, *Nat. Chem. Biol.*, 2018, **14**, 156.
4. R. M. Schmaltz, S. R. Hanson and C. H. Wong, *Chem. Rev.*, 2011, **111**, 4259.
5. L. Guo, X. Chen, L. Xu, M. Xiao and L. Lu, *Appl. Environ. Microbiol.*, 2018, **84**, e00071-18.
6. L. X. Wang and W. Huang, *Curr. Opin. Chem. Biol.*, 2009, **13**, 592.
7. L. X. Wang, *Chem. Biol.*, 2009, **16**, 1026.
8. R. Kittl and S. G. Withers, *Carbohydr. Res.*, 2010, **345**, 1272.
9. L. F. Mackenzie, Q. Wang, R. A. J. Warren and S. G. Withers, *J. Am. Chem. Soc.*, 1998, **120**, 5583.
10. K. Yamamoto, *J. Biosci. Bioeng.*, 2001, **92**, 493.
11. L. X. Wang, *Carbohydr. Res.*, 2008, **343**, 1509.
12. M. Umekawa, C. Li, T. Higashiyama, W. Huang, H. Ashida, K. Yamamoto and L. X. Wang, *J. Biol. Chem.*, 2010, **285**, 511.
13. Y. W. Kim, S. S. Lee, R. A. Warren and S. G. Withers, *J. Biol. Chem.*, 2004, **279**, 42787.

14. H. Lin, H. Tao and V. W. Cornish, *J. Am. Chem. Soc.*, 2004, **126**, 15051.
15. A. Ben-David, G. Shoham and Y. Shoham, *Chem. Biol.*, 2008, **15**, 546.
16. J. H. Shim, H. M. Chen, J. R. Rich, E. D. Goddard-Borger and S. G. Withers, *Protein Eng., Des. Sel.*, 2012, **25**, 465.
17. P. M. Danby and S. G. Withers, *ACS Chem. Biol.*, 2016, **11**, 1784.
18. M. Jahn, J. Marles, R. A. Warren and S. G. Withers, *Angew. Chem., Int. Ed. Engl.*, 2003, **42**, 352.
19. Y. W. Kim, R. Zhang, H. Chen and S. G. Withers, *Chem. Commun.*, 2010, **46**, 8725.
20. C. Li, S. Zhu, C. Ma and L. X. Wang, *J. Am. Chem. Soc.*, 2017, **139**, 15074.
21. H. J. Gijsen, L. Qiao, W. Fitz and C. H. Wong, *Chem. Rev.*, 1996, **96**, 443.
22. H. H. Freeze, G. W. Hart and R. L. Schnaar, in *Essentials of Glycobiology [Internet]*, ed. A. Varki, R. D. Cummings, J. D. Esko, P. Stanley, G. W. Hart, M. Aebi, A. G. Darvill, T. Kinoshita, N. H. Packer, J. H. Prestegard, R. L. Schnaar and P. H. Seeberger, Cold Spring Harbor Laboratory Press, Cold Spring Harbor, NY, 3rd edn, 2017, ch. 5.
23. L. Li, Y. Liu, W. Wang, J. Cheng, W. Zhao and P. Wang, *Carbohydr. Res.*, 2012, **355**, 35.
24. L. Cai, W. Guan, M. Kitaoka, J. Shen, C. Xia, W. Chen and P. G. Wang, *Chem. Commun.*, 2009, 2944.
25. L. Cai, W. Guan, W. Wang, W. Zhao, M. Kitaoka, J. Shen, C. O'Neil and P. G. Wang, *Bioorg. Med. Chem. Lett.*, 2009, **19**, 5433.
26. W. Guan, L. Cai, J. Fang, B. Wu and P. George Wang, *Chem. Commun.*, 2009, 6976.
27. W. Guan, L. Cai and P. G. Wang, *Chem. - Eur. J.*, 2010, **16**, 13343.
28. Y. Li, H. Yu, Y. Chen, K. Lau, L. Cai, H. Cao, V. K. Tiwari, J. Qu, V. Thon, P. G. Wang and X. Chen, *Molecules*, 2011, **16**, 6396.
29. G. Zhao, W. Guan, L. Cai and P. G. Wang, *Nat. Protoc.*, 2010, **5**, 636.
30. L. Li, Y. Liu, Y. Wan, Y. Li, X. Chen, W. Zhao and P. G. Wang, *Org. Lett.*, 2013, **15**, 5528.
31. Y. Guo, J. Fang, T. Li, X. Li, C. Ma, X. Wang, P. G. Wang and L. Li, *Carbohydr. Res.*, 2015, **411**, 1.
32. H. Yu, H. Yu, R. Karpel and X. Chen, *Bioorg. Med. Chem.*, 2004, **12**, 6427.
33. S. Williams, Glycosyltransferases, in CAZypedia, available at URL, http://www.cazypedia.org/, accessed January 2018.
34. P. Stanley, N. Taniguchi and M. Aebi, in *Essentials of Glycobiology [Internet]*, ed. A. Varki, R. D. Cummings, J. D. Esko, P. Stanley, G. W. Hart, M. Aebi, A. G. Darvill, T. Kinoshita, N. H. Packer, J. H. Prestegard, R. L. Schnaar and P. H. Seeberger, Cold Spring Harbor Laboratory Press, Cold Spring Harbor, NY, 3rd edn, 2017, ch. 9.
35. H. Yu, V. Thon, K. Lau, L. Cai, Y. Chen, S. Mu, Y. Li, P. G. Wang and X. Chen, *Chem. Commun.*, 2010, **46**, 7507.
36. J. Yan, X. Chen, F. Wang and H. Cao, *Org. Biomol. Chem.*, 2013, **11**, 842.
37. *Handbook of Glycosyltransferases and Related Genes*, ed. N. Taniguchi, K. Honke, M. Fukuda, H. Narimatsu, Y. Yamaguchi and T. Angata, Springer Japan, Tokyo, 2nd edn, 2014.

38. N. O. Seto, M. M. Palcic, O. Hindsgaul, D. R. Bundle and S. A. Narang, *Eur. J. Biochem.*, 1995, **234**, 323.
39. F. V. Rao, J. R. Rich, B. Rakic, S. Buddai, M. F. Schwartz, K. Johnson, C. Bowe, W. W. Wakarchuk, S. Defrees, S. G. Withers and N. C. Strynadka, *Nat. Struct. Mol. Biol.*, 2009, **16**, 1186.
40. J. Fang, J. Li, X. Chen, Y. Zhang, J. Wang, Z. Guo, W. Zhang, L. Yu, K. Brew and P. G. Wang, *J. Am. Chem. Soc.*, 1998, **120**, 6635.
41. B. Ramakrishnan and P. K. Qasba, *J. Mol. Biol.*, 2001, **310**, 205.
42. I. Brockhausen, *Front. Immunol.*, 2014, **5**, 492.
43. H. Yu, S. Huang, H. Chokhawala, M. Sun, H. Zheng and X. Chen, *Angew. Chem., Int. Ed. Engl.*, 2006, **45**, 3938.
44. C. M. Nycholat, W. Peng, R. McBride, A. Antonopoulos, R. P. de Vries, Z. Polonskaya, M. G. Finn, A. Dell, S. M. Haslam and J. C. Paulson, *J. Am. Chem. Soc.*, 2013, **135**, 18280.
45. L. Li, Y. Liu, C. Ma, J. Qu, A. D. Calderon, B. Wu, N. Wei, X. Wang, Y. Guo, Z. Xiao, J. Song, G. Sugiarto, Y. Li, H. Yu, X. Chen and P. G. Wang, *Chem. Sci.*, 2015, **6**, 5652.
46. Z. Wu, Y. Liu, C. Ma, L. Li, J. Bai, L. Byrd-Leotis, Y. Lasanajak, Y. Guo, L. Wen, H. Zhu, J. Song, Y. Li, D. A. Steinhauer, D. F. Smith, B. Zhao, X. Chen, W. Guan and P. G. Wang, *Org. Biomol. Chem.*, 2016, **14**, 11106.
47. W. Jing and P. L. DeAngelis, *J. Biol. Chem.*, 2004, **279**, 42345.
48. J. Cheng, H. Yu, K. Lau, S. Huang, H. A. Chokhawala, Y. Li, V. K. Tiwari and X. Chen, *Glycobiology*, 2008, **18**, 686.
49. W. W. Wakarchuk, A. Cunningham, D. C. Watson and N. M. Young, *Protein Eng.*, 1998, **11**, 295.
50. C. P. Chiu, A. G. Watts, L. L. Lairson, M. Gilbert, D. Lim, W. W. Wakarchuk, S. G. Withers and N. C. Strynadka, *Nat. Struct. Mol. Biol.*, 2004, **11**, 163.
51. S. Bernatchez, M. Gilbert, M. C. Blanchard, M. F. Karwaski, J. Li, S. Defrees and W. W. Wakarchuk, *Glycobiology*, 2007, **17**, 1333.
52. G. Sugiarto, K. Lau, J. Qu, Y. Li, S. Lim, S. Mu, J. B. Ames, A. J. Fisher and X. Chen, *ACS Chem. Biol.*, 2012, **7**, 1232.
53. S. W. Lin, T. M. Yuan, J. R. Li and C. H. Lin, *Biochemistry*, 2006, **45**, 8108.
54. K. Schmolzer, T. Czabany, C. Luley-Goedl, T. Pavkov-Keller, D. Ribitsch, H. Schwab, K. Gruber, H. Weber and B. Nidetzky, *Chem. Commun.*, 2015, **51**, 3083.
55. J. B. McArthur, H. Yu, J. Zeng and X. Chen, *Org. Biomol. Chem.*, 2017, **15**, 1700.
56. Y. Xu, Y. Fan, J. Ye, F. Wang, Q. Nie, L. Wang, P. G. Wang, H. Cao and J. Cheng, *ACS Catal.*, 2018, **8**, 7222.
57. J. B. McArthur and X. Chen, *Biochem. Soc. Trans.*, 2016, **44**, 129.
58. A. Aharoni, K. Thieme, C. P. Chiu, S. Buchini, L. L. Lairson, H. Chen, N. C. Strynadka, W. W. Wakarchuk and S. G. Withers, *Nat. Methods*, 2006, **3**, 609.
59. G. Yang, J. R. Rich, M. Gilbert, W. W. Wakarchuk, Y. Feng and S. G. Withers, *J. Am. Chem. Soc.*, 2010, **132**, 10570.
60. D. C. Briggs and E. Hohenester, *Structure*, 2018, **26**, 801.

61. K. Matsuo and G. Atsumi, *J. Biosci. Bioeng.*, 2018, **126**, 371.
62. J. Tamura, T. Nakamura-Yamamoto, Y. Nishimura, S. Mizumoto, J. Takahashi and K. Sugahara, *Carbohydr. Res.*, 2010, **345**, 2115.
63. B. T. Kim, H. Kitagawa, J. Tamura, T. Saito, M. Kusche-Gullberg, U. Lindahl and K. Sugahara, *Proc. Natl. Acad. Sci. U. S. A.*, 2001, **98**, 7176.
64. I. Takahashi, N. Noguchi, K. Nata, S. Yamada, T. Kaneiwa, S. Mizumoto, T. Ikeda, K. Sugihara, M. Asano, T. Yoshikawa, A. Yamauchi, N. J. Shervani, A. Uruno, I. Kato, M. Unno, K. Sugahara, S. Takasawa, H. Okamoto and A. Sugawara, *Biochem. Biophys. Res. Commun.*, 2009, **383**, 113.
65. R. D. Gordon, P. Sivarajah, M. Satkunarajah, D. Ma, C. A. Tarling, D. Vizitiu, S. G. Withers and J. M. Rini, *J. Mol. Biol.*, 2006, **360**, 67.
66. P. L. Lee, J. J. Kohler and S. R. Pfeffer, *Glycobiology*, 2009, **19**, 655.
67. R. Almeida, S. B. Levery, U. Mandel, H. Kresse, T. Schwientek, E. P. Bennett and H. Clausen, *J. Biol. Chem.*, 1999, **274**, 26165.
68. I. A. Gagarinov, T. Li, J. S. Torano, T. Caval, A. D. Srivastava, J. A. Kruijtzer, A. J. Heck and G. J. Boons, *J. Am. Chem. Soc.*, 2017, **139**, 1011.
69. A. R. Prudden, L. Liu, C. J. Capicciotti, M. A. Wolfert, S. Wang, Z. Gao, L. Meng, K. W. Moremen and G. J. Boons, *Proc. Natl. Acad. Sci. U. S. A.*, 2017, **114**, 6954.
70. O. Blixt, I. van Die, T. Norberg and D. H. van den Eijnden, *Glycobiology*, 1999, **9**, 1061.
71. Y. Li, M. Xue, X. Sheng, H. Yu, J. Zeng, V. Thon, Y. Chen, M. M. Muthana, P. G. Wang and X. Chen, *Bioorg. Med. Chem.*, 2016, **24**, 1696.
72. W. Peng, J. Pranskevich, C. Nycholat, M. Gilbert, W. Wakarchuk, J. C. Paulson and N. Razi, *Glycobiology*, 2012, **22**, 1453.
73. C. Dumon, C. Bosso, J. P. Utille, A. Heyraud and E. Samain, *ChemBioChem*, 2006, **7**, 359.
74. H. Bakker, P. S. Schoenmakers, C. A. Koeleman, D. H. Joziasse, I. van Die and D. H. van den Eijnden, *Glycobiology*, 1997, **7**, 539.
75. E. Bettler, E. Samain, V. Chazalet, C. Bosso, A. Heyraud, D. H. Joziasse, W. W. Wakarchuk, A. Imberty and A. R. Geremia, *Glycoconjugate J.*, 1999, **16**, 205.
76. K. J. Williams, K. M. Halkes, J. P. Kamerling and P. L. DeAngelis, *J. Biol. Chem.*, 2006, **281**, 5391.
77. P. L. DeAngelis, L. C. Oatman and D. F. Gay, *J. Biol. Chem.*, 2003, **278**, 35199.
78. P. Mattila, H. Salminen, L. Hirvas, J. Niittymaki, H. Salo, R. Niemela, M. Fukuda, O. Renkonen and R. Renkonen, *J. Biol. Chem.*, 1998, **273**, 27633.
79. M. Fukuda and M. Suzuki-Anekoji, in *Handbook of Glycosyltransferases*, ed. N. Taniguchi, K. Honke, M. Fukuda, H. Narimatsu and Y. Yamaguchi, Springer Japan, Tokyo, 2nd edn, 2014, ch. 33, pp. 347–354.
80. I. Brockhausen, T. Dowler and H. Paulsen, *Biochim. Biophys. Acta*, 2009, **1790**, 1244.
81. S. Zeng, A. Dinter, D. Eisenkratzer, M. Biselli, C. Wandrey and E. G. Berger, *Biochem. Biophys. Res. Commun.*, 1997, **237**, 653.

82. Y. Gao, R. P. Aryal, T. Ju, R. D. Cummings, G. Gahlay, D. L. Jarvis, K. L. Matta, J. Z. Vlahakis, W. A. Szarek and I. Brockhausen, *Biochim. Biophys. Acta*, 2013, **1830**, 4274.
83. M. Kaneko, G. Alvarez-Manilla, M. Kamar, I. Lee, J. K. Lee, K. Troupe, W. Zhang, M. Osawa and M. Pierce, *FEBS Lett.*, 2003, **554**, 515.
84. G. Alvarez-Manilla, K. Troupe, M. Fleming, E. Martinez-Uribe and M. Pierce, *Glycobiology*, 2010, **20**, 166.
85. K. Y. Do, N. Fregien, M. Pierce and R. D. Cummings, *J. Biol. Chem.*, 1994, **269**, 23456.
86. J. W. Dennis, N. Taniguchi and M. Pierce, in *Handbook of Glycosyltransferases*, ed. N. Taniguchi, K. Honke, M. Fukuda, H. Narimatsu and Y. Yamaguchi, Springer Japan, Tokyo, 2nd edn, 2014, ch. 21, pp. 233–246.
87. A. D. Calderon, J. Zhou, W. Guan, Z. Wu, Y. Guo, J. Bai, Q. Li, P. G. Wang, J. Fang and L. Li, *Org. Biomol. Chem.*, 2017, **15**, 7258.
88. K.-i. Inamori, M. Pierce and N. Taniguchi, in *Handbook of Glycosyltransferases*, ed. N. Taniguchi, K. Honke, M. Fukuda, H. Narimatsu and Y. Yamaguchi, Springer Japan, Tokyo, 2nd edn, 2014, ch. 22, pp. 247–256.
89. J. Ye, X. W. Liu, P. Peng, W. Yi, X. Chen, F. Wang and H. Cao, *ACS Catal.*, 2016, **6**, 8140.
90. R. Woodward, W. Yi, L. Li, G. Zhao, H. Eguchi, P. R. Sridhar, H. Guo, J. K. Song, E. Motari, L. Cai, P. Kelleher, X. Liu, W. Han, W. Zhang, Y. Ding, M. Li and P. G. Wang, *Nat. Chem. Biol.*, 2010, **6**, 418.
91. E. Bettler, A. Imberty, B. Priem, V. Chazalet, A. Heyraud, D. H. Joziasse and R. A. Geremia, *Biochem. Biophys. Res. Commun.*, 2003, **302**, 620.
92. X. Chen, W. Zhang, J. Wang, J. Fang and P. G. Wang, *Biotechnol. Prog.*, 2000, **16**, 595.
93. D. Adlercreutz, J. T. Weadge, B. O. Petersen, J. O. Duus, N. J. Dovichi and M. M. Palcic, *Carbohydr. Res.*, 2010, **345**, 1384.
94. W. Yi, J. Shao, L. Zhu, M. Li, M. Singh, Y. Lu, S. Lin, H. Li, K. Ryu, J. Shen, H. Guo, Q. Yao, C. A. Bush and P. G. Wang, *J. Am. Chem. Soc.*, 2005, **127**, 2040.
95. D. M. Su, H. Eguchi, W. Yi, L. Li, P. G. Wang and C. Xia, *Org. Lett.*, 2008, **10**, 1009.
96. J. Zhang, P. Kowal, J. Fang, P. Andreana and P. G. Wang, *Carbohydr. Res.*, 2002, **337**, 969.
97. R. Steffensen, K. Carlier, J. Wiels, S. B. Levery, M. Stroud, B. Cedergren, B. Nilsson Sojka, E. P. Bennett, C. Jersild and H. Clausen, *J. Biol. Chem.*, 2000, **275**, 16723.
98. S. Koizumi, T. Endo, K. Tabata and A. Ozaki, *Nat. Biotechnol.*, 1998, **16**, 847.
99. X. Bai, D. Zhou, J. R. Brown, B. E. Crawford, T. Hennet and J. D. Esko, *J. Biol. Chem.*, 2001, **276**, 48189.
100. M. B. Vester-Christensen, L. Hansen and H. Clausen, in *Handbook of Glycosyltransferases*, ed. N. Taniguchi, K. Honke, M. Fukuda, H. Narimatsu and Y. Yamaguchi, Springer Japan, Tokyo, 2nd edn, 2014, ch. 10, pp. 101–108.

101. M. Amado, R. Almeida, F. Carneiro, S. B. Levery, E. H. Holmes, M. Nomoto, M. A. Hollingsworth, H. Hassan, T. Schwientek, P. A. Nielsen, E. P. Bennett and H. Clausen, *J. Biol. Chem.*, 1998, **273**, 12770.
102. M. B. Vester-Christensen, L. Hansen and H. Clausen, in *Handbook of Glycosyltransferases*, ed. N. Taniguchi, K. Honke, M. Fukuda, H. Narimatsu and Y. Yamaguchi, Springer Japan, Tokyo, 2nd edn, 2014, ch. 7, pp. 73–80.
103. X. W. Liu, C. Xia, L. Li, W. Y. Guan, N. Pettit, H. C. Zhang, M. Chen and P. G. Wang, *Bioorg. Med. Chem.*, 2009, **17**, 4910.
104. W. Yi, R. S. Perali, H. Eguchi, E. Motari, R. Woodward and P. G. Wang, *Biochemistry*, 2008, **47**, 1241.
105. M. Randriantsoa, S. Drouillard, C. Breton and E. Samain, *FEBS Lett.*, 2007, **581**, 2652.
106. T. Antoine, B. Priem, A. Heyraud, L. Greffe, M. Gilbert, W. W. Wakarchuk, J. S. Lam and E. Samain, *ChemBioChem*, 2003, **4**, 406.
107. H. Malekan, G. Fung, V. Thon, Z. Khedri, H. Yu, J. Qu, Y. Li, L. Ding, K. S. Lam and X. Chen, *Bioorg. Med. Chem.*, 2013, **21**, 4778.
108. K. Lau, V. Thon, H. Yu, L. Ding, Y. Chen, M. M. Muthana, D. Wong, R. Huang and X. Chen, *Chem. Commun.*, 2010, **46**, 6066.
109. O. Blixt, J. Brown, M. J. Schur, W. Wakarchuk and J. C. Paulson, *J. Org. Chem.*, 2001, **66**, 2442.
110. C. M. Nycholat, R. McBride, D. C. Ekiert, R. Xu, J. Rangarajan, W. Peng, N. Razi, M. Gilbert, W. Wakarchuk, I. A. Wilson and J. C. Paulson, *Angew. Chem., Int. Ed. Engl.*, 2012, **51**, 4860.
111. K. Furukawa, H. Clausen and T. Sato, in *Handbook of Glycosyltransferases*, ed. N. Taniguchi, K. Honke, M. Fukuda, H. Narimatsu and Y. Yamaguchi, Springer Japan, Tokyo, 2nd edn, 2014, ch. 6, pp. 63–72.
112. Z. Wang, M. Gilbert, H. Eguchi, H. Yu, J. Cheng, S. Muthana, L. Zhou, P. G. Wang, X. Chen and X. Huang, *Adv. Synth. Catal.*, 2008, **350**, 1717.
113. M. Li, X. W. Liu, J. Shao, J. Shen, Q. Jia, W. Yi, J. K. Song, R. Woodward, C. S. Chow and P. G. Wang, *Biochemistry*, 2008, **47**, 378.
114. L. Engels and L. Elling, *Glycobiology*, 2014, **24**, 170.
115. N. Pettit, T. Styslinger, Z. Mei, W. Han, G. Zhao and P. G. Wang, *Biochem. Biophys. Res. Commun.*, 2010, **402**, 190.
116. D. B. Stein, Y. N. Lin and C. H. Lin, *Adv. Synth. Catal.*, 2008, **350**, 2313.
117. C. Zhao, Y. Wu, H. Yu, I. M. Shah, Y. Li, J. Zeng, B. Liu, D. A. Mills and X. Chen, *Chem. Commun.*, 2016, **52**, 3899.
118. B. Tefsen, C. M. van Stijn, M. van den Broek, H. Kalay, J. C. Knol, C. R. Jimenez and I. van Die, *Carbohydr. Res.*, 2009, **344**, 1501.
119. R. Kannagi, in *Handbook of Glycosyltransferases*, ed. N. Taniguchi, K. Honke, M. Fukuda, H. Narimatsu and Y. Yamaguchi, Springer Japan, Tokyo, 2nd edn, 2014, ch. 51, pp. 559–572.
120. T. Kudo and H. Narimatsu, in *Handbook of Glycosyltransferases*, ed. N. Taniguchi, K. Honke, M. Fukuda, H. Narimatsu and Y. Yamaguchi, Springer Japan, Tokyo, 2nd edn, 2014, ch. 52, pp. 573–580.

121. S. Nishihara, H. Iwasaki, M. Kaneko, A. Tawada, M. Ito and H. Narimatsu, *FEBS Lett.*, 1999, **462**, 289.
122. K. Nguyen, I. van Die, K. M. Grundahl, Z. S. Kawar and R. D. Cummings, *Glycobiology*, 2007, **17**, 586.
123. W. Wang, T. Hu, P. A. Frantom, T. Zheng, B. Gerwe, D. S. Del Amo, S. Garret, R. D. Seidel 3rd and P. Wu, *Proc. Natl. Acad. Sci. U. S. A.*, 2009, **106**, 16096.
124. C. Meng, A. Sasmal, Y. Zhang, T. Gao, C. C. Liu, N. Khan, A. Varki, F. Wang and H. Cao, *Angew. Chem., Int. Ed. Engl.*, 2018, **57**, 9003.
125. S. Wang, Q. Zhang, C. Chen, Y. Guo, M. R. Gadi, J. Yu, U. Westerlind, Y. Liu, X. Cao, P. G. Wang and L. Li, *Angew. Chem., Int. Ed. Engl.*, 2018, **57**, 9268.
126. S. S. Shivatare, S. H. Chang, T. I. Tsai, S. Y. Tseng, V. S. Shivatare, Y. S. Lin, Y. Y. Cheng, C. T. Ren, C. C. Lee, S. Pawar, C. S. Tsai, H. W. Shih, Y. F. Zeng, C. H. Liang, P. D. Kwong, D. R. Burton, C. Y. Wu and C. H. Wong, *Nat. Chem.*, 2016, **8**, 338.
127. C. Dumon, E. Samain and B. Priem, *Biotechnol. Prog.*, 2004, **20**, 412.
128. L. Zhang, K. Lau, J. Cheng, H. Yu, Y. Li, G. Sugiarto, S. Huang, L. Ding, V. Thon, P. G. Wang and X. Chen, *Glycobiology*, 2010, **20**, 1077.
129. K. T. Huang, B. C. Wu, C. C. Lin, S. C. Luo, C. Chen, C. H. Wong and C. C. Lin, *Carbohydr. Res.*, 2006, **341**, 2151.
130. A. Duffels, L. G. Green, R. Lenz, S. V. Ley, S. P. Vincent and C. H. Wong, *Bioorg. Med. Chem.*, 2000, **8**, 2519.
131. K. Witte, P. Sears, R. Martin and C. H. Wong, *J. Am. Chem. Soc.*, 1997, **119**, 2114.
132. R. Kannagi, in *Handbook of Glycosyltransferases*, ed. N. Taniguchi, K. Honke, M. Fukuda, H. Narimatsu and Y. Yamaguchi, Springer Japan, Tokyo, 2nd edn, 2014, ch. 50, pp. 549–558.
133. T. Li, M. Huang, L. Liu, S. Wang, K. W. Moremen and G. J. Boons, *Chem. - Eur. J.*, 2016, **22**, 18742.
134. A. Khaled, T. Ivannikova and C. Auge, *Carbohydr. Res.*, 2004, **339**, 2641.
135. T. Kudo and H. Narimatsu, in *Handbook of Glycosyltransferases*, ed. N. Taniguchi, K. Honke, M. Fukuda, H. Narimatsu and Y. Yamaguchi, Springer Japan, Tokyo, 2nd edn, 2014, ch. 48, pp. 531–540.
136. D. A. Rasko, G. Wang, M. M. Palcic and D. E. Taylor, *J. Biol. Chem.*, 2000, **275**, 4988.
137. S. Rabbani, V. Miksa, B. Wipf and B. Ernst, *Glycobiology*, 2005, **15**, 1076.
138. A. D. Calderon, Y. Liu, X. Li, X. Wang, X. Chen, L. Li and P. G. Wang, *Org. Biomol. Chem.*, 2016, **14**, 4027.
139. Q. Yang, R. Zhang, H. Cai and L. X. Wang, *J. Biol. Chem.*, 2017, **292**, 14796.
140. A. Bastida, A. Fernandez-Mayoralas, R. Gomez Arrayas, F. Iradier, J. C. Carretero and E. Garcia-Junceda, *Chem. - Eur. J.*, 2001, **7**, 2390.
141. Z. Wang, Z. S. Chinoy, S. G. Ambre, W. Peng, R. McBride, R. P. de Vries, J. Glushka, J. C. Paulson and G. J. Boons, *Science*, 2013, **341**, 379.

142. M. Ogata, K. I. Hidari, T. Murata, S. Shimada, W. Kozaki, E. Y. Park, T. Suzuki and T. Usui, *Bioconjugate Chem.*, 2009, **20**, 538.
143. K. Sujino, R. J. Jackson, N. W. Chan, S. Tsuji and M. M. Palcic, *Glycobiology*, 2000, **10**, 313.
144. M. Gilbert, J. R. Brisson, M. F. Karwaski, J. Michniewicz, A. M. Cunningham, Y. Wu, N. M. Young and W. W. Wakarchuk, *J. Biol. Chem.*, 2000, **275**, 3896.
145. M. Gilbert, M. F. Karwaski, S. Bernatchez, N. M. Young, E. Taboada, J. Michniewicz, A. M. Cunningham and W. W. Wakarchuk, *J. Biol. Chem.*, 2002, **277**, 327.
146. M. Gilbert, A. M. Cunningham, D. C. Watson, A. Martin, J. C. Richards and W. W. Wakarchuk, *Eur. J. Biochem.*, 1997, **249**, 187.
147. M. Gilbert, R. Bayer, A. M. Cunningham, S. DeFrees, Y. Gao, D. C. Watson, N. M. Young and W. W. Wakarchuk, *Nat. Biotechnol.*, 1998, **16**, 769.
148. N. Fierfort and E. Samain, *J. Biotechnol.*, 2008, **134**, 261.
149. M. Izumi, G. J. Shen, S. Wacowich-Sgarbi, T. Nakatani, O. Plettenburg and C. H. Wong, *J. Am. Chem. Soc.*, 2001, **123**, 10909.
150. H. Yu, H. Chokhawala, R. Karpel, H. Yu, B. Wu, J. Zhang, Y. Zhang, Q. Jia and X. Chen, *J. Am. Chem. Soc.*, 2005, **127**, 17618.
151. G. Sugiarto, K. Lau, Y. Li, Z. Khedri, H. Yu, D. T. Le and X. Chen, *Mol. BioSyst.*, 2011, **7**, 3021.
152. L. Wen, D. Liu, Y. Zheng, K. Huang, X. Cao, J. Song and P. G. Wang, *ACS Cent. Sci.*, 2018, **4**, 451.
153. H. Tsukamoto, Y. Takakura, T. Mine and T. Yamamoto, *J. Biochem.*, 2008, **143**, 187.
154. G. J. Shen, A. K. Datta, M. Izumi, K. M. Koeller and C. H. Wong, *J. Biol. Chem.*, 1999, **274**, 35139.
155. S. L. Marcus, R. Polakowski, N. O. Seto, E. Leinala, S. Borisova, A. Blancher, F. Roubinet, S. V. Evans and M. M. Palcic, *J. Biol. Chem.*, 2003, **278**, 12403.
156. J. A. Alfaro, R. B. Zheng, M. Persson, J. A. Letts, R. Polakowski, Y. Bai, S. N. Borisova, N. O. Seto, T. L. Lowary, M. M. Palcic and S. V. Evans, *J. Biol. Chem.*, 2008, **283**, 10097.
157. W. Yi, J. Shen, G. Zhou, J. Li and P. G. Wang, *J. Am. Chem. Soc.*, 2008, **130**, 14420.
158. W. Yi, Q. Yao, Y. Zhang, E. Motari, S. Lin and P. G. Wang, *Biochem. Biophys. Res. Commun.*, 2006, **344**, 631.
159. K. Y. Do, S. I. Do and R. D. Cummings, *Glycobiology*, 1997, **7**, 183.
160. J. Shao, J. Zhang, P. Kowal, Y. Lu and P. G. Wang, *Biochem. Biophys. Res. Commun.*, 2002, **295**, 1.
161. N. Vadaie and D. L. Jarvis, *J. Biol. Chem.*, 2004, **279**, 33501.
162. T. Uyama, H. Kitagawa, J. Tanaka, J. Tamura, T. Ogawa and K. Sugahara, *J. Biol. Chem.*, 2003, **278**, 3072.
163. R. S. Houliston, S. Bernatchez, M. F. Karwaski, R. E. Mandrell, H. C. Jarrell, W. W. Wakarchuk and M. Gilbert, *Glycobiology*, 2009, **19**, 153.

164. Y. Tone, H. Kitagawa, K. Imiya, S. Oka, T. Kawasaki and K. Sugahara, *FEBS Lett.*, 1999, **459**, 415.
165. M. Fondeur-Gelinotte, V. Lattard, S. Gulberti, R. Oriol, G. Mulliert, M. W. Coughtrie, J. Magdalou, P. Netter, M. Ouzzine and S. Fournel-Gigleux, *Glycobiology*, 2007, **17**, 857.
166. Y. Tone, L. C. Pedersen, T. Yamamoto, T. Izumikawa, H. Kitagawa, J. Nishihara, J. Tamura, M. Negishi and K. Sugahara, *J. Biol. Chem.*, 2008, **283**, 16801.
167. S. Kakuda, Y. Sato, Y. Tonoyama, S. Oka and T. Kawasaki, *Glycobiology*, 2005, **15**, 203.
168. I. Morita, Y. Kizuka, S. Kakuda and S. Oka, *J. Biochem.*, 2008, **143**, 719.
169. B. T. Kim, K. Tsuchida, J. Lincecum, H. Kitagawa, M. Bernfield and K. Sugahara, *J. Biol. Chem.*, 2003, **278**, 9116.
170. N. Sugiura, S. Shimokata, T. Minamisawa, J. Hirabayashi, K. Kimata and H. Watanabe, *Glycoconjugate J.*, 2008, **25**, 521.
171. B. S. Tracy, F. Y. Avci, R. J. Linhardt and P. L. DeAngelis, *J. Biol. Chem.*, 2007, **282**, 337.
172. C. A. Munro, S. Bates, E. T. Buurman, H. B. Hughes, D. M. Maccallum, G. Bertram, A. Atrih, M. A. Ferguson, J. M. Bain, A. Brand, S. Hamilton, C. Westwater, L. M. Thomson, A. J. Brown, F. C. Odds and N. A. Gow, *J. Biol. Chem.*, 2005, **280**, 1051.
173. N. Shibata and Y. Okawa, *Biol. Pharm. Bull.*, 2010, **33**, 895.
174. P. H. Tam, G. S. Besra and T. L. Lowary, *ChemBioChem*, 2008, **9**, 267.
175. G. M. Watt, L. Revers, M. C. Webberley, I. B. Wilson and S. L. Flitsch, *Carbohydr. Res.*, 1997, **305**, 533.
176. G. M. Watt, L. Revers, M. C. Webberley, I. B. H. Wilson and S. L. Flitsch, *Angew. Chem., Int. Ed.*, 1997, **36**, 2354.
177. O. Kanie, S. C. Crawley, M. M. Palcic and O. Hindsgaul, *Carbohydr. Res.*, 1993, **243**, 139.
178. O. Kanie, S. C. Crawley, M. M. Palcic and O. Hindsgaul, *Bioorg. Med. Chem.*, 1994, **2**, 1231.
179. I. Brockhausen and P. Stanley, in *Essentials of Glycobiology [Internet]*, ed. A. Varki, R. D. Cummings, J. D. Esko, P. Stanley, G. W. Hart, M. Aebi, A. G. Darvill, T. Kinoshita, N. H. Packer, J. H. Prestegard, R. L. Schnaar and P. H. Seeberger, Cold Spring Harbor Laboratory Press, Cold Spring Harbor, NY, 3rd edn, 2017, ch. 10.
180. L. Li, Y. Liu, T. Li, W. Wang, Z. Yu, C. Ma, J. Qu, W. Zhao, X. Chen and P. G. Wang, *Chem. Commun.*, 2015, **51**, 10310.
181. T. I. Tsai, H. Y. Lee, S. H. Chang, C. H. Wang, Y. C. Tu, Y. C. Lin, D. R. Hwang, C. Y. Wu and C. H. Wong, *J. Am. Chem. Soc.*, 2013, **135**, 14831.
182. H. Yu, Y. Li, Z. Wu, L. Li, J. Zeng, C. Zhao, Y. Wu, N. Tasnima, J. Wang, H. Liu, M. R. Gadi, W. Guan, P. G. Wang and X. Chen, *Chem. Commun.*, 2017, **53**, 11012.
183. W. Li, R. Yu, B. Ma, Y. Yang, X. Jiao, Y. Liu, H. Cao, W. Dong, L. Liu, K. Ma, T. Fukuda, Q. Liu, T. Ma, Z. Wang, J. Gu, J. Zhang and N. Taniguchi, *J. Immunol.*, 2015, **194**, 2596.

184. X. Wang, M. Mathieu and R. J. Brezski, *Protein Cell*, 2018, **9**, 63.
185. A. Varki, R. L. Schnaar and R. Schauer, in *Essentials of Glycobiology [Internet]*, ed. A. Varki, R. D. Cummings, J. D. Esko, P. Stanley, G. W. Hart, M. Aebi, A. G. Darvill, T. Kinoshita, N. H. Packer, J. H. Prestegard, R. L. Schnaar and P. H. Seeberger, Cold Spring Harbor Laboratory Press, Cold Spring Harbor, NY, 3rd edn, 2017, ch. 15.
186. B. Qu, M. Ziak, C. Zuber and J. Roth, *Proc. Natl. Acad. Sci. U. S. A.*, 1996, **93**, 8995.
187. L. K. Mahal, N. W. Charter, K. Angata, M. Fukuda, D. E. Koshland Jr and C. R. Bertozzi, *Science*, 2001, **294**, 380.
188. O. Blixt and N. Razi, *Methods Enzymol.*, 2006, **415**, 137.
189. H. Yu, Y. Li, J. Zeng, V. Thon, D. M. Nguyen, T. Ly, H. Y. Kuang, A. Ngo and X. Chen, *J. Org. Chem.*, 2016, **81**, 10809.
190. R. D. Cummings, *Mol. BioSyst.*, 2009, **5**, 1087.
191. U. Lindahl, J. Couchman, K. Kimata and J. D. Esko, in *Essentials of Glycobiology [Internet]*, ed. A. Varki, R. D. Cummings, J. D. Esko, P. Stanley, G. W. Hart, M. Aebi, A. G. Darvill, T. Kinoshita, N. H. Packer, J. H. Prestegard, R. L. Schnaar and P. H. Seeberger, Cold Spring Harbor Laboratory Press, Cold Spring Harbor, NY, 3rd edn, 2017, ch. 17.
192. A. S. Ramirez, J. Boilevin, C. W. Lin, B. Ha Gan, D. Janser, M. Aebi, T. Darbre, J. L. Reymond and K. P. Locher, *Glycobiology*, 2017, **27**, 726.
193. S. T. Li, N. Wang, X. X. Xu, M. Fujita, H. Nakanishi, T. Kitajima, N. Dean and X. D. Gao, *FASEB J.*, 2018, **32**, 2492.
194. Z. Xiao, Y. Guo, Y. Liu, L. Li, Q. Zhang, L. Wen, X. Wang, S. M. Kondengaden, Z. Wu, J. Zhou, X. Cao, X. Li, C. Ma and P. G. Wang, *J. Org. Chem.*, 2016, **81**, 5851.
195. C. H. Wong, S. L. Haynie and G. M. Whitesides, *J. Org. Chem.*, 1982, **47**, 5416.
196. X. Chen, J. Fang, J. Zhang, Z. Liu, J. Shao, P. Kowal, P. Andreana and P. G. Wang, *J. Am. Chem. Soc.*, 2001, **123**, 2081.
197. H. Yu and X. Chen, *Org. Biomol. Chem.*, 2016, **14**, 2809.
198. A. Santra, Y. Li, H. Yu, T. J. Slack, P. G. Wang and X. Chen, *Chem. Commun.*, 2017, **53**, 8280.
199. S. Li, S. Wang, X. Fu, X. W. Liu, P. G. Wang and J. Fang, *Carbohydr. Polym.*, 2017, **178**, 221.
200. T. Endo, S. Koizumi, K. Tabata, S. Kakita and A. Ozaki, *Carbohydr. Res.*, 1999, **316**, 179.
201. T. Endo, S. Koizumi, K. Tabata and A. Ozaki, *Appl. Microbiol. Biotechnol.*, 2000, **53**, 257.
202. X. Chen, Z. Liu, J. Zhang, W. Zhang, P. Kowal and P. G. Wang, *ChemBioChem*, 2002, **3**, 47.
203. C. Dumon, B. Priem, S. L. Martin, A. Heyraud, C. Bosso and E. Samain, *Glycoconjugate J.*, 2001, **18**, 465.
204. B. Priem, M. Gilbert, W. W. Wakarchuk, A. Heyraud and E. Samain, *Glycobiology*, 2002, **12**, 235.
205. D. Huang, K. Yang, J. Liu, Y. Xu, Y. Wang, R. Wang, B. Liu and L. Feng, *Metab. Eng.*, 2017, **41**, 23.

206. F. Baumgärtner, L. Seitz, G. A. Sprenger and C. Albermann, *Microb. Cell Fact.*, 2013, **12**, 40.
207. Y. W. Chin, N. Seo, J. H. Kim and J. H. Seo, *Biotechnol. Bioeng.*, 2016, **113**, 2443.
208. T. Antoine, A. Heyraud, C. Bosso and E. Samain, *Angew. Chem., Int. Ed. Engl.*, 2005, **44**, 1350.
209. S. Drouillard, H. Driguez and E. Samain, *Angew. Chem., Int. Ed. Engl.*, 2006, **45**, 1778.
210. L. Ban, N. Pettit, L. Li, A. D. Stuparu, L. Cai, W. Chen, W. Guan, W. Han, P. G. Wang and M. Mrksich, *Nat. Chem. Biol.*, 2012, **8**, 769.

CHAPTER 4

Chemical Synthesis of N-Glycans

YU-HSUAN LIH, SACHIN S. SHIVATARE AND CHUNG-YI WU*

The Genomics Research Center, Academia Sinica, Taiwan
*E-mail: cyiwu@gate.sinica.edu.tw

4.1 Introduction

Glycosylation is the most common and fundamental post-translational protein modification that nature uses to expand structural microheterogeneity and to regulate important biological processes such as cell adhesion, protein clearance, host–pathogen interaction, development, and immune response.[1] The *N*-linked protein glycosylation is found in both eukaryotes and prokaryotes.[2] More than 50% of human proteins are glycosylated through *N*-glycosidic covalent attachments between oligosaccharide building units and asparagine (Asn) residues.[3] Sugar residues are linked to one another as glycan chains in five different glycosidic linkages, typically between carbon 1 and 4 of the sugar molecules that are mostly *N*-acetylglucosamine (GlcNAc), mannose (Man), galactose (Gal), *N*-acetylneuraminic acid (Neu5Ac), and fucose (Fuc). In mammalian tissues, glycans are mainly in conjugate forms such as glycoproteins, glycolipids, and proteoglycans.

In general, *N*-glycans are categorized into three major types, high-mannose, hybrid, and complex types (Figure 4.1).[4] All three types have a common $Man_3GlcNAc_2$ core that is linked to a peptide backbone through an asparagine side chain.[5] In high-mannose series, mono-(Man), di-(Manα1–2Man), and tri-(Manα1–2Manα1–2Man) saccharides can be attached to an

Chemical Biology No. 11
Synthetic Glycomes
Edited by Wanyi Guan, Lei Li and Peng George Wang

Published by the Royal Society of Chemistry, www.rsc.org

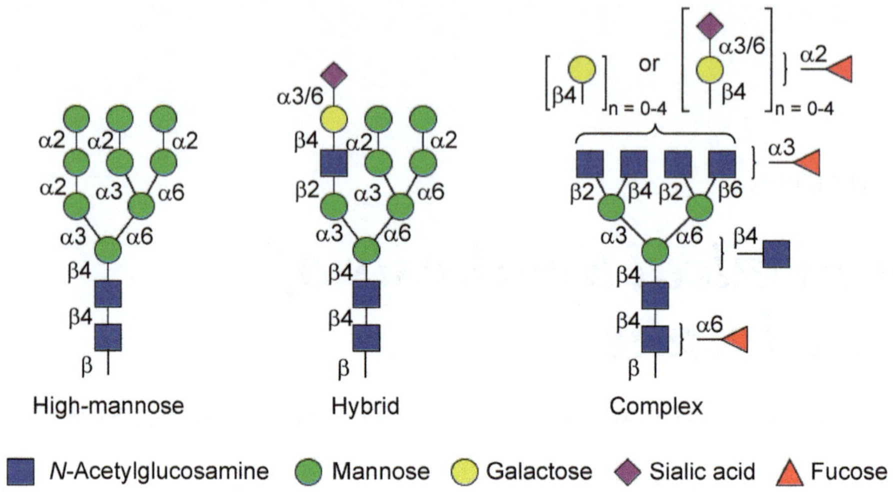

Figure 4.1 Structural representation of high-mannose, hybrid, and complex type *N*-glycans.

α1–3 branch of core trisaccharide (Manβ1–4GlcNAcβ1–4GlcNAcβ1-Asn), whereas an α1–6 branch can be occupied by branched tri-, tetra-, and pentamannose units. Thereupon, attachment of various mannose residues forms $Man_{5/6/7/8/9}GlcNAc_2$ glycans. In hybrid series, the α1–6 branch carries forward the high-mannose units; however, the α1–3 branch is attached to complex type antennae formed by sequential addition of GlcNAc (β1–2), galactose (β1–4), and Neu5Ac (α2–3/6) residues catalyzed by various glycosyltransferases. In complex series, glycans are generally highly branched and presented in bi-, tri-, and tetra-antennary forms containing α2–3/6-Neu5Ac, core bisected β1–4-*N*-acetylglucosamine, core α1–6-fucose, terminal α1–2/3-fucose, and extended poly-*N*-acetyllactosamine (LacNAc) residues. Particularly, in biantennary complex *N*-glycans, both α1–3 and α1–6 branches of core trisaccharide are elongated through GlcNAcβ1–2Man. In triantennary complex *N*-glycans, two forms exist. One form with GlcNAcβ1–2Man elongation at the α1–3 branch and GlcNAcβ1–2(GlcNAcβ1–6)-Man elongation at the α1–6 branch, while another form with GlcNAcβ1–2(GlcNAcβ1–4)Man elongation at the α1–3 branch and GlcNAcβ1–2Man elongation at the α1–6 branch. In tetraantennary complex *N*-glycans, the α1–3 branch is elongated through GlcNAcβ1–2(GlcNAcβ1–4)-Man and the α1–6 branch is elongated through GlcNAcβ1–2(GlcNAcβ1–6)-Man.

4.2 Chemical Synthesis of *N*-Glycans

The structural heterogeneity associated with *N*-glycans not only promotes their importance in many biological processes but also renders their chemical synthesis difficult. Extraction and purification of *N*-glycans from natural sources are often tedious, requiring a wide variety of analysis and separation procedures. Chemical synthesis of carbohydrates is one way to obtain pure and well-defined oligosaccharides for functional and biophysical studies. Synthesis of oligosaccharides is more complicated than that of proteins and nucleic acids due to difficulties associated with the regioselective protection of poly-hydroxyl groups, challenges related to the stereoselective assembly of glycosidic linkages, and problems arising from the scale-up of orthogonally protected monosaccharide units. Reaction of a glycosyl donor bearing a labile leaving group at the anomeric carbon with a nucleophilic acceptor carrying strategically positioned free hydroxyl group often produces a mixture of isomers due to the absence of neighboring group participation. Therefore, it is necessary to install suitable protecting groups on the remaining hydroxyl groups for overall reactivity tuning to 'match' the donor–acceptor pair and for further deprotection and glycosylation or functional group modifications. This section summarizes recent developments in chemical synthesis of *N*-glycans.

4.2.1 Chemical Synthesis of Core Trisaccharide in *N*-Glycans

As mentioned previously, *N*-glycans bear common trisaccharide core structure **1** (Figure 4.2) that consists of two *N*-acetylglucosaminides and one mannoside with all β-linkages. The most challenging part of the chemical synthesis is the β-configuration on mannoside.

There are two major strategies for constructing β-mannoside, direct glycosylation of mannosyl donor and indirect reactions through epimerization of glucoside to mannoside. Direct β-mannosylation can be achieved utilizing triflate-mediated mannosyl donor masked by 4,6-benzylidene, which provides torsional rigidity to the pyranose ring for glycosylation stereoselectivity[6–9] (Scheme 4.1a). In indirect methods, glucosyl donors are used to form β-glucoside by the neighboring group effect; then, the β-mannoside is obtained after epimerization.[10–15] The neighboring participating group on C-2″ of glucose residue is replaced by triflate as the leaving group. After intramolecular attack[11,13,14] or nucleophilic substitution with tri-*n*-butylammonium acetate,[12,15] the desired β-mannoside is produced (Scheme 4.1b). The sequential C-2″ oxidation and reduction can also accomplish epimerization.[10]

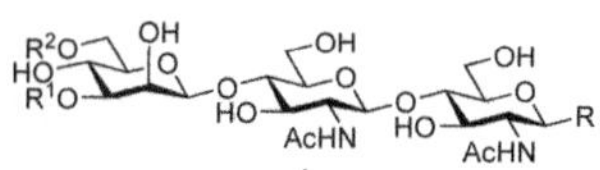

1

Figure 4.2 Structure of core trisaccharide **1** of *N*-glycans (R = reducing end residue; R^1 = 3″-armed glycan residue; R^2 = 6″-armed glycan residue).

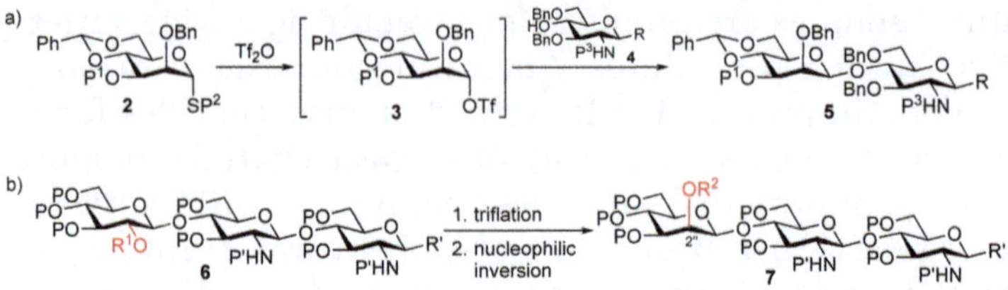

Scheme 4.1 Construction of β-mannoside in core trisaccharide. (a) Direct β-mannosylation through particular mannosyl donor (R = protecting group or sugar; P = protecting groups; Bn = benzyl; Ph = phenyl). (b) Indirect β-construction through epimer glucose (R′ = reducing end residue; R^1 = Ac or Lev (levulinoyl); R^2 = H or Ac; P = protecting groups).

To build the trisaccharide, the β-mannoside can be glycosylated with the other two β-GlcNAc units one by one[6,7] or with pre-coupled di-*N*-acetylglucosaminide.[8,9,11,12,14,15] Glycosylation at C-4 of GlcNAc is generally achieved through 4,6-benzylidene protection, which can undergo selective ring-opening at either site.

To install glycan arms, the mannoside in core trisaccharide is usually masked by 4,6-benzylidene and selectively protected by a temporary group at the 3-hydroxyl group. Typically, the C-3 hydroxyl group is exposed first to introduce 3″-armed glycan; then, either regioselective ring-opening of 4,6-benzylidene is achieved by BH_3/Bu_2BOTf to release C-6 hydroxyl group, or complete reduction of 4,6-benzylidene ring is allowed to generate 4,6-diol for selective 6″-armed glycan installation.

4.2.2 Chemical Synthesis of High-mannose Type *N*-Glycans

Two approaches often used in high-mannose type *N*-glycan synthesis are consecutive installation of oligomannoside[16–18] and attachment of pre-constructed saccharide arms.[17,19–22] Using $Man_9GlcNAc_2$ as an example, in a consecutive synthetic pathway, mannosyl donors **9** and **10** with temporary protecting groups on C-2 and C-3,6, respectively, are coupled to core trisaccharide to give pentasaccharide. Trimannosylation of triol acceptor (**11**) is performed twice to produce a nonamannoside skeleton; then, global deprotection produces $Man_9GlcNAc_2$.[17] Note that the glycosylation is conducted under radical conditions to activate ethylthiomannosyl and phenylthiomannosyl donors (Scheme 4.2).

The nonamannoside can also be constructed through α-selective coupling of trisaccharide and pentasaccharide to the 3″- and 6″-arms of core trisaccharide, respectively. Mannoside building blocks with leaving groups of distinct reactivities or activation conditions are required for building the glycan arms. Hence, chloromannosyl donors (**14**) are activated by AgOTf to build

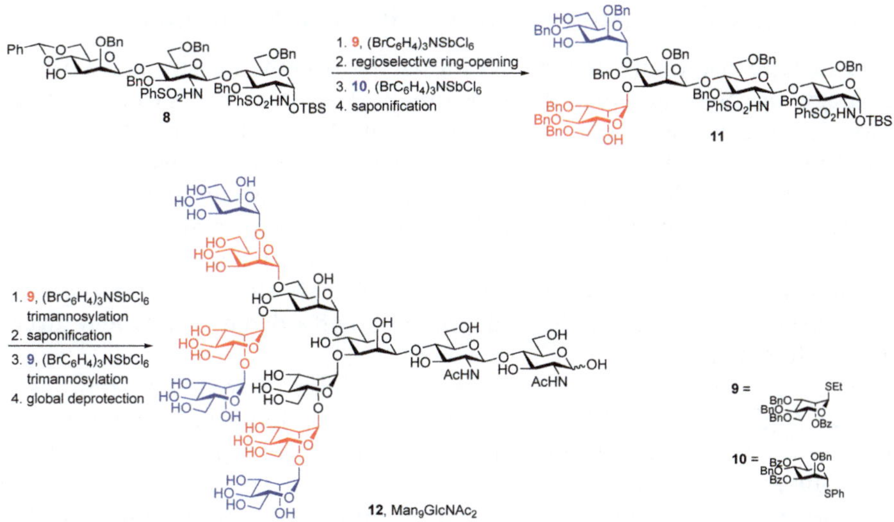

Scheme 4.2 Consecutive synthetic pathway of $Man_9GlcNAc_2$ (TBS = *tert*-butyldimethylsilyl; Bz = benzoyl).

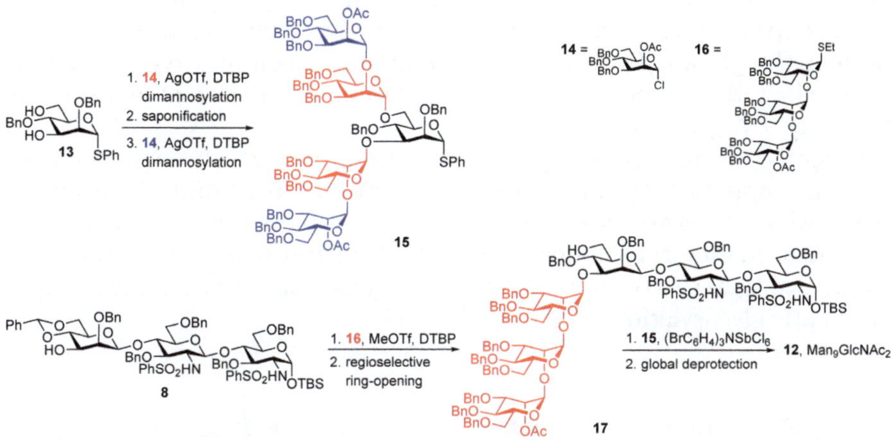

Scheme 4.3 Tri- and pentasaccharide-based synthetic pathway of $Man_9GlcNAc_2$ (DTBP = 2,6-di-*tert*-butylpyridine; Tf = trifluoromethanesulfonyl).

tri- and pentamannoside arms (**16** and **15**), and these arms are coupled with core trisaccharide (**8**) by MeOTf-mediated and radical activation glycosylation, respectively.[17] Even though more monosaccharide building blocks are involved in this approach, it is considered to be more efficient in glycosylation (Scheme 4.3).

In addition to $Man_9GlcNAc_2$ glycan, $Man_3GlcNAc_2$,[23] Man_4,[24] Man_5,[24–26] Man_6,[26] $Man_6GlcNAc_2$,[27] Man_8,[28,29] and Man_9[16,20,29] glycans are routinely synthesized in the same approaches.

4.2.3 Chemical Synthesis of Hybrid Type *N*-Glycans

Hybrid type *N*-glycans can be synthesized through the similar synthetic strategy of high-mannose type *N*-glycan. Danishefsky's group coupled monomannoside with core trisaccharide at C-3″ under radical activation condition; then, the resulting tetrasaccharide acceptor was glycosylated with *N*-phthalimidolactosaminide **18** to form mono-armed hexasaccharide. Radical activation condition was also used in the second glycosylation, but the yield was relatively lower. After regioselective unmasking of 6″-hydroxyl group from benzylidene acetal, the hexasaccharide **19** was coupled with trimannosyl donor **20**; then, global deprotection afforded hybrid type *N*-glycan **21**[30] (Scheme 4.4).

Wong and Wu's groups developed a modular approach to synthesize diverse *N*-glycans.[15,31] A library of glycan building blocks were prepared, and Figure 4.3 shows the building blocks used in hybrid type *N*-glycan synthesis. This strategy starts by treating core trisaccharide **26** that has exposed mono-hydroxyl group with trisaccharide fluoride donor **27** under AgOTf/Cp_2HfCl_2 activation to build C-3″ complex glycan arm. Acid-mediated benzylidene ring-opening is followed to form 4″,6″-diol acceptor **28**, and regioselective glycosylation with **24** or **25** is performed under $(BrC_6H_4)_3NSbCl_6$ activation to build C-6″ tri- or penta-mannosyl glycan arm. Further global deprotection produces hybrid type *N*-glycan **29** and **30** (Scheme 4.5). Besides the glycans in Scheme 4.5, biantennary complex type glycan can also be installed at the 3″-arm by glycosylation with donor **23** to form other hybrid type *N*-glycans.[31]

Recently, Wang's group synthesized the glycan **21** through chemoenzymatic approach,[32] in which ethylthio-*N*-phthalimidoglucosaminylmannoside and ethylthio-trimannoside were successively glycosylated with core trisaccharide *via N*-iodosuccinimide (NIS) and silver trifluoromethanesulfonate (AgOTf) activation. The terminal galactoside was installed through enzymatic glycosylation.

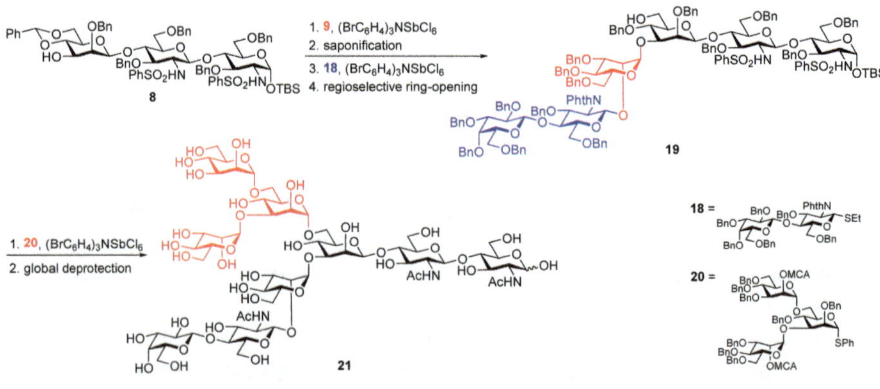

Scheme 4.4 Method for synthesis of hybrid type *N*-glycan **21** developed by Danishefsky's group (MCA = monochloroacetate; Phth = phthalimido).

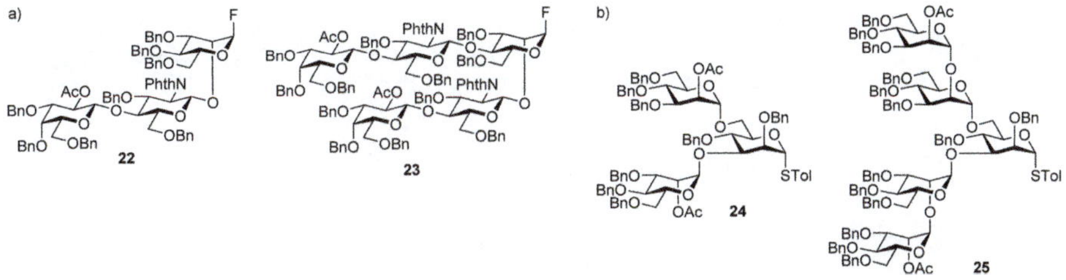

Figure 4.3 Glycan building blocks from Wong and Wu's groups. (a) Target donors to 3″-arm. (b) Target donors to 6″-arm (Tol = 4-tolyl).

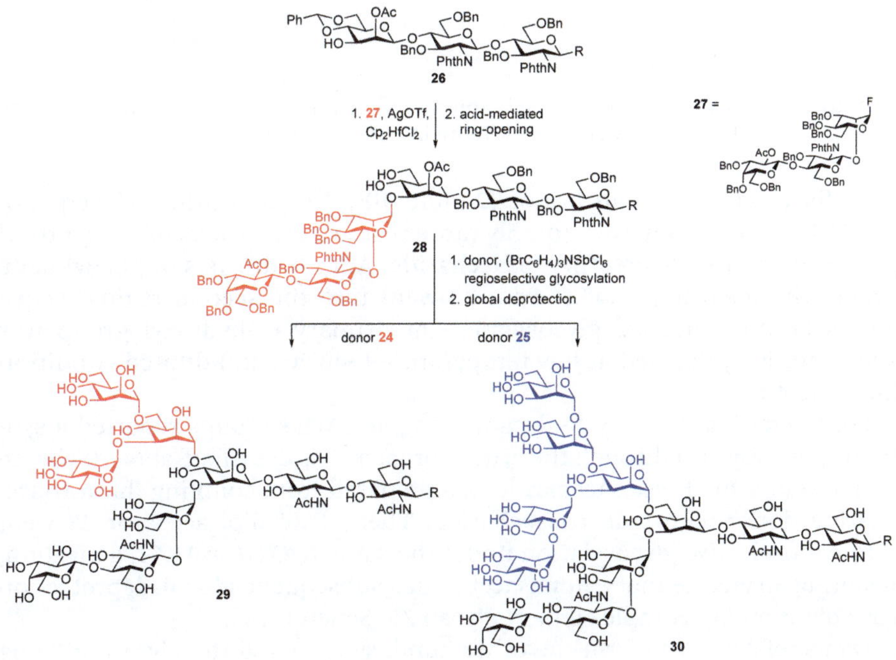

Scheme 4.5 Synthetic pathway of hybrid type *N*-glycan **29** and **30** developed by Wong and Wu's groups (R = reducing end residue; Cp_2HfCl_2 = bis(cyclopentadienyl)hafnium dichloride).

4.2.4 Chemical Synthesis of Complex Type *N*-Glycans

In this section, the synthesis of bi-, tri-, and tetra-antennary type *N*-glycans will be described separately.

4.2.4.1 Synthesis of Symmetric Biantennary Complex Type N-*Glycans*

Earlier research from Unverzagt's group demonstrated arm-approached synthesis of biantennary complex type *N*-glycan[33] with two sequential regioselective glycosylation. Core trisaccharide **31** with 2″,3″-diol was treated

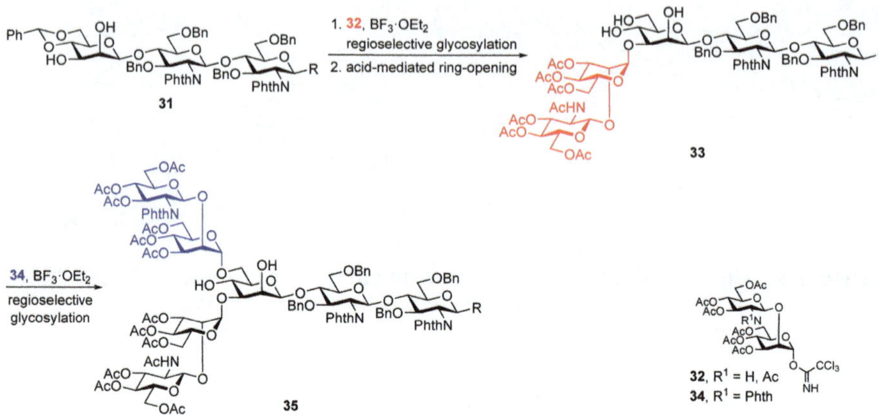

Scheme 4.6 Synthetic pathway of biantennary complex type *N*-glycan developed by Unverzagt's group (R = reducing end residue).

with disaccharide imidate donor **32** under BF_3–OEt_2 activation. The equatorial 3″-hydroxyl group was notably more reactive than the axial 2″-hydroxyl group, giving the desired pentasaccharide. After hydrolysis of benzylidene acetal, the resulting 2″,4″,6″-triol **33** went into the second regioselective glycosylation. Selective glycosylation at primary 6″-hydroxyl group was successfully performed at low temperature (−40 °C) and diluted condition (Scheme 4.6).

For longer biantennary *N*-glycan, Wong and Wu's groups reported a synthetic pathway also through the arm-approach.[15] As described above (Scheme 4.5), mono-armed hexasaccharide was prepared from coupling fluorotrisaccharide donor with core trisaccharide. Then, 4″,6″-diol acceptor **28** went into regioselective glycosylation under the same AgOTf/Cp_2HfCl_2 activation, resulting in biantennary nonasaccharide. Subsequent global deprotection gave biantennary complex type *N*-glycan **36** (Scheme 4.7).

Danishefsky's group, on the other hand, synthesized the glycan utilizing core pentasaccharide $Man_3GlcNAc_2$ with two C-2 hydroxyl groups exposed on terminal mannosides. The pentasaccharide **39** was prepared by coupling diol acceptor **37** with mannosyl donor **38** under radical activation. Later, MeOTf-mediated di-*N*-acetyllactosamination and global deprotection were performed to afford the desired biantennary *N*-glycan **40**[34,35] (Scheme 4.8). However, the coupling yield was relatively low when *N*-acetyllactosamine residue acted as the donor.

4.2.4.2 *Synthesis of Symmetric Triantennary Complex Type* N-*Glycans*

As described in Section 4.1, triantennary complex type *N*-glycans include two forms, 2,2,6-armed and 2,4,2-armed in abbreviation. These *N*-glycans are synthesized by similar synthetic strategy of biantennary *N*-glycans.

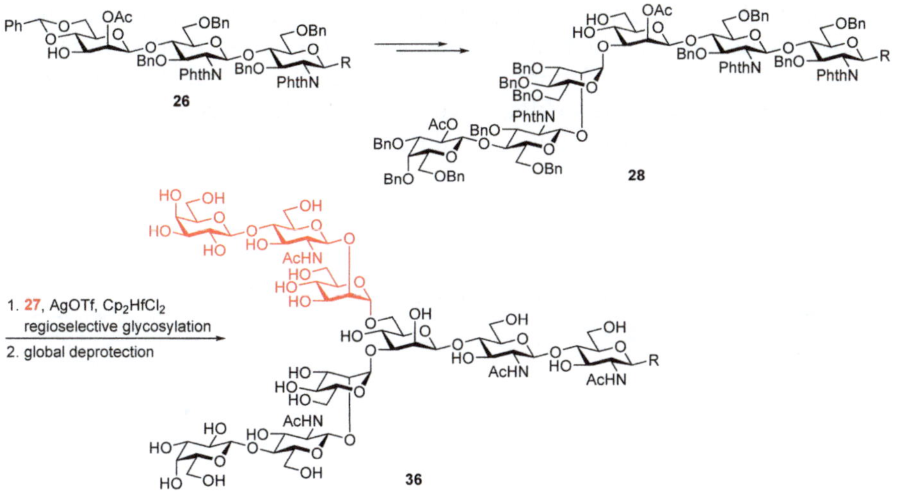

Scheme 4.7 Synthetic pathway of biantennary complex type *N*-glycan **36** developed by Wong and Wu's groups (R = reducing end residue).

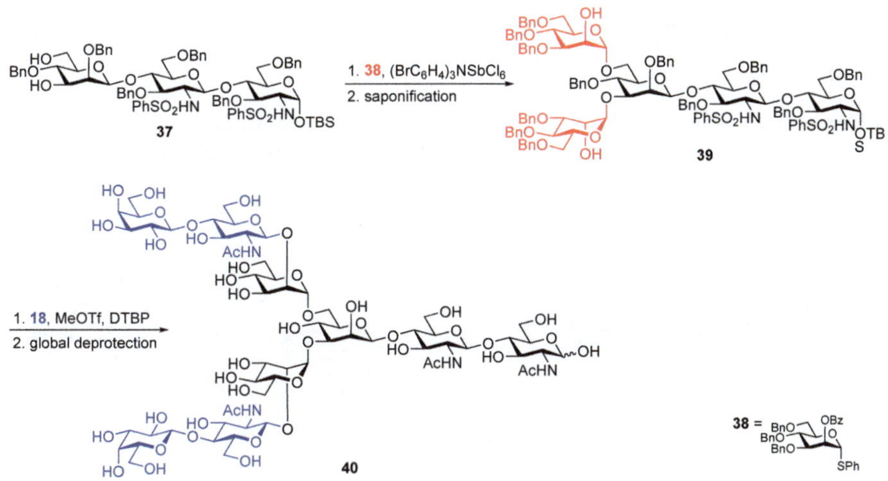

Scheme 4.8 Synthetic pathway of biantennary complex type *N*-glycan **40** developed by Danishefsky's group.

4.2.4.2.1 Synthesis of Symmetric 2,2,6-Armed Triantennary Complex Type *N*-Glycans. Logically, 2,2,6-triantennary glycan can be derived from synthesis of biantennary glycan. Unverzagt's group coupled disaccharide imidate **34** to the C-3″ of the core trisaccharide through regioselective glycosylation. After protecting group management, the prepared 2,6-elongated trisaccharide imidate donor **42** was selectively glycosylated to the primary hydroxyl group of diol acceptor **41** under BF_3–OEt_2 activation. Then, the skeleton of 2,2,6-triantennary glycan was obtained[36,37] (Scheme 4.9).

Similarly, Wong and Wu's groups utilized mono-armed hexasaccharide **28** that was synthesized from trisaccharide **26** to build 2,2,6-armed complex type *N*-glycan. Pre-constructed pentasaccharide, 2,6-*N*-acetyllactosaminyl mannoside, was fluorinated to be donor **44**. Regioselective glycosylation was activated by AgOTf/Cp_2HfCl_2; then, global deprotection gave 2,2,6-triantennary *N*-glycan **45**[15] (Scheme 4.10).

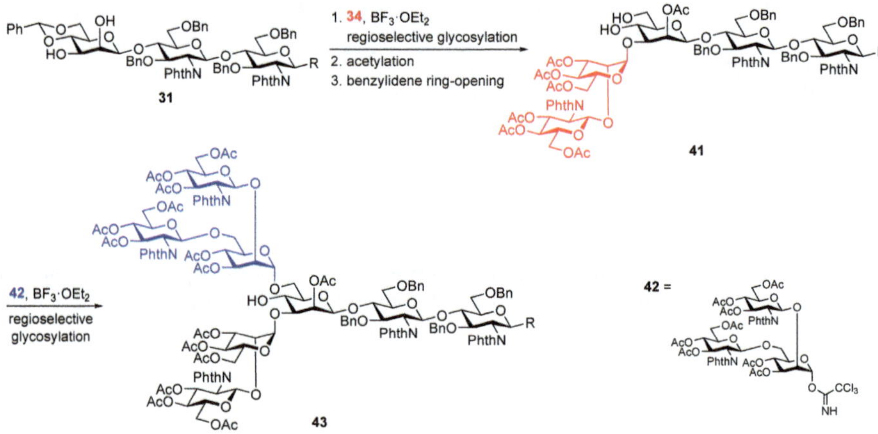

Scheme 4.9 Synthetic pathway of 2,2,6-triantennary complex type *N*-glycan developed by Unverzagt's group (R = reducing end residue).

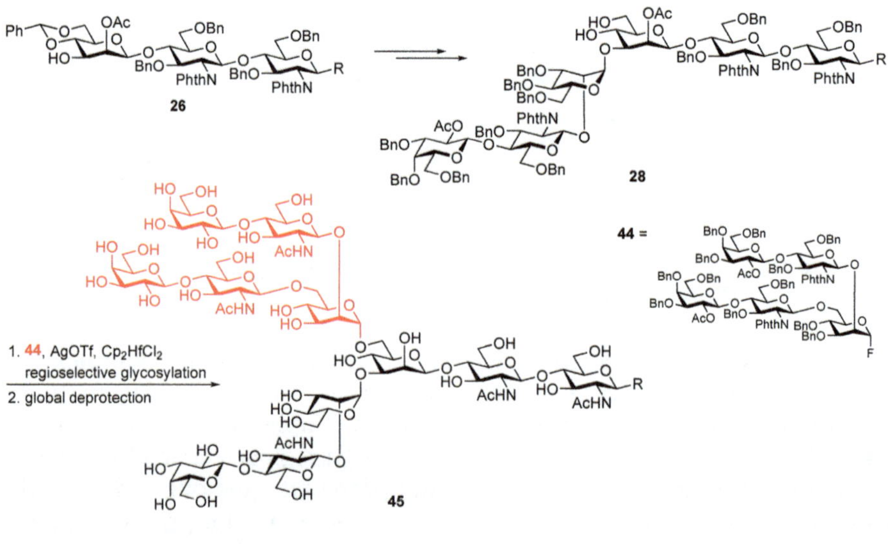

Scheme 4.10 Synthetic pathway of 2,2,6-triantennary complex type *N*-glycan **45** developed by Wong and Wu's groups (R = reducing end residue).

4.2.4.2.2 Synthesis of Symmetric 2,4,2-Armed Triantennary Complex Type *N*-Glycans. To synthesize 2,4,2-armed triantennary *N*-glycan, 2,4-armed saccharide residue is required. Similar to the synthesis of biantennary glycans described above (Scheme 4.6), Unverzagt's group coupled imidate-equipped 2,4-armed trisaccharide **46** with core trisaccharide through BF_3–OEt_2 activation. After modifying protecting groups, hexasaccharide acceptor **47** went through the second glycosylation to furnish 2,4,2-triantennary *N*-glycan skeleton[36,37] (Scheme 4.11).

Likewise, Wong and Wu's groups built 2,4-armed pentasaccharide and converted the reducing end to fluoride as the donor. The pentasaccharide donor **49** was first coupled with core trisaccharide at C-3″ under AgOTf/Cp_2H-fCl_2 activation; then, hydrolysis of benzylidene provided octasaccharide diol **50** as acceptor for the next glycosylation. The 6″-armed trisaccharide was installed smoothly in the same condition; then, global deprotection afforded the desired 2,4,2-triantennary *N*-glycan **51**[15] (Scheme 4.12).

Danishefsky's group, in contrast, prepared three hydroxyl group-exposed pentasaccharide **53** by coupling core trisaccharide **8** with C-2,4 and C-2 temporary-protected mannosides, respectively. Further MeOTf-activated tri-*N*-acetyllactosamination and global deprotection of the pentasaccharide gave the triantennary *N*-glycan **54**[34,35] (Scheme 4.13).

4.2.4.3 *Synthesis of Symmetric Tetraantennary Complex Type* N-*Glycans*

To synthesize tetraantennary complex type *N*-glycan, Unverzagt's group used the known 2,4- and 2,6-armed trisaccharide imidates to glycosylate with core trisaccharide. The arms were successfully installed at C-3″ and C-6″, respectively, to form tetraantennary *N*-glycan skeleton **55**[36,37] (Scheme 4.14).

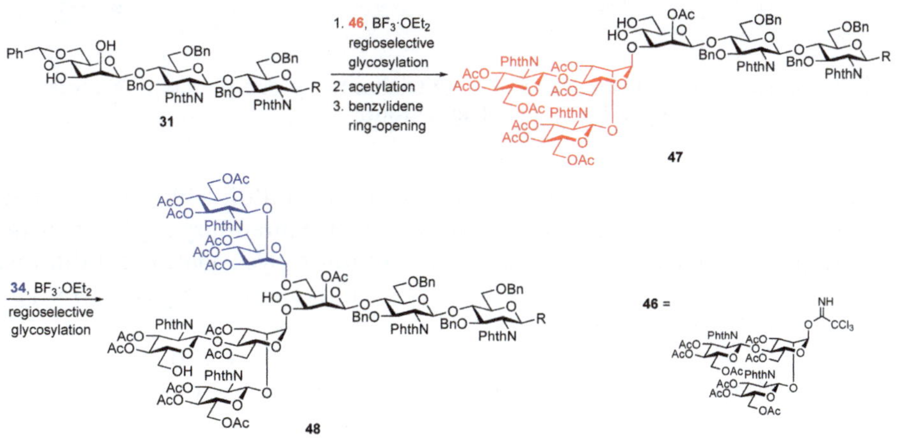

Scheme 4.11 Synthetic pathway of 2,4,2-triantennary complex type *N*-glycan developed by Unverzagt's group (R = reducing end residue).

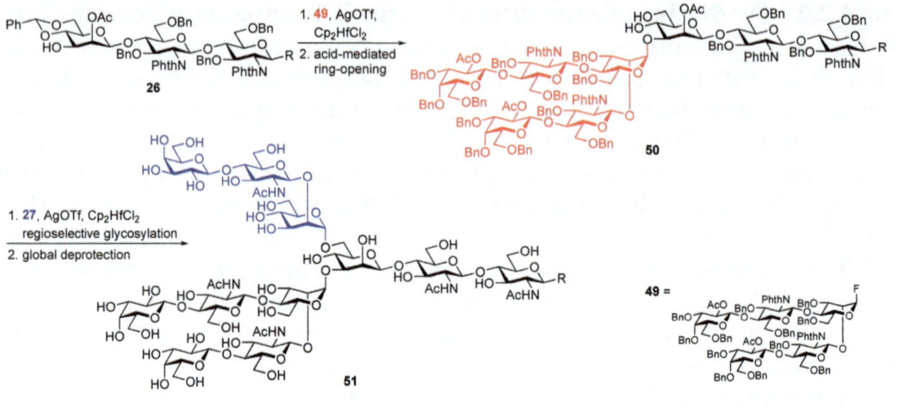

Scheme 4.12 Synthetic pathway of 2,4,2-triantennary complex type *N*-glycan **51** developed by Wong and Wu's groups (R = reducing end residue).

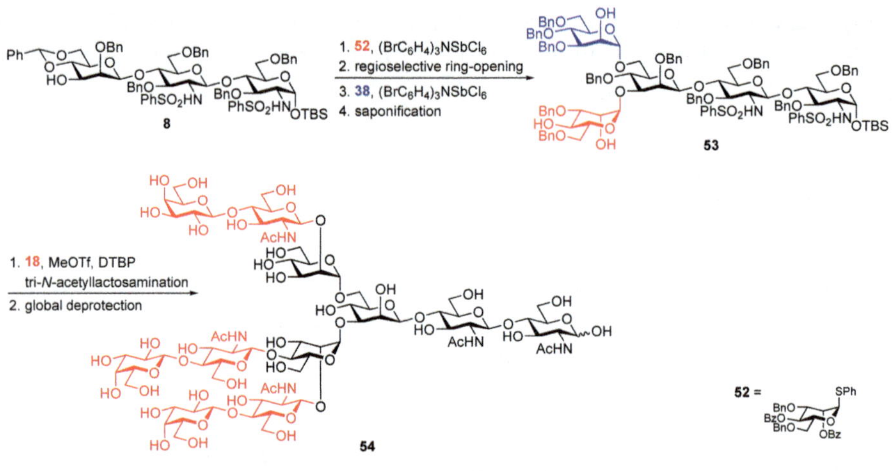

Scheme 4.13 Synthetic pathway of 2,4,2-triantennary complex type *N*-glycan **54** developed by Danishefsky's group.

Wong and Wu's groups utilized pentasaccharide building blocks described above (Schemes 4.10 and 4.12) to synthesize tetraantennary *N*-glycan. Symmetric glycan arms were introduced under common glycosylation conditions; then, global deprotection produced the desired tetraantennary complex type *N*-glycan **56**[15] (Scheme 4.15).

Using a different strategy, Danishefsky's group installed two terminal mannosides with four selectively exposed hydroxyl groups first to form core pentasaccharide **58**. The four glycan arms were equipped concurrently through tetra-*N*-acetyllactosamination activated by MeOTf; then, global deprotection afforded the tetraantennary *N*-glycan **59**[34,35] (Scheme 4.16).

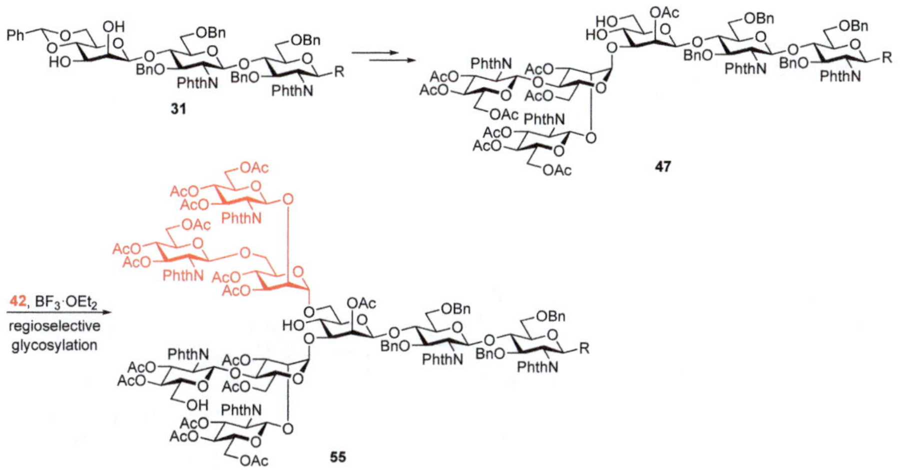

Scheme 4.14 Synthetic pathway of tetraantennary complex type *N*-glycan developed by Unverzagt's group (R = reducing end residue).

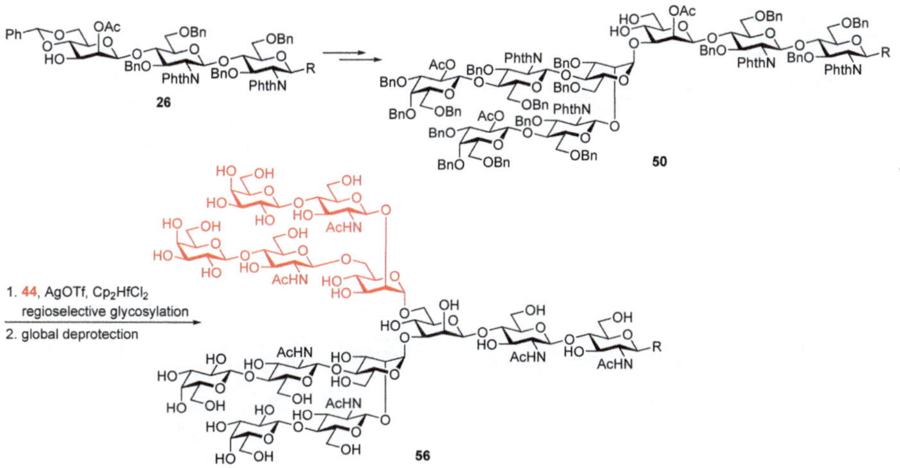

Scheme 4.15 Synthetic pathway of tetraantennary complex type *N*-glycan **56** developed by Wong and Wu's groups (R = reducing end residue).

4.2.5 Synthesis of Neu5Ac-containing Complex Type *N*-Glycans

Due to the α-glycosidic bond and deprotection methods, chemical synthesis of Neu5Ac-containing *N*-glycans is considered trickier and more time-consuming than that of glycans without Neu5Ac. Danishefsky's group prepared Neu5Ac-containing trisaccharide glycan arm and core pentasaccharide with different temporary protecting groups beforehand. To build

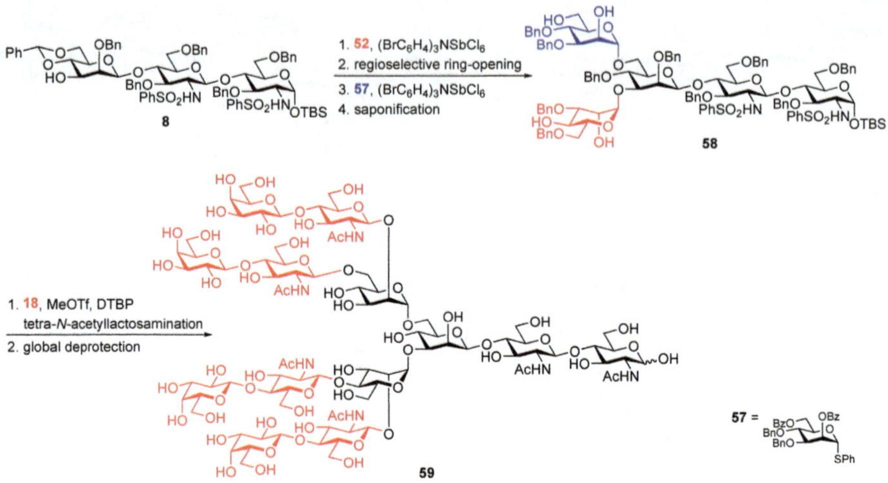

Scheme 4.16 Synthetic pathway of tetraantennary complex type *N*-glycan **59** developed by Danishefsky's group.

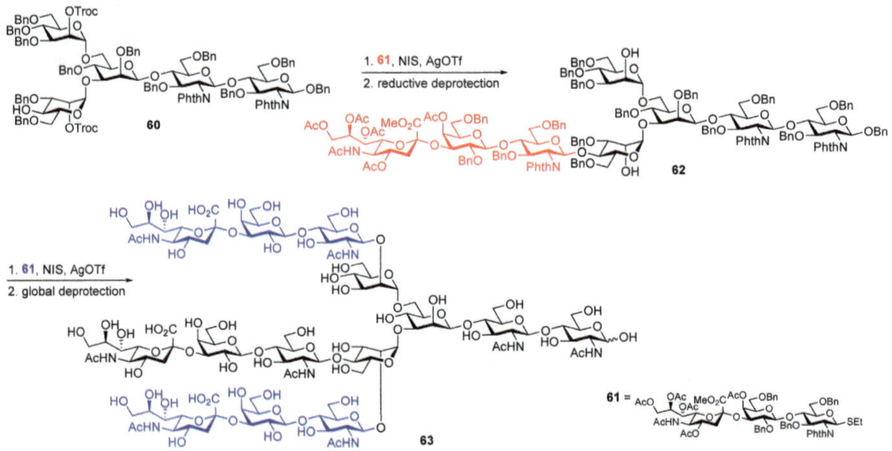

Scheme 4.17 Synthetic pathway of sialylated triantennary *N*-glycan **63** developed by Danishefsky's group (Troc = 2,2,2-trichloroethoxycarbonyl).

2,4,2-armed fully sialylated triantennary *N*-glycan **63**, thiotrisaccharide donor **61** was first glycosylated to C-4 of 3″-armed mannoside, the most sterically hindered position. Later, the other two arms were installed concurrently; then, global deprotection produced sialylated *N*-glycan **63**[9] (Scheme 4.17).

On the other hand, Wong and Wu's groups developed a modular synthesis method that includes Neu5Ac-containing *N*-glycan synthesis. First, Neu5Ac-containing glycan building blocks **64–66** (Figure 4.4) were prepared, and several sialylated *N*-glycans were synthesized based on these modules. Taking tri-sialylated tetraantennary *N*-glycan **69** as an illustration; under the same

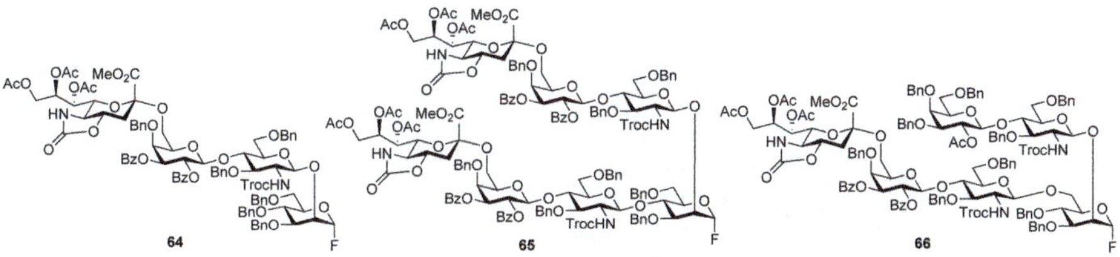

Figure 4.4 Neu5Ac-containing glycan building blocks from Wong and Wu's groups.

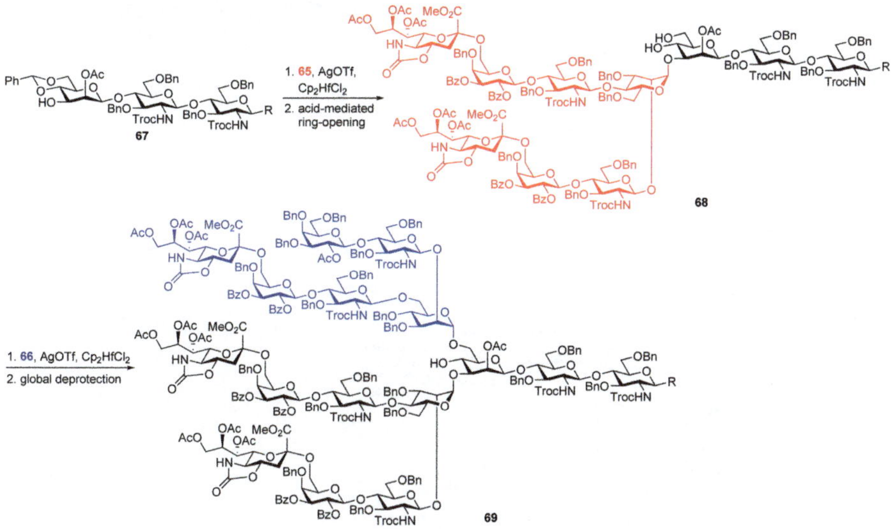

Scheme 4.18 Synthetic pathway of sialylated tetraantennary complex type *N*-glycan **69** developed by Wong and Wu's groups (R = reducing end residue).

modular strategy described previously (Section 4.2.4), sequential glycosylation and deprotection successfully introduced glycan arms with Neu5Ac[38] (Scheme 4.18). Note that global deprotection of sialylated glycans was tricky due to the existence of both acid (–COOH) and base (–NH_2) protecting groups. Ester would be converted to amide during the deprotection of amine protecting groups, and therefore ester should be hydrolyzed first. To circumvent the difficulty, a sequence of deprotection conditions were carefully optimized, and the more efficient chemoenzymatic approaches also have gained considerable popularity.[15,32,39,40]

4.2.6 Synthesis of Asymmetric Complex Type *N*-Glycans

Although synthesis of asymmetric *N*-glycan is much more complex, glycan modules reported by Wong and Wu's groups can be used to construct diverse asymmetric *N*-glycans directly by coupling of pre-constructed asymmetric glycan arms[38] (Scheme 4.19a). Instead of a total chemical approach,

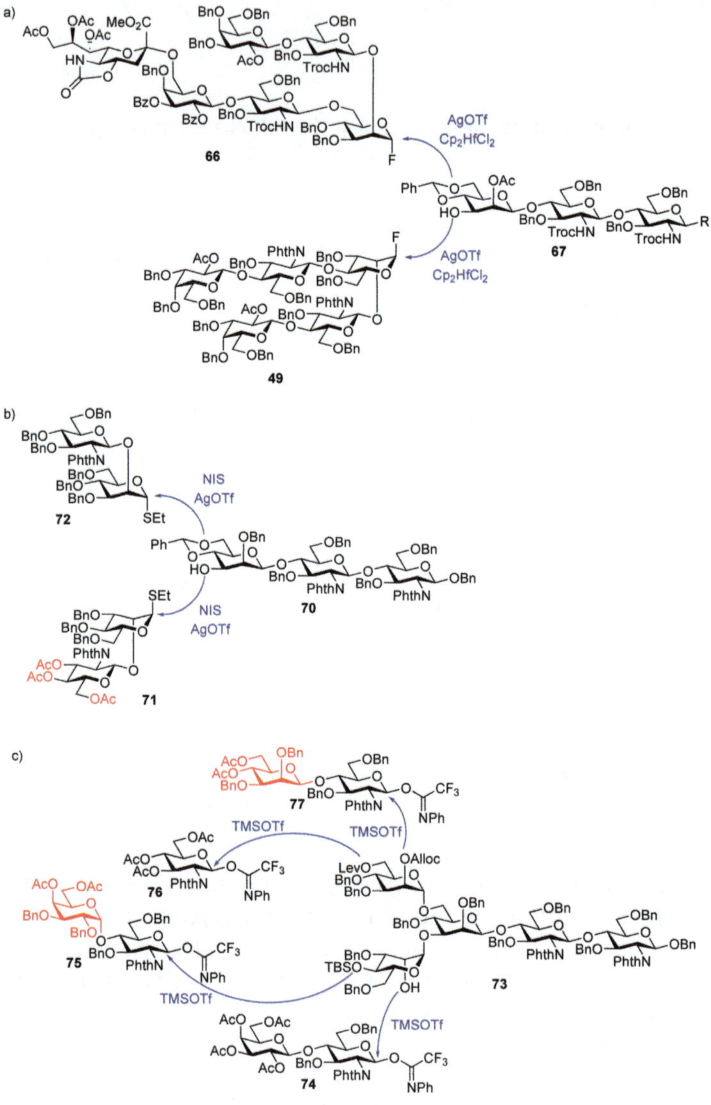

Scheme 4.19 Synthetic strategy of asymmetric complex type *N*-glycans. (a) Strategy from Wong and Wu's groups (R = reducing end residue). (b) Strategy from Wang's group. (c) Strategy from Boons' group (Alloc = allyloxycarbonyl).

the asymmetric syntheses are often managed by chemoenzymatic strategies, combining chemical and enzymatic approaches. Generally, protecting groups that can be selectively kept and further removed under aqueous phase[32,41] (*e.g.* acetyl group) or unnatural sugar residues[40,42] (*e.g.* α-galactoside and β-mannoside) are installed at specific positions as blocks through chemical approaches. Wang's group utilized acetyl groups to block the sugar side from being a substrate for galactosyltransferase (Scheme 4.19b). Global deprotection kept the acetyl groups, which were removed after selective glycan elongation by enzymes.[32] Boons' group employed unnatural glycan linkages (indicated in red in Scheme 4.19c) as blocks that are unrecognizable to galactosyltransferase or sialyltransferase. Fully deprotected glycans went into asymmetric enzymatic elongation. Unnatural sugar blocking sites were also differentiable, and they could be cleaved by α-galactosidase or β-mannosidase.[40]

4.2.7 Synthesis of Bisecting Type *N*-Glycans

Bisecting *N*-glycans exhibiting steric effects present a challenge in synthesis. The bisected *N*-acetylglucosaminide could be introduced in either an early stage[43–46] or a late stage.[47] Li's group reported an early stage installation strategy[45] (Scheme 4.20), in which the bisected GlcNAc was introduced through NIS/TfOH activation before 6″-armed glycan was attached. Less steric hindrance of glycosylation might be encountered through this method.

In contrast, Unverzagt's group developed a late stage introduction of bisected GlcNAc[48] (Scheme 4.21), in which the bisected GlcNAc was introduced after all the antennae were installed, including di-, tri- and tetra

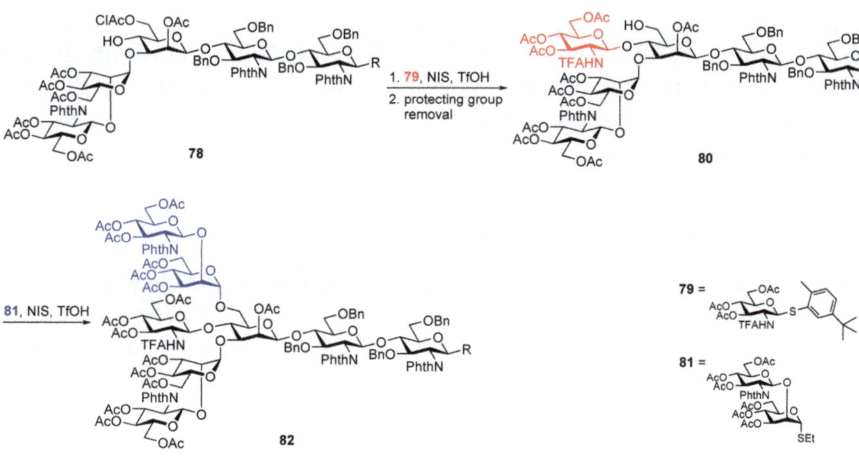

Scheme 4.20 Synthetic strategy of bisecting type *N*-glycan by early stage introduction (R = reducing end residue; AcCl = chloroacetyl; TFA = trifluoroacetamido).

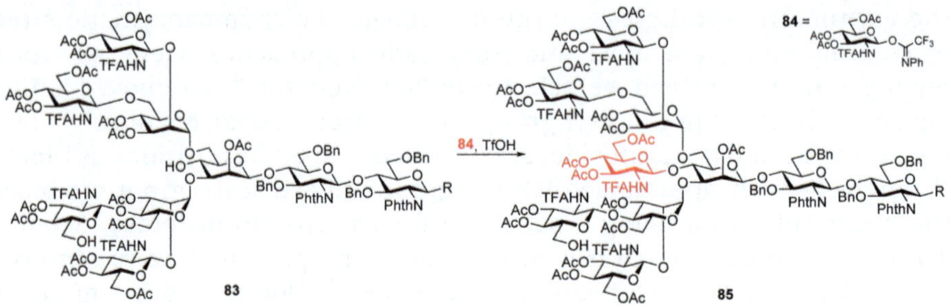

Scheme 4.21 Synthetic strategy of bisecting type *N*-glycan by late stage introduction (R = reducing end residue).

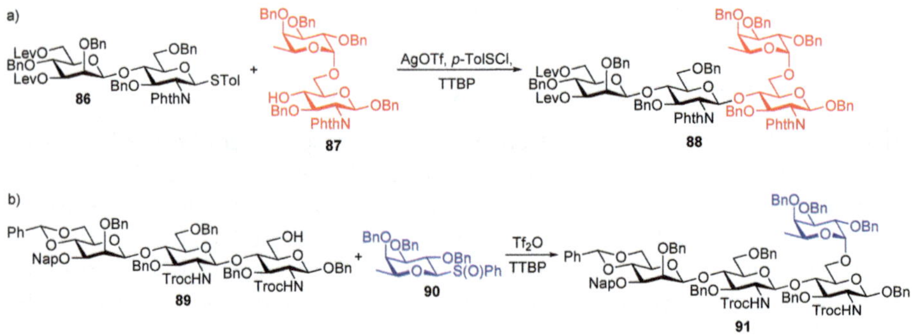

Scheme 4.22 Preparation of core-fucosylated core tetrasaccharide. (a) [2 + 2] strategy (PMB = *p*-methoxybenzyl; TTBP = 2,4,6-tri-*tert*-butylpyrimidine). (b) [3 + 1] strategy (Nap = 2-naphthylmethyl).

antennae. Protected GlcNAc with *N*-Phenylimidate as a leaving group was utilized as donor (**84**) and activated by triflic acid. Trifluoroacetamido protecting groups were considered important in this approach, showing higher reactivity than bulky phthalimido groups. Hence, a single route can apply to multiantennary glycans.

4.2.8 Synthesis of Core-fucosylated *N*-Glycans

Synthesis of core-fucosylated *N*-glycans could also be accomplished in either an early stage[41,46,49–51] or a late stage[52,53] introduction. In recent reports, the early stage introduction, *i.e.* to install the fucoside before glycan antennae, was the dominating approach. Fucose-containing core tetrasaccharide was constructed through [2 + 2] strategy[50] or [3 + 1] strategy[41] (Scheme 4.22). These tetrasaccharide then went into glycan elongation utilizing strategies described above. Core-fucosylated Neu5Ac-containing[49,51] and bisecting[46] *N*-glycans were also achievable by this strategy.

4.3 Discussion

Among the synthetic strategies, building block-based approaches are applied frequently. Unverzagt,[37,48,54] Wong and Wu,[15,31,38] and Wang[32] took trisaccharide $ManGlcNAc_2$ as the core structure while Boons[40–42] and Danishefsky[9,34,35] chose pentasaccharide $Man_3GlcNAc_2$ as the main skeleton. When trisaccharide (**92**) is the acceptor, mannose residue is the relative donor. In contrast, GlcNAc residue is the donor when pentasaccharide (**93** or **94**) is the acceptor. Glycosylation yields with GlcNAc as the donor are generally lower than that of mannose as the donor, but more diverse building blocks are required for the trisaccharide-based pathway. In other words, the diversity of *N*-glycans derived from core trisaccharide depends on pre-constructed branched glycans. Core pentasaccharide, on the other hand, already exposes hydroxyl groups on required positions; thus, the required building blocks are simple. In asymmetric *N*-glycan synthesis, core pentasaccharide with different protecting groups at elongation sites (indicated in red in Figure 4.5b) provides an easier approach that allows selective introduction of various glycan arms at specific sites. The donors for elongation here carry better reactivities than the branched glycan donors for core trisaccharide. It is worth mentioning that for Neu5Ac-containing *N*-glycan synthesis, the core trisaccharide-based pathway needs pre-installed Neu5Ac in the donor building blocks while the core pentasaccharide-based pathway enables late introduction.

4.4 Summary

Chemical synthesis of *N*-glycans is complicated and difficult; nevertheless, several strategies have been developed to accomplish the task. Overall, the chemical syntheses were performed through two major pathways, the consecutive approach and building block-based approach. The later approach is

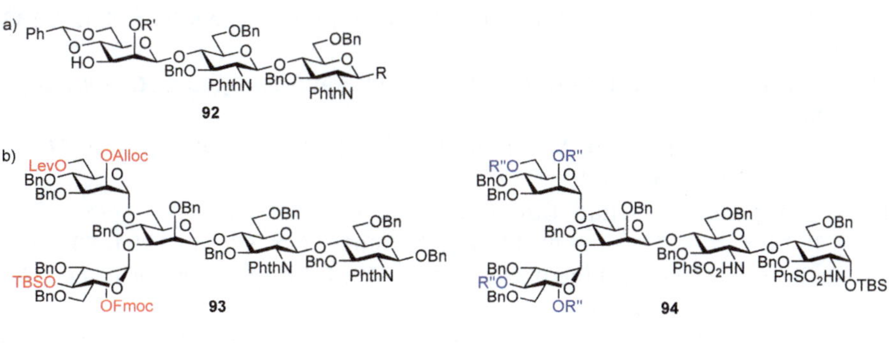

Figure 4.5 Structures of core tri- and penta-saccharide acceptors. (a) Trisaccharide (R = reducing end residue; R′ = H or Ac or Bn). (b) Pentasaccharide (R″ = H or Bn; Fmoc = fluorenylmethyloxycarbonyl).

more favorable and includes pathways through core trisaccharide or pentasaccharide. Both strategies have some pros and cons. Recently, synthesis of asymmetric complex type *N*-glycans has gained attention. Those *N*-glycans with high complexity can be synthesized either through chemical approach or chemoenzymatic approach.

References

1. E. Bieberich, *Adv. Neurobiol.*, 2014, **9**, 47.
2. C. Boscher, J. W. Dennis and I. R. Nabi, *Curr. Opin. Cell Biol.*, 2011, **23**, 383.
3. E. Weerapana and B. Imperiali, *Glycobiology*, 2006, **16**, 91R.
4. J. Breitling and M. Aebi, *Cold Spring Harbor Perspect. Biol.*, 2013, **5**, a013359.
5. R. Kornfeld and S. Kornfeld, *Annu. Rev. Biochem.*, 1985, **54**, 631.
6. Z.-W. Guo, Y. Nakahara, Y. Nakahara and T. Ogawa, *Bioorg. Med. Chem.*, 1997, **5**, 1917.
7. V. Y. Dudkin and D. Crich, *Tetrahedron Lett.*, 2003, **44**, 1787.
8. V. Y. Dudkin, J. S. Miller and S. J. Danishefsky, *Tetrahedron Lett.*, 2003, **44**, 1791.
9. M. A. Walczak and S. J. Danishefsky, *J. Am. Chem. Soc.*, 2012, **134**, 16430.
10. Z.-G. Wang, X. Zhang, M. Visser, D. Live, A. Zatorski, U. Iserloh, K. O. Lloyd and S. J. Danishefsky, *Angew. Chem., Int. Ed.*, 2001, **40**, 1728.
11. C. Unverzagt, *Chem. - Eur. J.*, 2003, **9**, 1369.
12. G. M. Watt and G.-J. Boons, *Carbohydr. Res.*, 2004, **339**, 181.
13. Z. Lu, N. Ding, W. Zhang, P. Wang and Y. Li, *Tetrahedron Lett.*, 2011, **52**, 3320.
14. D. Ott, J. Seifert, I. Prahl, M. Niemietz, J. Hoffman, J. Guder, M. Mönnich and C. Unverzagt, *Eur. J. Org. Chem.*, 2012, **2012**, 5054.
15. S. S. Shivatare, S.-H. Chang, T.-I. Tsai, C.-T. Ren, H.-Y. Chuang, L. Hsu, C.-W. Lin, S.-T. Li, C.-Y. Wu and C.-H. Wong, *J. Am. Chem. Soc.*, 2013, **135**, 15382.
16. D. M. Ratner, O. J. Plante and P. H. Seeberger, *Eur. J. Org. Chem.*, 2002, **2002**, 826.
17. X. Geng, V. Y. Dudkin, M. Mandal and S. J. Danishefsky, *Angew. Chem., Int. Ed.*, 2004, **43**, 2562.
18. L. Jiang and T. H. Chan, *Can. J. Chem.*, 2005, **83**, 693.
19. J. R. Merritt, E. Naisang and B. Fraser-Reid, *J. Org. Chem.*, 1994, **59**, 4443.
20. P. Grice, S. V. Ley, J. Pietruszka and H. W. M. Priepke, *Angew. Chem., Int. Ed. Engl.*, 1996, **35**, 197.
21. P. Grice, S. V. Ley, J. Pietruszka, H. M. I. Osborn, H. W. M. Priepke and S. L. Warriner, *Chem. - Eur. J.*, 1997, **3**, 431.
22. J. Ramos-Soriano, M. C. de la Fuente, N. de la Cruz, R. C. Figueiredo, J. Rojo and J. J. Reina, *Org. Biomol. Chem.*, 2017, **15**, 8877.

23. I. Matsuo, Y. Nakahara, Y. Ito, T. Nukada, Y. Nakahara and T. Ogawa, *Bioorg. Med. Chem.*, 1995, **3**, 1455.
24. J. J. Bailey and D. R. Bundle, *Org. Biomol. Chem.*, 2014, **12**, 2193.
25. C. Uriel, A. M. Gómez, J. C. López and B. Fraser-Reid, *Org. Biomol. Chem.*, 2012, **10**, 8361.
26. R. Jiang, X.-m. Liang, S.-h. Jin, H.-z. Lu, Y.-h. Dong, D.-q. Wang and J.-j. Zhang, *Synthesis*, 2016, **48**, 213.
27. Y. Liu, Y. M. Chan, J. Wu, C. Chen, A. Benesi, J. Hu, Y. Wang and G. Chen, *ChemBioChem*, 2011, **12**, 685.
28. I. Matsuo, T. Miyazaki, M. Isomura, T. Sakakibara and K. Ajisaka, *J. Carbohydr. Chem.*, 1998, **17**, 1249.
29. Y. Zhu, L. Chen and F. Kong, *Carbohydr. Res.*, 2002, **337**, 207.
30. M. Mandal, V. Y. Dudkin, X. Geng and S. J. Danishefsky, *Angew. Chem., Int. Ed.*, 2004, **43**, 2557.
31. V. S. Shivatare, S. S. Shivatare, C.-C. D. Lee, C.-H. Liang, K.-S. Liao, Y.-Y. Cheng, G. Saidachary, C.-Y. Wu, N.-H. Lin, P. D. Kwong, D. R. Burton, C.-Y. Wu and C.-H. Wong, *J. Am. Chem. Soc.*, 2018, **140**, 5202.
32. L. Li, Y. Liu, C. Ma, J. Qu, A. D. Calderon, B. Wu, N. Wei, X. Wang, Y. Guo, Z. Xiao, J. Song, G. Sugiarto, Y. Li, H. Yu, X. Chen and P. G. Wang, *Chem. Sci.*, 2015, **6**, 5652.
33. C. Unverzagt, *Angew. Chem., Int. Ed. Engl.*, 1994, **33**, 1102.
34. V. Y. Dudkin, J. S. Miller and S. J. Danishefsky, *J. Am. Chem. Soc.*, 2004, **126**, 736.
35. V. Y. Dudkin, J. S. Miller, A. S. Dudkina, C. Antczak, D. A. Scheinberg and S. J. Danishefsky, *J. Am. Chem. Soc.*, 2008, **130**, 13598.
36. C. Unverzagt, *Angew. Chem., Int. Ed. Engl.*, 1997, **36**, 1989.
37. C. Unverzagt, G. Gundel, S. Eller, R. Schuberth, J. Seifert, H. Weiss, M. Niemietz, M. Pischl and C. Raps, *Chem. - Eur. J.*, 2009, **15**, 12292.
38. S. S. Shivatare, S.-H. Chang, T.-I. Tsai, S. Y. Tseng, V. S. Shivatare, Y.-S. Lin, Y.-Y. Cheng, C.-T. Ren, C.-C. D. Lee, S. Pawar, C.-S. Tsai, H.-W. Shih, Y.-F. Zeng, C.-H. Liang, P. D. Kwong, D. R. Burton, C.-Y. Wu and C.-H. Wong, *Nat. Chem.*, 2016, **8**, 338.
39. C. Unverzagt and J. Seifert, *Tetrahedron Lett.*, 2000, **41**, 4549.
40. I. A. Gagarinov, T. Li, J. S. Toraño, T. Caval, A. D. Srivastava, J. A. W. Kruijtzer, A. J. R. Heck and G.-J. Boons, *J. Am. Chem. Soc.*, 2017, **139**, 1011.
41. T. Li, M. Huang, L. Liu, S. Wang, K. W. Moremen and G.-J. Boons, *Chem. - Eur. J.*, 2016, **22**, 18742.
42. Z. Wang, Z. S. Chinoy, S. G. Ambre, W. Peng, R. McBride, R. P. de Vries, J. Glushka, J. C. Paulson and G.-J. Boons, *Science*, 2013, **341**, 379.
43. S. Weiler and R. R. Schmidt, *Tetrahedron Lett.*, 1998, **39**, 2299.
44. H. Weiss and C. Unverzagt, *Angew. Chem., Int. Ed.*, 2003, **42**, 4261.
45. G. Wang, W. Zhang, Z. Lu, P. Wang, X. Zhang and Y. Li, *J. Org. Chem.*, 2009, **74**, 2508.
46. P. Wang, J. Zhu, Y. Yuan and S. J. Danishefsky, *J. Am. Chem. Soc.*, 2009, **131**, 16669.

47. S. Eller, C. Raps, M. Niemietz and C. Unverzagt, *Tetrahedron Lett.*, 2010, **51**, 2648.
48. M. Mönnich, S. Eller, T. Karagiannis, L. Perkams, T. Luber, D. Ott, M. Niemietz, J. Hoffman, J. Walcher, L. Berger, M. Pischl, M. Weishaupt, C. Wirkner, R. G. Lichtenstein and C. Unverzagt, *Angew. Chem., Int. Ed.*, 2016, **55**, 10487.
49. B. Wu, Z. Hua, J. D. Warren, K. Ranganathan, Q. Wan, G. Chen, Z. Tan, J. Chen, A. Endo and S. J. Danishefsky, *Tetrahedron Lett.*, 2006, **47**, 5577.
50. B. Sun, B. Srinivasan and X. Huang, *Chem. - Eur. J.*, 2008, **14**, 7072.
51. M. Nagasaki, Y. Manabe, N. Minamoto, K. Tanaka, A. Silipo, A. Molinaro and K. Fukase, *J. Org. Chem.*, 2016, **81**, 10600.
52. J. Seifert and C. Unverzagt, *Tetrahedron Lett.*, 1996, **37**, 6527.
53. R. Schuberth and C. Unverzagt, *Tetrahedron Lett.*, 2005, **46**, 4201.
54. X. Schratt and C. Unverzagt, *Tetrahedron Lett.*, 2005, **46**, 691.

CHAPTER 5

Chemoenzymatic Synthesis of N-Glycans

LEI LI*[a], WANYI GUAN[b], ZHIGANG WU[c] AND PENG G. WANG[a]

[a]Department of Chemistry, Georgia State University, Atlanta, GA 30303, USA; [b]College of Life Science, Hebei Normal University, Shijiazhuang, Hebei 050024, China; [c]College of Bioscience and Bioengineering, Hebei University of Science and Technology, Shijiazhuang, Hebei 050018, China
*E-mail: lli22@gsu.edu

5.1 Introduction

Asparagine-linked glycosylation (*N*-glycosylation) is the most ubiquitous and structurally varied form of post-translational modifications (PTM) of proteins in eukaryotes. *N*-glycan, which is linked to the β-amide nitrogen of the asparagine (Asn) residues, modulate the protein's structure and functions. They were demonstrated to play essential roles in protein folding and stability/degradation, cell adhesion and trafficking, cell signaling, host–pathogen interactions, *etc.*[1] Abnormal expression of *N*-glycans is usually associated with diseases such as cancer and atherosclerosis. Accordingly, glycan/glycoprotein biomarkers have been developed.[2,3] However, a clear understanding of the structure–function relationships of such glycans is often hampered by the inaccessibility of sufficient structurally well-defined *N*-glycoforms.

Unlike precise template-directed transcription/translation of nucleic acids/proteins, glycans are synthesized by enzymes, such as glycosyltransferases (GTs) and glycoside hydrolases (GHs). Their activity and localization, as

Chemical Biology No. 11
Synthetic Glycomes
Edited by Wanyi Guan, Lei Li and Peng George Wang

Published by the Royal Society of Chemistry, www.rsc.org

well as the availability of donor substrates can contribute to the final glycoforms. Particularly, the synthesis of *N*-glycans in eukaryotes is an extremely tedious and complex process that involves 50 plus steps and large numbers of enzymes.[1] As a result, *N*-glycans found in nature possess inherited complexity, diversity and for the majority of glycoforms, low abundancy. Such properties greatly hampered the isolation, separation and purification of *N*-glycoforms.

All eukaryotic *N*-glycans share a core pentasaccharide [Manα1–6(Manα1–3)Manβ1–4GlcNAcβ1–4GlcNAc], and are classified into three types based on glycans linked to the core: (1) high-mannose, in which only mannose residues are attached to the core (Figure 5.1A); (2) complex, in which GlcNAc residues are attached to the core, and then extended (Figure 5.1B); and (3) hybrid, in which only mannose residues are attached to the α1–6-Man branch, and GlcNAc initiated glycans are attached to the α1–3-Man branch (Figure 5.1C). The complex type represents the major portion of *N*-glycans, and are most structurally varied. As shown in Figure 5.1D, typical complex type *N*-glycans can be asymmetric, in which each branch may contain varied structures. Core-fucosylation, bisecting, sulfation, and multi-antennary are also commonly identified in mammalian glycomes.

A number of chemical methodologies have been developed to generate specific *N*-glycans, mainly focused on simple, symmetric ones (*e.g.*, Figure 5.1A–C). In each case, usually one or a few structures were generated. Recent advances in chemoenzymatic synthesis, on the other hand, have revolutionized *N*-glycan preparation, and enabled batch production of such structures

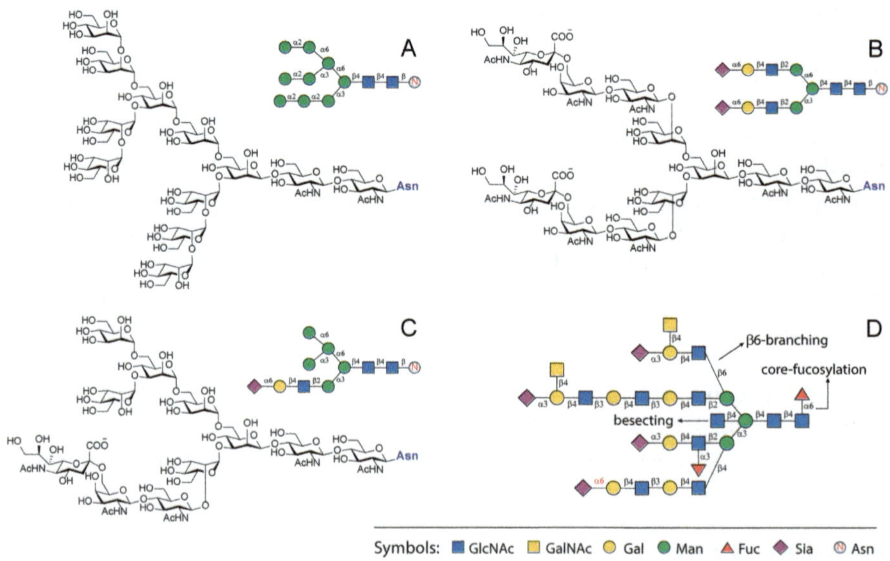

Figure 5.1 Chemical structures and symbol representations of the three classes of *N*-glycans (A–C) and a typical complex *N*-glycan found on mature glycoproteins (D).

(especially asymmetric ones) in a facile and efficient manner. This chapter summarizes enzymes that were applied in the synthesis of *N*-glycans, and recently reported chemoenzymatic and enzymatic strategies to access complex *N*-glycans.

5.2 Enzymes Employed in the Preparation of *N*-Glycans

In contrast to chemical methodologies, enzyme-catalyzed glycan synthesis eliminates tedious protection/deprotection steps and enables high regio- and stereo-selectivity. Many enzymes were applied in the synthesis of glycoconjugates and have been well-summarized previously.[4] Specifically, about 30 enzymes were employed in the synthesis of *N*-glycans to date (Table 5.1), which belong to three classes: GTs, GHs and glycosynthases (Figure 5.2).

GTs comprise a class of natural enzymes that generate glycosidic bonds. They transfer monosaccharide or oligosaccharide moieties from sugar donors onto acceptors in a stereo-specific manner. For example, as shown in Figure 5.2A, a β1–4-galactosyltransferase transfers a Gal residue from the donor UDP-Gal onto the C4-hydroxyl group of the acceptor GlcNAc-OR, forming a β1–4-linkage. To date, all GTs employed in *N*-glycan synthesis are Leloir type, which utilize sugar nucleotides as sugar donors, such as UDP-Gal, GDP-L-Fuc, CMP-Neu5Ac, *etc.*

GHs (also named glycosidases) are enzymes that cleave specific glycosidic bonds, releasing two sugar moieties. For example, a β-galactosidase cleaves the β1–4-linkage of the disaccharide lactose, releasing two monosaccharides, Gal and Glc (Figure 5.2B). Many glycosidases are involved in the biosynthetic pathway of *N*-glycans, trimming and breaking down nascent structures for further extension and modification to form various *N*-glycoforms.[1] In a few reports, certain glycosidases were employed in *N*-glycan preparation to generate novel core structures for enzymatic extension.[5–7]

Glycosynthases are not naturally occurring enzymes, but mutants of glycosidases that could promote glycosidic bond formation in the presence of a suitable activated glycosyl donor.[8,9] Glycosynthases are devoid of product hydrolysis activity due to the point mutation at the critical nucleophilic residue. For example, *endo*-β-*N*-acetylglucosaminidases (ENGases) are a class of *endo*-glycosidases that cleave the β1–4-glycosidic bond in the *N,N*′-diacetylchitobiose core of *N*-glycans. Many GH85 family ENGases were found to possess transglycosylation activity, which were engineered to corresponding glycosynthases by a single point amino acid mutation. Such glycosynthases could catalyze highly efficient transglycosylation to form *N*-glycans in the presence of a reaction intermediate mimic, glycan-oxazoline (Figure 5.2C).[10,11] This class of glycosynthases had been extensively applied in the synthesis of homogenous glycopeptides and glycoproteins.[11]

Table 5.1 List of enzymes employed in the synthesis of *N*-glycans.

Enzyme	Enzyme name (Source)	Linkage synthesized	Reference
Glycosyltransferase			
α2–3-Sialyltransferase	ST3Gal-III (rat)	Neu5Ac**α2–3**Galβ1–3/4GlcNAc	14,16–21,30,39
	ST3Gal-IV (human)	Neu5Ac**α2–3**Galβ1–4GlcNAc	7,22,23
	23ST (*Vibrio* sp. JT-FAJ-16)	Neu5Ac**α2–3**Gal	22,29
	PmST1 (*Pasteurella multocida*)	Neu5Ac**α2–3**Gal	26
	PmST1 E271F/R313Y (*P. multocida*)	Neu5Ac**α2–3**Gal	5,27,28
		Neu5Gc**α2–3**Gal	5,27
α2–6-Sialyltransferase	ST6GAL-I (human)	Neu5Ac**α2–6**Galβ1–4GlcNAc	13–23
	Pd2,6ST (*Photobacterium damselae*)	Neu5Ac**α2–6**Gal	5,24–29
		Neu5Gc**α2–6**Gal	5,6,24,27
α2–8-Sialyltransferase	CjCstII (*Campylobacter jejuni*)	Neu5Ac**α2–8**Neu5Acα2–3/6Gal	5
		Neu5Gc**α2–8**Neu5Gcα2–3/6Gal	5
		Neu5Gc**α2–8**Neu5Acα2–3/6Gal	5
		Neu5Ac**α2–8**Neu5Gcα2–3/6Gal	5
α1–3-Galactosyltransferease	A3GalT (bovine)	Gal**α1–3**Galβ1–4GlcNAc	5
β1–4-Galactosyltransferease	B4GalT1 (bovine milk)	Gal**β1–4**GlcNAc	6,7,13–20,22,23,25–30,33
	NmLgtB (*Neisseria meningitidis*)	Gal**β1–4**GlcNAc	5,21,31,39,40
α1–2-Fucosyltransferease	FUT-I (human)	Fuc**α1–2**Galβ1–4GlcNAc	7
α1–3-Fucosyltransferease	AtFucTA (*Arabidopsis thaliana*)	GlcNAcβ1–4(Fuc**α1–3**)GlcNAc	35
	Hp3FT (*Helicobacter pylori*)	Galβ1–4(Fuc**α1–3**)GlcNAc	5,6,22,27–29
	FUT-V (human)	Galβ1–4(Fuc**α1–3**)GlcNAc	23
	CeFUT6 (*Caenorhabditis elegans*)	Galβ1–4(Fuc**α1–3**)GlcNAc	35

α1–6-Fucosyltransferease	CeFUT8 (*C. elegans*)	GlcNAcβ1–4(Fuc**α1–6**)GlcNAc	26,35
	FUT8 (human)	GlcNAcβ1–4(Fuc**α1–6**)GlcNAc	6,28
β1–4-*N*-Acetylgalacosaminyltransferase	B4GalT1 Y289L (bovine)	GalNAc**β1–4**GlcNAc	37
β1–3-*N*-Acetylglucosaminyltransferase	Hp3GlcNAcT (*H. pylori*)	GlcNAc**β1–3**Galβ1–4GlcNAc	21,22,40
	NmLgtA (*N. meningitidis*)	GlcNAc**β1–3**Galβ1–4GlcNAc	5,39
	β3-GnT-II (human)	GlcNAc**β1–3**Galβ1–4GlcNAc	5,7
β1–6-*N*-Acetylglucosaminyltransferase	MGAT-V (human)	GlcNAc**β1–6**Man	6
Glycosidase and Glycosynthase			
endo-β-*N*-Acetylglucosaminidases (ENGases)	*Endo*-A (*Arthrobacter protophormiae*)	Manβ1–4GlcNAc**β1–4**GlcNAc	33
	Endo-M-N175A (*Mucor hiemalis*)	Manβ1–4GlcNAc**β1–4**GlcNAc	33
	Endo-M-N175Q (*M. hiemalis*)	Manβ1–4GlcNAc**β1–4**GlcNAc	34
β-*N*-Acetyl-glucosaminidase	bGlcNAcA (*Canavalia ensiformis*)	Digest GlcNAcβ1–2Man	26
α-Fucosidase	FucA1 (human)	Digest Fucα1–6GlcNAc	23
Sialidase	BiNanH2 (*Bifidobacterium infantis*)	Digest Sia-Gal	5,6
Sialidase	VcNanH (*Vibrio cholerae*)	Digest Sia-Gal	39
β-Mannosidase	βManA (*Helix pomatia*)	Release terminal β-Man	7
α-Galactosidase	αGalA (green coffee bean)	Release terminal α-Gal	7
β-Galactosidase	LacZ (*Escherichia coli*)	Release terminal β-Gal	6
N-Glycosidase F	PNGaseF (*Flavobacterium meningoseptica*)	Release intact *N*-glycans	6

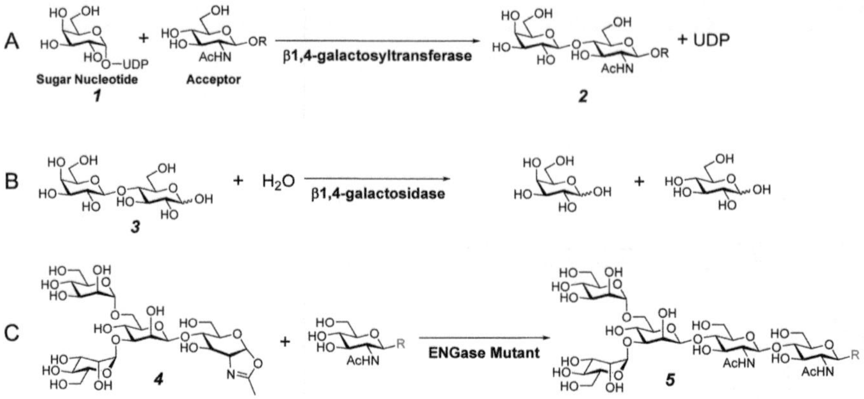

Figure 5.2 Enzyme-catalyzed glycosylation and deglycosylation. (A) GTs-catalyzed glycosylation, taking β1–4-galactosyltransferase as an example; (B) Glycosidase-catalyzed digestion, taking β-galactosidase as an example; (C) Transglycosylation catalyzed by ENGase-derived glycosynthases, which transfers an intermediate mimic **4** (Man_3GlcNAc-oxazoline) onto a GlcNAc residue to form an intact *N*-glycan.

5.2.1 Sialyltransferases

Sialyltransferases (SiaTs) transfer Neu5Ac residues or derivatives from cytidine monophosphate (CMP) activated donors to acceptors (usually to Gal or GalNAc residues). SiaTs are among the earliest and most widely applied GTs in *N*-glycan synthesis. This is partly due to the fact that Sia residues are generally terminal sugars, as well as particular difficulties in chemical synthesis of Sia-containing structures because of their hindered and charged tertiary anomeric center.[12] In 1996, Unverzagt first utilized human α2–6-SiaT (ST6GAL-I, purchased from Sigma) to synthesize sialylated bi-antennary *N*-glycan (Figure 5.1B).[13] Since then, the enzyme was widely applied in chemo-enzymatic synthesis of various *N*-glycans with terminal α2–6-linked Neu5Ac residues.[13–23] Recently, another robust bacterial α2–6-SiaT from *Photobacterium damselae* (Pd2–6ST) was applied in the synthesis of large numbers of *N*-glycans.[5,24–29] Unlike ST6GAL-I, Pd2,6ST catalyzes the transfer of Neu5Ac onto not only terminal Gal residues, but also internal ones, enabled generation of *N*-glycans with multisialylated poly-*N*-acetyllactosamine extensions (see Figure 5.3A for an example).[5,24] The relaxed donor substrate specificity of Pd2,6ST also allowed the generation of Neu5Gc-containing *N*-glycans in the presence of the sugar donor CMP-Neu5Gc.[5,6,24,27]

Animal cell surfaces are often capped with α2–3-sialylated glycans. To synthesize *N*-glycans with α2–3-sialylation, rat α2–3-SiaT (ST3Gal-III) was first introduced by Unverzagt and co-authors.[14,16] The enzyme was able to transfer Neu5Ac onto terminal Gal residues on bi-antennary *N*-glycans. Core-fucosylation, bisecting, and lack of reducing end *N,N′*-diacetylchitobiose core did not affect its activity.[16,17,20] Moreover, ST3Gal-III could recognize chemically

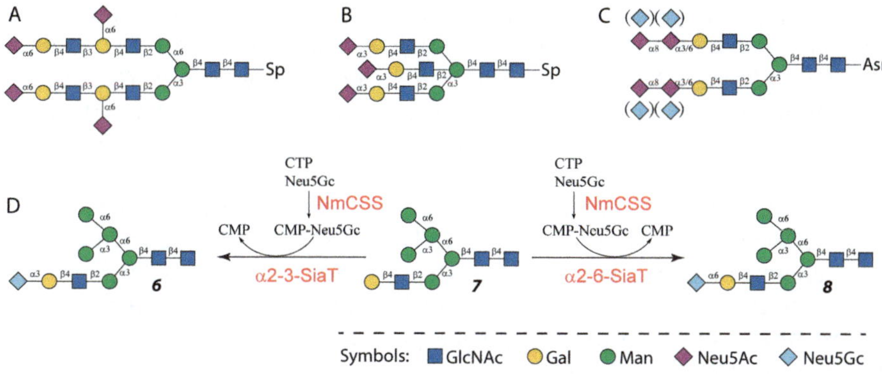

Figure 5.3 Examples of chemoenzymatically synthesized *N*-glycans (A, B, C) and one-pot multi-enzyme system for the synthesis of Neu5Gc-containing *N*-glycans (D).

prepared β1–2-branched structures, forming a natural tri-antennary *N*-glycoform (Figure 5.3B).[30] On the other hand, Boons utilized human α2–3-SiaT (ST3Gal-III) expressed in Chinese hamster ovary (CHO) cells to attach terminal Neu5Ac.[7,22,23] Bacterial α2–3-SiaTs had also been employed given their high activity and expression level. For example, Wong expressed an α2–3-SiaT from *Vibrio* sp. JT-FAJ-16 (23ST), which efficiently catalyzed the synthesis of numbers of symmetric complex *N*-glycans,[25] as well as oligosaccharide building blocks for modular assembly of *N*-glycans.[29] In another example, an α2–3-SiaT from *Pasteurella multocida* (PmST1) was utilized to attach α2–3-Neu5Ac onto isotope-labelled bi-antennary *N*-glycans.[26] Given that PmST1 also exhibits α2–3-sialidase activities,[31] which may yield unexpected product digestion, Wang utilized a double mutated PmST1 (E271F/R313Y, reduced sialidase activity) for chemoenzymatic synthesis of their *N*-glycan libraries.[5,27,28]

Alpha2–8-sialyation is commonly found on gangliosides, but also identified on *N*-glycans, *e.g.*, polysialic acid (PSA) on two specific *N*-glycans of neural cell adhesion molecule.[32] The only example was reported by Wu *et al.*, in which an α2–3/8-SiaT from *Campylobacter jejuni* (CjCstII) was employed to fulfill α2–8-sialyation, yielding 8 bi-antennary *N*-glycans terminated with α2–8-Neu5Ac/Gc (Figure 5.3C).[5] In many cases, to reduce synthetic cost, CMP-Neu5Ac and CMP-Neu5Gc donors were *in situ* generated by a CMP-sialic acid synthetase from *Neisseria meningitidis* (NmCSS) and then transferred by SiaTs in one-pot multi-enzyme (OPME) systems (Figure 5.3D).[5,27]

5.2.2 Galactosyltransferases

Galactosyltransferases (GalTs) can be used to transfer Gal residues from UDP-Gal to various acceptors.[4] In the chemoenzymatic synthesis of *N*-glycans, GalTs are employed to generate LacNAc motifs. The β1–4-GalT from bovine (B4GALT1) had been used in almost all reports to generate the most adjacent LacNAc motif to the pentasaccharide core,[6,7,13–20,22,23,25–30,33] as it was found that bacterial β1–4-GalTs shows reduced activity towards GlcNAc residues attached to the pentasaccharide core (unpublished data). On the other hand, a β1–4-GalT from *Neisseria meningitidis* (NmLgtB) was commonly used for efficient synthesis of poly-LacNAc extensions on *N*-glycans.[21,24,34]

Few other GalTs were tested and/or employed for the synthesis of *N*-glycans. We proved that bovine α1–3-GalT (A3GalT) could efficiently catalyze the generation of alpha-gal (Galα1–3Gal) terminated *N*-glycans,[5] whereas LgtC, a highly efficient α1–4-GalT from *N. meningitidis* that transfers Gal to lactose or LacNAc to form Gb3 epitopes, cannot recognize LacNAc motifs on *N*-glycans (unpublished data).

5.2.3 Fucosyltransferases

Fucosyltransferases (FucTs) transfer L-fucose residues from GDP-L-Fuc to Gal and GlcNAc residues of acceptors. L-Fuc is a C6 deoxy monosaccharide that endows unique binding properties to underlying glycans. L-Fuc is commonly found on *N*-glycans, *O*-glycans, human milk oligosaccharides, *etc*. FucTs have been integrated into chemoenzymatic synthesis of *N*-glycans with core-fucosylation, Lewis X (Le^X) and ABO blood antigens. In one example, a human α1–2-FucT (FUT-I) was incorporated in a chemoenzymatic strategy to produce asymmetric *N*-glycans that contain a blood H antigen (Fucα1–2Galα1–4GlcNAc).[7] In other examples, α1–6-FucTs from *Caenorhabditis elegans* (CeFUT8) and human (FUT8) were employed to fulfill core-fucosylation.[28,35] It is worth noting that the α1–6-FucTs requires a free β1–2-linked GlcNAc on the α1–3-Man branch of *N*-glycans for proper activity. We have developed an enzymatic strategy to overcome this hurdle, starting with partially protected *N*-glycan core structures (Figure 5.4). While α1–6-linked core-fucosylation is exclusively found in mammals, α1–3-linked core-fucose have been observed in plants and insects. The α1–3-FucT from *Arabidopsis thaliana* (AtFucTA) is

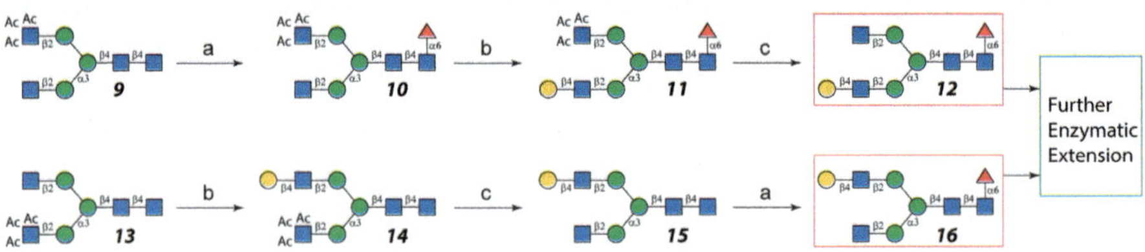

Figure 5.4 Chemoenzymatic synthesis of asymmetric core-fucosylated *N*-glycans. Starting with two chemically prepared *N*-glycan core structures (**9** and **10**) that harbor one acetylated GlcNAc residue at different branches to prevent galactosylation, step-wise synthesis was performed to afford two key intermediates, **12** and **16**. (a) FUT8-catalyzed core-fucosylation in the presence of GDP-L-Fuc; (b) B4GalT1-catalyzed galactosylation in the presence of UDP-Gal; (c) incubate with 30% ammonium hydroxide:H_2O (1:10) for 6 h to remove acetylation.

the only enzyme employed to synthesize α1–3-core-fucosylated *N*-glycans.[35] In addition, AtFucTA was able to recognize α1–6-core-fucosylated *N*-glycans, yielding bi-core-fucosylated structures. Similar to FUT8, AtFucTA requires a free β1–2-linked GlcNAc on the α1–3-Man branch of *N*-glycans for proper activity.

Le^X [Galβ1–4(Fucα1–3)GlcNAc-R] and sialyl-Le^X [Neu5Acα2–3Galβ1–4(Fucα1–3)GlcNAc-R] epitopes are common terminal glycan motifs of *N*-glycans that endow functions in involved biological processes, such as human fertilization.[36] Both bacterial (*e.g.*, Hp3FT from *Helicobacter pylori*) and mammalian α1–3-FucTs (*e.g.*, human FUT-V and *C. elegans* FUT6) have been utilized to generate such *N*-glycans.[7,22,23,26,27,37] While all FucTs could efficiently catalyze the generation of terminal Le^X or sialyl-Le^X epitopes, it was found that bacterial Hp3FT could also fucosylate internal LacNAc motifs to yield unique internal Le^X epitopes in *N*-glycans.[5]

5.2.4 *N*-Acetylhexosaminyltransferases

Complex *N*-glycans are initiated with GlcNAc residues attached to the core pentasaccharide (Figure 5.1D), which are fulfilled by *N*-acetylglucosaminyltransferases (GlcNAcTs). For example, in humans, MGAT5 catalyzes the attachment of a GlcNAc onto the α1–6-Man *via* a β1–6-linkage, initiating the β1–6-branching. Altered expression of β1–6-branching usually associates with diseases, for example, impairing TGF-β-mediated epithelial-to-mesenchymal transition in human lung cancer.[38] Heterogeneous expressed MGAT5 had been employed to identify, separate and synthesize tri- and tetra-antennary *N*-glycans.[6]

Poly-LacNAc extended *N*-glycans are commonly observed in mammals. Paulson proved that recent H3N2 and 2009 pandemic H1N1 human influenza viruses share binding specificity for human-type receptors with such extended glycan chains. Enzymatic synthesis has been performed using several β1–3-GlcNAcTs (*e.g.*, Hp3GlcNAcT from *H. pylori*, NmLgtA from *N. meningitidis*, β3-GnT-II from human), which attach GlcNAc onto terminal Gal *via* a β1–3-linkage, and β1–4GalT (NmLgtB), which attach Gal onto terminal GlcNAc *via* a β1–4-linkage.[5,21,24] Besides these GlcNAcTs, in a very recent report, Reichardt integrated a mutant of bovine milk GalT (B4GalT1 Y289L, which possesses β1–4-GalNAcT activity) in the chemoenzymatic synthesis of *N*-glycans containing LacDiNAc (GalNAcβ1–4GlcNAc).[37]

5.2.5 Glycosidases

The application of GHs is limited to digesting naturally extracted or chemically synthesized structures to afford *N*-glycan cores for further enzymatic extension. For example, Wu *et al.* utilized a sialiadase from *Bifidobacterium infantis* (BiNanH2) together with pronase E to digest sialylglycopeptide (SGP) from egg yolk (Figure 5.5), yielding a bi-antennary *N*-glycan $Gal_2GlcNAc_2Man_3GlcNAc_2$-Asn (Figure 5.5A, compound **17**).[5] In another example, Calderon *et al.* employed

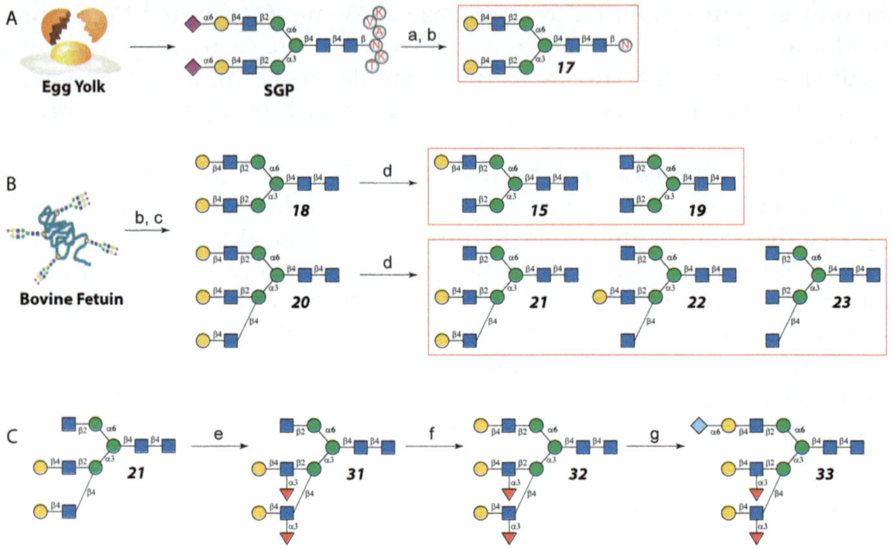

Figure 5.5 Enzymatic digestion of natural extracted glycan/glycopeptides to afford *N*-glycan cores (A, B) and enzymatic extension (C). (a) pronase E; (b) sialidase BiNanH2; (c) PNGaseF; (d) β-galactosiase LacZ; (e) Hp3FT, GDP-L-Fuc; (f) B4GalT1, UDP-Gal; (g) Pd2,6ST, NmCSS, CTP, Neu5Gc.

PNGaseF to release intact *N*-glycans from bovine fetuin, followed by sialidase digestion to remove all Sia residues. Taking advantage of branch preference of a β-galactosidase from *Escherichia coli* (LacZ), Gal residues on these *N*-glycans were then selectively digested to afford a number of (a)symmetric *N*-glycoforms (Figure 5.5B, compounds **15, 19, 21, 22, 23**).[6]

Boons' group, on the other hand, chemically prepared an unnatural asymmetric *N*-glycan, which contains different terminal monosaccharides (*e.g.*, β1–4-linked Man, α1–4-linked Gal) on each branch. After selective removal of the monosaccharide on the desired branch using corresponding *exo*-glycosidases, the branch can be further extended by GTs-catalyzed reactions (see Section 5.4.3 for details).[7]

5.2.6 Glycosynthases

ENGase-derived glycosynthases have been extensively applied in chemoenzymatic syntheses of homogenous glycopeptides and glycoproteins, which was well-summarized previously.[11] It was found that these glycosynthases have extremely relaxed substrate specificities toward acceptors,[34] and can transfer glycan-oxazolines onto almost any glycans with a terminal GlcNAc residue. The property has been applied to the preparation of non-natural *N*-glycan cluster structures.[33,34] As shown in Figure 5.6, Huang synthesized a panel of *N*-glycan cluster structures (compounds **26, 27, 28, 29, 30**) by repeated transferring of glycan-oxazolines (compounds **4, 25**) onto *N*-glycan acceptors.[33]

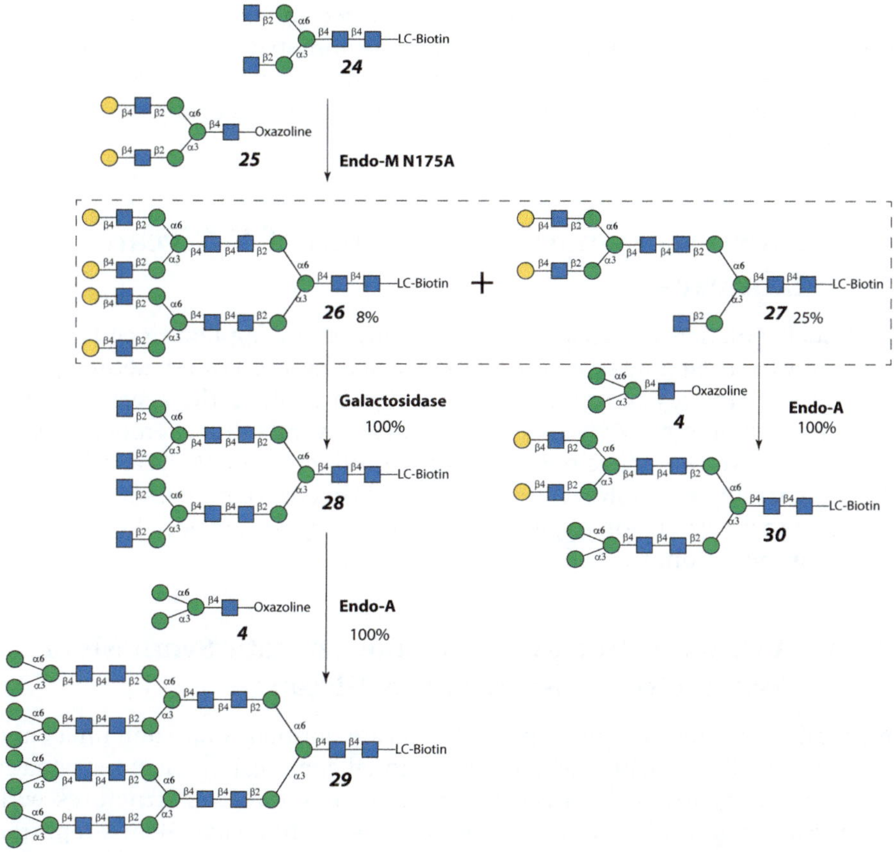

Figure 5.6 Glycosynthase-catalyzed synthesis of *N*-glycan clusters.

In another example, Wu *et al.* successfully transferred two $Man_2GlcNAc$-oxazoline isomers onto a number of asymmetric *N*-glycans, and branch-biased glycan-protein binding and spatial inhibition caused by a bulky group on the other branch of *N*-glycan were observed.[34]

5.3 Enzymatic Preparation of *N*-Glycans

Given the complexity of *N*-glycans and their biosynthetic pathway, total enzymatic synthesis of *N*-glycans is extremely challenging. As a result, chemoenzymatic strategies are usually employed. Nevertheless, a few pure enzymatic strategies have been developed, which involve enzymatic digestion of natural *N*-glycans to generate one or a few cores, followed by enzymatic extension to expand the cores to various complex *N*-glycans. For example, both Paulson's and Wang's groups utilized sialidase to digest SGP extracted from egg yolk, yielding a universal bi-antennary *N*-glycan core (compound **17**), which was then extended to complex symmetric *N*-glycan structures catalyzed by a number of GTs.[5,21,24,39,40] Moreover, as described in Section 5.2.5, we developed an enzymatic approach to generate asymmetric *N*-glycan cores from fetuin glycans, which can be further extended.[6] For example, compound **21** was extended to complex *N*-glycans **31**, **32**, **33** by GTs-catalyzed reactions (Figure 5.5C).[6]

5.4 Chemoenzymatic Preparation of *N*-Glycan Libraries

Chemoenzymatic strategies had been developed and applied since 1996, but have only been capable of synthesizing one or a few *N*-glycan structures, until the general strategy reported by Boons, which enabled the access of complex type asymmetric *N*-glycans.[22] Since then, several other chemoenzymatic strategies have been developed for batch production of *N*-glycans.[7,23,26,27,41] All these strategies involve chemical synthesis of core or precursor structures and enzymatic extension to generate various *N*-glycans, especially asymmetrically branched ones.

5.4.1 A General Strategy for Chemoenzymatic Synthesis of Asymmetric Complex Type *N*-Glycans

Naturally occurring complex type *N*-glycans are normally multi-antennary and asymmetrically branched, whereas traditional chemical or chemoenzymatic strategies were focused on the preparation of symmetric structures with identical glycan moieties on each branch. As the first case, Boons reported a general strategy to access these asymmetrically branched *N*-glycans,[22] in which five building blocks (Figure 5.7A) were prepared for the assembly of various complex *N*-glycan precursors for further enzymatic extension. A core pentasaccharide (compound **38**) with orthogonal protecting groups at potential branching positions (C2, C4 of the α1–3-Man, and C2, C6 of the α1–6-Man) serves as a universal base structure. The orthogonal protecting groups levulinoyl (Lev), fluorenylmethyloxycarbonate (Fmoc), allyloxycarbonate (Alloc) and 2-naphthylmethyl (Nap) allow selective attachment of specific sugar moieties (compounds **34**, **35**, **36**, **37**) to form bi-, tri- and tetra-antennary *N*-glycan precursors. For example, as shown in Figure 5.7B, to prepare a tri-antennary precursor **44**, the Fmoc protecting group on **38** was selectively removed to afford glycosyl acceptor **39**, which was coupled with donor **34**, followed by selective removal of the Nap protecting group, affording **41**. The glycosyl donor **35** was then attached onto **41** to form the ennea-saccharide **42**. After selective removal of the Lev protecting group on C6 of the α1–6-Man residue, donor **36** was attached. Final global deprotection yielded the precursor **45**.

Since each branch of the precursor contains a distinct terminal, they can be selectively extended by GTs-catalyzed reaction. As demonstrated in Figure 5.7C, the β2-branch on the α1–3-Man was first attached with an α2–3-Neu5Ac

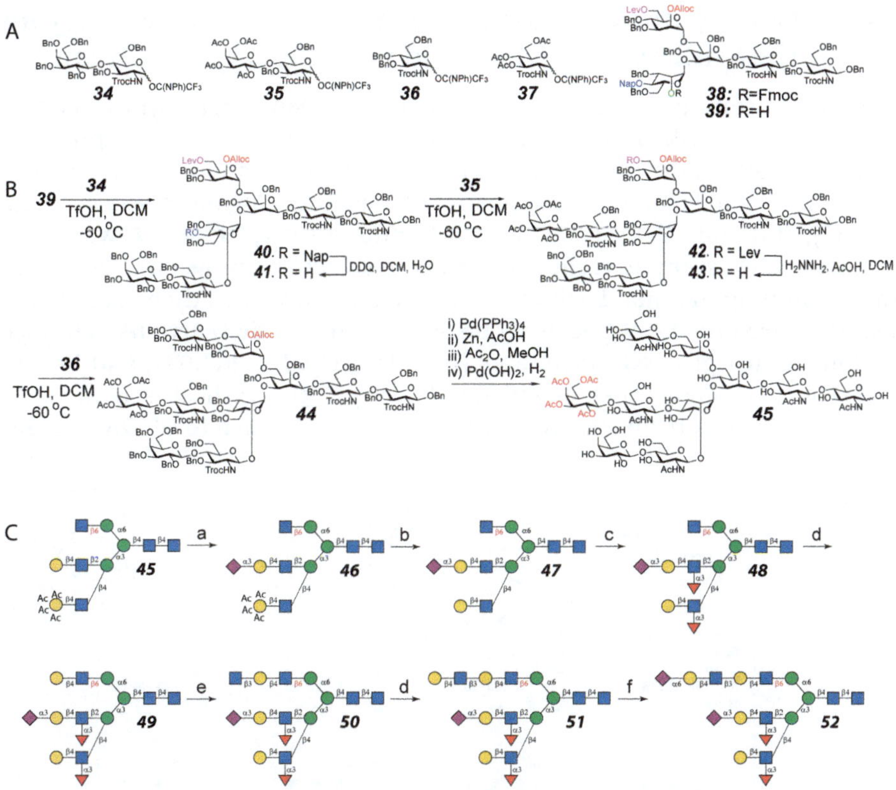

Figure 5.7 The general strategy for chemoenzymatic synthesis of asymmetric complex type *N*-glycans. (A) Five building blocks for convergent chemical assembly of various *N*-glycan precursors; (B) Chemical assembly of a tri-antennary *N*-glycan precursor **44**; (C) Enzymatic synthesis of asymmetrically branched *N*-glycans from the precursor. (a) ST3Gal-IV, CMP-Neu5Ac; (b) NH_4OH, H_2O; (c) Hp3FT, GDP-L-Fuc; (d) B4GalT1, UDP-Gal; (e) β3-GnT-II, UDP-GlcNAc; (f) ST6GAL-I, CMP-Neu5Ac.

catalyzed by ST3Gal-IV, whereas the acetylated Gal on the β4-branch renders sialyation. After removal of the acetyl groups, the β4-branch was then extended by α1–3-fucosylation catalyzed by Hp3FT to afford **48**. Finally, the β6-branch was enzymatically extended by B4GalT1, β3-GnT-II and ST6GAL-I catalyzed reactions in a step-wise manner, yielding the target compound **52**. It is worth noting that all these tri-antennary *N*-glycans contain a β6-branch but are missing the β2-branch on the α1–6-Man residue, which is quite unnatural as the attachment of the β6-branch requires the presence of the β2-branch in the natural *N*-glycan biosynthetic pathway.[42,43] Nevertheless, for the first time, the strategy enabled chemoenzymatic synthesis of multi-antennary asymmetric *N*-glycans. Using these glycans it was found that glycan-protein recognition is modulated by the presentation of minimal epitopes on complex *N*-glycans.

5.4.2 Core Synthesis/Enzymatic Extension (CSEE) Strategy for the Synthesis of Hybrid and Complex Type *N*-Glycans

Another outstanding example for the facile chemoenzymatic synthesis of *N*-glycans was described by Li and co-workers, in which the authors prepared 73 structurally well-defined *N*-glycans, including hybrid type, complex type, and 63 positional isomers.[27] As illustrated in Figure 5.8A, one glycosyl acceptor (**53**), and four oligosaccharyl thioether donors (**54, 55, 56, 57**) were used to chemically assemble eight *N*-glycan core structures in a convergent fragment coupling manner. Each of these core structures was then extended to between five and 17 *N*-glycan sequences by enzymatic reactions catalyzed by four robust glycosyltransferases (B4GalT1, Hp3FT, Pd2,6ST, and PmST1 E271F/R313Y). Figure 5.8B elucidated the enzymatic synthesis scheme of 17 *N*-glycans from core **60**. Firstly, a Gal residue was attached to GlcNAc residue

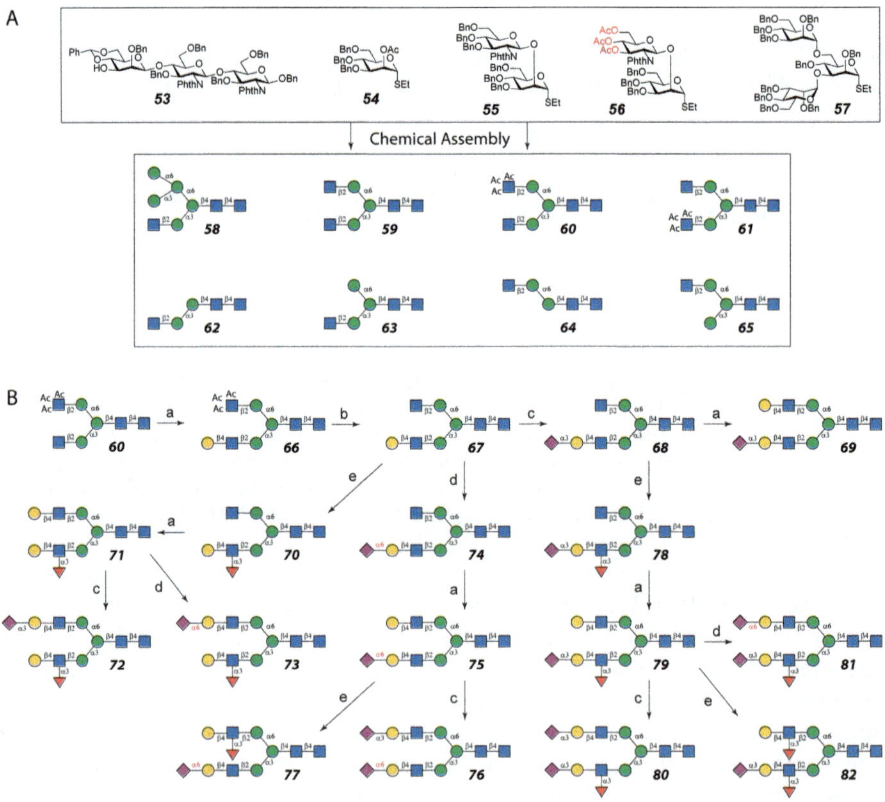

Figure 5.8 The core synthesis/enzymatic extension (CSEE) strategy for the preparation of hybrid and complex type *N*-glycan. (A) Convergent chemical assembly of eight *N*-glycan precursor cores from five building blocks; (B) Enzymatic synthesis of 17 asymmetrically branched *N*-glycans from precursor **60**. (a) B4GalT1, UDP-Gal; (b) NH_4OH, H_2O; (c) PmST1 E271F/R313Y, CMP-Neu5Ac; (d) Pd2,6ST, CMP-Neu5Ac; (e) Hp3FT, GDP-L-Fuc.

on the α1–3-Man branch of **60** by B4GalT1, galactosylation on the α1–6-Man branch was avoided by peracetylation of the corresponding GlcNAc residue. After HPLC purification and de-acetylation, an asymmetric *N*-glycan **67** was obtained. The structure can then be used to generate 15 asymmetrically branched *N*-glycans in a strictly controlled sequential manner, by firstly extending the α1–3-Man branch and then the other (Figure 5.8B). Such synthetic routes were designed according to the substrate specificity of each GT to avoid undesired glycosylation. This work represents the first batch synthesis of *N*-glycans, and provides an HPLC-based approach for milligram-scale separation of such structures. Using similar enzymatic strategies, we were able to synthesize numbers of Neu5Gc-terminated, core-fucosylated, and bisected *N*-glycans (unpublished data).

5.4.3 Enzymatic Synthesis of Complex Type *N*-Glycans from One Chemically Prepared Precursor

Most recently, Gagarinov and coworkers chemically prepared an unnatural tri-antennary *N*-glycan, which serves as a common precursor to enzymatically synthesize asymmetric bi-, tri- and tetra-antennary *N*-glycans.[7] The precursor (compound **83**) has different saccharide moieties attached to each branch, GlcNAc and unnatural Manβ1–4GlcNAc disaccharide on the β6- and β2-branch of the α1–6-Man, respectively, LacNAc and unnatural Galα1–4GlcNAc disaccharide on the β2- and β4-branch of the α1–3-Man, respectively. GTs and specific GHs can be employed to generate various complex type *N*-glycans starting with the precursor. For example, to generate tetra-antennary *N*-glycans, the β2-branch of the α1–3-Man was extended first, followed by the step-wise enzymatic extension of the β6-branch of the α1–6-Man to afford **86** (Figure 5.9A). The unnatural α1–4-Gal on the β4-branch of the α1–3-Man was then selectively removed by treatment with α-galactosidase from green coffee bean (αGalA), and the branch was extended subsequently by B4GalT1 and ST3Gal-IV catalyzed reactions. The so-yielded **88** was then treated with β-mannosidase from *Helix pomatia* (βManA) to remove the unnatural β1–4-Man on the β2-branch of the α1–6-Man (Figure 5.9A). Further GTs-catalyzed glycosylation can be performed to extend this branch.

To generate tri-antennary *N*-glycans, the intermediate **85** can be first treated with βManA and β-*N*-acetylglucosaminidase from *Canavalia ensiformis* (βGlcNAcA) to remove the β2-branch of the α1–6-Man, followed by treatment of αGalA (Figure 5.9B). The so-yielded compound **90** is ready for step-wise extension to afford various tri-antennary *N*-glycans. Alternatively, **85** can be first treated with αGalA and βGlcNAcA to remove the β4-branch of the α1–3-Man, and then treated with βManA, yielding the tri-antennary *N*-glycan **91** for further enzymatic extension (Figure 5.9B). On the other hand, the bi-antennary *N*-glycan core **67** could be obtained by first treating the common precursor **83** with αGalA and βGlcNAcA to remove both β4- and β6-branches, and then with βManA to release the unnatural β1–4-Man (Figure 5.9C).

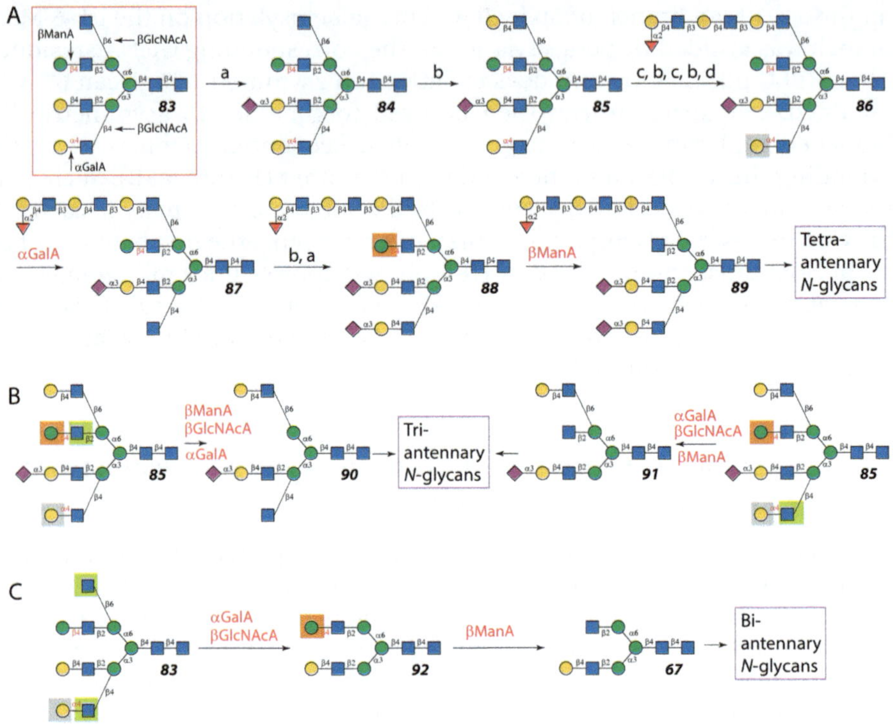

Figure 5.9 Enzymatic synthesis of tetra- (A), tri- (B) and bi-antennary (C) complex type *N*-glycans from a universal chemically prepared precursor **83**. (a) ST3Gal-IV, CMP-Neu5Ac; (b) B4GalT1, UDP-Gal; (c) β3-GnT-II, UDP-GlcNAc; (d) FUT-I, GDP-L-Fuc.

5.5 Modular Synthesis of All Types of *N*-Glycans

Distinct from common chemoenzymatic strategies in *N*-glycan preparation, which involve firstly chemical synthesis of core or precursor structures, and then enzymatic extension, Shivatare and coworkers developed a modular synthesis strategy to access all types of *N*-glycans.[29] Based on the retrosynthetic disconnection of different types of *N*-glycan structures (Figure 5.10), the authors designed four mannose/oligomannose modules which are synthesized chemically (Figure 5.11A), and over 30 other symmetric and asymmetric modules initiated with GlcNAcβ1–2Man disaccharide or GlcNAcβ1–4/6(GlcNAcβ1–2)Man trisaccharide which are synthesized chemoenzymatically (Figure 5.11B). This set of building blocks and trisaccharide cores with crucial β-mannoside linkage (*e.g.* compound **102**) can be used as starting materials for modular synthesis of various *N*-glycans. For example, as shown in Figure 5.11C, chemoenzymatically prepared modules **98** and **100** were firstly peracetylated, and the *p*-methoxyphenyl (PMP) group was transformed into fluoride, affording peracetylated glycosyl fluoride donors **99** and **101**. The trisaccharide core **102** was then coupled with donor **99** to form hexasaccharide **103**, followed by cleavage of the benzylidene to

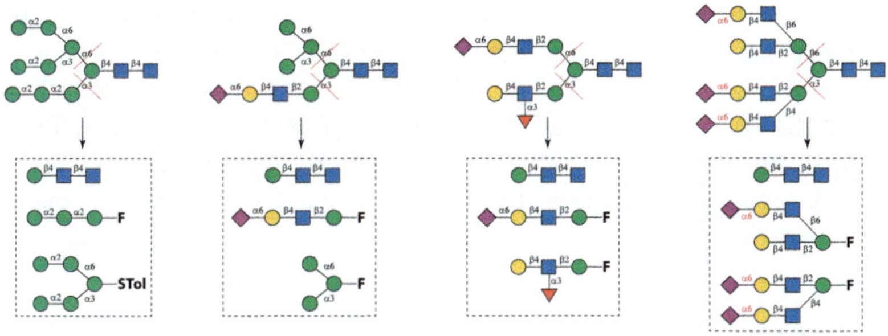

Figure 5.10 Retrosynthetic disconnections of high-mannose, hybrid and complex type *N*-glycans, showing the building blocks required for assembly.

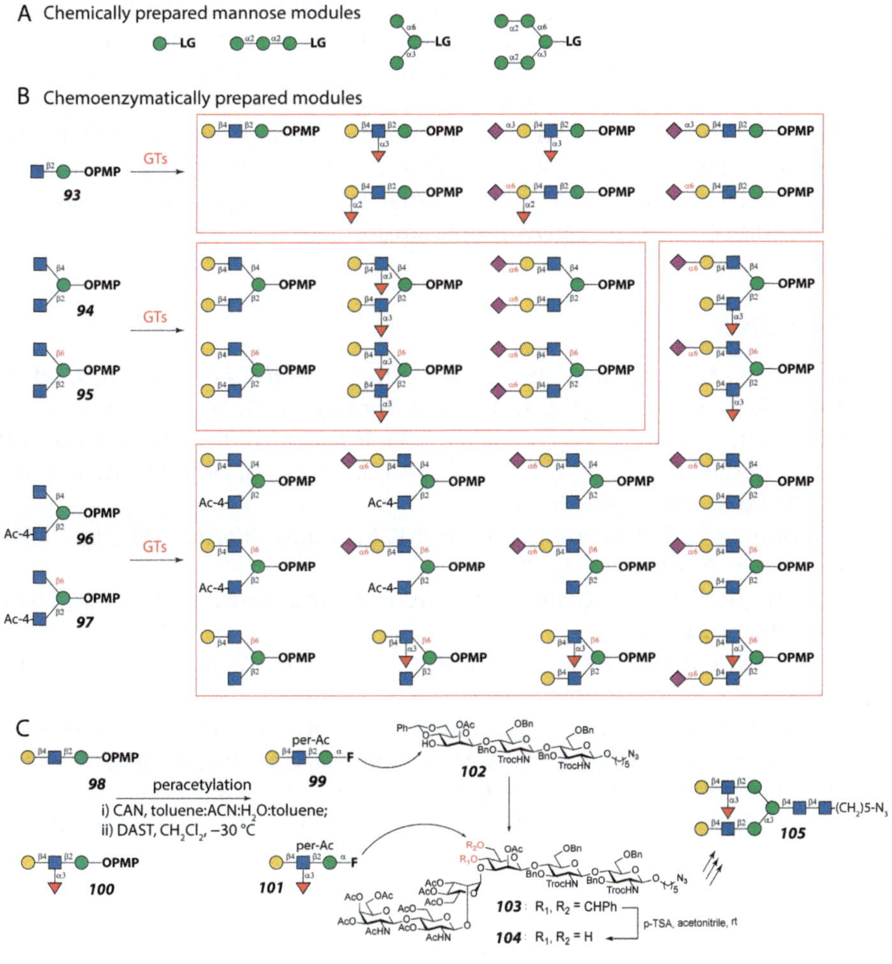

Figure 5.11 Modular chemoenzymatic synthesis of all types of *N*-glycans. (A) Chemily prepared mannose modules; (B) Chemoenzymatically prepared modules; (C) Proof-of-concept demonstration of the modular *N*-glycan synthesis.

expose the C4, C6 hydroxyl groups on the β-mannose. Finally, the glycosyl donor **101** was stereo-specifically installed at the C6-position to give decasaccharide, which was deprotected to afford glycan **105**. Theoretically, this strategy is capable of preparing all types of complex *N*-glycans with necessary modules. In addition, asymmetrically branched glycans (*e.g.*, **105**) prepared here could serve as precursors for enzymatic extension, further expanding *N*-glycan libraries.

5.6 Summary

Recent advances in chemoenzymatic synthesis enabled rapid and batch production of various glycan structures, including *N*-glycans, *O*-glycans, glycosaminoglycans, human milk oligosaccharides, *etc*. The application of various synthetically useful enzymes, especially GTs, enables the preparation of extremely complex structures and simplifies the synthetic routes, promoting chemoenzymatic synthesis to a brand-new level. On the other hand, these chemoenzymatic strategies still require expertise in chemical synthesis to generate cores or precursors, and a larger pool of synthetically useful enzymes is urgently needed. Machine-driven automatic synthesis that incorporates these chemoenzymatic or modular synthesis approaches may play an important role in future glycan preparation.

References

1. P. Stanley, N. Taniguchi and M. Aebi, in *Essentials of Glycobiology*, ed. A. Varki, R. D. Cummings, J. D. Esko, P. Stanley, G. W. Hart, M. Aebi, A. G. Darvill, T. Kinoshita, N. H. Packer, J. H. Prestegard, R. L. Schnaar and P. H. Seeberger, Cold Spring Harbor Laboratory Press, Cold Spring Harbor, NY, 3rd edn, 2015, pp. 99–111.
2. B. Adamczyk, T. Tharmalingam and P. M. Rudd, *Biochim. Biophys. Acta, Gen. Subj.*, 2012, **1820**, 1347.
3. N. Taniguchi and Y. Kizuka, *Advances in Cancer Research*, Elsevier, 2015, vol. 126, pp. 11–51.
4. R. M. Schmaltz, S. R. Hanson and C. H. Wong, *Chem. Rev.*, 2011, **111**, 4259.
5. Z. Wu, Y. Liu, C. Ma, L. Li, J. Bai, L. Byrd-Leotis, Y. Lasanajak, Y. Guo, L. Wen, H. Zhu, J. Song, Y. Li, D. A. Steinhauer, D. F. Smith, B. Zhao, X. Chen, W. Guan and P. G. Wang, *Org. Biomol. Chem.*, 2016, **14**, 11106.
6. A. D. Calderon, J. Zhou, W. Guan, Z. Wu, Y. Guo, J. Bai, Q. Li, P. G. Wang, J. Fang and L. Li, *Org. Biomol. Chem.*, 2017, **15**, 7258.
7. I. A. Gagarinov, T. Li, J. S. Toraño, T. Caval, A. D. Srivastava, J. A. Kruijtzer, A. J. Heck and G.-J. Boons, *J. Am. Chem. Soc.*, 2017, **139**, 1011.
8. G. Perugino, A. Trincone, M. Rossi and M. Moracci, *Trends Biotechnol.*, 2004, **22**, 31.

9. L. F. Mackenzie, Q. Wang, R. A. J. Warren and S. G. Withers, *J. Am. Chem. Soc.*, 1998, **120**, 5583.
10. M. Umekawa, W. Huang, B. Li, K. Fujita, H. Ashida, L. X. Wang and K. Yamamoto, *J. Biol. Chem.*, 2008, **283**, 4469.
11. C. Li and L. X. Wang, *Chem. Rev.*, 2018, **118**, 8359.
12. O. Schwardt, T. Visekruna, S. Rabbani and B. Ernst, *CHIMIA Int. J. Chem.*, 2006, **60**, 234.
13. C. Unverzagt, *Angew. Chem., Int. Ed.*, 1996, **35**, 2350.
14. S. André, C. Unverzagt, S. Kojima, X. Dong, C. Fink, K. Kayser and H.-J. Gabius, *Bioconjugate Chem.*, 1997, **8**, 845.
15. C. Unverzagt, *Tetrahedron Lett.*, 1997, **38**, 5627.
16. J. Seifert and C. Unverzagt, *Tetrahedron Lett.*, 1997, **38**, 7857.
17. C. Unverzagt and J. Seifert, *Tetrahedron Lett.*, 2000, **41**, 4549.
18. C. Unverzagt, S. André, J. Seifert, S. Kojima, C. Fink, G. Srikrishna, H. Freeze, K. Kayser and H.-J. Gabius, *J. Med. Chem.*, 2002, **45**, 478.
19. S. André, C. Unverzagt, S. Kojima, M. Frank, J. Seifert, C. Fink, K. Kayser, C. W. von der Lieth and H. J. Gabius, *Eur. J. Biochem.*, 2004, **271**, 118.
20. S. Hanashima, S. Manabe and Y. Ito, *Angew. Chem., Int. Ed.*, 2005, **117**, 4290.
21. W. Peng, R. P. de Vries, O. C. Grant, A. J. Thompson, R. McBride, B. Tsogtbaatar, P. S. Lee, N. Razi, I. A. Wilson and R. J. Woods, *Cell Host Microbe*, 2017, **21**, 23.
22. Z. Wang, Z. S. Chinoy, S. G. Ambre, W. Peng, R. McBride, R. P. de Vries, J. Glushka, J. C. Paulson and G.-J. Boons, *Science*, 2013, **341**, 379.
23. T. Li, M. Huang, L. Liu, S. Wang, K. W. Moremen and G. J. Boons, *Chem. - Eur. J.*, 2016, **22**, 18742.
24. C. M. Nycholat, W. Peng, R. McBride, A. Antonopoulos, R. P. de Vries, Z. Polonskaya, M. Finn, A. Dell, S. M. Haslam and J. C. Paulson, *J. Am. Chem. Soc.*, 2013, **135**, 18280.
25. S. S. Shivatare, S.-H. Chang, T.-I. Tsai, C.-T. Ren, H.-Y. Chuang, L. Hsu, C.-W. Lin, S.-T. Li, C.-Y. Wu and C.-H. Wong, *J. Am. Chem. Soc.*, 2013, **135**, 15382.
26. B. Echeverria, J. Etxebarria, N. Ruiz, A. Hernandez, J. Calvo, M. Haberger, D. Reusch and N.-C. Reichardt, *Anal. Chem.*, 2015, **87**, 11460.
27. L. Li, Y. Liu, C. Ma, J. Qu, A. D. Calderon, B. Wu, N. Wei, X. Wang, Y. Guo, Z. Xiao, J. Song, G. Sugiarto, Y. Li, H. Yu, X. Chen and P. G. Wang, *Chem. Sci.*, 2015, **6**, 5652.
28. A. D. Calderon, Y. Liu, X. Li, X. Wang, X. Chen, L. Li and P. G. Wang, *Org. Biomol. Chem.*, 2016, **14**, 4027.
29. S. S. Shivatare, S.-H. Chang, T.-I. Tsai, S. Y. Tseng, V. S. Shivatare, Y.-S. Lin, Y.-Y. Cheng, C.-T. Ren, C.-C. D. Lee, S. Pawar, C.-S. Tsai, H.-W. Shih, Y.-F. Zeng, C.-H. Liang, P. D. Kwong, D. R. Burton, C.-Y. Wu and C.-H. Wong, *Nat. Chem.*, 2016, **8**, 338.
30. I. Prahl and C. Unverzagt, *Angew. Chem., Int. Ed.*, 2002, **41**, 4259.
31. H. Yu, H. Chokhawala, R. Karpel, H. Yu, B. Wu, J. Zhang, Y. Zhang, Q. Jia and X. Chen, *J. Am. Chem. Soc.*, 2005, **127**, 17618.

32. L. Bonfanti, *Prog. Neurobiol.*, 2006, **80**, 129.
33. W. Huang, D. Wang, M. Yamada and L.-X. Wang, *J. Am. Chem. Soc.*, 2009, **131**, 17963.
34. Z. Wu, Y. Liu, L. Li, X.-F. Wan, H. Zhu, Y. Guo, M. Wei, W. Guan and P. G. Wang, *Org. Biomol. Chem.*, 2017, **15**, 8946.
35. S. Serna, S. Yan, M. Martin-Lomas, I. B. Wilson and N.-C. Reichardt, *J. Am. Chem. Soc.*, 2011, **133**, 16495.
36. P. C. Pang, P. C. Chiu, C. L. Lee, L. Y. Chang, M. Panico, H. R. Morris, S. M. Haslam, K. H. Khoo, G. F. Clark, W. S. Yeung and A. Dell, *Science*, 2011, **333**, 1761.
37. B. Echeverria, S. Serna, S. Achilli, C. Vivès, J. Pham, M. Thépaut, C. H. Hokke, F. Fieschi and N.-C. Reichardt, *ACS Chem. Biol.*, 2018, **13**, 2269.
38. N. Li, H. Xu, K. Fan, X. Liu, J. Qi, C. Zhao, P. Yin, L. Wang, Z. Li and X. Zha, *J. Cell. Mol. Med.*, 2014, **18**, 1975.
39. C. M. Nycholat, R. McBride, D. C. Ekiert, R. Xu, J. Rangarajan, W. Peng, N. Razi, M. Gilbert, W. Wakarchuk, I. A. Wilson and J. C. Paulson, *Angew. Chem., Int. Ed.*, 2012, **51**, 4860.
40. W. Peng, J. Pranskevich, C. Nycholat, M. Gilbert, W. Wakarchuk, J. C. Paulson and N. Razi, *Glycobiology*, 2012, **22**, 1453.
41. Z. S. Chinoy, F. Friscourt, C. J. Capicciotti, P. Chiu and G. J. Boons, *Chem. - Eur. J.*, 2018, **24**, 7970.
42. O. Kanie, S. C. Crawley, M. M. Palcic and O. Hindsgaul, *Carbohydr. Res.*, 1993, **243**, 139.
43. O. Kanie, S. C. Crawley, M. M. Palcic and O. Hindsgaul, *Bioorg. Med. Chem.*, 1994, **2**, 1231.

CHAPTER 6

Chemoenzymatic Synthesis of α-Dystroglycan O-Mannose Glycans

CAICAI MENG[a], CHANG-CHENG LIU[a,b] AND HONGZHI CAO*[a,b]

[a]National Glycoengineering Research Center and School of Pharmaceutical Science, Shandong University, Jinan 250012, China; [b]State Key Laboratory of Microbial Technology, Shandong University, Jinan 250100, China
*E-mail: hzcao@sdu.edu.cn

6.1 Introduction

The *O*-mannosyl glycan is a type of *O*-glycan which is initiated by a covalent attachment of mannose to the serine (Ser) or threonine (Thr) residues of proteins *via* α-glycosidic linkage. For many years, *O*-mannosyl glycans were considered to be a specific fungal protein modification. However, about 40 years ago, in an analysis of a proteoglycan enriched fraction of brain lysate, *O*-mannosyl glycans were first reported in mammalian system, and up to 30% of brain *O*-glycans are *O*-mannose-initiated glycans,[1] among which α-dystroglycan (α-DG) is the most well studied *O*-mannosylated mammalian protein.

Dystroglycan (DG) is a highly glycosylated basement membrane receptor involved in a variety of physiological processes including maintenance of the skeletal muscle-cell membrane integrity and establishment of the structure and function of the central nervous system. DG is encoded by

Chemical Biology No. 11
Synthetic Glycomes
Edited by Wanyi Guan, Lei Li and Peng George Wang

Published by the Royal Society of Chemistry, www.rsc.org

a single gene (*dystroglycan 1*, *DAG1*) and is cleaved into two subunits as cell-surface α-DG and transmembrane β-DG.[2] The α-DG is an integral glycoprotein of the dystrophin-glycoprotein complex (DGC) that serves as a connector between the actin cytoskeleton and the extracellular matrix (ECM) and acts as a receptor for laminin-G domain-containing ECM proteins such as laminin, agrin, perlecan in muscles, neurexin in the brain, and pikachurin in the retina in a glycosylation dependent manner.[2] Interestingly, the α-DG is a heavily glycosylated membrane protein. The molecular weight of α-DG core protein is estimated to be about 74 kDa, however, α-DG is commonly observed as a diffuse set of bands ranging from 150–200 kDa when separated by SDS-PAGE due to extensive glycosylation. The molecular mass of α-DG varies dramatically in different tissues which may be due to tissue-specific glycosylation, which probably have unique or variable functions in different tissues.[3] Specifically, the α-DG undergoes *N*-glycosylation, mucin-type *O*-GalNAc-linked and *O*-mannose-linked *O*-glycosylation, and the *O*-mannosyl glycan is believed to be responsible for the muscle function related ligands binding.

In addition to serving as a cell surface receptor for extracellular matrix proteins, α-DG is also the receptor of some arenaviruses including lymphocytic choriomeningitis virus (LCMV), Lassa virus (LASV), and clade C New World arenaviruses.[4] The high binding affinity to α-DG with LCMV and LASV was directly related to the viral tropism, the ability to infiltrate the white pulp of the spleen, and subsequently the ability of a virus to establish a persistent infection.[5] Recent studies also demonstrated that post-translational modification by the glycosyltransferase LARGE is crucial for α-DG's function as a cellular receptor for ECM proteins and arenaviruses.[6] The α-DG–arenavirus interaction is understood to be the key to early events of arenavirus infection. Successful strategies to intervene in this crucial interaction can prevent the virus infection.

Defects in the biosynthesis of *O*-mannose glycans often result in hypoglycosylation of α-DG and play critical roles in multiple forms of congenital muscular dystrophies, cancer progression and metastasis, and inhibited entry and infection of certain arenaviruses. In this chapter, we will mainly discuss structural and functional characteristics of *O*-mannose initiated glycan structures, enzymes underlying the biosynthesis pathway of *O*-mannose glycan, the diseases that are related with aberrant *O*-mannosylation and recent advances in the synthesis of *O*-mannose glycans.

6.2 Biosynthetic Pathway and Functional Roles of *O*-Mannose Glycans

6.2.1 Biosynthesis Pathway of *O*-Mannosyl Glycans

The *O*-mannose glycans are currently classified into three categories based on the linkage types of *N*-acetylglucosamine (GlcNAc) to the core mannose: core M1, GlcNAcβ1–2Man; core M2, GlcNAcβ1–2(GlcNAcβ1–6) Man; and core

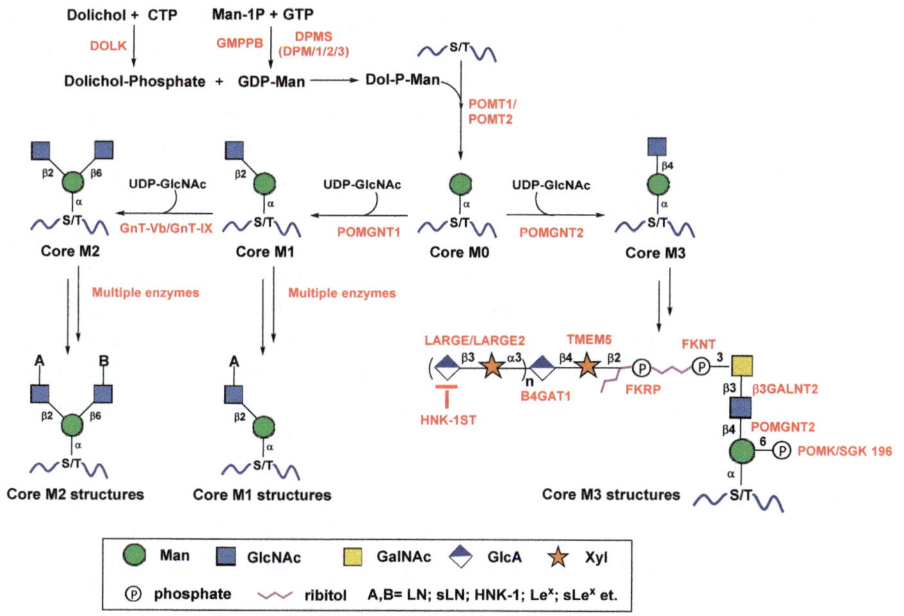

Scheme 6.1 Biosynthetic pathway of *O*-mannose glycans.

M3, GlcNAcβ1–4Man (Scheme 6.1).[7] The *O*-mannosylation of proteins is initiated in endoplasmic reticulum (ER) by the transfer of mannose from dolichol monophosphate-activated mannose (Dol-P-Man or DPM) to Ser and Thr residues by protein *O*-mannosyltransferases 1 (POMT1) and 2 (POMT2).[8] The resulting structure of Ser or Thr residue linked with only core α-mannose is defined as core M0.[9] The POMT1 and POMT2 are homologous to yeast protein *O*-mannosyltransferases, which have to be coexpressed to form a functional enzyme heterocomplex.[8] Also, the POMT1 and POMT2 are expressed in all human tissues, but POMT1 is highly expressed in fetal brain, testis, and skeletal muscle, while POMT2 is predominantly expressed in testis. The regulatory mechanisms of POMTs for protein *O*-mannosylation in each tissue are still unknown. The Dol-P-Man is synthesized by the DPM synthase complex (comprised of DPM1, DPM2 and DPM3) which catalyzes the transfer of mannose from guanidine diphosphate mannose (GDP-Man) to dolichol-phosphate.[10] *DOLK* encodes the dolichol kinase responsible for dolichol-phosphate formation and GMPPB catalyzes GDP-Man synthesis from GTP and mannose-1-phosphate.[11,12]

Subsequent addition of GlcNAc at C2-OH of core mannose is catalyzed by protein *O*-mannose-β1–2-*N*-acetylglucosaminyl transferase 1 (POMGnT1), resulting in the formation of linear core M1 structures (GlcNAcβ1–2Manα),[13] which can be further extended to form the classical tetrasaccharide (Neu5Acα2–3Galβ1–4GlcNAcβ1–2Man), Lewis X epitope (Galβ1–4(Fucα1–3) GlcNAcβ1–2Man), or a human natural killer-1 epitope (HNK-1, 3S-GlcAβ1–3Galβ1–4GlcNAcβ1–2Man). Core M1 structures account up to 15% of total brain protein O-glycans.[14] The most abundant and first identified O-mannose glycan structure is Neu5Acα2–3Galβ1–4GlcNAcβ1–2Man which was previously thought to be a requirement for laminin-binding activity.[15] However, more recent studies suggested that laminin binding most likely does not depend on this structure, as removing all detectable sialic acid and galactose moieties from this tetrasaccharide did not decrease but rather enhanced the laminin-binding activity.[16]

The addition of a second GlcNAc to the core mannose of core M1 structure at C6-OH by β1–6-*N*-acetylglucosaminyl transferase (GnT-Vb/GnT-IX) forms a branched core M2 glycan GlcNAcβ1–2(GlcNAcβ1–6)Manα. No β1–6-branching O-mannose glycans were observed in POMGnT1-deficient mice, suggesting that the activity of POMGnT1 precedes and is necessary for the activity of GnT-Vb involving in branching of O-mannose structures. The core M1 and M2 structures can be further elongated by β1–4-linked Gal and capped by sialic acid, glucuronic acid (GlcA), fucose (Fuc) or sulfate groups to generate the diverse and complex structures. Recent studies demonstrated that GnT-Vb is also involved in the formation of HNK-1 structure on 2,6-branched O-mannosyl glycan by RPTPβ and phosphacan, a secreted splicing variant of RPTPβ.[17] The core M3 glycans is initiated by β1–4-*N*-acetylglucosaminyltransferase 2 (POMGnT2) in the ER, which catalyses the transfer of GlcNAc to the core mannose to generate GlcNAcβ1–4Man. The resulting disaccharide is further glycosylated by β1–3-*N*-acetylgalactosaminyl transferase 2 (β3GALNT2) by transferring GalNAc from UDP-GalNAc to the GlcNAc residue to form GalNAcβ1–3GlcNAc. Moreover, SGK196/POMK is a specific O-mannose kinase, which is responsible for the phosphorylation of C6-OH of the core O-mannose to give the phospho-core M3 structure GalNAcβ1–3GlcNAcβ1–4Man6P.[18] Before the formation of the phospho-core M3 structure, the biosynthesis process occurs in ER. After mannosyl phosphorylation, the tandem Rbo5P structure is attached to the GalNAc residue by FKTN, FKRP, and ISPD in the Golgi.[19] Furthermore, ISPD is a mammalian CDP-Rbo synthase that uses Rbo5P and CTP as substrates, indicating that cellular levels of Rbo5P and CTP may affect Rbo5P modification. However, the biosynthetic pathway of Rbo5P is currently unknown in mammals. Both FKTN and FKRP are Rbo5P transferases that sequentially catalyze the transfer of Rbo5P from CDP-Rbo to phospho-core M3 structure and have strict substrate specificity. Then, TMEM5 transfer Xyl in a β1–2-linkage to the ribitol moiety of the FKRP products.[20] In addition, B4GAT1 is a β1–4 glucuronyltransferase, which transfers glucuronic acid (GlcA) residue to the terminal xylose of the TMEM5 product to form GlcAβ1–4Xylβ1–2Rbo1–P–5Rbo1–P–core M3.[21] The TMEM5 and B4GAT1 are the priming enzymes for the LARGE-dependent functional glycosylation of α-DG.[9] The GlcAβ1–4Xyl disaccharide serves as acceptor for LARGE to polymerize the glycosaminoglycan (GAG)-like repeating disaccharide of Xyl and GlcA (–3GlcAβ1–3Xylα1–). The LARGE is a bifunctional enzyme with both β1–3-glucuronosyltransferase and α1–3-xyrosyltransferase activities.[22] This novel glycosaminoglycan (GAG)-like polysaccharide with repeating disaccharide units of –3Xylα1–3GlcAβ1– was recently named as matriglycan, and has been only identified on α-dystroglycan (α-DG), which is responsible for the laminin-binding activity.[23] Human natural killer-1 sulfotransferase (HNK-1ST) can suppress glycosylation and ligand binding activity of α-DG, resulting in a reduction in cell migration of melanoma. The HNK-1ST is capable of transferring a sulfate group to the C3-OH of terminal GlcA of matriglycan and, blocking the xylosylation to the terminal GlcA of martiglycan, therefore, can stop the further chain elongation catalyzed by LARGE. So, HNK-1ST can be a suppressive factor for α-DG function, acting as a "molecular brake" to generate properly glycosylated α-DG.[24]

LARGE2 is a paralogue of LARGE, which catalyzes the same glycosylation reactions as LARGE and is a bifunctional glycosyltransferase exhibiting xylosyltransferase and glucuronyltransferase activities as well as polymerizing activity.[25,26] LARGE2 and LARGE have different pH profiles. LARGE is widely expressed, and the highest levels are found in the brain, heart, and skeletal muscle. In contrast, LARGE2 is highly expressed in the kidney and placenta but not in the brain. The differential expression of LARGE and LARGE2 suggest that the enzymes might be of distinctive importance in different organs. Recent study demonstrated that LARGE2 but not LARGE can modify proteoglycans (PGs) with the GAG-like polysaccharide composed of xylose and glucuronic acid to confer laminin binding.[27] To date, 24 genes have been identified as being involved in the protein *O*-mannosylation, and it is likely that additional novel genes will be identified in future.[28,29]

6.2.2 Dystroglycanopathies and Implicated Genes

Defects in the proper glycosylation of α-DG result in the loss of receptor function and cause numerous forms of congenital muscular dystrophies (CMDs) that are accompanied by a variety of brain and eye malformations collectively classified as dystroglycanopathies. Several genes have been proved to be associated with proper glycosylation of α-DG, including POMT1, POMT2, POMGnT1, LARGE, Fukutin, FKRP, and ISPD. Mutations in those genes will cause various dystroglycanopathies, such as consecutive weakness, skeletal muscle wasting, and mental retardation.

Walker–Warburg syndrome (WWS) is an autosomal recessive developmental disorder mainly caused by POMT1 and POMT2 mutations and associated with congenital muscular dystrophy, neuronal migration defects, and ocular abnormalities.[30] WWS is a relatively severe phenotype as infants with WWS rarely survive after 12 months. The expression of both POMT1 and POMT2 is prerequisite for normal mannosyltransferase activity, and mutations in either gene can cause WWS.

POMGnT1 enzyme is a type II transmembrane glycosyltransferase found in the Golgi apparatus and responsible for the formation of the GlcNAcβ1–2Man linkage of *O*-mannosyl glycan. This enzyme is indispensable in the formation of Core M1 and Core M2 *O*-mannosyl glycans, which are predominantly expressed in neural tissue. MEB is an autosomal recessive disorder characterized by congenital muscular dystrophy, ocular abnormalities, and brain malformation (type II lissencephaly). Previous study demonstrated that muscle–eye–brain disease (MEB) is mainly caused by mutations in the gene encoding POMGnT1.[13] Patients with MEB have a similar but milder clinical phenotype as patients with WWS. A child born with MEB are expected to live 12 or even 16 years, which is significantly longer than a WWS patients.[31]

Fukuyama-type congenital muscular dystrophy (FCMD) is mainly caused by mutations in the *fukutin* (*FKTN*) gene, which encodes Rob5P transferases that catalyze the transfer of Rbo5P from CDP-Rbo to phospho-core M3 structure prior to FKRP. Mutations in *FKTN* impede further elongation of phospho-core M3, resulting in a deficiency of functional Core M3 structures. Patients with FCMD show a wide range of variability in dystroglycanopathy disease phenotypes. FCMD is prevalent in the Japanese population with a carrier frequency of 1 in 88 and an estimated incidence of approximately of 1 in 10 000 births.[32] Theoretically, FKRP is not associated with the activity of POMT1, POMT2 or POMGNT1. However, the *FKTN* mutations could induce reduction of POMGNT1 activity in the Golgi apparatus, suggesting that the expression of FKTN has some undiscovered relationship with POMGnT1 activity.[33]

Fukutin-related protein (FKRP) is a homolog of FKTN and also a Rob5P transferase that catalyzes the transfer of Rbo5P from CDP-Rbo to the product of FKTN. Moreover, CMD type 1C (MDC1C) characterized by severe weakness, muscle degeneration and cardiomyopathy, is caused by a defect in *FKRP* gene. *FKRP* mutations are also responsible for a milder and more common form of muscular dystrophy termed limb–girdle muscular dystrophy type 2I (LGMD2I) that are characterized by limb–girdle weakness in childhood or adulthood and normal intelligence.[2,34]

Since the gene mutations and hypoglycosylation of α-DG cause α-dystroglycanopathies, exogenous gene insertion and enhanced glycosylation may provide a novel protocol for curing these diseases in the future.

6.2.3 Roles of *O*-Mannosylation in Tumor Metastasis

The interaction between epithelial cells and the extracellular matrix is crucial for tissue architecture and function. Dystroglycan is a membrane receptor responsible for mediating interactions between cells and basement membranes in various epithelia. During cancer progression, primitive cancer cells escape from this control by abolishing the binding affinities of their cell membrane receptors. In a variety of adenocarcinomas, the tumor aggressiveness is associated with reduced expression of DG protein including hypoglycosylation of α-DG and/or truncation β-DG. In aggressive prostate and breast carcinoma cell lines, laminin-binding glycans dramatically decrease, although the amount of α-DG and β-DG is maintained. The decrease of laminin-binding glycans and consequently increased cell migration are associated with the decreased expression of β3-*N*-acetylglucosaminyltransferase-1 (β3GnT1), which was later renamed β4GAT1 and plays a critical role in the synthesis of core M3 *O*-mannosyl glycan on α-DG in collaboration with LARGE or LARGE2.[21] Forced expression of β3GnT1 restored the laminin-binding glycans, and consequently resulted in reduced cell migration and decreased tumor formation.[35] Some studies also showed that the functional glycosylation of α-DG is substantially reduced in epithelia-derived cancer metastases. High levels of α-DG correlated strongly with cell polarization capability, which is related to normal cell growth in the presence of the basement membrane.[36] In breast, cervical, and lung cancer cell lines, the expression of LARGE was down-regulated, resulting in loss of laminin binding, which correlates with cancer progression. Ectopic restoration of LARGE expression in epithelium-derived cancer cell lines restores the normal glycosylation and laminin binding of α-DG, leading to enhanced cell adhesion and reduced cell migration.[37] In prostate cancer cells, decreased functional glycosylation of α-DG is clinically correlated with reduced LARGE2 but not LARGE. Restoring expression of LARGE2 rescues dystroglycan function and reduces the invasive and proliferative potential of cancer cells.[38] The re-expression of the LARGE/LARGE2 gene may thus represent a future avenue for the treatment of cancer.

Epithelial cadherin (E-cadherin) is also a cell surface glycoprotein, mediating interactions between cells and basement membranes. Reduced expression of E-cadherin and impairment of its functions are associated with tumor development and progression. Recent study has demonstrated that *O*-mannosylation of E-cadherin was reduced in human gastric cancer, affecting E-cadherin-mediated cellular adhesion. A reduced expression of *O*-mannosyl glycans concomitantly with increased modification of branched complex *N*-glycans is a signature of non-functional E-cadherin in human gastric carcinomas.[39]

6.2.4 Function of Core M1 and M2 Structures

Despite astonishing progress in the research into the biological activity of core M3 *O*-mannose glycans, little is known about the functional roles of core M1 and M2 structures.

Core M1 structures account for over 15% of brain *O*-glycans and are far more abundant than core M3 structures in the mucin-like domain of α-DG.[14] As mentioned above, defects in POMGnT1 lead to loss of core M1 glycans, relating to muscle-eye-brain disease (MEB).[13] The role of POMGnT1 in the modification of functional *O*-mannose glycans is not clear. Generally, POMGnT1 does not catalyze any reactions involved in the formation of core M3; however, in POMGnT1-KO mice and MEB patients, the Xyl-GlcA repeats on the *O*-mannosyl glycan of α-DG are impaired, suggesting that POMGnT1 or its product core M1 plays regulatory roles in postphosphoryl modification of core M3. Recent crystal structure study has shown that the stem domain of POMGnT1 recognized β-linked GlcNAc of *O*-mannosyl glycans (including GlcNAcβ1–2Man and GalNAcβ1–3GlcNAcβ1–4Man6P glycan).

This interaction may facilitate the clustering of core M1 and recruit other enzymes that interact with POMGnT1 (*e.g.* fukutin) for further functional modification.[40] The core M1 structures also serve as a precursor for core M2 structures.

The core M2 structures account for no more than 5% of brain *O*-glycans and are primarily found in brain and prostate tissue.

6.2.5 *O*-Mannosylated Substrates Beyond α-DG

The *O*-mannosyl glycan related mutations of glycosyltransferases are associated with brain malformations. However, the roles of *O*-mannosyl glycans and the proteins undergoing this modification in brain pathology are largely unknown. Recent studies[14] revealed that α-DG occupies only a minor proportion of the total *O*-mannosyl glycans in the mammalian brain. Glycomic analysis and mass spectrometry studies[14] of mammalian brain *O*-mannose glycans showed that the *O*-mannose-initiated glycan have no observable change when LARGE expression is lost or between α-DG-deficient mice and wild-type mice, indicating there must be additional proteins besides α-DG that are modified by *O*-mannosyl glycans in the mammalian brain.[14] Brain-specific *O*-mannosyl glycans are characterized by β1–2- and β1–6-branching of *O*-linked mannose (core M1 and core M2), and the branching GlcNAc transferase, GnT-IX, is expressed relatively more specifically and highly in the brain.[41,42]

In the *POMGnT1*-knockout mouse model, the receptor protein tyrosine phosphatase β (RPTPβ) and its secreted variant (phosphacan) shifted to a lower molecular weight, and the *O*-mannose glycan epitopes are either absent or substantially reduced, indicating that the RPTPβ/phosphacan is a substrate for protein *O*-mannosylation in the brain.[43] GnT-Vb activity promotes the addition of the *O*-mannosyl-linked HNK-1 in neuron-specific RPTPB modification, which promotes galectin-1 binding and RPTPβ retention levels on the cell surface. GnT-Vb-mediated glycosylation can inhibit RPTPβ phosphatase activity and subsequently increase tyrosine phosphorylation levels of β-catenin, resulting in reducing cell–cell adhesion and promoting migration and demonstrating that GnT-Vb is a regulator of RPTPβ signaling that influences cell–cell and cell–matrix interactions in the development of the nervous system.[24] The RPTPβ also plays a pivotal role in the recovery of demyelinating disease. A recent study showed that GnT-Vb-deficient mice, which lack branched *O*-mannosyl glycans, exhibited enhanced remyelination with reduced astrogliosis. Therefore, the branched *O*-mannosyl glycans of RPTPβ may be the key in inhibiting the remyelination process. However, the detailed molecular mechanism needs to be further studied.[44]

Clausen and coworkers identified a large superfamily of cadherins, protocadherins, and plexin of cell membrane receptors as the major carrier proteins of *O*-mannose glycans using the "Simple Cells" strategy.[45] The *O*-mannose glycans are only found at conserved positions of extracellular cadherin domains, and these glycosites appear highly conserved during evolution, suggesting that they serve important biological functions.[45] Strahl and coworkers also identified E-cadherin as an *O*-mannosylated glycoprotein, and this modification was crucial for E-cadherin-mediated cell adhesion.[46] E-cadherin is essential for the formation of adherens-junctions in the preimplantation embryo, and *O*-mannosylation-deficient embryos die at the morula-to-blastocyst transition. *O*-mannosylation of E-cadherin also mediates cell–cell adhesion in tissue morphogenesis and maintenance, as well as tumor invasion and metastasis. Moreover, dysregulation of E-cadherin is a molecular signature in cancer. The mechanism that regulates E-cadherin function may be glycosylation modifications such as *N*-glycosylation and *O*-mannosylation. The reduced levels of *O*-mannosyl glycans together with the presence of GnT-V-mediated branched *N*-glycosylation lead to E-cadherin cellular mislocalization and an incorrect assembly of adherens-junctions, thus contributing to tumor progression and metastases.[39] The prevention of *N*-glycosylation at Asn554 by Asn mutation facilitates the biological function recovery related *O*-mannosylation modification of E-cadherin. Recent studies[47] demonstrated that the *O*-mannosylations of cadherins and protocadherins are not catalyzed by POMT1 and/or POMT2 enzymes as in α-DG. Moreover, the *O*-mannose glycans on cadherins do not appear to be elongated. These results suggest the existence of a novel undiscovered *O*-mannosylation machinery in mammalian cells for the large cadherin superfamily and other proteins.[47] Using gene-knockout strategy combined with *O*-mannose glycoproteomics, four putative protein *O*-mannosyltransferases genes were identified as the requirements for the *O*-mannosylation of cadherins and protocadherins.[48]

Recently, a handful of additional mammalian glycoproteins were identified to contain *O*-mannosyl modifications, including CD24, recombinant IgG2,[49] neurofascin 186,[50] lecticans (aggrecan, brevican, neurocan, and versican),[51] a disulfide-isomerase (PDIA3), and a membrane protein of unknown function (KIAA1549).[45] Most of these *O*-mannosyl-modified proteins are identified in the brain, implicating the association between aberrant *O*-mannosylation and brain malformations. However, there is still much to learn about the roles of *O*-mannosyl glycans in brain pathology.

Proteins can acquire differential glycosylation patterns, and these heterogeneous glycoforms are not generated randomly. The variation of *O*-mannose glycans across tissue and cell types likely reflect important differences in functions. However, the proteins undergoing *O*-mannosylation modification, the functional roles of this modification, and the genes and enzymes underlying the biosynthesis pathway of *O*-mannosylation remain to be characterized. Therefore, it is urgent to acquire a library of structurally well-defined *O*-mannose glycans to understand the complexity of glycans related to various biochemical processes. The chemical and chemoenzymatic methods developed recently have facilitated the synthesis of structurally defined oligosaccharides and glycoconjugates. The previous synthesis mainly focused on the core M1 tetrasaccharide structure Neu5Acα2–3Galβ1–4GlcNAcβ1–2Manα[52–58] or the phosphorylated core M3 trisaccharide.[59] But a recent study reported a highly efficient diversity-oriented chemoenzymatic strategy for the rapid assembly of core M1 and core M2 *O*-mannose glycans.[60]

6.3 Synthesis of *O*-Mannose Glycans

6.3.1 Chemical or Chemoenzymatic Synthesis of Core M1 Structures

Core M1 tetrasaccharide Neu5Acα2–3Galβ1–4GlcNAcβ1–2ManαSer/Thr was the first characterized *O*-mannose glycan from the bovine peripheral nerve and the most abundant *O*-mannosyl glycan on α-DG from a range of different species and tissues.[28] This tetrasaccharide was previously proposed to play important roles in muscle function *via* binding with laminin. Therefore, previous efforts were mainly focused on the synthesis of this tetrasaccharide core M1 structure.

The first chemical approach for the synthesis of amino acid appended core M1 tetrasaccharide structures **8a** and **8b** was reported by Ito and his co-workers in 1999.[56] The tetrasaccharide glycoconjugates **8a** and **8b** were convergent assembled from four monosaccharide building blocks **1–4** (Scheme 6.2). After chemical glycosylation of thioglycoside donor **1** with galactoside acceptor **2**, protecting group manipulation gave disaccharide Schmidt donor **5**. On the other hand, after the glycosylation of thioglycoside **3** with mannoside acceptor **4**, deprotection of the Phth group from amine and conversion of the amine into azide provided the disaccharide acceptor **6**. The glycosylation of disaccharide donor **5** with disaccharide acceptor **6** gave the tetrasaccharide intermediate. The resulting tetrasaccharide intermediate was converted into trichloroacetimidate donor and glycosylated with serine acceptor **7a** or threonine acceptor **7b** to give the full protected tetrasaccharide glycoconjugate. The late stage installation of the amino acid aglycon was designed to prevent the epimerization of amino acid moiety during the protecting group manipulations. After global deprotection, the final serine- or threonine-containing core M1 tetrasaccharides **8a** and **8b** were obtained in 17% total yields in the longest linear sequence of 18 steps. A stepwise chemical synthesis approach was also developed by Ito's group for the synthesis of disaccharide and trisaccharide core M1 structures.[54]

The threonine-appended core M1 trisaccharides **12** and **13** were chemically synthesized by Westerlind's group (Scheme 6.3).[58] Threonine aglycon was installed at monosaccharide stage *via* glycosylation between trichloroacetimidate donor **9** and threonine acceptor **10**. The threonine-linked mannoside acceptor **11** was glycosylated with thioglycoside **14**, and the resulting trisaccharide product was then converted into protected trisaccharide derivative **12** in six steps for further SPPS-based glycopeptide synthesis. Following a similar approach, the disaccharide building block **13** was also prepared for the glycopeptide synthesis.

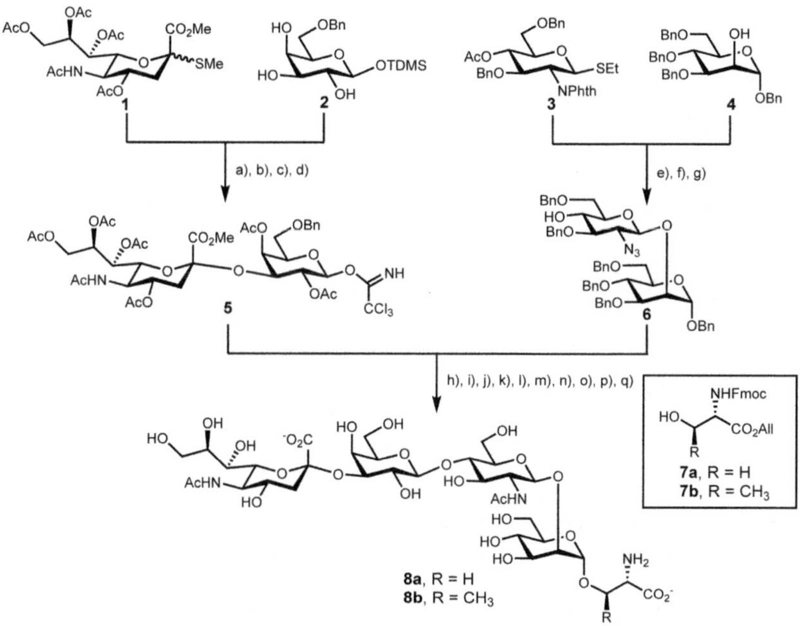

Scheme 6.2 Chemical synthesis of core M1 tetrasaccharides by Ito's group. Reaction conditions: (a) NIS, TfOH, CH_3CN, 3 Å MS, −40 °C; (b) Ac_2O, pyridine, 57% for 2 steps; (c) TBAF, HOAC, THF, 92%; (d) CCl_3CN, DBU, CH_2Cl_2, 0 °C, 93%; (e) NIS, TfOH, CH_3CN, 3 Å MS, −40 °C, 95%; (f) Ethylenediamine, *n*-BuOH, 100 °C; (g) TfN_3, DMAP, CH_2Cl_2, 82% for 2 steps; (h) TMSOTf, CH_2Cl_2, 4 Å MS, −20 °C, 83%; (i) 1,3-propanedithiol, DIPEA, pyridine/H_2O (7 : 3), then Ac_2O, 95% for two steps; (j) $Pd(OH)_2/C$, H_2, MeOH, HOAc, (k) Ac_2O, pyridine, 0 °C, 95% for 2 steps; (l) H_2NNH_2-HOAc, DMF, 0 °C; (m) CCl_3CN, DBU, CH_2Cl_2, 0 °C, 91% for 2 steps; (n) TMSOTf, CH_2Cl_2/toluene (1 : 1), 4 Å MS, −20 °C, 85% for using **7a** as donor, 84% for using **7b** as donor; (o) $Pd(Ph_3P)_4$, dimedone, THF, 90%; (p) Morpholine, $CHCl_3$, (q) NaOMe/MeOH, then NaOH (5 mM), MeOH, 90% for 3 steps.

The first chemoenzymatic synthesis of core M1 tetrasaccharide **23** was achieved by Ajisaka and coworkers in 1999 (Scheme 6.4).[55] Disaccharide **19** was prepared in 0.5% yield from GlcNAc and mannose catalyzed by *N*-acetylglucosaminidase from *Bacillus cirulans*. Then, disaccharide **19** was chemically converted into thioglycoside **20** in three steps in 92% yield. A β-galactosidase from *Bifidobacterium bifidum* was used for the β1–4-galactosylation to afford the trisaccharide **21** in 15% yield. The thioglycoside **21** was fully acetylated as donor for the glycosylation of a serine acceptor to give the serine-appended trisaccharide intermediate, which was deprotected to afford the core M1 trisaccharide **22** in 43% yield in 4 steps. The trisaccharide **22** can be recognized by a α2–3-sialyltransferase

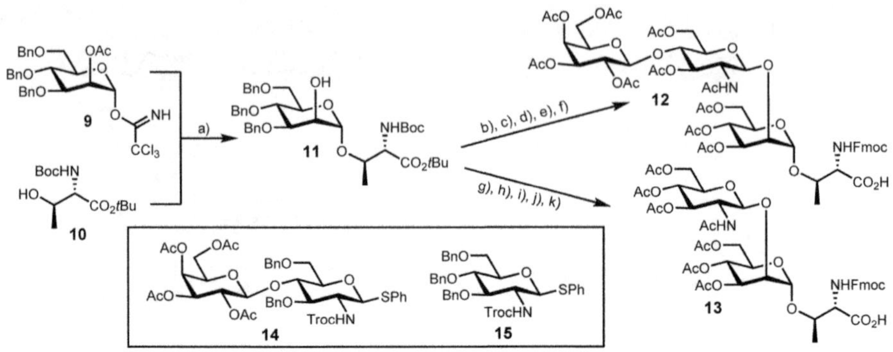

Scheme 6.3 Chemical synthesis of core M1 structures **12**, **13** and core M2 structure. Reaction conditions: (a) TMSOTf, Et_2O, −40 °C, then NaOMe/MeOH, 48% for 2 steps; (b) **14**, NIS, TfOH, CH_2Cl_2, −18 °C, 78%; (c) Zn, HOAc, 40 °C, then Ac_2O, pyridine, 69% for 2 steps; (d) Pd/C, H_2, MeOH, HOAc, then Ac_2O, pyridine; (e) TFA, CH_2Cl_2, anisole, 53% for 3 steps; (f) Fmoc-OSu, $NaHCO_3$, dioxane, 91%; (g) **15**, NIS, TfOH, CH_2Cl_2, −40 °C, 57%; (h) Zn, HOAc, then Ac_2O, pyridine, 67% for 2 steps; (i) Pd/C, H_2, MeOH, HOAc, then Ac_2O, pyridine; (j) TFA, CH_2Cl_2, anisole, 90% for 3 steps; (k) Fmoc-OSu, $NaHCO_3$, dioxane, 99%.

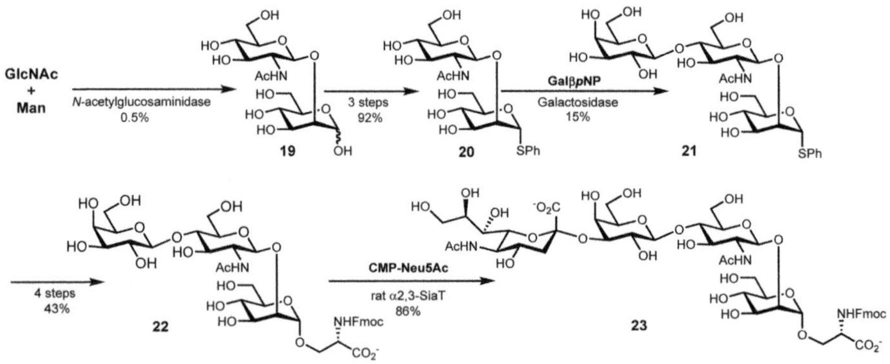

Scheme 6.4 Chemoenzymatic synthesis of core M1 tetrasaccharide **23** by Ajisaka's group.

from a rat to give core M1 tetrasaccharide **23** in 86% yield in the presence of CMP-Neu5Ac as donor. This chemoenzymatic strategy provided a short route towards core **M1** structures **22** and **23**; however, the total yields for either **22** or **23** is less than 0.25% due to the low yields of two glycosidases catalyzed reactions.

In 2003, Whitfield and co-workers reported another chemoenzymatic approach for the synthesis of core M1 tetrasaccharide **29** (Scheme 6.5).[53] The enzymatically synthesized trisaccharide building block **26** was used as the key building block for their synthesis, in which *N*-trichloroethoxycarbonyl (*N*-Troc) protected thioglycoside **24** was first galactosylated by a fusion

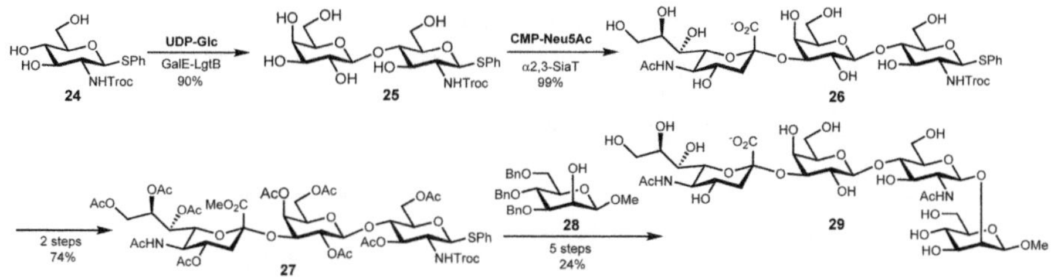

Scheme 6.5 Chemoenzymatic synthesis of core M1 tetrasaccharide **29** by Whitfield's group.

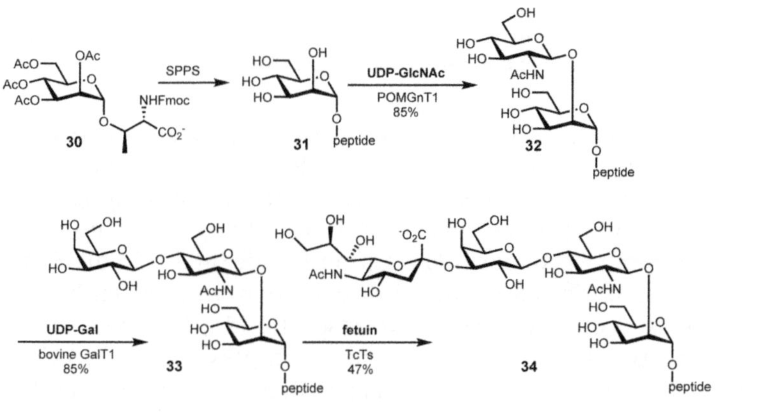

Scheme 6.6 Chemoenzymatic synthesis of core M1 tetrasaccharide glycopeptide **34** by Flitsch's group.

enzyme of GalE and LgtB in the presence of UDP-Glc to provide the LacNAc derivative **25** in 90% yield. The resulting LacNAc disaccharide **25** was then sialylated by *Neisseria meningitides* α2–3-sialyltransferase to give the sialoside **26**, which was converted to fully protected trisaccharide **27** in 74% yield in three steps. The trisaccharide **27** severed as donor for t chemical glycosylation of mannoside **28** to give the fully protected tetrasaccharide intermediate. The core M1 tetrasaccharide **29** was finally achieved after global deprotection.

The first chemoenzymatic synthesis of core M1 *O*-manosyl glycopeptide was achieved by Flitsch's group (Scheme 6.6).[52] The synthesis commenced with the solid phase peptide synthesis (SPPS) of mannosyl peptides **31**. Then, the β1–2-linked GlcNAc was introduced to the glycopeptide **31** using human POMGnT1 in the presence of UDP-GlcNAc as donor, providing the disaccharide glycopeptide **32** in 85% yield. The recombinant bovine β1–4-galactosyltransferase (GalT1) was used to introduce the β1–2-linked galactose to yield trisaccharide **33** in 87% yield. Finally, a *trans*-sialidase from

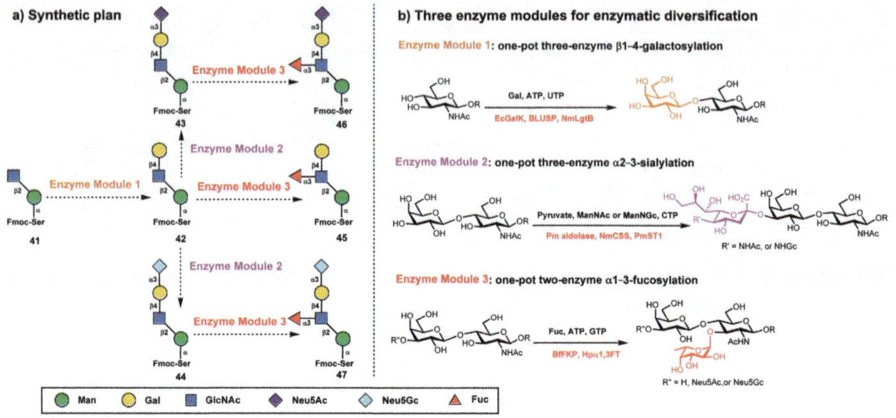

Scheme 6.7 (a) Retrosynthetic plan, and (b) enzyme modules for chemoenzymatic synthesis of core M1 structure **41–47** by Cao's group.

Trypanozoma cruzi (TcTS) was used for α2–3-sialylation of trisaccharide **33** to afford the target core M1 tetrasaccharide glycopeptide **34**. However, TcTs can also hydrolysis the sialosidic linkage, so the **34** was only achieved in 47% yield. A "one-pot" approach was also developed by the same group for the solid phase synthesis of core M1 tetrasaccharide on a gold platform.

The above-mentioned chemical or chemoenzymatic synthesis methods mainly focused on the core M1 tetrasaccharide; however, less attention was paid to the synthesis of the more complex Neu5Gc-, Le^{x}- and sLe^{x}-containing structures. To tackle this problem, a diversity-oriented chemoenzymatic strategy for the systematically synthesis of *O*-mannose core M1 structures was developed (Scheme 6.7).[60] Three enzyme modules were designed to target three glycosidic linkages, Galβ1–4-, Siaα2–3-, and Fucα1–3- presented in *O*-mannose glycan branches. The enzyme module 1 is a one-pot three-enzyme system that was designed for β1–4-galactosylation to form the Galβ1–4GlcNAc (LacNAc) moiety. In this module, the galactose was activated by an *Escherichia coli* galactokinase (EcGalK) and a *Bifidobacterium longum* UDP-sugar pyrophosphorylase (BLUSP) to generate UDP-Gal in the presence of ATP and UTP. The *in situ* generated UDP-Gal was taken by a *N. meningitidis* β1–4-galactosyltransferase (NmLgtB) to form the desired Galβ1–4GlcNAc linkage. The enzyme module 2 is another one-pot three-enzyme system, which is designed for the α2–3-sialylation to form both Neu5Acα2–3Gal and Neu5Gcα2–3Gal units. In this module, the *Pasteurella multocida* sialic acid aldolase (Pm aldolase) is responsible for converting ManNAc/ManNGc to Neu5Ac/Neu5Gc in the presence of sodium pyruvate. The sialic acid was then converted to CMP-Sia (CMP-Neu5Ac or CMP-Neu5Gc) catalyzed by *N. meningitidis* CMP-sialic acid synthetase (NmCSS) in the presence of CTP, and the *in situ* generated CMP-sialic acid was utilized by *P. multocida* multifunctional α2–3-sialyltransferase (PmST1) to form the desired α2–3-linked sialoside. The enzyme module 3 is a one-pot two-enzyme system, which is applied for the α1–3-fucosylation to form the Fucα1–3GlcNAc linkage of Le^x or sLe^x. In this module, the L-fucose was converted to GDP-Fuc in the presence of ATP, GTP, and a bifunctional L-fucokinase/GDP-fucose pyrophosphorylase (FKP) from *Bacteroides fragilis*. The resulting GDP-Fuc was then taken by a recombinant *Helicobacter pylori* α1–3-fucosyltransferase (Hpα1–3FTΔ52) to form Le^x and sLe^x moiety. To overcome the enzyme limitation, the serine-appended disaccharide **41** was chosen for modular enzymatic diversification.

The chemical synthesis of serine-appended disaccharide **41** (Scheme 6.8) starts from the coupling of mannoside acceptor **35** with trichloroacetimidate donor **36** to give the disaccharide **37** in 95% yield. To prevent the epimerization of amino acid moiety under multi-step deprotection conditions, the protecting groups of **37** were changed into acetyl groups, and the late stage glycosylation strategy developed by Ito *et al.*[56] was adopted for installation of the serine aglycon. The *tert*-butyl (*t*Bu) and *O*-acetyl groups were removed under mild conditions to provide disaccharide intermediates for further glycopeptide synthesis or enzymatic diversification.

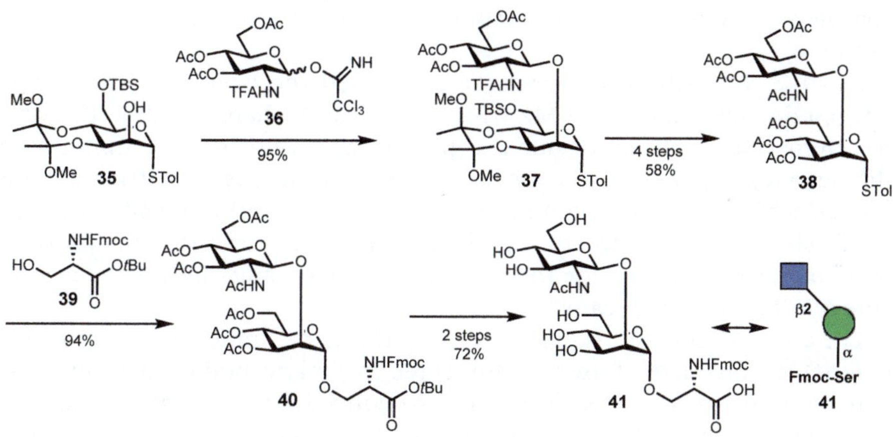

Scheme 6.8 Chemical synthesis of disaccharide **41**.

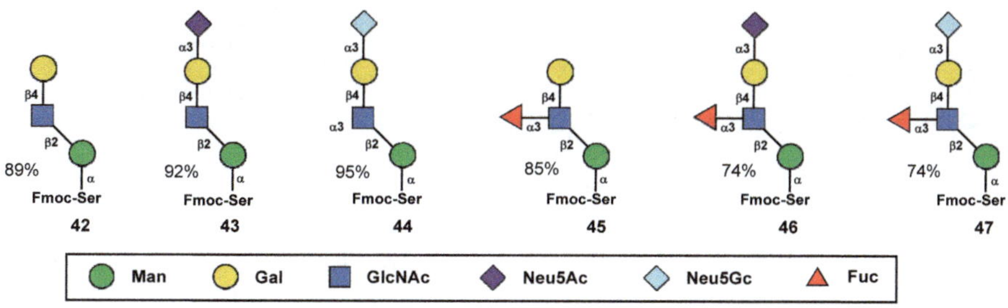

Scheme 6.9 Chemical or chemoenzymatic synthesis of core M2 structures.

The chemically synthesized disaccharide **41** was used for the enzymatic modular assembly of core M1 structure **42–47** (Scheme 6.9). All the *O*-mannose glycans are tolerated by corresponding bacterial glycosyltransferases to afford the glycosylation products at preparative scale in good yields. Each enzyme module contains a glycosyltransferase and enzyme(s) for *in situ* generation of nucleotide-activated donor (UDP-Gal, CMP-Sia acid, or GDP-Fuc) for each module, and all the enzymes are from bacteria and can be overexpressed in convenient *E. coli* systems. Therefore, this enzymatic modular assembly strategy provides a practical and efficient approach for the parallel synthesis of mammalian *O*-mannose glycans.

6.3.2 Chemical or Chemoenzymatic Synthesis of Core M2 *O*-Mannose Glycans

The core M1 *O*-mannose glycans comprise less than 10 possible structures with known branching structures. In contrast, about 20 core M2 structures have been identified, and the theoretical structures of core M2 *O*-mannose glycan could be more than 50. However, the synthesis of these branched complex structures is very challenging and has not been extensively explored.

In 2017, Westerlind and coworkers reported the chemical synthesis of a threonine-appended core M2 trisaccharide **50** (Scheme 6.10).[57] The double glycosylation of mannoside acceptor **48** with thioglycoside donor **49** is the key step for the assembly of core M2 trisaccharide. The trisaccharide **50** was achieved in 70% yield in four steps. The trisaccharide **50** is the first synthesized core M2 structure and was utilized in the construction of *O*-mannosylated peptides for the preparation of specific antibodies against core M2 *O*-mannose glycans.

The core M2 *O*-mannose glycans share the same branching structures as core M1 glycans. Therefore, the same three enzymatic modular assemblies of core M1 structures[60] were adopted for the collective synthesis of 49 putative Core M2 structures **56–105** (Scheme 6.11).[61] The synthesis commenced with chemical synthesis of three core structures **56–58**, which were designed as

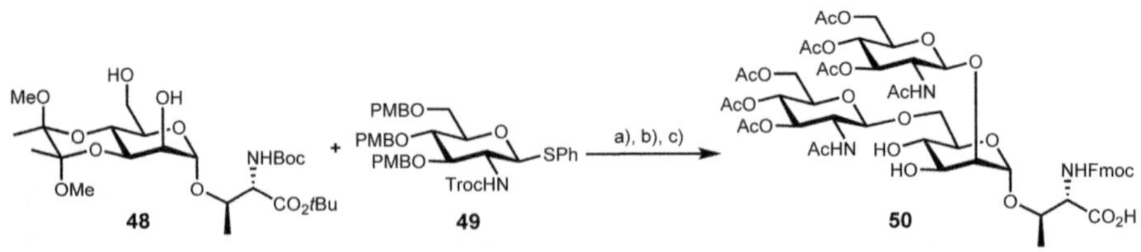

Scheme 6.10 Chemical synthesis of core M2 structures **50**. Reaction conditions: (a) NIS, TfOH, CH_2Cl_2, −50 °C, 90%; m) CAN, CH_3CN/H_2O (9 : 1), then Ac_2O, pyridine, 91%; (b) Zn, HOAc, then Ac_2O, pyrindine, 85%; (c) TFA/CH_2Cl_2 (3 : 1), 62%.

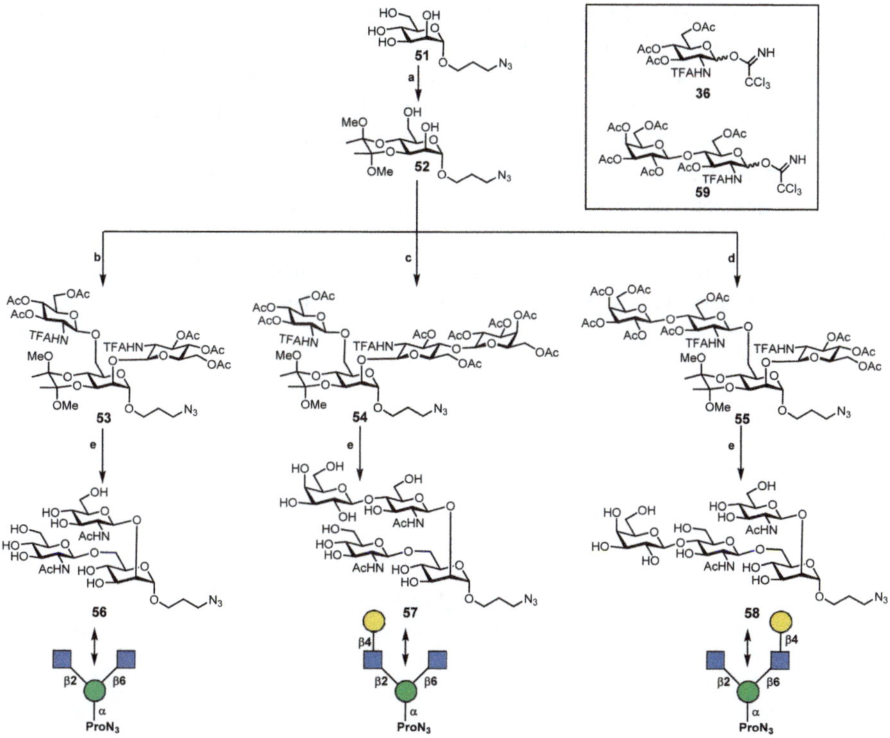

Scheme 6.11 Chemical synthesis of core M2 structures **56–58**. Reagents and conditions: (a) 2,3-butanedione, trimethyl orthoformate, (±)-camphorsulfonic acid, 110 °C, 12 h, 58%; (b) **36**, TMSOTf, 4 Å MS, −50 °C to −20 °C, 85%; (c) **36** (1.0 equiv.), TMSOTf, 4 Å MS, −50 °C, 1 h, then **59** (1.5 equiv.), −20 °C, 1 h, 65% for 2 steps; (d) **59** (1.0 equiv.), TMSOTf, 4 Å MS, −50 °C, 1 h, then **36** (1.5 equiv.), −20 °C, 1 h, 74% for 2 steps; (e) (1) TFA/H_2O (9 : 1, v/v); (2) NaOMe/MeOH, then 1.0 M LiOH; (3) Ac_2O/MeOH, **56** (1.13 g) 84% for 3 steps; **57** (1.58 g) 91% for 3 steps; **58** (1.41 g) 90% for 3 steps.

the key intermediates for further enzymatic diversification without the need of human POMGnT1, the enzyme naturally responsible for the attachment of a GlcNAc residue onto *O*-mannosylated protein in the α-DG biosynthesis. The chemically synthesized core M2 structure **56** was designed for the enzymatic synthesis of six symmetrical core M2 structures **60–65** by synchronous enzymatic extension at both C2- and C6-branched GlcNAc. The chemically synthesized core M2 structures **57–58** were designed for the enzymatic assembly of **40** asymmetrical core M2 structures **66–105**.

A highly convergent strategy was developed for the chemical synthesis of core structures **56–58** (Scheme 6.11). The mannoside **52** was used as the common acceptor for the synthesis of all three core structures **56–58**. Double glycosylation of **52** with trichloroacetimidate donor **36** gave the protected trisaccharide **53** in 85% yield. Taking advantage of the difference in reactivity between the primary C6-OH and the secondary C2-OH of **52**, a one-pot sequential glycosylation approach was applied for the synthesis of asymmetrical tetrasaccharide isomers **54** and **55** by simply changing the glycosylation sequence of donor **36** and **59**. The unified three-step deprotection approach was applied for the global deprotection of **53–55** to provide the core structures of **56–58**. This highly convergent chemical synthesis provides easy access tocore structures at gram-scale for the next step enzymatic diversification.

The trisaccharide **56** was used for the enzymatic synthesis of core M2 symmetrical structures **60–65** (Scheme 6.12). Using enzyme module 1,

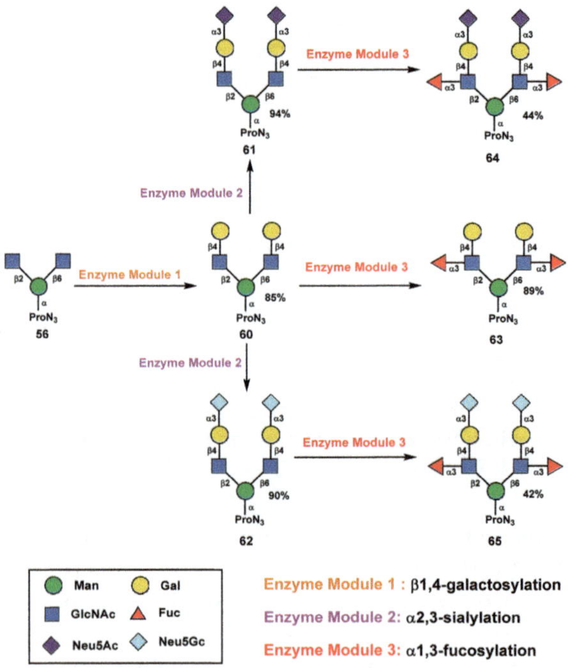

Scheme 6.12 Enzymatic modular assembly of core M2 symmetrical structures **60–65**.

β1–4-galactosylation of trisaccharide **52** provided the symmetrical LacNAc-branched core M2 structure **60** in 85% yield. Then, pentasaccharide **60** was utilized as acceptor by enzyme module 2 for parallel α2–3-sialylation to achieve the Neu5Acα2–3- or Neu5Gcα2–3-linked core M2 structures **61–62** in 94% and 90% yield, respectively. Finally, **60–62** were used as acceptors by enzyme module 3 for parallel α1–3-fucosylation to provide core M2 symmetrical structures **63–65** (Scheme 6.12).

The chemically synthesized asymmetrical core M2 tetrasaccharide **57** was used for the diversity-oriented enzymatic modular assembly of asymmetrical core M2 structures **66–87** (Scheme 6.13). First, the tetrasaccharide **57** was elongated at C2-branched GlcNAc by using enzyme module 2 and 3 sequentially to afford core M2 structures **66–69** with two different sialic acid forms (Neu5Ac or Neu5Gc). Both **68** and **69** were then elongated at C6-branched GlcNAc by

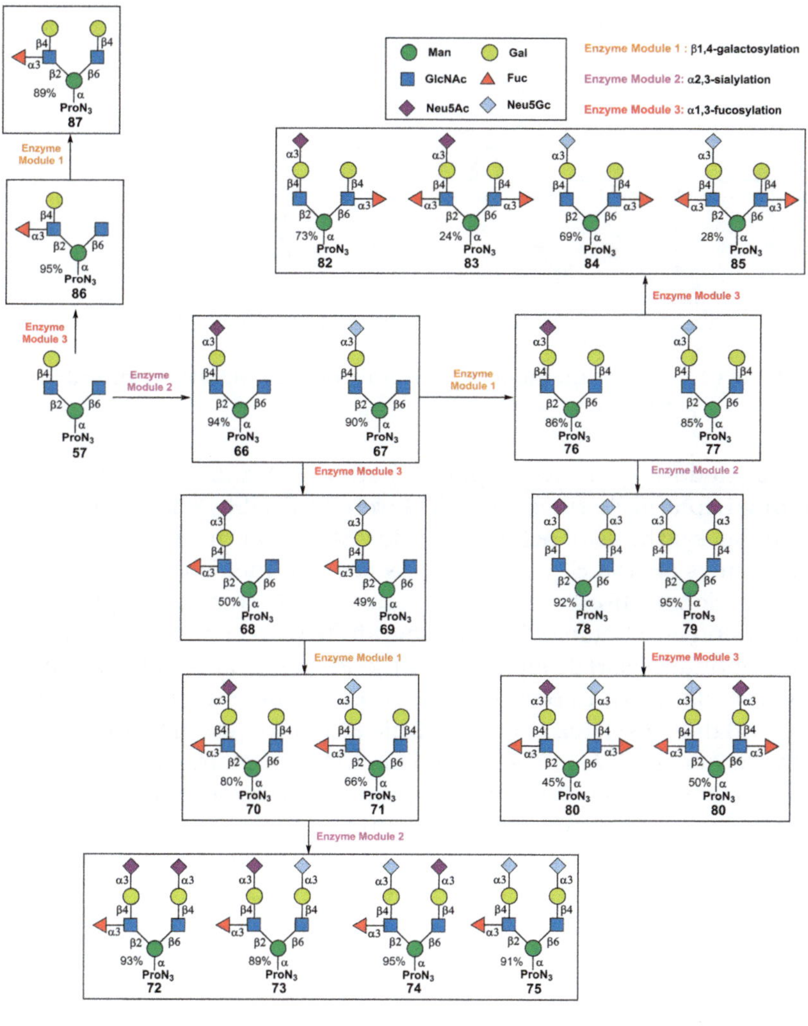

Scheme 6.13 Enzymatic modular assembly of core M2 asymmetrical structures.

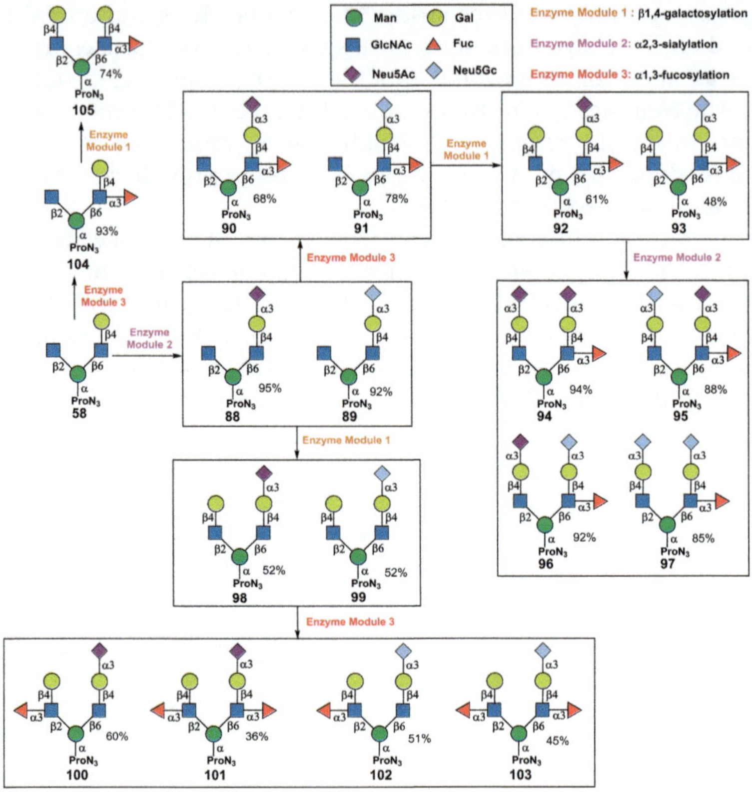

Scheme 6.14 Enzymatic modular assembly of core M2 asymmetrical structures **88–105.**

enzyme module 1 to give heptasaccharides **70** and **71**, respectively. The resulting heptasaccharides **70** and **71** were utilized by enzyme module 2 for the parallel synthesis of octasaccharides **72–75.** On the other hand, the pentasaccharides **66** and **67** also served as key intermediates for the enzymatic diversification to afford the asymmetrical core M2 *O*-mannose glycans **76–85.** Then, α1–3-fucosylation of hexasaccharide **76** with enzyme module **3** yielded both mono-fucosylated heptasaccharide **82** (73%) and di-fucosylated octasaccharide **83** (24%) in one-pot. The mono-fucosylated heptasaccharide **84** (69%) and di-fucosylated octasaccharide **85** (28%) were also prepared in one-pot by enzyme module 3. The asialo-core M2 asymmetrical structures **86** and **87** were prepared from chemically synthesized **57** by sequential glycosylation of enzyme module 3 and 1 (Scheme 6.13).

The chemically synthesized asymmetrical tetrasaccharide **58** was used for the diversity-oriented enzymatic assembly of core M2 asymmetrical structures **88–105** (Scheme 6.14).

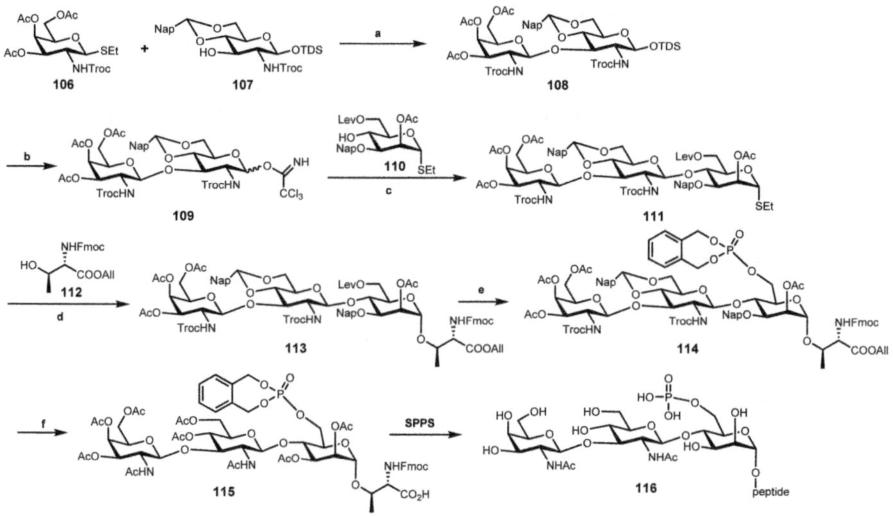

Scheme 6.15 Chemical synthesis of core M3 structure by Boons' group. Reaction conditions: (a) NIS, TMSOTf, CH_2Cl_2, 0 °C, 91%; (b) TBAF, THF-HOAc, then CCl_3CN, Cs_2CO_3, DMF, 71% for 2 steps; (c) TMSOTf, CH_2Cl_2, −35 °C, 86%; (d) NIS, AgOTf, CH_2Cl_2, −20 °C, then H_2NNH_2-HOAc, CH_2Cl_2-MeOH; (e) *N,N*-diethyl-1,5-dihydro-2,4,3-benzodioxaphosphepin-3-amine, 1H-tetrazole, CH_2Cl_2, then mCPBA oxidation at −20 °C, 51% for 4 steps; (f) DDQ, CH_2Cl_2-H_2O, then Zn, HOAc, finally Ac_2O, pyridine, 56% for 3 steps.

6.3.3 Chemical Synthesis of Phosphorylated Core M3 Trisaccharide

In 2010, Campbell's group identified a phosphorylated core M3 *O*-mannosyl glycan that is required for laminin binding on α-DG.[62] To investigate the biosynthetic pathway and enzymes involved in the assembly of this special phosphorylated core M3 trisaccharide structure, Boons' group developed a chemical synthetic approach for the total synthesis of phosphorylated *O*-mannosyl glycopeptides (Scheme 6.15).[59] First, trisaccharide **111** was constructed from building blocks **106**, **107** and **110**. Then, thiogalactoside **106** was coupled with glycosyl acceptor **107** in the presence of NIS/TMSOTf to provide the disaccharide **108** in 91% yield. The anomeric dimethylthexylsilyl (TDS) ether of **108** was removed and converted into corresponding trichloroacetimidate **109**, which was used as glycosyl donor and coupled with thiomannosyl acceptor **110** in the presence of TMSOTf to give trisaccharide **111** in 86% yield. Trisaccharide **111** was coupled with properly protected threonine **112** in the presence of

NIS/AgOTf to furnish glycosylated amino acid **113** in 72% yield. Then the Lev ester of **113** was selectively removed for introducing phosphotriester to afford **114** in 51% yield in four steps. After protecting group manipulation, the trisaccharide **114** was converted into trisaccharide derivative **115** in 56% yield in three steps, and the **115** was further applied for the solid phase peptide synthesis of core M3 *O*-mannose glycopeptides.[59]

6.4 Conclusion

Due to the important roles of protein *O*-mannosylation in many biological processes, chemical or chemoenzymatic strategies have been developed to synthesize structurally well-defined *O*-mannose glycans in sufficient amounts. These systematically synthesized *O*-mannose glycans are invaluable tools for exploring the biosynthetic pathways and functional roles of this novel *O*-glycan family.

Abbreviations

Ser	serine
Thr	threonine
DG	dystroglycan
α-DG	α-dystroglycan
DGC	dystrophin-glycoprotein complex
ECM	extracellular matrix
ER	endoplasmic reticulum
Dol-P-Man or DPM	dolichol monophosphate-activated mannose
POMT1/2	protein *O*-mannosyltransferases 1/2
GDP-Man	guanidine diphosphate mannose
POMGnT1	protein *O*-mannose-β1–2-*N*-acetylglucosaminyl transferase 1
GnT-Vb/GnT-IX	β1–6-*N*-acetylglucosaminyl transferase
POMGnT2	β1–4-*N*-acetylglucosaminyltransferase 2
β3GALNT2	β1–3-linked *N*-acetylgalactosamine
Gal	galactose
GlcA	glucuronic acid
PGs	proteoglycans
HNK-1ST	human natural killer-1 sulfotransferase
CMDs	congenital muscular dystrophies
WWS	Walker–Warburg syndrome
MEB	muscle–eye–brain disease
FCMD	Fukuyama-type congenital muscular dystrophy
FKRP	fukutin-related protein
FKTN	*fukutin*
MDC1C	CMD type 1C

LGMD2I	limb–girdle muscular dystrophy type 2I
MDC1D	CMD type 1D
β3GnT1	β1–3-*N*-acetylglucosaminyltransferase-1
E-cadherin	epithelial cadherin
MEB	muscle-eye-brain disease
RPTPβ	receptor protein tyrosine phosphatase β
cdhs	cadherins
pcdhs	protocadherins
NIS	*N*-iodosuccinimide
TfOH	trifluoromethanesulfonic acid
N-Phth	*N*-phthaloyl
N-DMP	2,5-dimethylpyrrole
N-Troc	trichloroethoxycarbonyl
OPME	one-pot multienzyme
TMSOTf	trimethylsilyl trifluoromethane-sulfonate
AgOTf	silver trifluoromethanesulfonate
Neu5Ac	*N*-acetylneuraminic acid
TDS	dimethylthexylsilyl

Acknowledgement

The support from National Natural Science Foundation of China (Grant Nos. 21672128, 21877072, 21807064), State Key Laboratory of Microbial Technology (M2016-06), Department of Science and Technology of Shandong Province (2016GGH4502, 2016GSF121002, ZR201709190252) and Shandong University (2018JC053) are acknowledged.

References

1. J. Finne, T. Krusius, R. K. Margolis and R. U. Margolis, *J. Biol. Chem.*, 1979, **254**, 10295.
2. R. Barresi and K. P. Campbell, *J. Cell Sci.*, 2006, **119**, 199.
3. T. Endo, *J. Biochem.*, 2015, **157**, 1.
4. W. Cao, *Science*, 1998, **282**, 20791.
5. S. C. Smelt, P. Borrow, S. Kunz, W. Cao, A. Tishon, H. Lewicki, K. P. Campbell and M. B. Oldstone, *J. Virol.*, 2001, **75**, 448.
6. S. Kunz, J. M. Rojek, M. Kanagawa, C. F. Spiropoulou, R. Barresi, K. P. Campbell and M. B. Oldstone, *J. Virol.*, 2005, **79**, 14282.
7. J. L. Praissman and L. Wells, *Biochemistry*, 2014, **53**, 3066.
8. H. Manya, A. Chiba, A. Yoshida, X. Wang, Y. Chiba, Y. Jigami, R. U. Margolis and T. Endo, *Proc. Natl. Acad. Sci. U. S. A.*, 2004, **101**, 500.
9. M. O. Sheikh, S. M. Halmo and L. Wells, *Glycobiology*, 2017, **27**, 806.
10. Y. Maeda, S. Tanaka, J. Hino, K. Kangawa and T. Kinoshita, *EMBO J.*, 2000, **19**, 2475.

11. D. J. Lefeber, A. P. de Brouwer, E. Morava, M. Riemersma, J. H. Schuurs-Hoeijmakers, B. Absmanner, K. Verrijp, W. M. van den Akker, K. Huijben, G. Steenbergen, J. van Reeuwijk, A. Jozwiak, N. Zucker, A. Lorber, M. Lammens, C. Knopf, H. van Bokhoven, S. Grunewald, L. Lehle, L. Kapusta, H. Mandel and R. A. Wevers, *PLoS Genet.*, 2011, **7**, e1002427.
12. K. J. Carss, E. Stevens, A. R. Foley, S. Cirak, M. Riemersma, S. Torelli, A. Hoischen, T. Willer, M. van Scherpenzeel, S. A. Moore, S. Messina, E. Bertini, C. G. Bonnemann, J. E. Abdenur, C. M. Grosmann, A. Kesari, J. Punetha, R. Quinlivan, L. B. Waddell, H. K. Young, E. Wraige, S. Yau, L. Brodd, L. Feng, C. Sewry, D. G. MacArthur, K. N. North, E. Hoffman, D. L. Stemple, M. E. Hurles, H. van Bokhoven, K. P. Campbell, D. J. Lefeber, UK10K Consortium, Y. Y. Lin and F. Muntoni, *Am. J. Hum. Genet.*, 2013, **93**, 29.
13. A. Yoshida, K. Kobayashi, H. Manya, K. Taniguchi, H. Kano, M. Mizuno, T. Inazu, H. Mitsuhashi, S. Takahashi and M. Takeuchi, *Dev. Cell*, 2001, **1**, 717.
14. S. H. Stalnaker, K. Aoki, J. M. Lim, M. Porterfield, M. Liu, J. S. Satz, S. Buskirk, Y. Xiong, P. Zhang, K. P. Campbell, H. Hu, D. Live, M. Tiemeyer and L. Wells, *J. Biol. Chem.*, 2011, **286**, 21180.
15. A. Chiba, K. Matsumura, H. Yamada, T. Inazu, T. Shimizu, S. Kusunoki, I. Kanazawa, A. Kobata and T. Endo, *J. Biol. Chem.*, 1997, **272**, 2156.
16. A. C. Combs and J. M. Ervasti, *Biochem. J.*, 2005, **390**, 303.
17. J. Morise, Y. Kizuka, K. Yabuno, Y. Tonoyama, N. Hashii, N. Kawasaki, H. Manya, Y. Miyagoe-Suzuki, S. Takeda, T. Endo, N. Maeda, H. Takematsu and S. Oka, *Glycobiology*, 2014, **24**, 314.
18. T. Yoshida-Moriguchi, T. Willer, M. E. Anderson, D. Venzke, T. Whyte, F. Muntoni, H. Lee, S. F. Nelson, L. Yu and K. P. Campbell, *Science*, 2013, **341**, 896.
19. M. Kanagawa, K. Kobayashi, M. Tajiri, H. Manya, A. Kuga, Y. Yamaguchi, K. Akasaka-Manya, J. I. Furukawa, M. Mizuno, H. Kawakami, Y. Shinohara, Y. Wada, T. Endo and T. Toda, *Cell Rep.*, 2016, **14**, 2209.
20. H. Manya, Y. Yamaguchi, M. Kanagawa, K. Kobayashi, M. Tajiri, K. Akasaka-Manya, H. Kawakami, M. Mizuno, Y. Wada, T. Toda and T. Endo, *J. Biol. Chem.*, 2016, **291**, 246187.
21. T. Willer, K. Inamori, D. Venzke, C. Harvey, G. Morgensen, Y. Hara, D. Beltran Valero de Bernabe, L. Yu, K. M. Wright and K. P. Campbell, *eLife*, 2014, **3**, e03941.
22. J. L. Praissman, D. H. Live, S. Wang, A. Ramiah, Z. S. Chinoy, G.-J. Boons, K. W. Moremen and L. Wells, *eLife*, 2014, **3**, e03943.
23. K. Inamori, T. Yoshida-Moriguchi, Y. Hara, M. E. Anderson, L. Yu and K. P. Campbell, *Science*, 2012, **335**, 93.
24. N. Nakagawa, H. Manya, T. Toda, T. Endo and S. Oka, *J. Biol. Chem.*, 2012, **287**, 30823.
25. A. Ashikov, F. F. Buettner, B. Tiemann, R. Gerardy-Schahn and H. Bakker, *Glycobiology*, 2013, **23**, 303.
26. K. Inamori, Y. Hara, T. Willer, M. E. Anderson, Z. Zhu, T. Yoshida-Moriguchi and K. P. Campbell, *Glycobiology*, 2013, **23**, 295.

27. K. I. Inamori, A. M. Beedle, D. B. de Bernabe, M. E. Wright and K. P. Campbell, *Glycobiology*, 2016, **26**, 1284.
28. L. Wells, *J. Biol. Chem.*, 2013, **288**, 6930.
29. T. Yoshida-Moriguchi and K. P. Campbell, *Glycobiology*, 2015, **25**, 702.
30. K. Akasaka-Manya, H. Manya and T. Endo, *Biochem. Biophys. Res. Commun.*, 2004, **325**, 75.
31. U. Hehr, G. Uyanik, C. Gross, M. C. Walter, A. Bohring, M. Cohen, B. Oehl-jaschkowitz, L. M. Bird, G. M. Shamdeen and U. Bogdahn, *Neurogenetics*, 2007, **8**, 279.
32. U. Yis, G. Uyanik, P. B. Heck, M. Smitka, H. Nobel, F. Ebinger, E. Dirik, L. Feng, S. H. Kurul, K. Brocke, A. Unalp, E. Ozer, H. Cakmakci, C. Sewry, S. Cirak, F. Muntoni, U. Hehr and D. J. Morris-Rosendahl, *Neuromuscular Disord.*, 2011, **21**, 20.
33. H. Xiong, K. Kobayashi, M. Tachikawa, H. Manya, S. Takeda, T. Chiyonobu, N. Fujikake, F. Wang, A. Nishimoto, G. E. Morris, Y. Nagai, M. Kanagawa, T. Endo and T. Toda, *Biochem. Biophys. Res. Commun.*, 2006, **350**, 935.
34. M. Brockington, Y. Yuva, P. Prandini, S. C. Brown, S. Torelli, M. A. Benson, R. Herrmann, L. V. B. Anderson, R. Bashir and J. M. Burgunder, *Hum. Mol. Genet.*, 2001, **10**, 2851.
35. X. Bao, M. Kobayashi, S. Hatakeyama, K. Angata, D. Gullberg, J. Nakayama, M. N. Fukuda and M. Fukuda, *Proc. Natl. Acad. Sci. U. S. A.*, 2009, **106**, 12109.
36. J. Muschler, D. Levy, R. Boudreau, M. Henry, K. Campbell and M. J. Bissell, *Cancer Res.*, 2002, **62**, 71029.
37. D. B. de Bernabe, K. Inamori, T. Yoshida-Moriguchi, C. J. Weydert, H. A. Harper, T. Willer, M. D. Henry and K. P. Campbell, *J. Biol. Chem.*, 2009, **284**, 11279.
38. A. K. Esser, M. R. Miller, Q. Huang, M. M. Meier, D. B.-V. de Bernabe, C. S. Stipp, K. P. Campbell, C. F. Lynch, B. J. Smith, M. B. Cohen and M. D. Henry, *J. Biol. Chem.*, 2013, **288**, 2132.
39. S. Carvalho, T. Oliveira, M. F. Bartels, E. Miyoshi, M. Pierce, N. Taniguchi, F. Carneiro, R. Seruca, C. A. Reis and S. Strahl, *Oncotarget*, 2016, **7**, 65231.
40. N. Kuwabara, H. Manya, T. Yamada, H. Tateno, M. Kanagawa, K. Kobayashi, K. Akasaka-Manya, Y. Hirose, M. Mizuno, M. Ikeguchi, T. Toda, J. Hirabayashi, T. Senda, T. Endo and R. Kato, *Proc. Natl. Acad. Sci. U. S. A.*, 2016, **113**, 92805.
41. K.-i. Inamori, T. Endo, Y. Ide, S. Fujii, J. Gu, K. Honke and N. Taniguchi, *J. Biol. Chem.*, 2003, **278**, 43102.
42. K. L. Abbott, R. T. Matthews and M. Pierce, *J. Biol. Chem.*, 2008, **283**, 33026.
43. C. A. Dwyer, E. Baker, H. Hu and R. T. Matthews, *Neuroscience*, 2012, **220**, 47.
44. K. Kanekiyo, K. Inamori, S. Kitazume, K. Sato, J. Maeda, M. Higuchi, Y. Kizuka, H. Korekane, I. Matsuo, K. Honke and N. Taniguchi, *J. Neurosci.*, 2013, **33**, 10037.

45. M. B. Vester-Christensen, A. Halim, H. J. Joshi, C. Steentoft, E. P. Bennett, S. B. Levery, S. Y. Vakhrushev and H. Clausen, *Proc. Natl. Acad. Sci. U. S. A.*, 2013, **110**, 21018.
46. M. Lommel, P. R. Winterhalter, T. Willer, M. Dahlhoff, M. R. Schneider, M. F. Bartels, I. Renner-Mueller, T. Ruppert, E. Wolf and S. Strahl, *Proc. Natl. Acad. Sci. U. S. A.*, 2013, **110**, 21024.
47. I. S. B. Larsen, Y. Narimatsu, H. J. Joshi, Z. Yang, O. J. Harrison, J. Brasch, L. Shapiro, B. Honig, S. Y. Vakhrushev, H. Clausen and A. Halim, *J. Biol. Chem.*, 2017, **292**, 11586.
48. I. S. B. Larsen, Y. Narimatsu, H. J. Joshi, L. Siukstaite, O. J. Harrison, J. Brasch, K. M. Goodman, L. Hansen, L. Shapiro, B. Honig, S. Y. Vakhrushev, H. Clausen and A. Halim, *Proc. Natl. Acad. Sci. U. S. A.*, 2017, **114**, 11163.
49. T. Martinez, D. Pace, L. Brady, M. Gerhart and A. Balland, *J. Chromatogr. A*, 2007, **1156**, 183.
50. S. Pacharra, F. G. Hanisch and I. Breloy, *J. Proteome Res.*, 2012, **11**, 3955.
51. S. Pacharra, F. G. Hanisch, M. Muhlenhoff, A. Faissner, U. Rauch and I. Breloy, *J. Proteome Res.*, 2013, **12**, 1764.
52. R. Sardzik, A. P. Green, N. Laurent, P. Both, C. Fontana, J. Voglmeir, M. J. Weissenborn, R. Haddoub, P. Grassi, S. M. Haslam, G. Widmalm and S. L. Flitsch, *J. Am. Chem. Soc.*, 2012, **134**, 4521.
53. F. Yan, S. Mehta, E. Eichler, W. W. Wakarchuk, M. Gilbert, M. J. Schur and D. M. Whitfield, *J. Org. Chem.*, 2003, **68**, 2426.
54. J. Seifert, T. Ogawa, S. Kurono and Y. Ito, *Glycoconjugate J.*, 2000, **17**, 407.
55. I. Matsuo, M. Isomura and K. Ajisaka, *Tetrahedron Lett.*, 1999, **40**, 5047.
56. J. Seifert, T. Ogawa and Y. Ito, *Tetrahedron Lett.*, 1999, **40**, 6803.
57. J. Yu, O. C. Grant, C. Pett, S. Stahl, R. J. Woods and U. Westerlind, *Chem. - Eur. J.*, 2017, **23**, 3466.
58. J. Yu and U. Westerlind, *ChemBioChem*, 2014, **15**, 939.
59. K.-F. Mo, T. Fang, S. H. Stalnaker, P. S. Kirby, M. Liu, L. Wells, M. Pierce, D. H. Live and G.-J. Boons, *J. Am. Chem. Soc.*, 2011, **133**, 14418.
60. Y. Zhang, C. Meng, L. Jin, X. Chen, F. Wang and H. Cao, *Chem. Commun.*, 2015, **51**, 11654.
61. C. Meng, A. Sasmal, Y. Zhang, T. Gao, C.-C. Liu, N. Khan, A. Varki, F. Wang and H. Cao, *Angew. Chem., Int. Ed.*, 2018, **57**, 9003.
62. T. Yoshida-Moriguchi, L. Yu, S. H. Stalnaker, S. Davis, S. Kunz, M. Madson, M. B. A. Oldstone, H. Schachter, L. Wells and K. P. Campbell, *Science*, 2009, **327**, 88.

CHAPTER 7

Chemical Synthesis of Glycopeptides and Glycoproteins

NIKESH NARANG[a], ROHIT K. SHARMA[b] AND HIRONOBU HOJO*[c]

[a]Osaka University, Department of Chemistry, 1-1 Machikaneyama, Toyonaka, 560-0043, Japan; [b]Panjab University, Department of Chemistry & Centre of Advanced Studies, Sector-14, 160014, Chandigarh, India; [c]Osaka University, Institute for Protein Research, 3-2 Yamadaoka, Suita, 565-0871, Japan
*E-mail: hojo@protein.osaka-u.ac.jp

7.1 Introduction

The oligosaccharides of the glycoproteins provide ligands for cell adhesion and pathogen invasion along with other functions such as cell–cell interactions, tumor metastasis, immune surveillance, inflammatory reactions, auto immunity, *etc*. The biosynthesis of the glycan chain is controlled by various competitive enzymes leading to glycosylation of the same protein by different glycan units to form glycoforms of a protein.[1] Attempts have been made for the structural elucidation of glycoproteins using various biophysical techniques. However, the heterogeneity of the carbohydrates on the glycoproteins makes structural analysis complicated.[2]

Chemical Biology No. 11
Synthetic Glycomes
Edited by Wanyi Guan, Lei Li and Peng George Wang

Published by the Royal Society of Chemistry, www.rsc.org

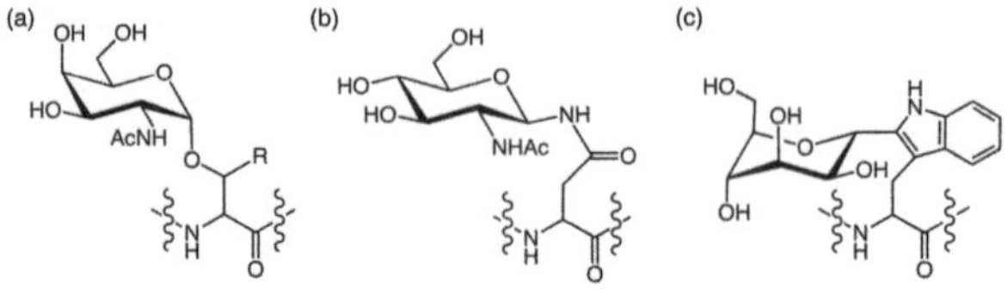

Figure 7.1 Pictorial representation of the reducing end structure of (a) *O*-linked, (b) *N*-linked and (c) *C*-linked carbohydrates on glycoproteins.

The development of a method for glycoprotein synthesis would provide a solution to this problem as it efficiently produces the homogeneous glycoproteins, which are essential for their structure–activity relationship (SARs) studies. In addition, it is important for regulatory therapeutic applications as well as for other purposes, such as the development of antifreeze agents, *etc.*

The major glycosidic linkages exhibited by natural systems have been *O*-linked (Figure 7.1(a)) and *N*-linked (Figure 7.1(b)) with glycan-protein linkages at the Ser/Thr and Asn, respectively. Correspondingly, the most common glycan motifs are GalNAcα-Ser/Thr, GlcNAcβ-Asn along with less common GlcNAcβ-Ser/Thr, GalNAcα-Asn and Glc-Asn.[3] Although GalNAcα-Asn and Glc-Asn are less pronounced, they are found in the nephritogenic glycopeptide[4] and in the halobacterial cell wall glycoprotein,[5] respectively. The *O*-linked glycoproteins can be classified based on the structure as core 1 to 8, where core 1 is often referred as the T antigen. All the core structures have the reducing end α-GalNAc in common, which is called the Tn antigen.[6] All the *N*-linked glycoproteins possess a common core pentasaccharide structure, *i.e.*, Manα1–3(Manα1–6)Manβ1–4GlcNAcβ1–4GlcNAcβ1, and are classified into the oligomannose, complex and hybrid-type.[7] Another linkage, which was recently discovered, is the *C*–*C* linkage (Figure 7.1(c)) which possesses mannosyl linkages at the C-2 carbon of the tryptophan residues.[8] The discussion in the following sections will be mainly based on different methodologies employed for the chemical synthesis of the *O*-linked and *N*-linked glycopeptides and glycoproteins.

7.2 Synthesis of Glycopeptide

7.2.1 Synthesis of Polypeptide Chain

The solid phase peptide synthesis (SPPS) (Figure 7.2(a)) is used as an efficient tool for the glycopeptide synthesis.[9] The SPPS method works on the principle that the *N*-protected amino acid is used for sequential chain elongation. The C-terminus of the peptide is bound to a resin for easy purification during the chain elongation step. As almost half of the natural amino acids possess a side chain with reactive functionalities, these groups are also orthogonally protected.

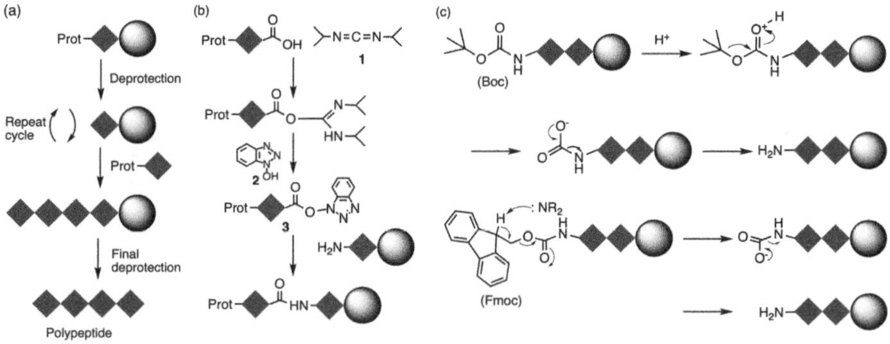

Figure 7.2 (a) Schematic diagram representing solid phase peptide synthesis (SPPS);(b)Representativemechanismoftheintroductionoftheaminoacidusedin the SPPS; (c) deprotection mechanism of Boc and Fmoc.

A representative reaction is initiated by activation of the carboxy group of the *N*-protected amino acid by an activator, such as diisopropylcarbodiimide (DIC) (**1**), which then reacts with 1-hydroxybenzotriazole (HOBt) (**2**) to yield the active ester **3** (Figure 7.2(b)). The amino group of the resin bound amino acid then reacts with the carbonyl carbon of the active ester (**3**) leading to formation of a peptide bond. Other coupling reagents, such as HBTU [*O*-(benzotriazol-1-yl)-*N,N,N′,N′*-tetramethyluronium hexafluorophosphate] and COMU [(1-cyano-2-ethoxy-2-oxoethylidenaminoxy)-dimethylamino-morpholino-carbenium, are also used in recent syntheses.[10,11] The deprotection procedure varies depending upon the nature of the protecting group employed. The most common amino protecting groups used are the *t*-butoxycarbonyl (Boc) and 9-fluorenylmethoxycarbonyl (Fmoc) groups, which are generally removed by trifluoroacetic acid (TFA) and piperidine, respectively. The resin cleavage along with the final deprotection for the Boc and Fmoc based SPPS is carried out using anhydrous HF and TFA. Since the Boc method employs acid treatments, especially at the end of the synthesis by HF, its application to the synthesis of glycopeptides is difficult, as glycosidic linkages are sensitive to strong acids. Thus Fmoc-SPPS may be suitable for the synthesis of glycopeptides (Figure 7.2(c)).

7.2.2 Synthesis of PSGL-1

For synthesis of the glycopeptides, the side chain hydroxy groups of the glycan chain are usually protected by an acetyl group due to the ease of preparation of acetyl-protected sugar units, as well as their glycosidic linkage-stabilizing effect on acids. As an example, the synthesis of the PSGL-1 glycopeptide is described. It has been reported that the carbohydrate recognizing sites of the P- and E-selectin play an important role in the acute inflammatory processes in which the P- and E-selectin receptors

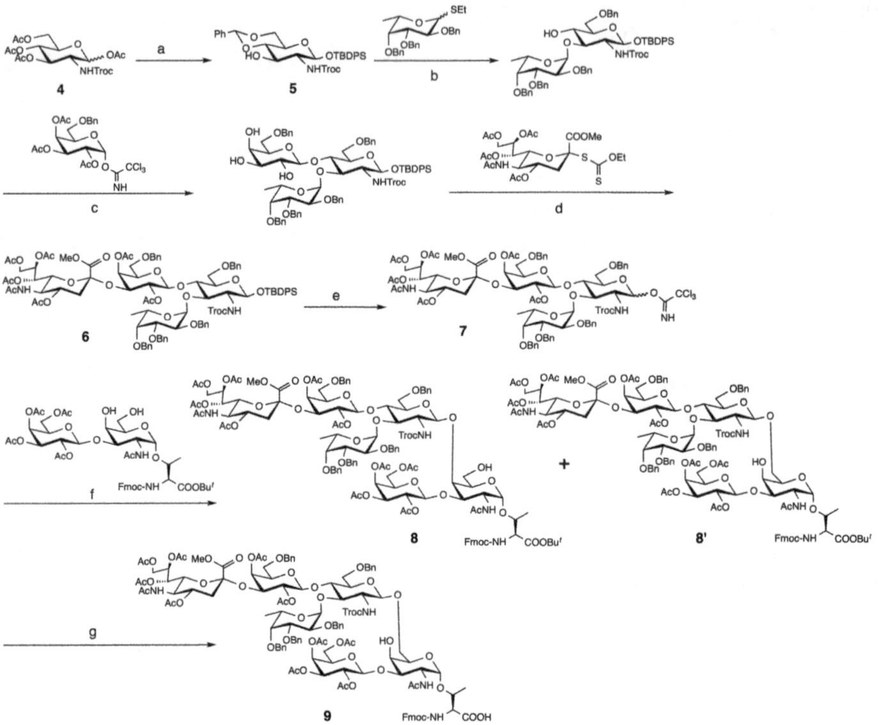

Figure 7.3 Synthesis scheme of glycan chain of PSGL-1. Reagents and conditions, (a): 1. NH_2–NH_2, AcOH, 2. TBDPS-Cl, imidazole, DMAP, 3. NaOMe, MeOH, 4. $PhCH(OMe)_2$, *p*-TsOH; (b): 1. $CuBr_2$, Bu_4NBr, 2. Et_3SiH, TfOH; (c): 1. TMSOTf, 2. NaOMe, MeOH; (d): 1. AgOTf, MeSBr, 2. Ac_2O; (e): 1. TBAF, AcOH, 2. Cl_3CCN; (f): TMSOTf (g): 1. (i) Zn, AcOH, (ii) Ac_2O, 2. (i) H_2, $Pd(OH)_2/C$, (ii) Ac_2O, 3. TFA/DCM (1/2), Et_3SiH.

are exposed on the activated endothelium. This is similar to the case with tumor metathesis making P-selectin intervening agents promising antitumor tools. Considering this, Kunz *et al.* synthesized the binding site of the P-selectin glycoprotein ligand PSGL-1.[12]

The synthesis of the glycan part was initiated with **4** to generate the acceptor **5** (Figure 7.3). **5** was then glycosylated with a series of sugars to obtain a tetrasaccharide **6**. The anomeric hydroxy group was deprotected using TBAF, then converted to the imidate using CCl_3CN generating **7**. **7** was then reacted with the Fmoc-Thr carrying core 1 sugar to yield two products **8** and **8′** and it was observed that **8** was obtained as the major quantity. However, since **8′** was obtained in sufficient quantity, it was deprotected and all the hydroxy groups were acetylated to obtain **9** for the SPPS.

The synthesis of the PSGL glycopeptide was then performed by Fmoc-SPPS (Figure 7.4), in which the glycosylated Thr was introduced using **9**. The peptide was detached from the resin using acetic acid and trifluoroethanol in DCM. Peptide deprotection was further performed using TFA/DCM/water/

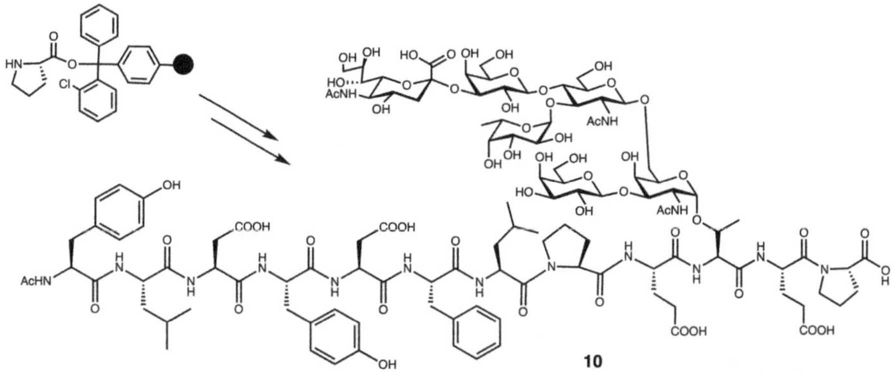

Figure 7.4 SPPS of PSGL-1. Reagents and conditions, 1. SPPS cycle, Fmoc-AA-OH, HBTU, HOBt, DIPEA; Fmoc removal by piperidine/NMP (1:4) 2. AcOH: trifluoroethanol: DCM(1:1:8) 3. TFA:DCM:water:TIS (20:20:1:1), 4. NaOMe, MeOH, 5. NaOH, H2O, 6. AcOH.

triisopropylsilane (20/20/1/1) followed by deprotection of the sugars to obtain the desired PSGL-1 (**10**). Due to the acid-stabilizing effect of the acetyl group, the highly acid-sensitive α-sialyl linkage was stable during the TFA treatment at the end of the Fmoc-SPPS.

7.2.3 Synthesis of CD52

Nakahara and our group have been using the benzyl-protection strategy for synthesis of the glycopeptide, in which the benzyl group is used for protection of the side chain hydroxy groups of the glycan chain. The benzyl protection strategy has several advantages over the commonly used acetyl protection: (i) the benzyl sugars retain a high reactivity for glycosylation facilitating synthesis of the complex glycan chains; (ii) benzyl groups can remain intact during the basic treatment used in the Fmoc based SPPS; (iii) the cleavage of the benzyl group does not require basic conditions which could lead to the racemization of amino acids, β-elimination and acetyl-group migration to the amino group of the growing polypeptide chains as in the case of acetyl protection; (iv) the benzyl group can be easily cleaved by mild hydrogenation or treatment with low TfOH.[13]

As an example, the chemical synthesis of CD52[14] is depicted here. CD52 is a GPI anchored antigen which is expressed in most human lymphocytes. It is the shortest cell surface glycoprotein with only 12 amino acid residues, but its molecular weight is 21–28 kDa through extensive glycosylation. First, for the synthesis of the Asn linked pentasaccharide **19**, sugars **11** and **12** were reacted over silver alumina silicate to yield **13** (α/β: 1/1.4) (Figure 7.5(a)). As desired, the anomeric fluoride was untouched. The disaccharide **13** was then stereoselectively conjugated with **14** and subjected to the iridium-catalyzed deallylation procedure to produce **15**, which was then reacted with **16** in the presence of silver trifluoromethanesulfonate as a promoter to yield **17**. The sugar **17** was subjected to dephthaloylation in the presence

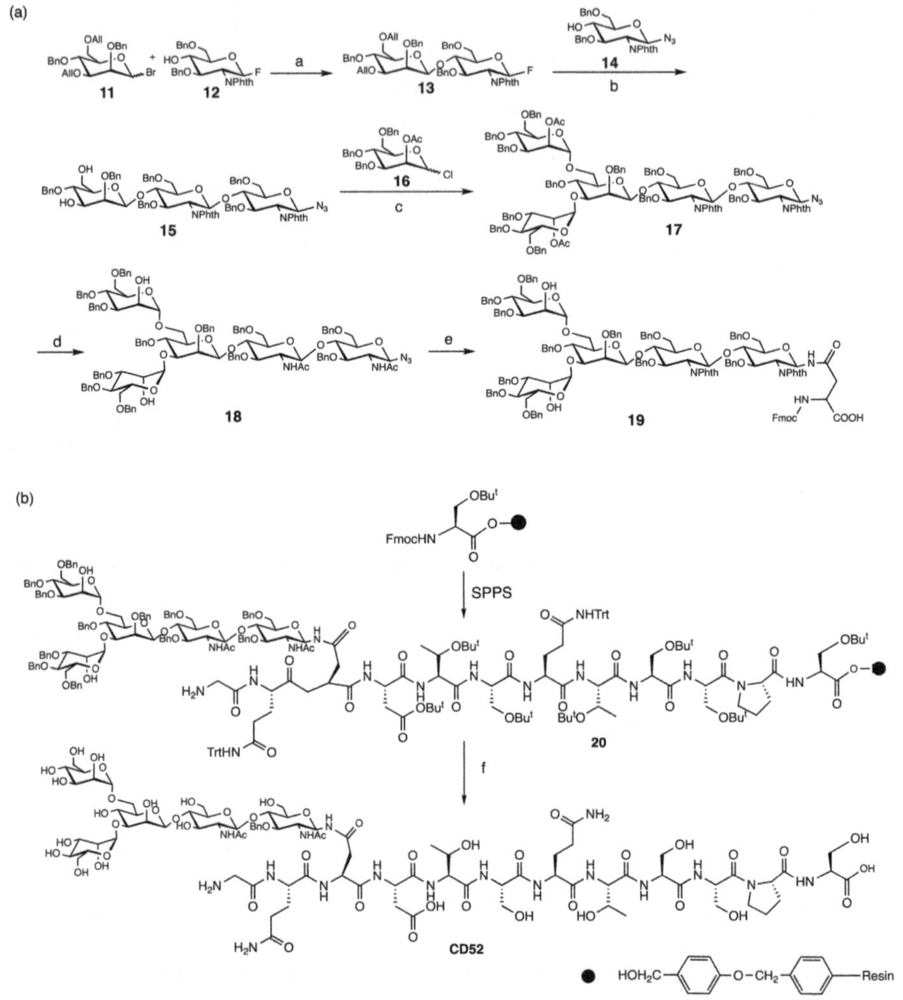

Figure 7.5 Synthesis scheme of glycan chain (a) and SPPS of CD52 (b). Reagents and conditions, (a): silver alumina-silicate, (b): 1. Cp_2HfCl_2, $AgClO_4$, 2. **14**, 3 $[Ir(COD)(PMePh_2)_2]PF_6$, 4. $HgO/HgCl_2$; (c): 1. AgOTf, 2. **16**, (d): 1. $NH_2CH_2CH_2NH_2$ (en), *n*-BuOH, 2. Ac_2O/MeOH, (e): 1. Lindlar's Catalyst, H_2/MeOH; 2. Fmoc-Asp-OBu^t, 3. TFA : DCM (1 : 2); (f): 1. 95% aq. TFA, 2.5% $HSCH_2CH_2SH$; 2. 20% $Pd(OH)_2$/C, H_2, EtOH, H_2O.

of ethylenediamine (en) followed by subsequent allylation with Ac_2O/MeOH yielding **18**. During this course of the reactions, the anomeric azide did not undergo any reaction. It was then reduced using Lindler's catalyst under a hydrogen atmosphere and conjugated to Fmoc-Asp-OBu^t. Finally, the product was treated with TFA : DCM (1 : 2) to yield **19**.

Fmoc-SPPS was employed for building the desired peptide segment CD52 (Figure 7.5(b)). During the SPPS cycle, the glycosylated amino acid **19** was manually introduced followed by SPPS elongation to obtain **20**. Final deprotection was then carried out using TFA. It was followed by cleavage of the benzyl groups of the glycan chain by hydrogenation using 20% $Pd(OH)_2/C$ as a catalyst to efficiently synthesize CD52.[14]

As application of the SPPS method is usually limited to around 50 amino acid residues, due to the increasing by-product as the chain length becomes longer, it poses a major challenge for the synthesis of glycoproteins with larger polypeptide chains ($n \geq 100$). Thus, more practical strategies are needed for the effective synthesis of larger and more complex glycopeptides.

7.3 Ligation Method

7.3.1 Basic Strategies

To efficiently obtain a longer polypeptide, methods to chemoselectively condense the peptide segments have been developed and are called ligation methods. The basic requirement of the ligation method is the presence of mutually reactive moieties. At present, the native chemical ligation (NCL) method,[15] developed by Kent, is mainly used for this purpose (Figure 7.6(a)). The method uses the chemoselective reaction between the thioester group of one peptide with the cysteine residue at the N-terminus of the other peptide. The ligation starts by the attack of the carbonyl carbon of the peptidyl thioester by the thiol group of the terminal cysteine. The thiolate ion departs, and as a result, the peptide segments are joined together by a thioester linkage. Following this, the intramolecular *S*-to-*N* acyl shift takes place leading to the formation of the desired peptide bond (Figure 7.6(a)). This high chemoselectivity makes the NCL reaction protection-free, which leads to the simple preparation of the peptide segments. However, the lower abundance of the cysteine limits the applicability of the method.

Therefore, to overcome this limitation, Dawson *et al.* exchanged the *N*-terminal alanine with cysteine making it capable of ligation. Once the ligation is complete, the cysteine residue is then subjected to desulfurization for conversion to the alanine residue (Figure 7.6(a)(i)). This approach widened the applicability of the ligation method to alanine as a ligation site.[16] This thiol modification of amino acids followed by desulfurization was later applied to other amino acids like Phe,[17] Thr,[18] Val,[19,20] *etc.*, and proved to be successful. However, the issue of using RANEY® Ni was that it tends to adsorb the polypeptides, especially in the case of the larger ones. This problem was later overcome by free radical desulfurization in a homogeneous solution (Figure 7.6(a)(ii)) developed by Danishefsky.[21]

Another ligation method, called the thioester method, was developed by Hojo and Aimoto in 1991. The method was designed based on the Boc-derived SPPS to obtain the peptide thioester. After all the free amino groups were protected, the Ag(I) selectively activates the thioester group (Figure 7.6(b)) which facilitates the nucleophilic attack of the free amino group of the other polypeptide chain leading to the condensation reaction of two peptide segments. After the ligation, the protecting groups are removed to obtain the desired

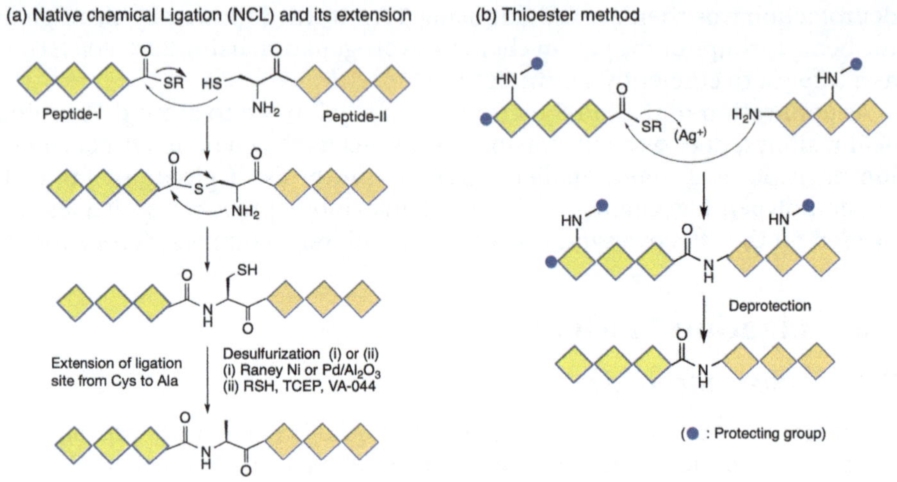

Figure 7.6 (a) Mechanistic illustration of the native chemical ligation method and its extension to the alanyl site followed by (i) Dawson's and (ii) Danishefsky's desulfurization; (b) mechanistic representation of the thioester method.

polypeptide.[22] The efficiency of the method was demonstrated by the synthesis of the DNA binding domain of the c-Myb protein (142–193) which is a 52 amino acid long polypeptide chain. It was found that if the more reactive aryl thioester is used, the ligation proceeds even in the absence of Ag^+ (Ag^+-free thioester method).[23]

7.3.2 Glycopeptide Thioester Synthesis

As already discussed, the basic requirement for ligation is the presence of mutually-reactive moieties. The most common reacting moiety used, which is prone to undergo nucleophilic attack, is the peptide thioester. However, in the case of glycopeptide thioesters, the following additional aspects need to be considered. The glycosidic linkages in the glycoprotein are acid-sensitive, except for the amide bond between the reducing end GlcNAc and Asn. Thus, the Boc method, which uses HF as the final deprotection reagent, is difficult to use for the synthesis of the glycosylated peptide thioester. Adding to it, the Fmoc-SPPS employs piperidine for periodic deprotection of the Fmoc group which cleaves the thioester linkage.

To overome these limitations, Aimoto's group used the less nucleophilic cocktail base composed of 2% hexamethyleneimine, 2% HOBt, and 25% 1-methylpyrrolidine in NMP/DMSO (1 : 1) for the Fmoc deprotection and successfully produced the thioester of a 25-mer peptide segment.[24] Furthermore, Guo's group also used DBU/HOBt (0.88/1) which efficiently cleaved Fmoc but some of the thioester linkage was also cleaved in this method.[25]

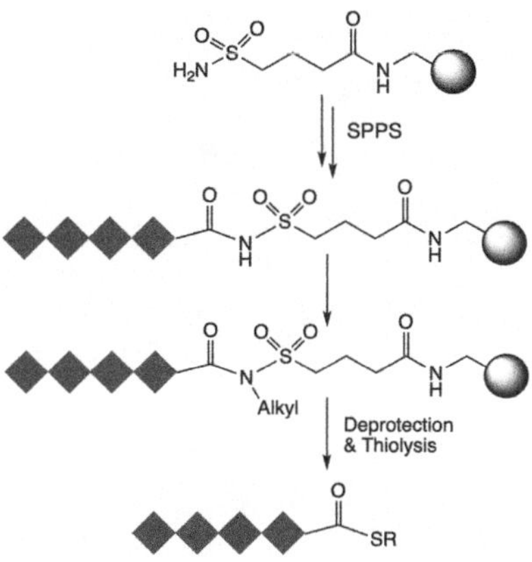

Figure 7.7 Peptide thioesterification using sulfonamide safety catch linker.

Another approach is the use of the sulfonamide linker (Figure 7.7).[26] In this method, the amino acid is conjugated to the sulfonamide linker and the peptide segment is elongated by the Fmoc-SPPS. The peptidyl sulfonamide bond is then alkylated at the nitrogen atom. This makes the carbonyl group of the C-terminal amino acid more prone to undergo nucleophilic attack by thiols to facilitate the peptide thioester synthesis. The drawback of this method is that peptidyl sulfonamide can undergo a number of side reactions during alkylation. The sulfonamide linker has been modified to a double linker strategy in which the additional linker is orthogonally stable and helps in easy release of the linker. The double linker strategy has been used to efficiently synthesize RNase C (1–124) as well as RNase B (30–68).[27,28]

In order to avoid the stepwise thioester cleavage, another approach can be generation of the thioester after synthesis of the entire polypeptide chain. Our group demonstrated the thioesterification after the peptide elongation using the *N*-alkylcysteine (NAC method) as a thioesterification device. In this method, the Fmoc-amino acid is elongated as depicted in Figure 7.8. After the formation of the desired peptide chain, the peptide was deprotected and the NAC moiety undergoes an *N*-to-*S* acyl shift followed by nucleophilic substitution by the appropriate thiols to obtain the peptide thioester.[29]

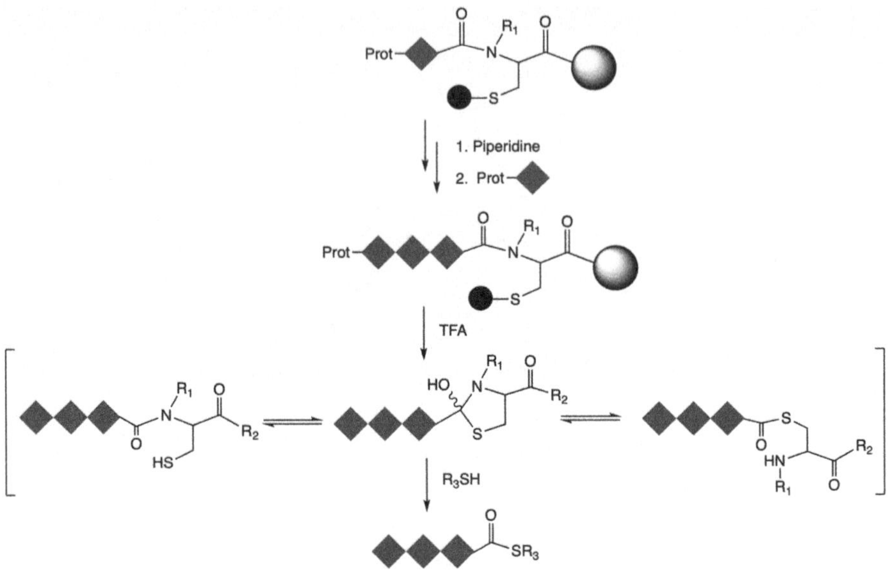

Figure 7.8 Schematic diagram of NAC method for the synthesis of peptide thioester by the Fmoc method.

The peptide thioester synthesis by the Fmoc method can now be made possible by various methods, which have been summarized by many authors.[30–32] Due to the development of these methods, reports about the synthesis of glycoproteins are increasing. The following sections depict the application of the solid phase synthesis in combination with ligation methods for the glycoprotein synthesis.

7.4 Synthesis of *O*-linked Glycoproteins

7.4.1 Synthesis of Antifreeze Glycoprotein

Antifreeze glycoproteins (AFGPs) have been known to be present in polar fish for sustaining sub-zero temperatures as a measure for adaptation in extreme cold environments. These AFGPs interact with the ice surface which leads to a decrease in the freezing point even up to 200–300 fold as compared to normally-used antifreeze agents.[33] It was revealed through AFGP gene characterization that it has *O*-linked sugars at the Thr residues. The typical AFGP structure consists of an iterative $(\text{Ala-Ala-Thr}^{G})_n$ sequence, where n varies from 4 to 50 and Thr^{G} is Gal-β(1–3)-GalNAcα-Thr.[34]

An efficient synthesis of the AFGP analogue was carried out by Nishimura's group as an improvement over their previous attempt at synthesizing AFGPs.[35] For this, the synthesis was initiated with the acetylated sugar **21** which was converted into its fluoride analogue **22** using the

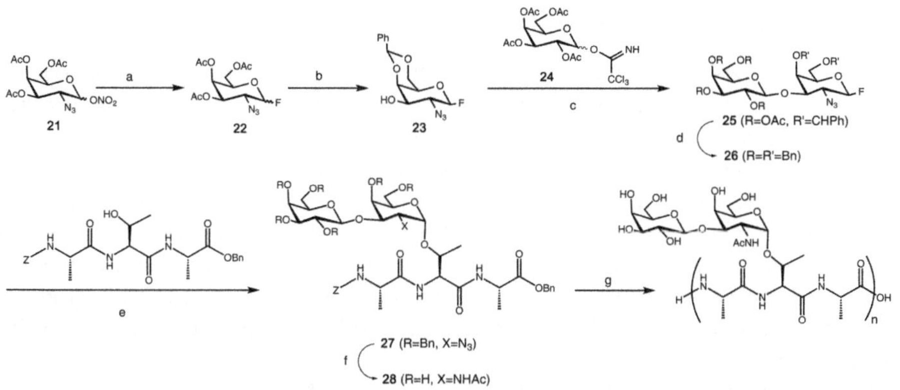

Figure 7.9 Schematic diagram for the synthesis of AFGP analogues; reagents and conditions,(a): $Et_3N/3HF$, NEt_3, CH_3CN, (b): 1. NaOMe, MeOH 2. $PhCH(OMe)_2$, CSA (c):TMSOTf (d): 1. NaOMe, MeOH 2. CSA, MeOH, 3. BnBr, NaH, (e): Cp_2ZrCl_2, $AgClO_4$ (f): 1. AcSH, pyridine 2. H_2, Pd/C, DMF, H_2O, AcOH, (g): 1. DPPA, Et_3N, DMF 2. 25 mM NaOH.

$Et_3N/3HF$ complex (Figure 7.9). This method also generated a hydroxy analogue which was converted to the desired fluoride **22** using (diethylamino) sulfur trifluoride (DAST). The β-isomer was then subjected to deacetylation followed by benzylidenation which resulted in the formation of **23**. The free hydroxy group of **23** was then glycosylated by the galactosyl imidate **24** to obtain the disaccharide **25**. The acetyl protections were then changed to benzyl protections (**25** to **26**) by first modifying them to the corresponding hydroxy analogues and then reacting them with benzyl bromide and sodium hydride.

The benzyl protected glycosyl donor **26** was next reacted with Z-Ala-Thr-Ala-OBn (*Z* = benzyloxycarbonyl) in the presence of $Cp_2ZrCl_2/AgClO_4$ (1 : 2) to yield the glycosylated tripeptide **27**. Thereafter, **27** was treated with AcSH in pyridine for conversion of the azido group to the actamido group. The following reactions included removal of all the protecting groups to generate the unprotected glycopeptide **28**. Finally, **28** was activated using diphenylphosphoryl azide (DPPA) to synthesize the proposed AFGP in quantitative yield as indicated by ^{1}H and ^{13}C NMR. It may be noted that the tripeptide sequence used for the synthesis of (Ala-Ala-Thr^G)$_n$ is Ala-Thr-Ala instead of Ala-Ala-Thr to avoid steric hindrance during the polymerization reaction.[36]

The antifreeze activity of the synthesized analogue was analyzed in which the bipyramidal structures clearly depicted the inhibition of ice crystals. These synthetic AFGP analogues were successfully used for the cryopreservation of islet cells and it was observed that these analogues elevate the reproducibility of islet cells from 63% to 85%.[37]

7.4.2 Synthesis of MUC2 Model

Mucins are typical *O*-glycosylated proteins. They are surface glycoproteins of epithelial cells densely clustered over a small region with characteristic tandem repeat units in it.[38] The difference in the carbohydrate structure on the cell-surface depending on the cells helps the pharmaceutical design to achieve the targeted drug delivery as well as in making biomarkers.[39] The structure elucidation of mucins is difficult due to the repeating units and the complexity of the glycan unit. This is one of the major motivating factors for the chemical synthesis of homogeneous glycopeptides as the biological activity can be later correlated with the natural counterpart to ascertain the exact structure.

Thus, the efficient convergent synthesis of MUC2 was performed which involves the challenging task of synthesizing the 23mer MUC2 unit carrying seven GalNAcs (**32**, Figure 7.10). Among the GalNAcs in the sequence, five of them are consecutively arranged making the coupling in SPPS more difficult due to the steric factors. The synthesis was initiated from Fmoc-Pro-Thr(Bn)-Gly-SCH_2CH_2CONH-resin (NH_2-resin: CLEAR amide-resin) by the Fmoc method using the less nucleophilic deFmoc reagent developed by Aimoto's group to obtain **30** (Figure 7.10). The benzyl protected

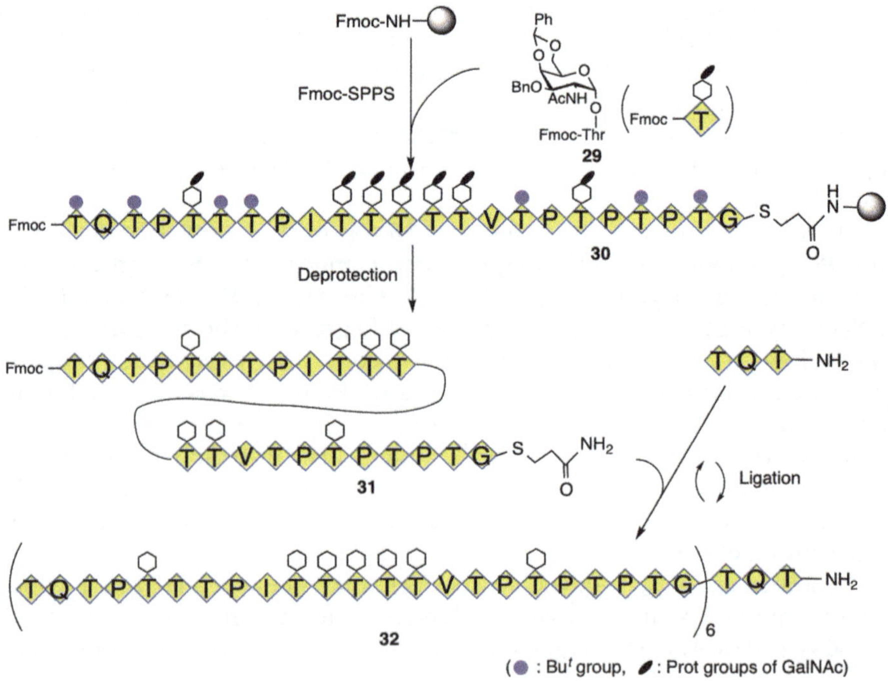

Figure 7.10 Schematic representation for synthesis of MUC2 using Fmoc-based thioester method.

α-GalNAc-Thr was introduced using **29**. Furthermore, the deprotection of **30** was performed using TFA followed by low TfOH to get **31**. The crude **31** was purified by RP-HPLC in 0.1% TFA in aqueous acetonitrile, followed by 20 mM ammonium acetate in aqueous acetonitrile. The ligation was then repeatedly performed using **31** to TQT-NH_2, followed by purification using RP-HPLC leading to the synthesis of the pure 23 kDa glycoprotein with 141 amino acid residues (**32**). The method provides the freedom to design the heterogeneously glycosylated peptide thioesters to synthesize the mucin model with a heterogeneous glycan unit over the tandem repeating peptide backbone. The advent of such designed glycoproteins will not only simulate the real mucin surface, but can also be used for the structural and functional analyses of mucins.[40]

7.4.3 Synthesis of Interleukin-2

The ligation method was also used for the synthesis of the *O*-glycosylated cytokine, *i.e.*, the human interleukin-2 (IL-2).[41] IL-2 consists of a polypeptide chain comprised of 133 amino acids in which the synthetic design with three ligation sites by the thioester method leads to four polypeptide segments: (1–27), (28–65), (66–98), and (99–133). Despite its globular structure, the C-terminal segment of IL-2 was highly insoluble. To overcome the solubility issue, two isopeptide structures,[42] which are reported to increase the solubility of the peptide, were introduced at Ala^{112}-Thr^{113} and Ile^{122}-Thr^{123}. Although it led to a partial increase in the solubility, the preparative purification was difficult. As a solution to this problem, the negatively-charged carboxylate of Glu was converted to the positively charged picolyl ester, reversing the polarity to efficiently increase the isoelectric point of the peptide.[43] Thus, the synthesis of the (99–133) segment employed Fmoc-SPPS in which Glu was introduced as Fmoc-Glu(OPic)-OH along with the introduction of the isodipeptide unit at Ala^{112}-Thr^{113} and Ile^{122}-Thr^{123}. As a result, the solubility of the peptide was highly increased and the desired peptide was easily purified by RP-HPLC in a 0.1% TFA solution of aqueous ACN.

The core 1 *O*-glycosylated sugar attached to Thr^3 in the (1–27) segment was synthesized using 4-methylbenzyl (MBn) protection.[44] Although the low TfOH treatment is efficient for removal of the benzyl group, partial decomposition of the glycosidic linkages is observed as the glycan chain becomes longer.[45] Therefore, we examined the use of MBn protection, which is more sensitive to acid. The synthesis was initiated from Fmoc-Gly-(Et)Cys(Trt)-[Lys(Boc)]$_2$-NH-resin (NH_2-resin: Rink amide MBHA-resin). The peptide chain was elongated through the automated Fmoc-SPPS except for the Lys(*i*Noc) (*i*Noc: 4-pyridylmethoxycarbonyl) and the glycosylated Thr residue, which were manually introduced. The obtained peptide resin was treated with TFA, which resulted in the complete deprotection of both the peptide and carbohydrate moiety. The crude peptide was then thioesterified using 3-hydroxythiophenol.

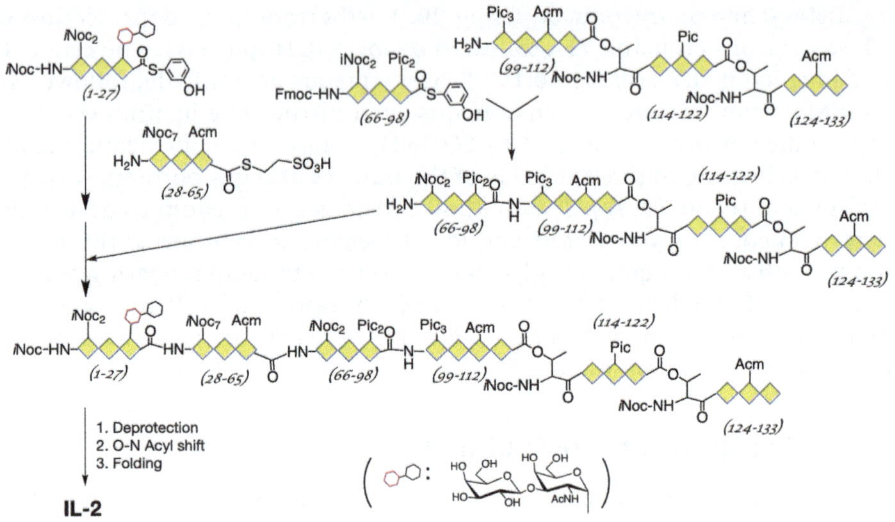

Figure 7.11 Synthetic scheme of IL-2.

After the synthesis of all four peptide segments, the Ag^+-free thioester method was applied for (66–98) and (99–133) to obtain the (66–133) segment. Based on the relative reactivity of the aryl and alkyl thioesters, Ag^+-free ligation was then performed for the aryl thioester of (1–27) and alkyl thioester of (28–65) to produce segment (1–65). Without purification, the (66–133) segment and Ag^+ was added to perform the Ag^+-induced condensation to obtain the whole segment (1–133). The Acm groups were then removed by $AgNO_3$ treatment in acetic acid, followed by *i*Noc and picolyl ester removal by Zn treatment in aqueous 3-mercaptopropionic acid. Finally, the isopeptide units were converted into native amino acids, and the oxidative folding reaction was performed to obtain the final IL-2 (1–133) segment (Figure 7.11). The synthesized IL-2 exhibited an α-helical structure which is similar to the non-glycosylated recombinant IL-2. The $CD4^+$ T cells were well proliferated with both the synthesized and recombinant IL-2 in the presence of the anti-CD-3 antibody along with production of IFN-γ which depicts the successful synthesis of the IL-2.

7.5 Synthesis of *N*-linked Glycoprotein

7.5.1 Strategies for *N*-linked Glycoprotein Synthesis

In contrast to the *O*-linked glycoproteins, for the synthesis of the *N*-linked glycoproteins, the glycan chain can be made available from natural glycoproteins, like fetuin and ribonuclease B, by mild hydrazinolysis. The sugars were then converted to glycosylamines which were then made to react with Fmoc-Asp(ODhbt)-OBu^t to produce glycosylated Asn derivatives. The

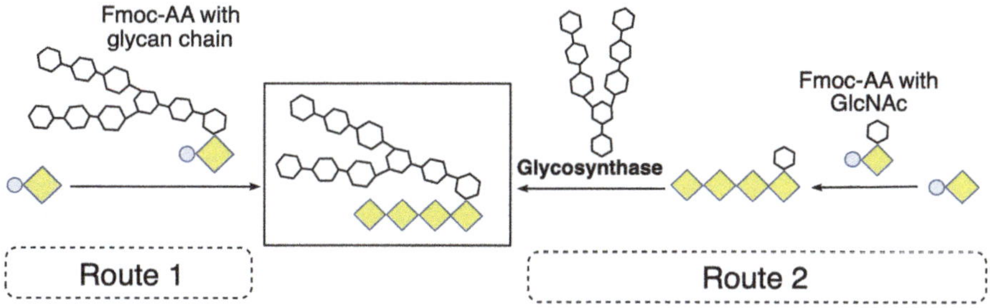

Figure 7.12 Scheme illustrating synthetic design of *N*-linked glycoprotein in SPPS.

obtained glycosylated Asn unit was successfully used for synthesis of the *N*-linked glycoprotein.[46] However, the more practical method for use of the natural *N*-linked sugar for glycoprotein synthesis was developed by Inazu *et al.* who successfully isolated the glycosylated Asn unit by the digestion of the glycoprotein obtained from egg yolk.[47] Kajihara *et al.* extended the method to prepare Fmoc-Asn units carrying various carbohydrate structures,[48] which was used for the synthesis of glycoproteins.

In addition, the enzymatic glycosyl chain transfer reactions have been carried out for the glycopeptide carrying the reducing end GlcNAc using endoglycosidases, such as endo-M.[49] This leads to two synthetic possibilities for the *N*-linked glycoproteins. First, the carbohydrate group can be added to the peptide chain by introducing a glycosylated amino acid during SPPS (Figure 7.12, Route 1). Alternatively, glycosynthase can be employed for transfer of the entire glycan chain to the GlcNAc on the peptide chain (Figure 7.12, Route 2).[50–52]

7.5.2 Synthesis of Human Interferon-β

Human interferon-β (IFN-β) is a cytokine, which is composed of 166 amino acid residues glycosylated at Asn^{80}, and exhibits immune surveillance, antiviral, and antitumor properties as well as a host cell defense. For the IFN-β's *in-vivo* antitumor activity, the oligosaccharide chain plays the major role. To obtain the IFN-β with homogeneous carbohydrate, its chemical synthesis was demonstrated.[53]

IFN-β was synthesized in disialylated and non-sialylated forms by the NCL method as shown in Figure 7.13. The ligation sites were $Cys^{68,89}$ which were used as precursors of the alanine residues. After the successive native chemical ligation of (Thz^{68}-Leu^{88}), (Cys^{89}-Asn^{166}) and (Met^{1}-Phe^{67}), metal-free desulfurization[21] was performed to retrieve the corresponding $Ala^{68,89}$ residues. Following this, the Acm groups were removed using silver acetate, and the benzyl ester of the oligosaccharide chain was subjected to basic hydrolysis. The polypeptide was then allowed to fold into a highly-helical secondary structure as depicted by the CD spectra.

The *in vitro* antitumor activity on the Daudi cell line (human Burkitt's lymphoma cell line) was checked with both sialyl and asialo IFN-β

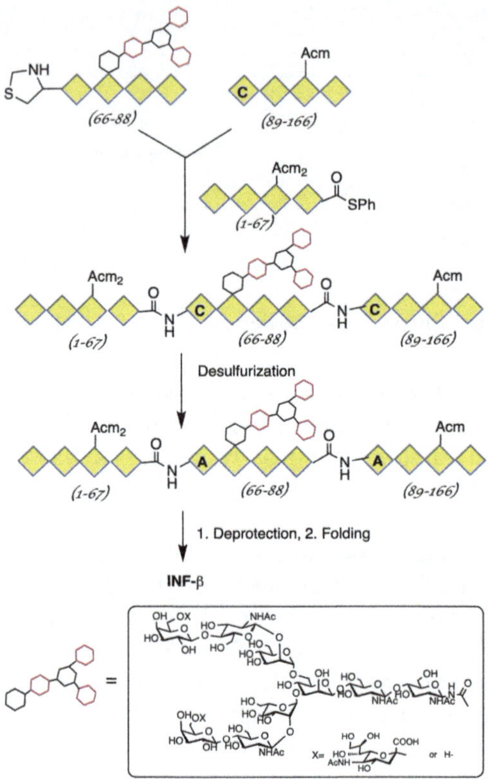

Figure 7.13 Synthetic scheme of human IFN-β.

showing similar activities. The sialyl IFN-β in tumor bearing mice completely suppressed the tumor growth in comparison to asialo IFN-β and the commercially-available IFN-β, as the asialo type is immediately metabolized by the liver. Thus, the synthesis becomes an appropriate application of SPPS for generating synthetic analogues of the biological motifs with an appreciable activity.

7.5.3 Synthesis of TIM-3

As an example of the synthesis of the *N*-linked glycoprotein *via* route 2 in Figure 7.12, synthesis of the TIM-3 Ig domain is discussed (Figure 7.14).[54] TIM-3 is a surface glycoprotein of T-cells, whose extracellular Ig domain binds to galectin-9 and promotes apoptosis which allows TIM-3 to play a significant role in immunotolerance. The synthetic scheme was planned by the thioester method to use Gly[31,68] as ligation sites splitting the peptide into three parts, (1–31), (32–68), (69–107) as depicted in Figure 7.14(a). The segment (1–31) aryl thioester was synthesized using the NAC method and the (32–68) thioester was synthesized using the Boc method.

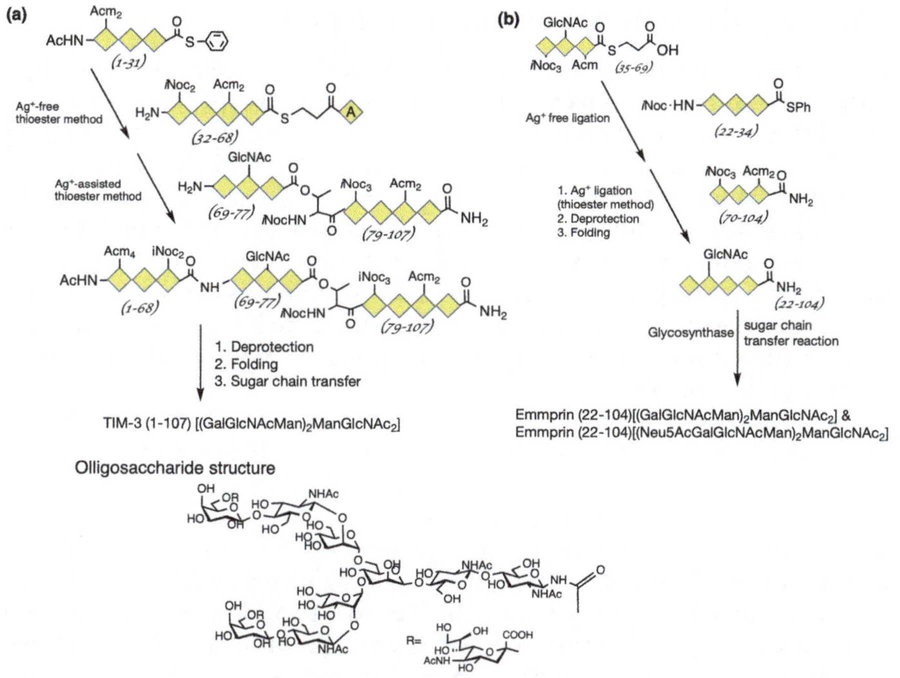

Figure 7.14 Synthetic scheme of Ig domain of TIM-3 (A) and emmprin using chemoenzymatic method (B).

Taking advantage of the higher reactivity of the aryl thioester in (1–31) segment, the ligation with the (32–68) segment was performed to chemoselectively produce the (1–68) thioester segment. Without purification, it was then reacted with GlcNAc carrying (69–107)-$CONH_2$ or GlcNAc carrying (69–107)-$CONH_2$ having the isopeptide structure at Val^{77}-Thr^{78} in one-pot. It is noted that the yield of the C-terminal segment without the isopeptide structure was four times lower than that of the C-terminal peptide with the isopeptide structure attributed to the role of the isopeptide units in increasing the solubility and thereby the respective yields.

The ligated *N*-acetylglucosaminylated H-(1–107)-$CONH_2$ with the isopeptide structure at Val^{77}-Thr^{78} was then subjected to deprotection of the Acm and *i*Noc groups. Thereafter, it was subjected to the *O*-to-*N* acyl shift and folding. After synthesis of the Ig domain of TIM-3 with the GlcNAc unit, the same was subjected to undergo glycosyl transfer with the synthetic sugar oxazoline[55] in the presence of glycosynthase which successfully yielded the final structure of the Ig domain of TIM-3(1–107). The anti-TIM-3 antibody test along with the β-sheet-rich structure inferred by the CD spectra confirmed proper folding of the synthetic glycoprotein. The synthetic glycoprotein can then be subjected to biological studies for a better understanding of the SARs with galectin-9 and the plausible apoptosis mechanism.

7.5.4 Synthesis of Ig Domain of Emmprin

The development of the chemoenzymatic method diversified the applicability of SPPS by conveniently synthesizing glycoprotein units with the GlcNAc unit and thereby transferring the glycan chain after the polypeptide assembly. Motivated by this, we resynthesized the Ig domain of emmprin which we previously synthesized by adding the glycosylated amino acid during SPPS.[56] Emmprin is a surface glycoprotein expressed on cancer cells and plays a vital role in tumor metastasis by stimulating fibroblasts to produce matrix metalloproteinases (MMPs). However, in the absence of the glycan unit, the MMPs activity becomes dormant. Also, as the number of glycan units increases, the MMP-production activity becomes stronger.

The peptide was divided into three segments, *i.e.*, (22–34), (35–69) and (70–104) (Figure 7.14(b)). For synthesis of the individual segments, the thioesters of (22–34) and (35–69) were prepared using the NAC method, and the benzyl protection strategy was used for protecting the GlcNAc. After the ligation of the *N*-terminal two segments by the Ag^+-ion free thioester method to obtain the (22–69) segment, the Ag^+-promoted second ligation was performed in one-pot yielding the (22–104) peptide segment which was isolated by the gel filtration method. Deprotection of the Acm and *i*Noc groups was then performed and the peptide segment was allowed to fold. The glycosynthase was used to introduce the undecasaccharide unit over the GlcNAc motif.[57]

7.6 Conclusion

Glycopeptides have been mainly prepared by the Fmoc-SPPS, which does not use strong acids during the synthesis. This is because the glycosidic linkages are labile to strong acids, such as anhydrous HF used as the final deprotection reagent in the Boc SPPS. To synthesize the larger glycopolypeptides or glycoproteins, the ligation methods, such as the native chemical ligation and thioester methods, have to be applied. However, the preparation of the key intermediate for these ligation methods, the peptide thioester, has been difficult by the Fmoc method, as the thioester linkage is sensitive to piperidine used for the Fmoc group removal. Due to the recent development of the methods for the peptide thioester synthesis by the Fmoc method, such as the NAC method, the glycosylated peptide thioesters have become easily available and have been used for the chemical or chemoenzymatic synthesis of various homogeneous *O*-linked and *N*-linked glycoproteins. The achieved glycoproteins significantly contribute to the structure elucidation, understanding the structure–activity relationship and therapeutics-based applications.

References

1. K. W. Moremen, M. Tiemeyer and A. V. Nairn, *Nat. Rev. Mol. Cell Biol.*, 2012, **13**, 448.
2. K. B. Lee, D. Loganathan, Z. M. Merchant and R. J. Linhardt, *Appl. Biochem. Biotechnol.*, 1990, **23**, 53.
3. D. P. Gamblin, E. M. Scanlan and B. G. Davis, *Chem. Rev.*, 2009, **109**, 131.
4. S. Shibata, T. Takeda and Y. Natori, *J. Biol. Chem.*, 1988, **263**, 12483.
5. F. Wieland, R. Heitzer and W. Schaefer, *Proc. Natl. Acad. Sci. U. S. A.*, 1983, **80**, 5470.
6. I. Brockhausen, H. Schachter and P. Stanley, O-GalNAc Glycans, in *Essentials of Glycobiology. 2nd Edition*, ed. A. Varki, R. D. Cummings and J. D. Esko, *et al.*, Cold Spring Harbor Laboratory Press, Cold Spring Harbor, NY, 2009, ch. 9.
7. P. Stanley, H. Schachter, N. Taniguchi and N. Glycans, in *Essentials of Glycobiology. 2nd Edition*, ed. A. Varki, R. D. Cummings and J. D. Esko, *et al.*, Cold Spring Harbor Laboratory Press, Cold Spring Harbor, NY, 2009, ch. 8.
8. R. G. Spiro, *Glycobiology*, 2002, **12**, 43R.
9. R. B. Merrifield, *J. Am. Chem. Soc.*, 1963, **85**, 2149.
10. T. I. A. Warhi, H. M. A. Hazimi and A. E. Faham, *J. Saudi Chem. Soc.*, 2012, **16**, 97.
11. R. S. Funosas, R. Prohens, R. Barbas, A. E. Faham and F. Albericio, *Chem. - Eur. J.*, 2009, **15**, 9394.
12. K. Baumann, D. Kowalczyk and H. Kunz, *Angew. Chem., Int. Ed.*, 2008, **47**, 3445.
13. J. P. Tam, W. F. Heath and R. B. Merrifield, *J. Am. Chem. Soc.*, 1986, **108**, 5242.
14. Z. W. Guo, Y. Nakahara, Y. Nakahara and T. Ogawa, *Bioorg. Med. Chem.*, 1997, **5**, 1917.
15. P. E. Dawson, T. W. Muir, I. Clark-Lewis and S. B. H. Kent, *Science*, 1994, **266**, 776.
16. L. Z. Yan and P. E. Dawson, *J. Am. Chem. Soc.*, 2001, **123**, 526.
17. D. Crich and A. Banerjee, *J. Am. Chem. Soc.*, 2007, **129**, 10064.
18. J. Chen, P. Wang, J. Zhu, Q. Wan and S. J. Danishefsky, *Tetrahedron*, 2010, **66**, 2277.
19. C. Haase, H. Rohde and O. Seitz, *Angew. Chem., Int. Ed.*, 2008, **47**, 6807.
20. J. Chen, Q. Wan, Y. Yuan, J. Zhu and S. J. Danishefsky, *Angew. Chem., Int. Ed.*, 2008, **47**, 8521.
21. Q. Wan and S. J. Danishefsky, *Angew. Chem., Int. Ed.*, 2007, **46**, 9248.
22. H. Hojo and S. Aimoto, *Bull. Chem. Soc. Jpn.*, 1991, **64**, 111.
23. H. Hojo, Y. Murasawa, H. Katayama, T. Ohira, Y. Nakahara and Y. Nakahara, *Org. Biomol. Chem.*, 2008, **6**, 1808.
24. X. Q. Li, T. Kawakami and S. Aimoto, *Tetrahedron Lett.*, 1998, **39**, 8669.

25. X. Z. Bu, G. Y. Xie, C. W. Law and Z. H. Guo, *Tetrahedron Lett.*, 2002, **43**, 2419.
26. B. J. Backes and J. A. Ellman, *J. Org. Chem.*, 1999, **64**, 2322.
27. C. Piontek, D. V. Silva, C. Heinlein, C. Pohner, S. Mezzato, P. Ring, A. Martin, F. X. Schmid and C. Unverzagt, *Angew. Chem., Int. Ed.*, 2009, **48**, 1941.
28. S. Mezzato, M. Schaffrath and C. Unverzagt, *Angew. Chem., Int. Ed.*, 2005, **44**, 1650.
29. H. Hojo, Y. Onuma, Y. Akimoto, Y. Nakahara and Y. Nakahara, *Tetrahedron Lett.*, 2007, **48**, 25.
30. J. Kang and D. Macmillan, *Org. Biomol. Chem.*, 2010, **8**, 1993.
31. F. Mende and O. Seitz, *Angew. Chem., Int. Ed.*, 2011, **50**, 1232.
32. L. Raibaut, N. Ollivier and O. Melnyk, *Chem. Soc. Rev.*, 2012, **41**, 7001.
33. A. Y. Avanov, *Mol. Biol.*, 1990, **24**, 473.
34. J. K. Bang, J. H. Lee, R. N. Murugan, S. G. Lee, H. Do, H. Y. Koh, H. E. Shim, H. C. Kim and H. J. Kim, *Mar. Drugs*, 2013, **11**, 2013.
35. T. Tsuda and S. I. Nishimura, *Chem. Commun.*, 1996, 2779.
36. Y. Tachibana, N. Matsubara, F. Nakajima, T. Tsuda, S. Tsuda, K. Monde and S. I. Nishimura, *Tetrahedron*, 2002, **58**, 10213.
37. S. Matsumoto, M. Matsusita, T. Morita, H. Kamachi, S. Tsukiyama, Y. Furukawa, S. Koshida, Y. Tachibana, S. I. Nishimura and S. Todo, *Cryobiology*, 2006, **52**, 90.
38. Y. S. Kim, J. Gum Jr and I. Brockhausen, *Glycoconjugate J.*, 1996, **13**, 693.
39. H. Kunz, *J. Pept. Sci.*, 2003, **9**, 563.
40. H. Hojo, Y. Matsumoto, Y. Nakahara, E. Ito, Y. Suzuki, M. Suzuki, A. Suzuki and Y. Nakahara, *J. Am. Chem. Soc.*, 2005, **127**, 13720.
41. Y. Asahina, S. Komiya, A. Ohagi, R. Fujimoto, H. Tamagaki, K. Katsuhiro Nakagawa, T. Sato, S. Akira, T. Takao, A. Ishii, Y. Nakahara and H. Hojo, *Angew. Chem., Int. Ed.*, 2015, **54**, 8226.
42. Y. Sohma, M. Sasaki, Y. Hayashi, T. Kimura and Y. Kiso, *Chem. Commun.*, 2004, 124.
43. J. Rizo, F. Albericio, G. Romero, C. G. Echeverria, J. Claret, C. Muller, E. Giralt and E. Pedroso, *J. Org. Chem.*, 1988, **53**, 5386.
44. Y. Asahina, R. Fujimoto, A. Suzuki and H. Hojo, *J. Carbohydr. Chem.*, 2015, **34**, 12.
45. A. Ueki, Y. Takano, A. Kobayashi, Y. Nakahara, H. Hojo and Y. Nakahara, *Tetrahedron*, 2010, **66**, 1742.
46. E. Meinjohanns, M. Meldal, H. Paulsen, R. A. Dwek and K. J. Bock, *J. Chem. Soc., Perkin Trans. 1*, 1998, 549.
47. T. Inazu, M. Mizuno, T. Yamazaki and K. Haneda, in *Peptide Science 1998*, ed. M. Kondo, Protein Research Foundation, Osaka, 1999, pp. 153–156.
48. Y. Kajihara, Y. Suzuki, N. Yamamoto, K. Sasaki, T. Sakakibara and L.R. Juneja, *Chem. -Eur. J.*, 2004, **10**, 971.
49. M. Mizuno, K. Haneda, R. Iguchi, I. Muramoto, T. Kawakami, S. Aimoto, K. Yamamoto and T. Inazu, *J. Am. Chem. Soc.*, 1999, **121**, 284.

50. M. Umekawa, W. Huang, B. Li, K. Fujita, H. Ashida, L. X. Wang and K. Yamamoto, *J. Biol. Chem.*, 2008, **283**, 4469.
51. W. Huang, C. Li, B. Li, M. Umekawa, K. Yamamoto, X. Zhang and L. X. Wang, *J. Am. Chem. Soc.*, 2009, **131**, 2214.
52. M. Umekawa, T. Higashiyama, Y. Koga, T. Tanaka, M. Noguchi, A. Kobayashi, S. Shoda, W. Huang, L. X. Wang, H. Ashida and K. Yamamoto, *Biochim. Biophys. Acta*, 2010, **1800**, 1203.
53. I. Sakamoto, K. Tezuka, K. Fukae, K. Ishii, K. Taduru, M. Maeda, M. Ouchi, K. Yoshida, Y. Nambu, J. Igarashi, N. Hayashi, T. Tsuji and Y. Kajihara, *J. Am. Chem. Soc.*, 2012, **134**, 5428.
54. Y. Asahina, S. Kamitori, T. Takao, N. Nishi and H. Hojo, *Angew. Chem., Int. Ed.*, 2013, **52**, 9733.
55. H. Hojo, H. Tanaka, M. Hagiwara, Y. Asahina, A. Ueki, H. Katayama, Y. Nakahara, A. Yoneshige, J. Matsuda, Y. Ito and Y. Nakahara, *J. Org. Chem.*, 2012, **77**, 9437.
56. H. Hojo, E. Haginoya, Y. Matsumoto, Y. Nakahara, K. Nabeshima, B. P. Toole and Y. Watanabe, *Tetrahedron Lett.*, 2003, **44**, 2961.
57. Y. Asahina, M. Kanda, H. Katayama, Y. Nakahara and H. Hojo, *Org. Biomol. Chem.*, 2013, **11**, 7199.

CHAPTER 8

Synthesis of Chondroitin Sulfate Oligosaccharides and Chondroitin Sulfate Glycopeptides

SHERIF RAMADAN[a,b], WEIZHUN YANG[a] AND XUEFEI HUANG*[a,c,d]

[a]Department of Chemistry, Michigan State University, 578 South Shaw Lane, East Lansing, MI 48824, USA; [b]Chemistry Department, Faculty of Science, Benha University, Benha, Qaliobiya 13518, Egypt; [c]Department of Biomedical Engineering, Michigan State University, East Lansing, MI 48824, USA; [d]Institute for Quantitative Health Science and Engineering, Michigan State University, East Lansing, MI 48824, USA

*E-mail: xuefei@chemistry.msu.edu, Website: https://www2.chemistry.msu.edu/faculty/huang/

8.1 Introduction to Chondroitin Sulfate (CS) and Chondroitin Sulfate Proteoglycan (CSPG)

Chondroitin sulfates (CSs) are linear anionic polysaccharides belonging to the glycosaminoglycan (GAG) family, a class of poly-anionic glycoconjugates.[1–3] CSs consist of a disaccharide repeating unit of D-glucuronic acid (GlcA) and 2-acetamido-2-deoxy-D-galactose (GalNAc) arranged in the sequence of

Chemical Biology No. 11
Synthetic Glycomes
Edited by Wanyi Guan, Lei Li and Peng George Wang

Published by the Royal Society of Chemistry, www.rsc.org

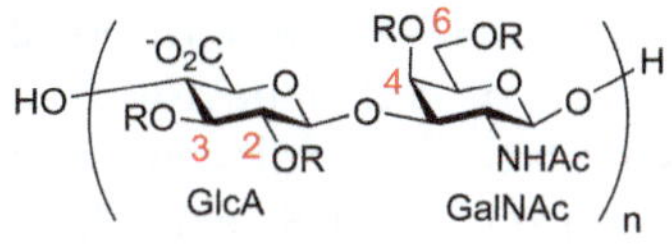

R= SO_3^- or H

A: GlcA-GalNAc($4SO_3^-$)
C: GlcA-GalNAc($6SO_3^-$)
D: GlcA($2SO_3^-$)-GalNAc($4SO_3^-$)
E: GlcA-GalNAc(4, $6SO_3^-$)
K: GlcA($3SO_3^-$)-GalNAc($4SO_3^-$)
L: GlcA($3SO_3^-$)-GalNAc($6SO_3^-$)
M: GlcA($3SO_3^-$)-GalNAc(4, $6SO_3^-$)
R: GlcA(2,$3SO_3^-$)-GalNAc

Figure 8.1 CS structures and major subtypes.

D-GlcA-β-(1–3)-D-GalNAc-β-(1–4). CS is often modified by *O*-sulfate groups on 4-*O* and 6-*O* of GalNAc and 2-*O* and 3-*O* of GlcA. The number and positions of the *O*-sulfates define the subtypes of CS (Figure 8.1). For instance, CS-A contains one sulfate group on 4-*O* of GalNAc, CS-C has one sulfate on 6-*O* of GalNAc, CS-D has 4-*O* on GalNAc and 2-*O* on GlcA sulfated, while CS-E has both 4-*O* and 6-*O* of GalNAc sulfated.

In nature, CS typically exists as part of a proteoglycan, with one or more CS chains covalently attached to serine residues in the core protein through a tetrasaccharide linker of D-GlcA-β-D-(1–3)-galactose(Gal)-β-D-(1–3)-Gal-β-D-(1–4)-xylose(Xyl). These types of macromolecules are referred to as CSPGs, which are widely expressed by animal cells in extracellular matrixes.

Many enzymes are involved in the biosynthesis of CSPGs, which occurs in the endoplasmic reticulum/Golgi apparatus.[4] The process is initiated by stepwise synthesis of the tetrasaccharide linkage region (common to all sulfated proteoglycans), starting by the addition of a Xyl moiety to a serine residue in the core protein through a xylosyl transferase. Subsequently, two Gal moieties are added by galactosyl transferases, followed by the addition of a GlcA residue by a glucuronyl transferase, completing the common tetrasaccharide linkage region. To the non-reducing end GlcA in the tetrasaccharide linkage region, the first GalNAc is transferred by the GalNAc transferase initiating the synthesis of chondroitin chain. The repeating disaccharide units of CS are polymerized by alternating addition of D-GlcA and D-GalNAc catalyzed by chondroitin sulfate synthases with both GlcA and GalNAc transferase activities[5] aided by a chondroitin polymerizing factor.[6] The sulfate groups are then introduced to the chondroitin backbone mainly at 2-*O* of GlcA and 4-*O* and/or 6-*O* of GalNAc residues by multiple sulfotransferases.[7] The sulfation reactions are often incomplete, which result in significant structural diversities of naturally existing CS/CSPG.

CSs and CSPGs have been implicated in many important events such as intercellular signaling,[2,8] regulation of cell behaviors,[9,10] development and repair of the central nervous system,[11–13] inflammation[14,15] and amyloid plaque formation.[16–18] Sulfation patterns of CSs can play important roles in directing biological functions. For example, while CS-A-A and CS-C-C tetrasaccharides did not affect neuron growth much, CS-E-E tetrasaccharide exhibited potent neuronal stimulations.[19] This was not due to the overall sulfation levels since another CS tetrasaccharide bearing the same number of *O*-sulfates as CS-E-E did not show much activity. These results supported the existence of sulfo code and highlighted the need for systematically varied structurally well-defined CS glycans.[20]

With the appreciation of their important biological functions, there is a great demand for CSs in homogeneous forms with defined structures. The high structural heterogeneity of naturally existing CSs and the difficulties in separating a significant amount from natural sources in pure form prompted many research groups to synthesize CSs for further biological and medicinal studies. Here in this chapter, we first review the chemical synthesis of CSs by focusing on the critical protective group strategies as well as methods to expedite building block preparation. This is followed by discussions on chemoenzymatic synthesis of CS oligosaccharides, the synthesis of CS conjugates, as well as the preparation of homogenous CS glycopeptides.

8.2 Chemical Synthesis of CS

8.2.1 General Synthetic Design for Chemical Synthesis of CS

Generation of CS oligosaccharides with various sulfate patterns has been pursued by multiple groups. For comprehensive discussions of all CS oligosaccharide syntheses, readers can refer to several excellent reviews.[21–23] Herein, we focus on some key aspects of the synthesis. Chondroitin backbones consist of repeating GlcA-GalNAc disaccharides that all are β-linked to each other. In chemical synthesis of these molecules, many protecting groups have been used to facilitate the β-stereoselectivities for both GalNAc and GlcA linkages. For GlcA building blocks, their 2-*O*-positions were typically decorated with a protective group such as benzoyl (Bz) capable of neighboring group participation. The amino group at 2-*N* position on GalNAc was masked either as azido group, which could be easily transformed into *N*-acetyl group under mild condition, or as a protective group capable of neighboring group participation. Another important issue in CS synthesis design was the differentiation of the hydroxyl groups to be sulfated, which include 4-*O* and/or 6-*O* on GalNAc and 2-*O* and/or 3-*O* on GlcA. Furthermore, methods are needed to expedite the preparation of building blocks for CS synthesis. These factors are reviewed and discussed in the following sections.

8.2.2 Strategies Addressing *N*-protective Groups

8.2.2.1 Azide Protective Group

In 1995, the Tamura group reported the synthesis of CS oligosaccharides with the azide bearing disaccharide **1** as a key intermediate (Scheme 8.1).[24] **1** was functionalized with chemoselectively cleavable protecting groups, *p*-methoxyphenyl (MP) and levulinoyl (Lev) groups at the reducing and the non-reducing end respectively. The MP group in **1** was converted to the trichloroacetimidate donor **2**, while the Lev group was cleaved to give acceptor **3**. Coupling between **2** and **3** afforded the tetrasaccharide **4** in 50% yield. Disaccharide **3** and tetrasaccharide **4** were subjected to azide reduction to acetamides. Subsequently, selective benzylidene opening,

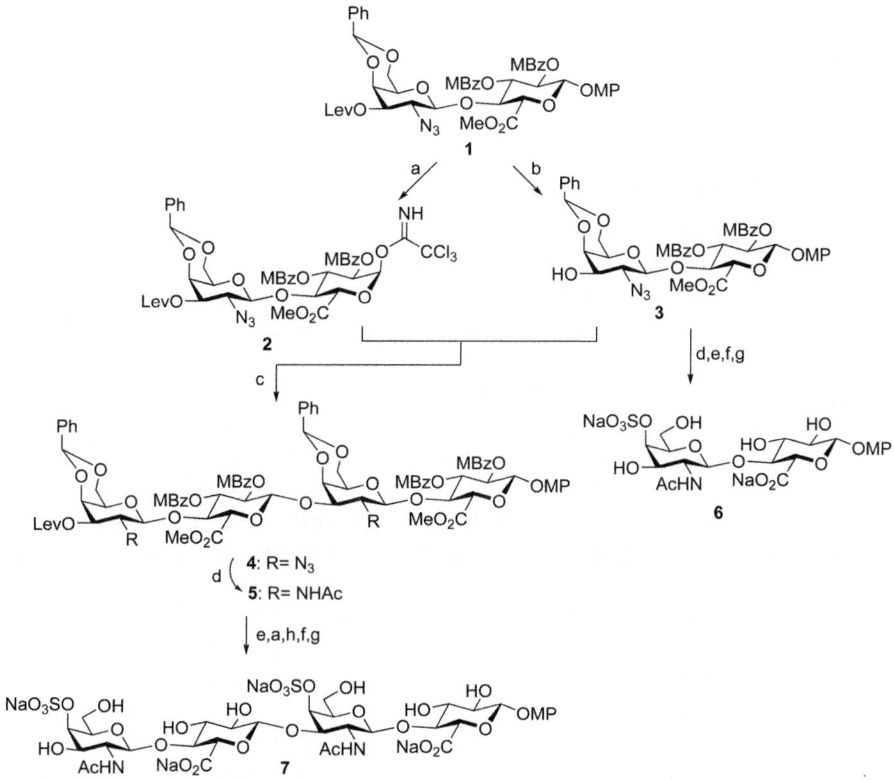

Scheme 8.1 Synthesis of CS-A di- and tetra-saccharide: (a) (i) cerium ammonium nitrate; (ii) CCl_3CN, DBU, DCM, 94%; (b) NH_2NH_2·AcOH, toluene/EtOH, quantitative; (c) $BF_3 \cdot Et_2O$ (1 equiv.), toluene, −20 °C, 50%; (d) AcSH, pyridine, 75%; (e) $NaBH_3CN$, THF, ethereal HCl, 3 Å MS, 87% for disaccharide, 51% for tetrasaccharide; (f) SO_3·TMA, DMF, 50 °C, quantitative for disaccharide and 92% for tetrasaccharide; (g) (i) 1 M LiOH, THF/H_2O; (ii) NaOMe/MeOH; (iii) H_2, $Pd(OH)_2$, H_2O, 82% for disaccharide and 70% for tetrasaccharide; (h) Ac_2O, pyridine, DMAP.

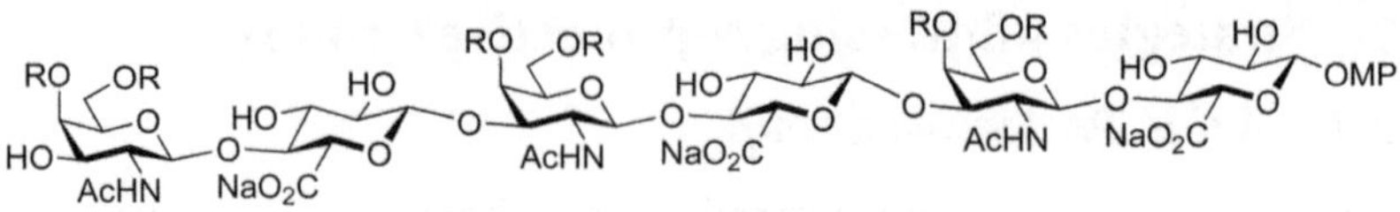

Figure 8.2 Hexasaccharide **9** and CS-E hexasaccharide **10**.

sulfation of the resulting free 4-OH groups followed by basic hydrolysis of ester groups and hydrogenolysis gave CS disaccharide **6** and tetrasaccharide **7** (Scheme 8.1).

Using the same strategy, Tamura and Tokuyoshi[25] attempted to synthesize hexasaccharide **9** (Figure 8.2) from building block **1**. However, they observed a significant decrease of the yield (<10%) for reducing the three azides in the protected hexasaccharide to acetamides, rendering it unsuitable for the synthesis of longer CSs.

8.2.2.2 N-*Acetamide Group*

With the difficulties in azide reduction in longer CS precursors, the *N*-acetamide group was explored (Scheme 8.2).[26] This was attractive as it would not require further synthetic manipulations to adjust *N*-protective groups. The glycosylation between acetamide containing building blocks **11** and **12** proceeded smoothly, which followed by Lev removal and a second glycosylation with **11** afforded the hexasaccharide **14** in good yield. Hexasaccharide **14** was deprotected giving unsulfated chondroitin hexasaccharide **9**. For CS-E hexasaccharide, after the benzylidene hydrolysis from **14**, the resulting diols were subjected to sulfation, followed by deprotection to produce CS-E **10** in good overall yield. In a similar manner, CS-E octasaccharide **15**[26] was synthesized (Figure 8.3), which was the longest CS oligosaccharide chemically assembled to date.

During the reaction of **11** and **12**, an imidate side product **16** was isolated from the glycosylation reaction mixture (Figure 8.4). This presumably resulted from nucleophilic attack of the oxygen atom of acetamide on the activated donor. This side reaction is not uncommon in glycosylations of acetamide containing building blocks.[27] While the imidate side product could be transformed to the desired glycoside, the conversion yields were generally not very high, thus hindering the wide application of acetamide in CS synthesis.

8.2.2.3 N-*Trichloroacetamide Group*

The Jacquinet group introduced the *N*-trichloroacetamido (TCA) group as a powerful stereocontrolling auxiliary in CS synthesis.[28] The TCA moiety could be transformed into *N*-acetamide through tributyltin hydride (TBTH)

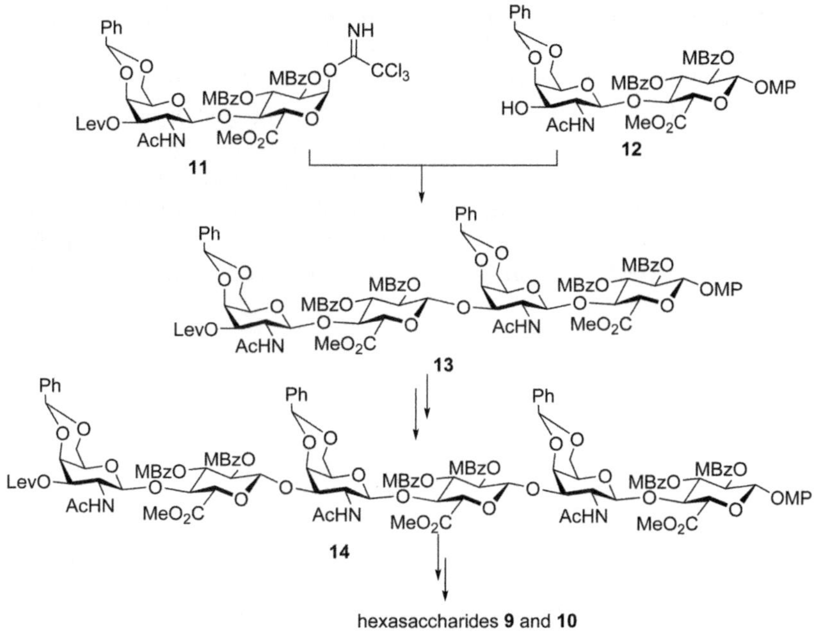

Scheme 8.2 Synthesis of CS-E hexasaccharide *via N*-acetamide-containing building blocks.

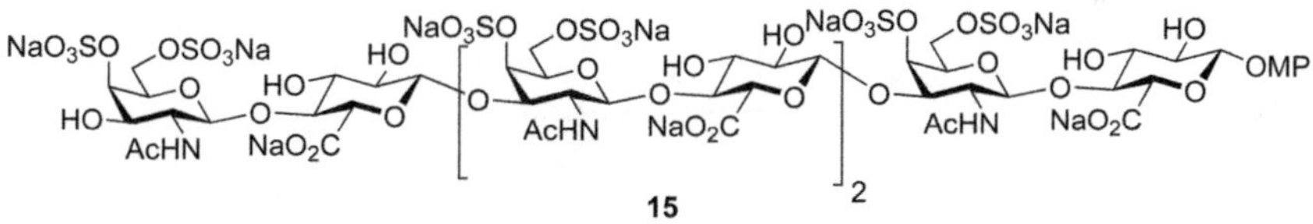

Figure 8.3 CS-E octasaccharide **15**.

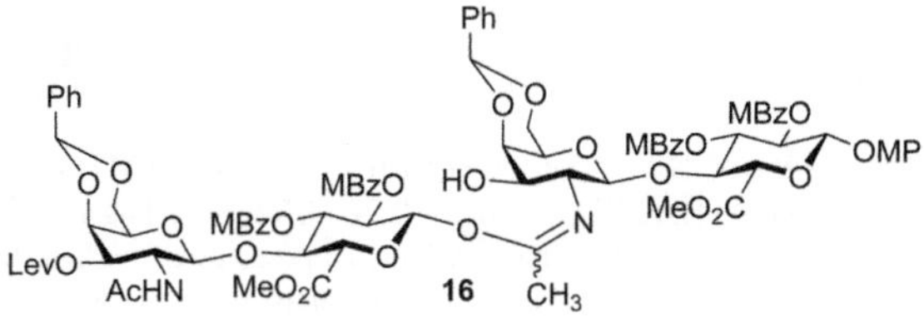

Figure 8.4 Imidate side product **16**.

reduction. The TCA group has been widely used for successful syntheses of many CS oligosaccharides.[12,19,29–32] As an example, Hsieh-Wilson's group reported a convergent strategy for the assembly of three common CS structures, *i.e.*, tetrasulfated CS-E **17**, disulfated CS-A **18** and CS-C **19**, in addition to the rare tetrasulfated CS-R **20** bearing sulfate groups at 2-*O* and 3-*O* of GlcA (Scheme 8.3).[12,33] These CS structures were generated from a core disaccharide **21**, where *N*-TCA on GalN and *O*-Bz at 2-*O* of GlcA were chosen to control the β-stereoselectivity during glycosylation. The key disaccharide **21** was obtained from donor **22** and acceptor **23**, which was then converted to donor **24** after transforming the allyl protecting group to trichloroacetimidate. Acceptor **25** was obtained from **21** *via* selective cleavage of the silyl protecting group. Coupling between **24** and **25** afforded the tetrasaccharide **26**, which underwent radical reduction of the *N*-TCA to

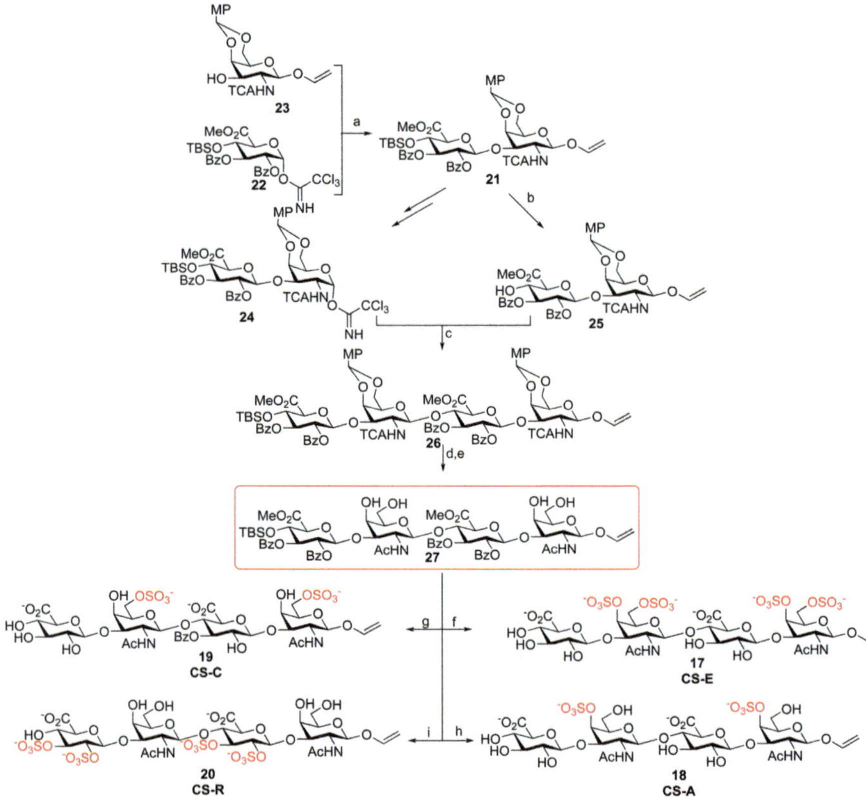

Scheme 8.3 Synthesis of CS-E, CS-C, CS-A and CS-R: (a) TMSOTf, DCM, −40 °C → −15 °C, 77%; (b) HF·pyridine, 0 °C, 85%; (c) TMSOTf, DCM, −75 °C → −20 °C, 44%; (d) Bu_3SnH, AIBN, benzene, 80 °C, 92%; (e) DDQ, H_2O, CH_3CN, 93%; (f) (i) SO_3·TMA (50 equiv.), DMF, 50 °C, 84%; (ii) HF·pyridine, 0 °C, 91%; (iii) 1 M LiOH, 30%, H_2O_2, 0 °C; (iv) NaOH, MeOH, H_2O, 70% (2 Steps); (g) (i) SO_3·TMA (5 equiv.), DMF, 50 °C, 61%; (ii) HF·pyridine, 0 °C, 89%; (iii) 1 M LiOH, 30%, H_2O_2, 0 °C; (iv) NaOH, MeOH, H_2O, 70%,Quant.; (h) (i) BzCN, pyridine, 93%; (ii) SO_3·TMA (20 equiv.), DMF, 50 °C, 84%; (iii) HF·Py., 0 °C, 90%; (iv) 1 M LiOH, 30%, H_2O_2, 0 °C; (v) NaOH, MeOH, H_2O, 70% (2 Steps); (i) (i) $PhCH(OMe)_2$, Cat. CSA, MeCN, 96%; (ii) 1 M LiOH, 30%, H_2O_2; (iii) NaOH, MeOH, H_2O, 56% (2 Steps); (iv) SO_3·TMA (20 equiv.), DMF, 50 °C; (v) 0.01 M Aq. AcOH, 17%.

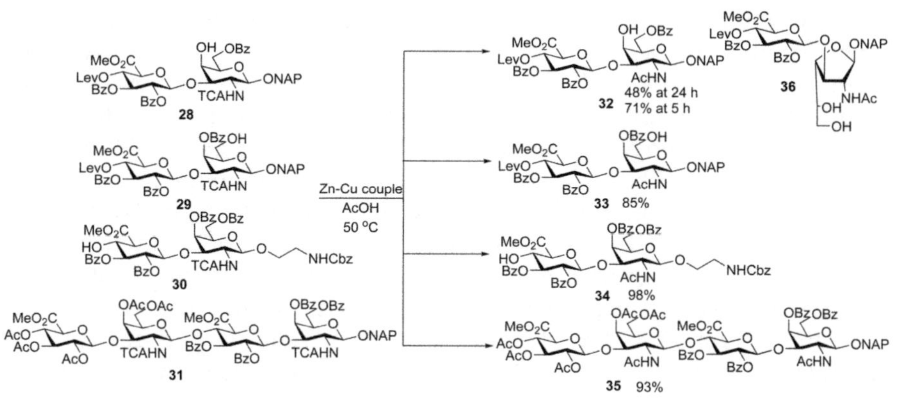

Scheme 8.4 *N*-TCA in chondroitin derivatives could be converted to *N*-acetamide using Zn–Cu couple.

N-acetyl, followed by cleavage of *p*-methoxybenzylidine leading to tetraol **27**. Extensive sulfation of **27** under vigorous condition generated the 4,6-*O* disulfated CS-E precursor, while milder sulfation reaction generated 6-*O* sulfated CS-C precursor. Subsequent silyl deprotection and saponification furnished CS-E **17** and CS-C tetrasaccharide **19**. CS-A tetrasaccharide **18** was synthesized *via* selective benzoylation of 6-OH of **27** followed by sulfation of the free hydroxyl groups at C4. Subsequent deprotection and saponification gave the desired CS-A tetrasaccharide **18**. Finally, CS-R tetrasaccharide **20** was obtained through the formation of benzylidene acetal, followed by saponification and sulfation of the resulting free hydroxyl groups then deprotection to produce CS-R **20**.

As an alternative to the TBTH mediated radical reduction method, the Jacquinet group reported the reduction of *N*-TCA using Zn–Cu couple in acetic acid,[34] which was successfully applied to the synthesis of chondroitin oligomers (Scheme 8.4). While for most reactions, the products were separated in good yields. However, the reaction on compound **28** gave furanoside **36** (22% yield), which may be due to zinc salts acting as a Lewis acid promoting the isomerization of pyranoside to furanoside. This side product was suppressed by decreasing the reaction time, improving the yield of the desired product **32** from 48% to 71% yield. The generality of the Zn–Cu reduction method is yet to be demonstrated in the synthesis of longer CS oligosaccharides.

8.2.2.4 N-Trifluoroacetamide

While multiple syntheses of CS oligomers have been accomplished using *N*-TCA as a protecting group,[12,19,29–32] there are some difficulties encountered with this group. For example, the formation of stable trichlorooxazoline side products from activated donors bearing *N*-TCA was reported by Huang[35] and Mulard.[36] Additionally, the conversion of *N*-TCA to the desired acetamide can be problematic. In some cases, radical reduction of *N*-TCA using TBTH/azobisisobutyronitrile produced significant amounts of mono- and dichloroacetamide side products.[37–40] Alternatively, basic hydrolysis of *N*-TCA could require long reaction time.[35,41,42] To address these challenges, Nieto and coworkers[43] reported *N*-trifluoroacetyl (*N*-TFA) group instead of *N*-TCA, which could be more readily cleaved under a mild condition to facilitate the final deprotection steps while still providing high β-selectivity in glycosylation.[44,45] They synthesized CS/dermatan sulfate-like tetrasaccharide **37** aided by *N*-TFA (Scheme 8.5). They found that the di-*tert*-butylsilylene bearing GalN **38** was an excellent glycosyl acceptor for coupling with uronic acid trichloroacetimidate donors **39** and **40,** leading to disaccharides **41** and **42.** Since the 4,6-di-*tert*-butylsilylene group of Gal and GalN donors led to the formation of

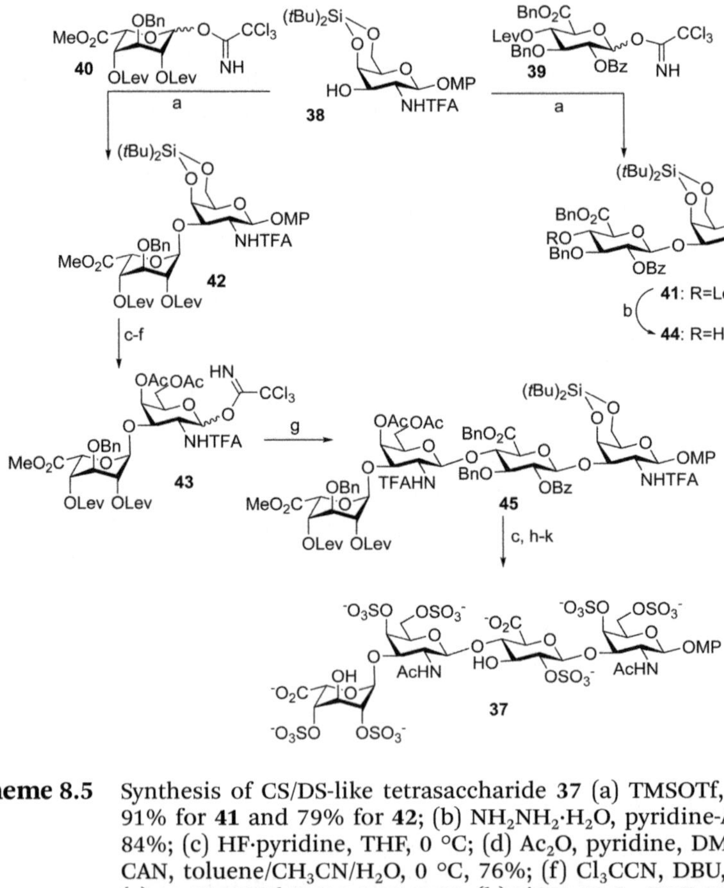

Scheme 8.5 Synthesis of CS/DS-like tetrasaccharide **37** (a) TMSOTf, DCM, 0 °C, 91% for **41** and 79% for **42**; (b) $NH_2NH_2 \cdot H_2O$, pyridine-AcOH, DCM, 84%; (c) HF·pyridine, THF, 0 °C; (d) Ac_2O, pyridine, DMAP, 89%; (e) CAN, toluene/CH_3CN/H_2O, 0 °C, 76%; (f) Cl_3CCN, DBU, DCM, 84%; (g) **44**, TMSOTf, DCM, 0 °C, 74%; (h) LiOH, H_2O_2, THF; NaOH, MeOH; (i) Ac_2O, MeOH, Et_3N; (j) $SO_3 \cdot Me_3N$, DMF, 100 °C, MW, 34% from **45**, 4 steps, 76%; (k) H_2, $Pd(OH)_2$, H_2O/MeOH, 82%.

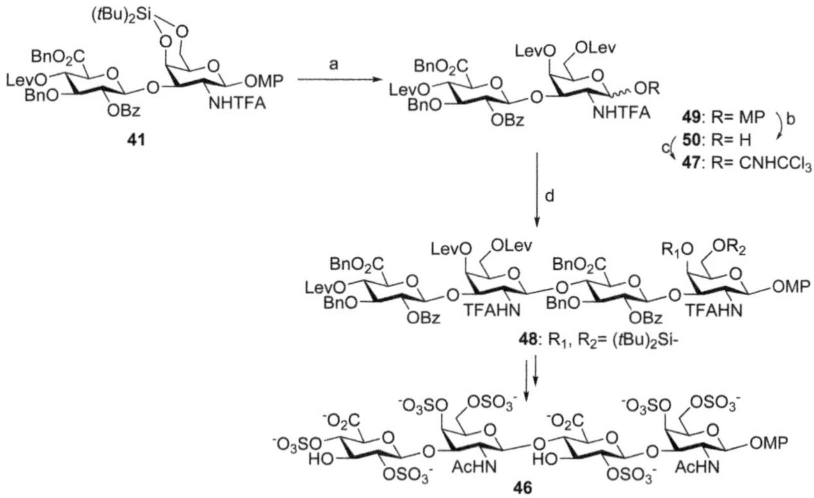

Scheme 8.6 Synthesis of CS tetrasaccharide **46** (a) HF·pyridine, THF, 0 °C; Lev_2O, pyridine, DMAP, 83%; (b) CAN, $DCM/CH_3CN/H_2O$, 81%; (c) Cl_3CCN, DBU, DCM, 70%; (d) **44**, TMSOTf, DCM, 0 °C, 71%.

1,2-*cis* glycosides despite the presence of participating groups at 2-*O*,[46] it was switched to di-*O*-acetate, followed by the removal of anomeric MP group and conversion of the newly formed hemiacetal to trichloroacetimidate donor **43**. Coupling of **43** with acceptor **44** afforded tetrasaccharide **45**. Tetrasaccharide **45** was then subjected to deprotection and sulfation under microwave (MW) irradiation,[47,48] generating the desired tetrasaccharide **37**.

The successful synthesis of **37** encouraged Nieto and his co-workers to apply *N*-TFA monomer in the synthesis of unnatural oversulfated CS tetrasaccharide **46** containing GlcA exclusively (Scheme 8.6).[49] Compound **41** was transformed to trichloroacetimidate donor **47** containing acyl groups on 4-*O* and 6-*O* of GalNAc moiety for better β-selectivity. Glycosylation between **47** and **44** gave the desired tetrasaccharide **48**, which was converted to the desired CS tetrasaccharide **46**.

8.2.2.5 N-*Tetrachlorophthalimide Group*

In addition to the TFA protecting group, the Nieto group investigated the utility of *N*-tetrachlorophthaloyl-(TCP) in CS synthesis (Scheme 8.7).[49] The TCP protecting group can facilitate the formation of 1,2-*trans* glycosidic linkages and can be cleaved under a mild condition. Glycosylation of GlcA trichloroacetimidate donor **51** and GalN acceptor **52** gave GlcA-GalN disaccharide **53**, while the coupling of donor **54** and acceptor **55** afforded GalN-GlcA disaccharide **56** in good yield. **56** was transformed to the tricholoroacetimidate donor **57**, which coupled with 5-(benzyloxycarbonylamino)-1-pentanol and gave the desired disaccharide **58**. The

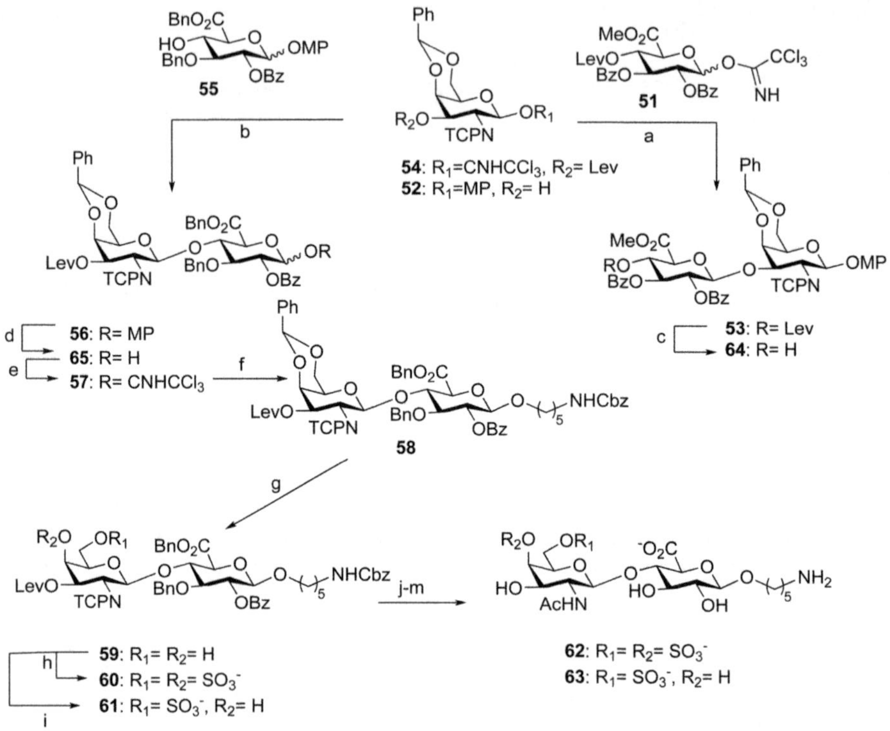

Scheme 8.7 Synthesis of CS disaccharides **62** and **63** using *N*-TCP bearing building blocks. (a) $BF_3{\cdot}Et_2O$, toluene, 54% + 27% recovered acceptor; (b) TMSOTf, DCM, 58% + 16% recovered acceptor; (c) $NH_2NH_2{\cdot}H_2O$, pyridine-AcOH, DCM; (d) CAN, $DCM/CH_3CN/H_2O$, 84%; (e) Cl_3CCN, K_2CO_3, DCM, 95%; (f) 5-(benzyloxycarbonylamino)-1-pen tanol, TMSOTf, DCM, 67%; (g) TFA, DCM, 0 °C, 99%; (h) $SO_3{\cdot}Me_3N$ (10 equiv.), DMF, 100 °C, 40 min, MW, 83%; (i) $SO_3{\cdot}Me_3N$ (2 equiv.), DMF, 50 °C, 30 min, MW, 84%; (j) LiOH, H_2O_2, THF; NaOH, MeOH; (k) $NH_2(CH_2)_2NH_2$, DMF, 100 °C, 90 min, MW; (l) Ac_2O, MeOH, Et_3N; (m) H_2, $Pd(OH)_2$, H_2O/MeOH, 65% for **62** for four steps, 78% for **63** for four steps.

benzylidene in **58** was cleaved affording diol **59** and the free hydroxyl groups were sulfated using microwave heating. The location of sulfations can be controlled by varying the amounts of sulfation agent producing two sulfated compounds **60** and **61**. Basic hydrolysis of the ester groups was followed by TCP cleavage with ethylenediamine and selective acetylation of the resulting amines. Finally, hydrogenolysis gave the final CS-E disaccharide **62** and CS-C disaccharide **63**.

Although *N*-TCP showed promising results in disaccharide synthesis, its application to the preparation of longer CS sequences was not successful. No desired tetrasaccharides GlcA-GalN-GlcA-GalN could be obtained from 2 + 2 couplings. Thus, further improvements are needed to broaden the utility of *N*-TCP.

8.2.2.6 Lactose Derived CS Mimetics

As the protective group for nitrogen adds a level of complexity to CS synthesis, the Mallet group reported the synthesis of CS-E mimetic containing lactose as the disaccharide repeating unit with the 2-*N*-acetamido group of GalNAc replaced by a hydroxyl group (Scheme 8.8).[50] In their strategy, lactoside building blocks **66** and **67** were derived from allyl lactoside **68**. Glycosylation of **66** and **67** followed by deprotection of the non-reducing end Fmoc produced the tetrasaccharide **69**. Repeating the sequence of glycosylation by **66** and Fmoc removal led to hexa- and octasaccharide **70** and **71**. These compounds were then deprotected, sulfated and functionalized with a biotin (compounds **72–75**) at the reducing end to aid in biological studies. While the synthesis was simplified using lactose as the building block, how well these analogs can mimic biological functions of CS-E remains to be established.

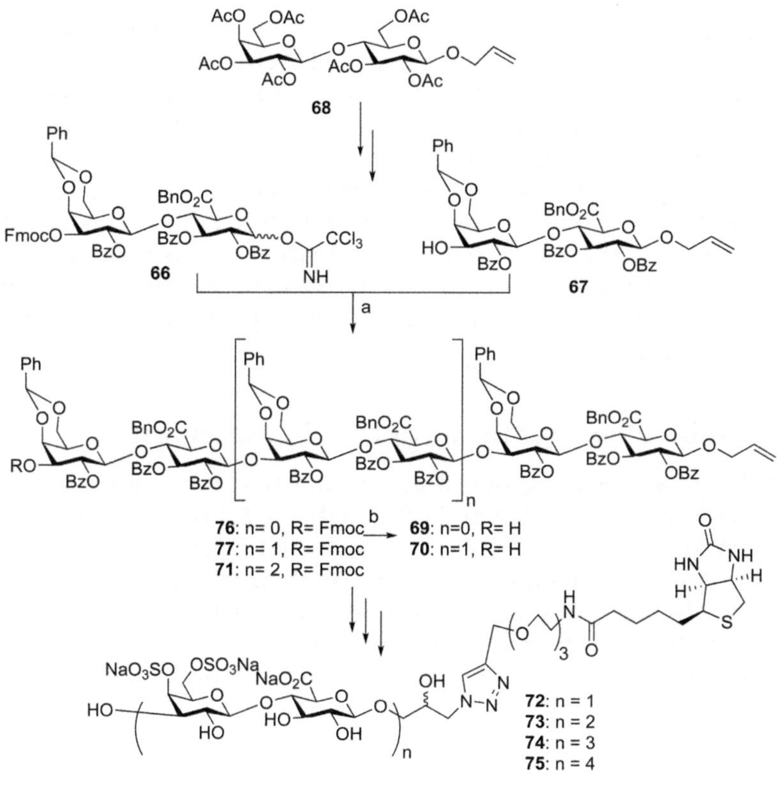

Scheme 8.8 Synthesis of tetra and hexasaccharide CS-E like analogs: (a) **66**, TMSOTf, MS 4 Å, DCM, rt, 68% for **76**, 62% for **77** and 40% for **71**; (b) NEt_3, DCM, rt, 74%.

8.2.3 Semi-synthesis of CS, Expediting Building Block Preparation Using Natural Sources

In 2006, the Jacquinet group reported an efficient semi-synthesis of size defined chondroitin oligomers from naturally existing CS polymers (Scheme 8.9).[30] They innovated a strategy based on previous work from Levene[51] and Meyer.[52] Acid hydrolysis of commercially available CS polymers (obtained from cartilage of shark and skate) resulted in desulfation and *N*-deacetylation, which was followed by esterification to afford disaccharide **78** having the sequence of GlcA-GalN. Disaccharide **78** was chemically transformed to the key intermediate **79** bearing *N*-TCA as the participating group and 2-naphthylmethyl (NAP) at the anomeric position that can be selectively removed for chain extension. Having **79** in hand, they prepared disaccharides **80, 81** and **82** with different protecting groups. Disaccharide **80** was transformed to acceptor **83** and to donor **84.** Glycosylation of acceptor **83** by donor **84** using Schmidt's procedure[53] with promoter TMSOTf gave tetrasaccharide **85.** The partially protected CS-C tetrasaccharide **86** was obtained *via* transformation of *N*-TCA to *N*-acetamide by radical reduction and acetylation.

The Jacquinet group also reported the synthesis of less common sulfo forms of CS oligomers, *i.e.*, CS-K, L and M (Scheme 8.10).[29] Disaccharide **81** was transformed to triol **92.** Exhaustive sulfation of triol **92** afforded 4,6,3′-trisulfated CS-M derivative **93** after deprotection, while controlling the amount of sulfating agent produced 6,3′-disulfated CS-L **94** in 75% yield with about 10% of trisulfated congener. On the other hand, selective benzoylation at 6-*O* of **92** followed by sulfation led to 4,3′-disulfated CS-K **95.** These results suggested that the reactivity of hydroxyl groups towards sulfation are of the order 6-*O*>3′-*O*>4-*O*. Similar selective protection and sulfation strategy has been extended to the preparation of CS-K and L tetrasaccharides **96** and **97.**

8.2.4 Solid Phase Synthesis of CS Oligosaccharides

In 2013, the Seeberger group reported the first automated solid-phase supported synthesis of CS-A and CS-C (Scheme 8.11).[54] For their protocol, they prepared glucuronic acid phosphate **103** bearing a Fmoc group at 3-*O* position, which could be cleaved for chain elongation. In addition, two GalN glycosyl phosphate building blocks **104** and **105** were synthesized containing Lev protecting groups strategically positioned either on the 4-*O* or 6-*O* position for future sulfation. A novel photolabile linker **106** was designed that was stable under both the basic condition used for removal of acyl type protecting groups and the acidic glycosylation condition, but readily cleavable under light irradiation. Hexasaccharide **107** was assembled using building blocks **104** and **103** in 16 steps with 13% overall yield (88% average yield per step). Using **105** and **103** as building blocks, the protected CS-A hexasaccharide **108** was produced in 8% yield

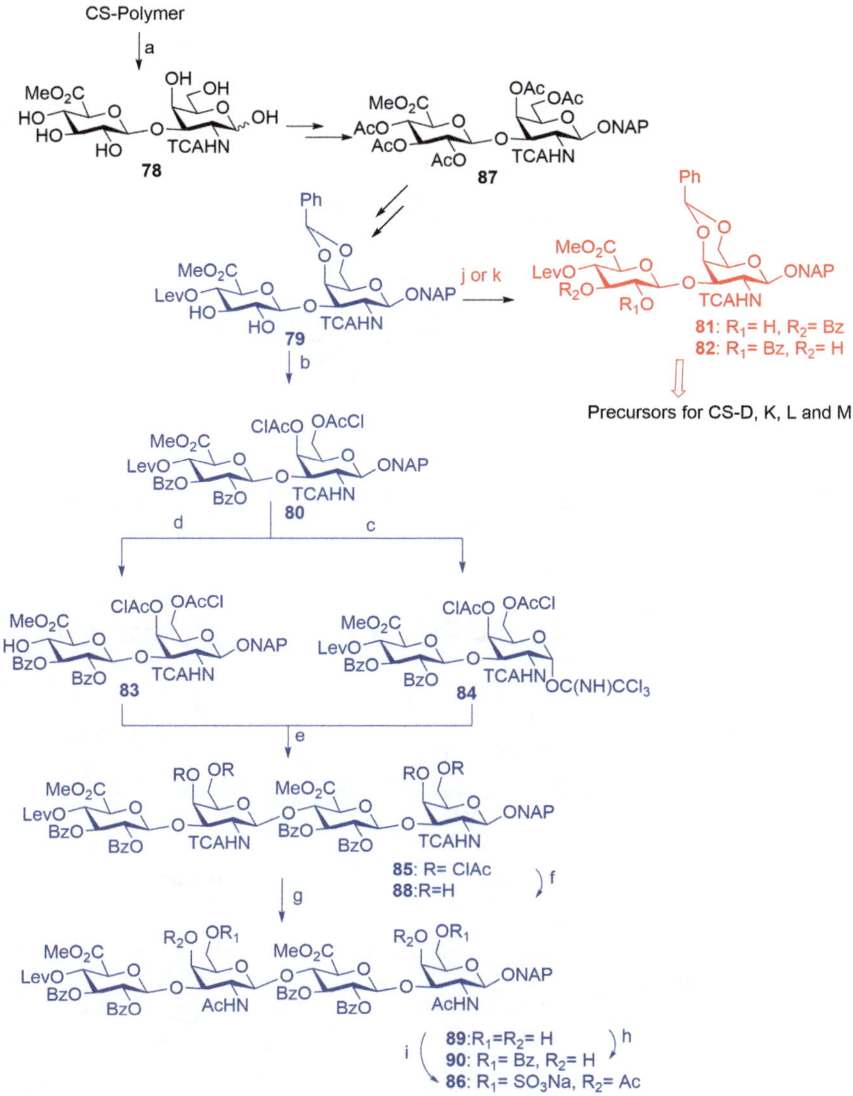

Scheme 8.9 Synthesis of chondroitin oligomers and sulfoform precursors: (a) (i) IR-120 [H^+] resin, H_2O; then 0.5 M H_2SO_4, 100 °C, 6 h; (ii) AcCl, MeOH, 0 °C, 72 h; (iii) Cl_3CCOCl, pyridine, 0 °C, 4 h; then DCM/MeOH/pyridine, rt, 3 h; (b) (i) PhCOCl pyridine, 0 °C, 1 h, 95%; (ii) 75% AcOH, 100 °C, 30 min; then $(ClAc)_2O$, pyridine, DCM, 0 °C, 1 h, 80%; (c) DDQ, DCM/MeOH, 24 h; then Cl_3CCN, DBU, DCM, 30 min, 66%; (d) hydrazine acetate, pyridine, 8 min, 89%; (e) TMSOTf, DCM, 30 min, 71% for **85**; (f) thiourea, pyridine/EtOH, 80 °C, 2 h, 90%; (g) Bu_3SnH, AIBN, DMAC, 80 °C, 2 h, 89%; (h) PhCOCN, pyridine, rt, 6 h, 88%; (i) $Me_3N{\cdot}SO_3$, DMF, 40 °C, 2 h; then Ac_2O, pyridine, rt, 16 h; then SP-C25 [Na+] resin, 81%; (m) PhCOCN, pyridine, rt, 6 h, 80%; (n) Bu_2SnO, dioxane/benzene, reflux, 7 h; then PhCOCl, rt, 8 h, 68% for **82**.

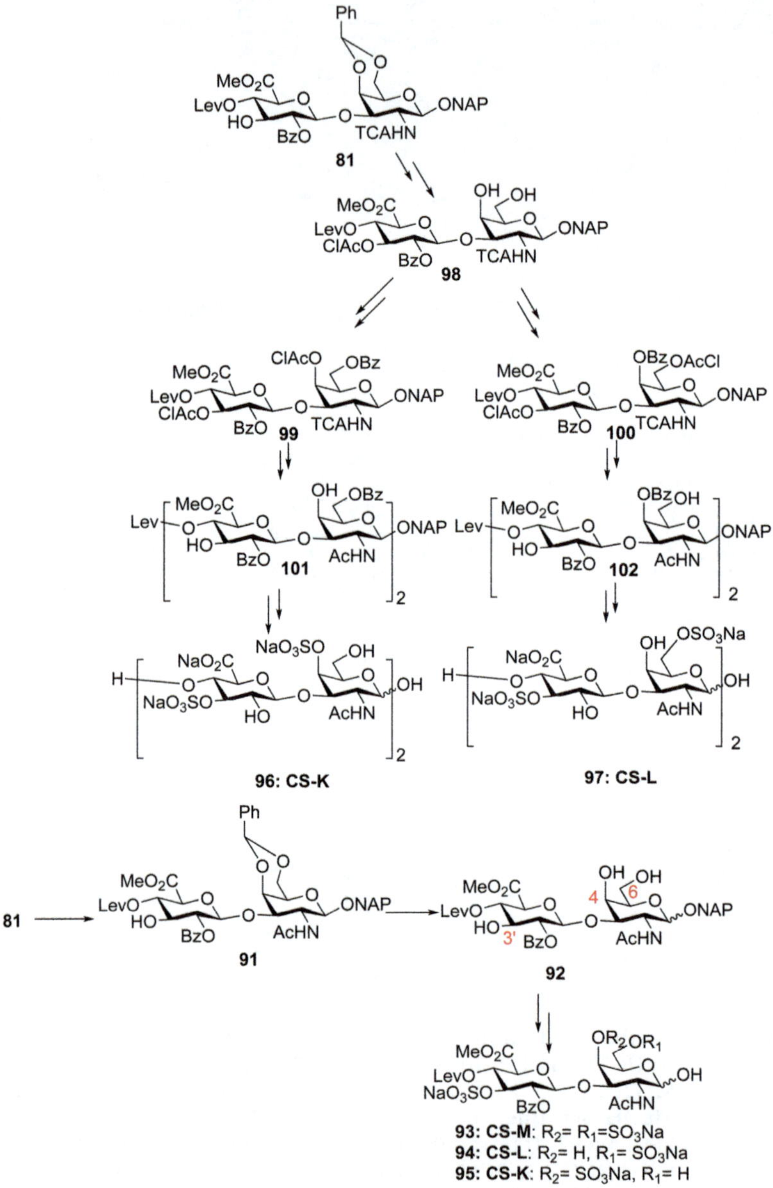

Scheme 8.10 Synthesis of CS-K, L and M disaccharides and CS-K and L tetrasaccharides from disaccharide building block **81**.

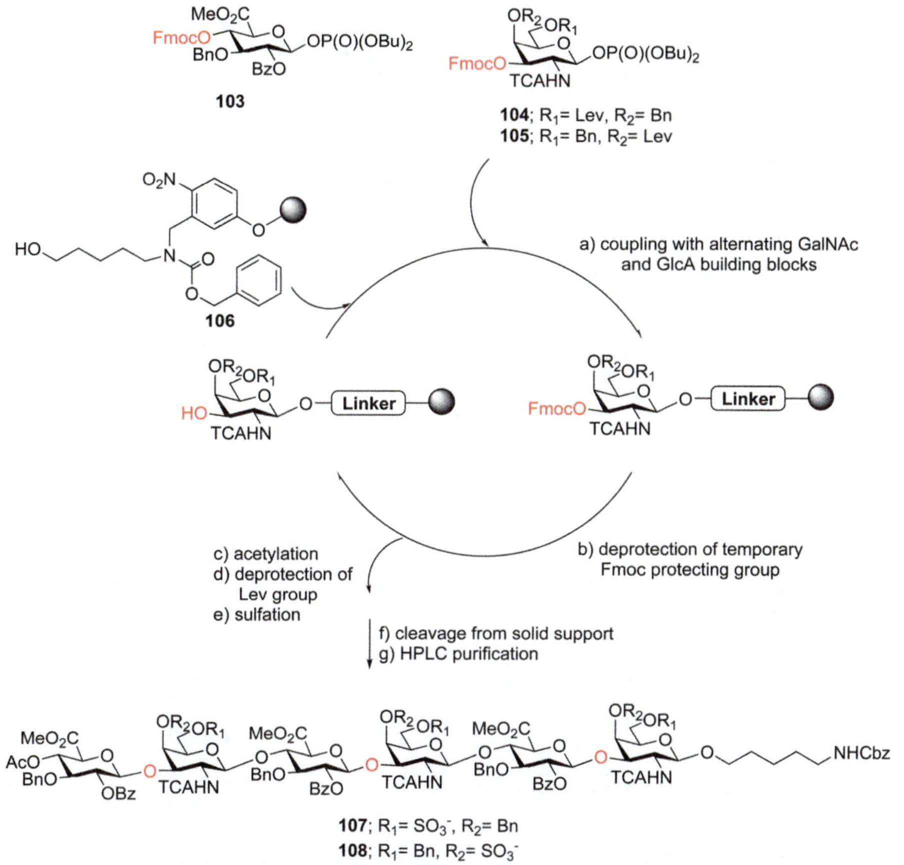

Scheme 8.11 Automated synthesis of CS-A and CS-C hexasaccharides **107** and **108**: (a) 3×3 equiv. building block, TMSOTf, DCM, −15 °C to 0 °C (15 min); (b) 3x20% piperidine in DMF, 25 °C (5 min); (c) 3x Ac_2O, pyridine, 25 °C (30 min); (d) 3x $N_2H_4{\cdot}H_2O$, pyridine, AcOH, DCM, 25 °C (60 min); (e) 3x $SO_3{\cdot}$ pyridine, pyridine, DMF, 50 °C, (3 h); (f) *hv*, DCM, 25 °C; (g) HPLC purification (**107**: 13% over 16 steps; **108**: 8% over 16 steps).

(86% average yield per step). While this automated solid phase synthesis provided a powerful strategy to access CS oligosaccharide backbones, further work is still needed to facilitate the deprotection on solid phase to produce the final CS oligosaccharides.

8.2.5 Chemical Synthesis of a CS Disaccharide Library and Biotinylated CS

Since the sulfate groups on CS can have important roles dictating their biological functions,[55–57] the Lam group developed a synthetic strategy to prepare all 16 theoretically possible sulfation motifs in CS disaccharides (Scheme 8.12).[58] In their study, benzyl ether and ester protecting groups

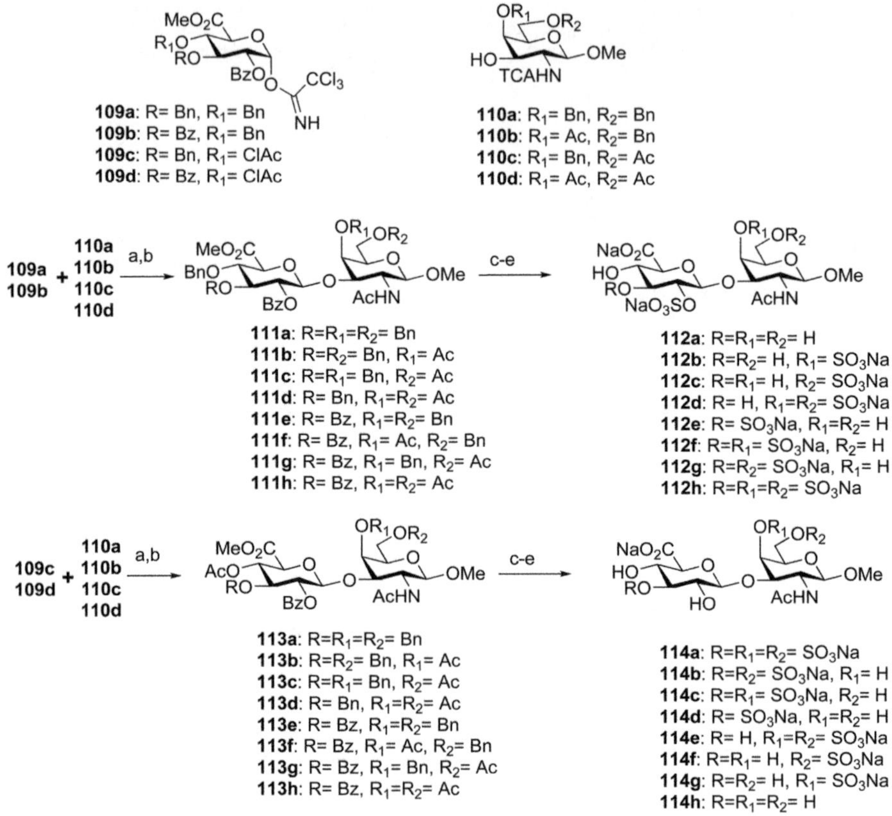

Scheme 8.12 Synthesis of CS disaccharide **112a–112h** and **114a–114h** (a) TMSOTf, DCM, 4 Å MS, −10 °C, 64–75%; (b) Bu_3SnH, AIBN, toluene, 100 °C, 69–80%; (c) LiOH, H_2O_2, THF/H_2O, −5 °C to rt; NaOH, MeOH, 0 °C to rt, 72–79%; (d) $SO_3 \cdot Et_3N$, DMF, 50 °C, 68–77%; (e) H_2, $Pd(OH)_2$, DCM/H_2O/MeOH, 85–95%.

were chosen to direct regioselective sulfation. Donors **109a–109d** and acceptors **110a–110d** were prepared.[59–61] Glycosylation between donors **109a–109d** and acceptors **110a–110d** were carried out with TMSOTf as the promoter. The resulting disaccharides were subjected to TCA reduction *via* TBTH mediated radical reduction. Basic hydrolysis of the ester groups on intermediates (**111a–111h**), followed by extensive sulfation of all free hydroxyl groups and global hydrogenolysis afforded eight CS disaccharides (**112a–112h**). The other eight CS disaccharides were obtained from intermediates **113a–113h**, where the benzyl ethers were removed *via* hydrogenolysis followed by sulfation of the resulting free hydroxyl groups and basic hydrolysis (**114a–114h**).

Biotinylated CS could be useful for biological assays and identifying new pharmaceutical targets. Jacquinet's group reported an efficient and stereocontrolled synthesis of homo 4-*O* and 6-*O* sulfated biotinylated chondroitin

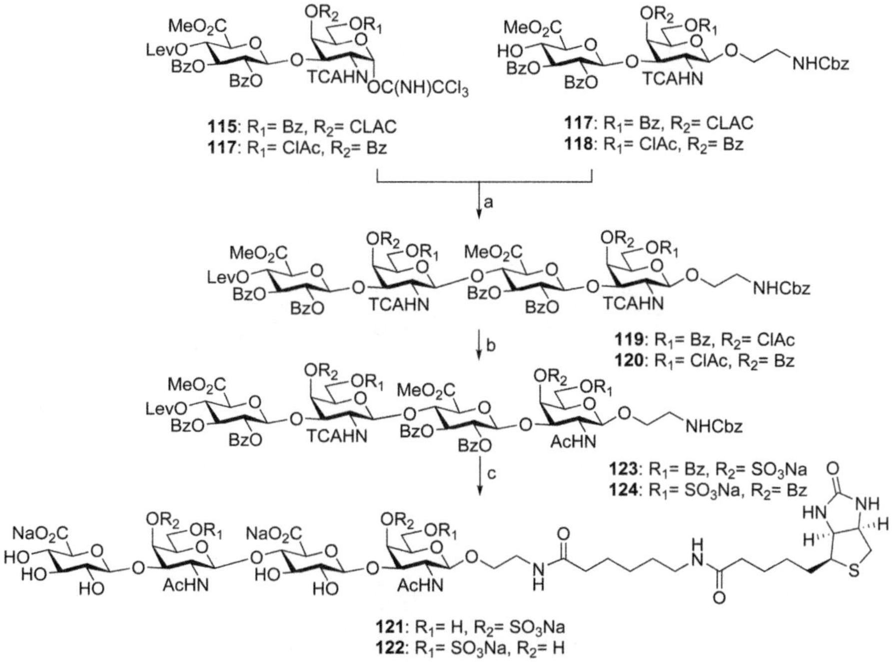

Scheme 8.13 Synthesis of biotinylated CS tetrasaccharides: (a) TMSOTf, DCM, 2 h, 55% for **119**, 63% for **120**; (b) (i) thiourea, pyridine/EtOH, 80 °C, 2 h; (ii) Bu_3SnH, AIBN, benzene/DMAC, 80 °C, 6 h; (iii) $Me_3N{\cdot}SO_3$, DMF, 60 °C; then LH-20; then Sephadex SP C-25 [Na+] resin, 81% for **123**, 71% for **124** (in 3 steps); (c) (i) $LiOH/H_2O_2$, THF/H_2O, −10 °C, rt, 16 h; then 4 M NaOH, MeOH, 8 h; (ii) H_2, Pd/C, H_2O; then *N*-hydroxysuccinimidoyl 6- (biotinylamido)hexanoate, Et_3N, DMF/H_2O, 1 h, 87% for **121**, 74% for **122**.

di- and tetrasaccharides.[31,62] For example, coupling between donor **115** with acceptor **116** and donor **117** with acceptor **118** gave tetrasaccharides **119** and **120** (Scheme 8.13). Selective dechloroacetylation, radical reduction of TCA, sulfation, catalytic hydrogenolysis and finally biotinylation provided the desired products **121** and **122**.

8.2.6 Chemical Synthesis of Heterologous Sulfated CS Oligosaccharides

Currently, most CS synthesis focused on the production of homologously sulfated CS oligosaccharides, *i.e.*, all disaccharide units in the oligosaccharides bear the same sulfation patterns. Suda's group, in 2015, reported a synthesis of CS tetrasaccharide analogs bearing homo- or hetero-dimeric sulfation patterns.[63] To access these CS compounds, they synthesized pentasaccharides **125–128**, which bear benzylidene, Lev and Bn at GalN to differentiate various hydroxyl groups for sulfation (Scheme 8.14). The Troc

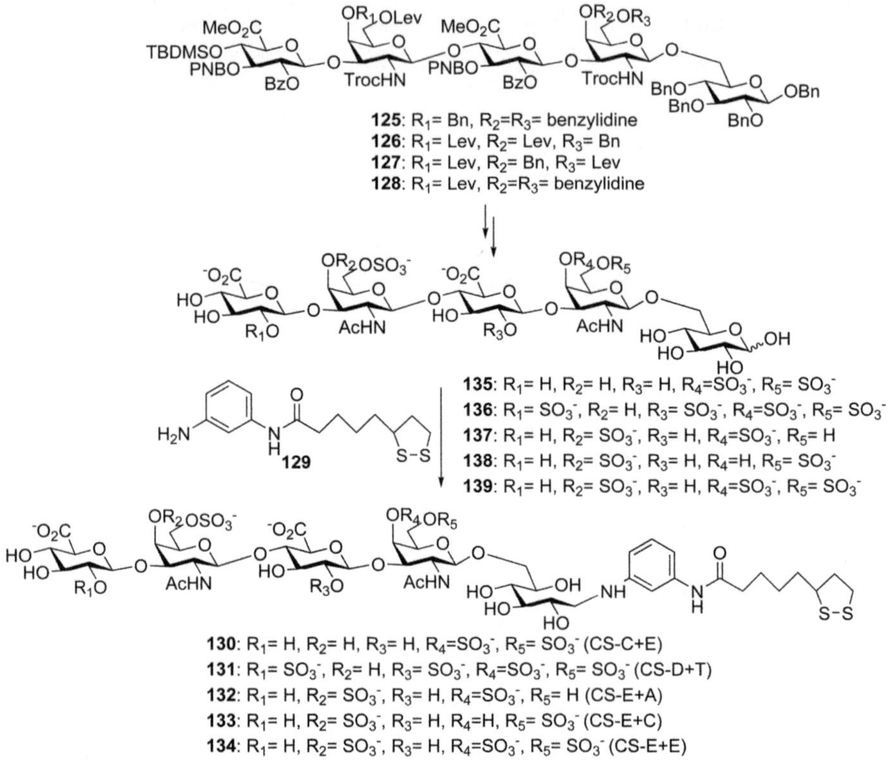

Scheme 8.14 Synthesis of homo- and hetero-dimeric CS tetrasaccharide analogs.

groups in these pentasaccharides were converted to NHAc using Zn–Cu couples in AcOH, followed by acetylation of the free amine. The Lev group was selectively deprotected and the resulting free hydroxyl group was sulfated. Finally, global deprotection was carried out, followed by reductive amination reaction with linker **129** affording tetrasaccharide analogs **130–134** containing two disaccharide units: CS-C + E, CS-D + T, CS-E + A, CS-E + C, or CS-E + E. Other examples of heterologously 4-*O* and 6-*O* sulfated CSs synthesized include the biotinylated tri- and tetrasaccharides **140–143** (Figure 8.5).[32]

8.2.7 Chemical Synthesis of CS Oligosaccharides with Linkage Region

Lopin-Bon and coworkers reported for the first time a stereocontrolled synthesis of biotinylated tetrasaccharides and pentasaccharides bearing a CS mono- or disaccharide linked to trisaccharide (GlcAβ1–3Galβ1–3Gal) of the truncated linkage region (Scheme 8.15).[64] The synthesis started from trisaccharides **144** and **145**,[65] in which the hydroxyl groups to be sulfated were protected with silylene acetal or chloroacetate groups. Trisaccharides **144** and

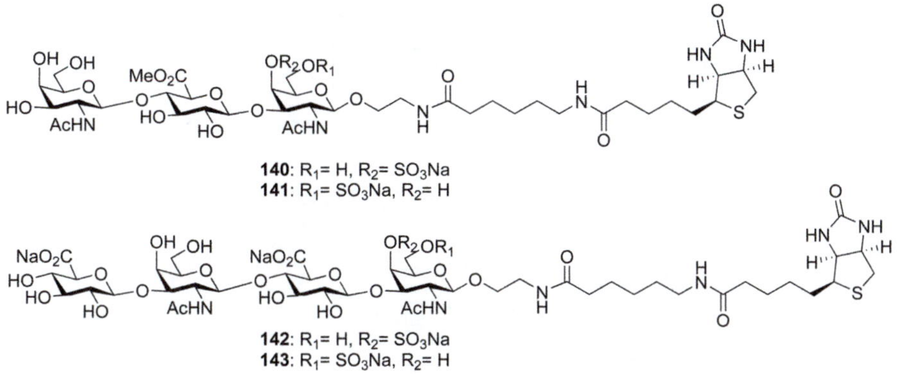

Figure 8.5 Biotinylated tri- and tetrasaccharides **140–143**.

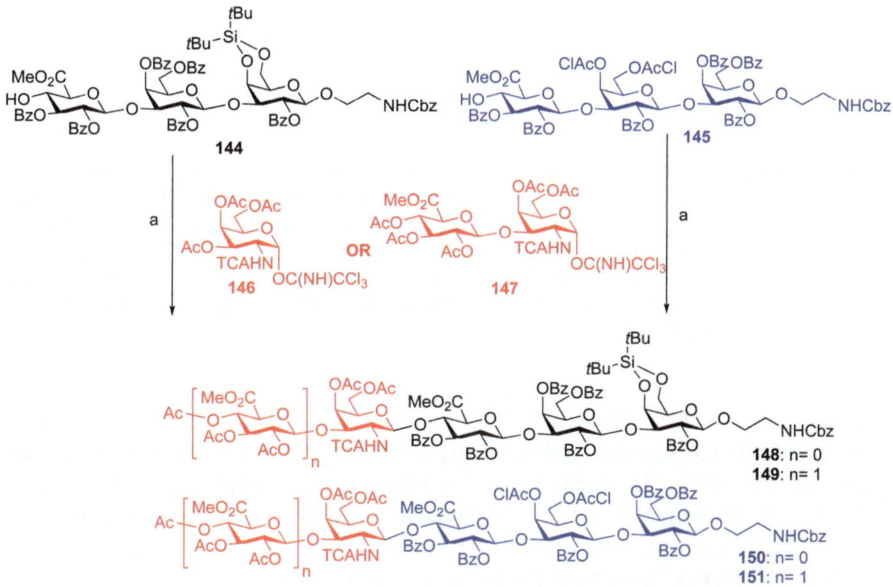

Scheme 8.15 Synthesis of tetra- and penta-saccharides containing the linkage region trisaccharide: (a) TMSOTf, DCM, rt, 30 min, 83% for **148**, 76% for **149**, 68% for **150**, 55% for **151**.

145 were coupled with monosaccharide donor **146** or disaccharide donor **147** following Schmidt's glycosylation procedure,[53] leading to the desired tetra- and pentasaccharides **148–151**.

For sulfation of the reducing end galactose, **148** and **149** were treated with Zn/Cu couple in acetic acid for *N*-TCA reduction[39] producing **152** and **153**, the silylene acetals of which were cleaved to give the diol **154** and **155** (Scheme 8.16). Regioselective 6-*O*-sulfation of **154** and **155**, followed by

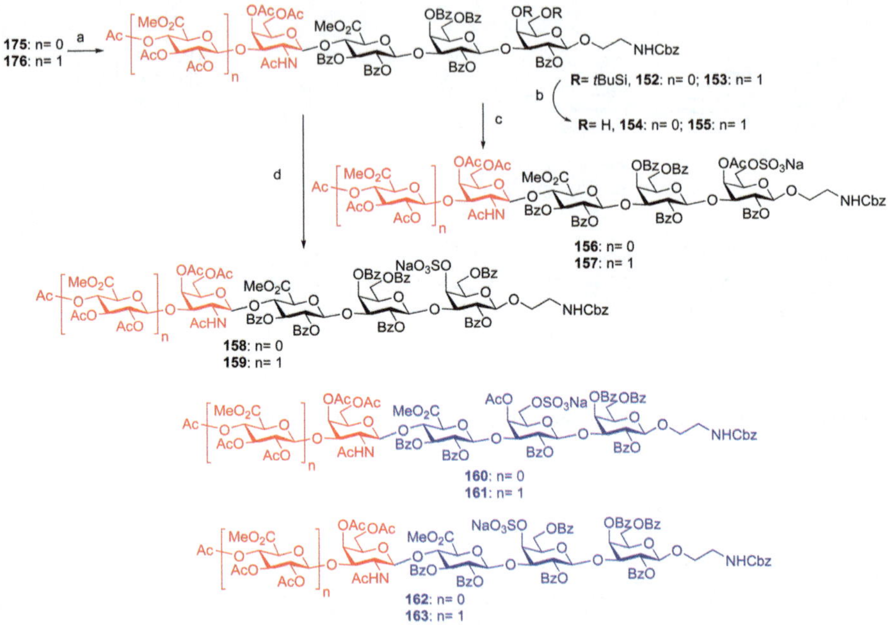

Scheme 8.16 Synthesis of tetra- and penta-saccharides bearing the linkage trisaccharide. (a) Zn/Cu couple, AcOH, 50 °C, 24 h, 92% for **152**, 86% for **153**; (b) NEt_3·3HF, THF, 0 °C, 4 h, 95% for **154**, 83% for **155**; (c) $Me_3N·SO_3$, DMF, 40 °C, 8–9 h; then Ac_2O, pyridine, rt, 16 h, 81% for **156**, 57% for **157**, 84% for **160**, 55% for **161**; (d) (i) BzCN, pyridine, rt, 3 h; (ii) $Me_3N·SO_3$, DMF, 60 °C, 40 h, 81% for **158**, 81% for **159**, 89% for **162**, 62% for **163**.

acetylation afforded **156** and **157**. Selective 6-*O*-benzoylation of **154** and **155** with benzoyl cyanide and subsequent sulfation afforded 4-*O*-sulfated products **158** and **159**. Similarly, tetra- and pentasaccharides **160–163** bearing sulfation at the penultimate galactose unit from the reducing end were prepared. Final deprotection and biotin conjugation led to biotinylated oligosaccharides **164–173** (Figure 8.6) in good yields.

8.3 Chemoenzymatic Synthesis of CS Oligosaccharides

In 2008, Sugiura *et al.* successfully prepared chondroitin oligosaccharide backbone up to 16-mers using two immobilized enzymes (Scheme 8.17).[66] Chondroitin polymerase K4CP, expressed in *Escherichia coli*, have two glycosyltransferase active domains and can transfer both D-GalNAc and D-GlcA to the non-reducing end of a chondroitin acceptor from donors uridine diphosphate (UDP)-D-GalNAc and UDP-D-GlcA. To bestow control

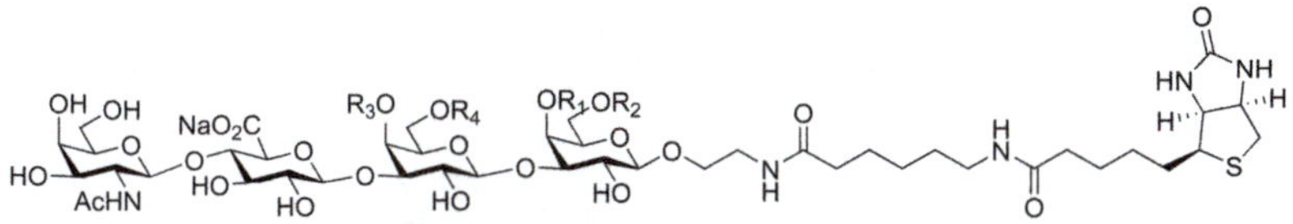

164: $R_1=R_2=R_3=R_4=$ H
165: $R_1=R_3=R_4=$ H, $R_2=$ SO_3Na
166: $R_2=R_3=R_4=$ H, $R_1=$ SO_3Na
167: $R_1=R_2=R_3=$ H, $R_4=$ SO_3Na
168: $R_1=R_2=R_4=$ H, $R_3=$ SO_3Na

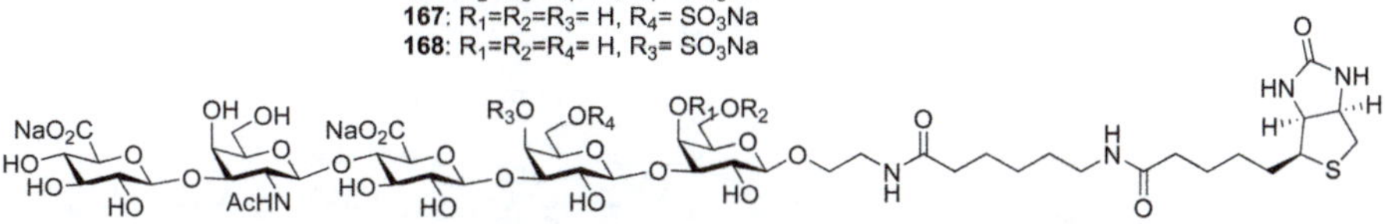

169: $R_1=R_2=R_3=R_4=$ H
170: $R_1=R_3=R_4=$ H, $R_2=$ SO_3Na
171: $R_2=R_3=R_4=$ H, $R_1=$ SO_3Na
172: $R_1=R_2=R_3=$ H, $R_4=$ SO_3Na
173: $R_1=R_2=R_4=$ H, $R_3=$ SO_3Na

Figure 8.6 Biotinylated oligosaccharides **164–173**.

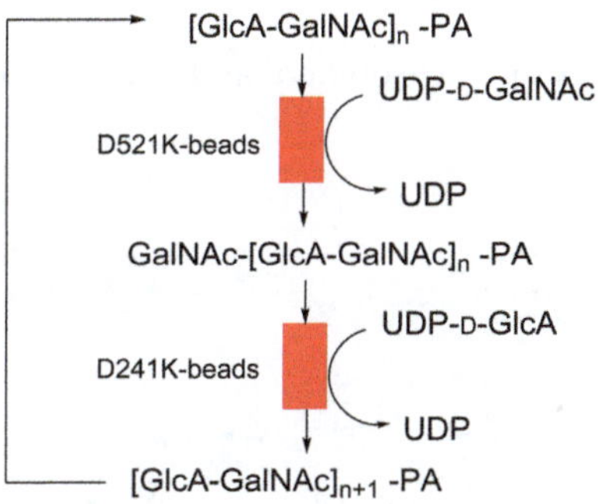

Scheme 8.17 Synthesis of chondroitin oligosaccharide using K4CP mutant-immobilized beads.

of backbone length, mutations of this enzyme were achieved, *i.e.*, D-241K for transferring the D-GlcA residue only and D-521K for transferring the D-GalNAc moiety only. These enzyme mutants were immobilized on agarose beads to facilitate product purification. Alternating usage of these enzymes elongated a pyridylamino (PA)-functionalized chondroitin up to 16-mers (Scheme 8.17).

In 2017, the Liu group reported for the first time the synthesis of 4-*O*-sulfated and 6-*O*-sulfated CS oligosaccharides **175–188** using an enzyme-based method (Scheme 8.18).[67] They synthesized the non-sulfated chondroitin backbone in gram scale (Scheme 8.19A) using bacterial glycosyltransferase called KfoC, which transfers both UDP-D-GalNAc and UDP-D-GlcA to an acceptor. They were also able to prepare the donor UDP-GalNAc using *N*-acetylhexosamine kinase (NAHK)[68] and mammalian

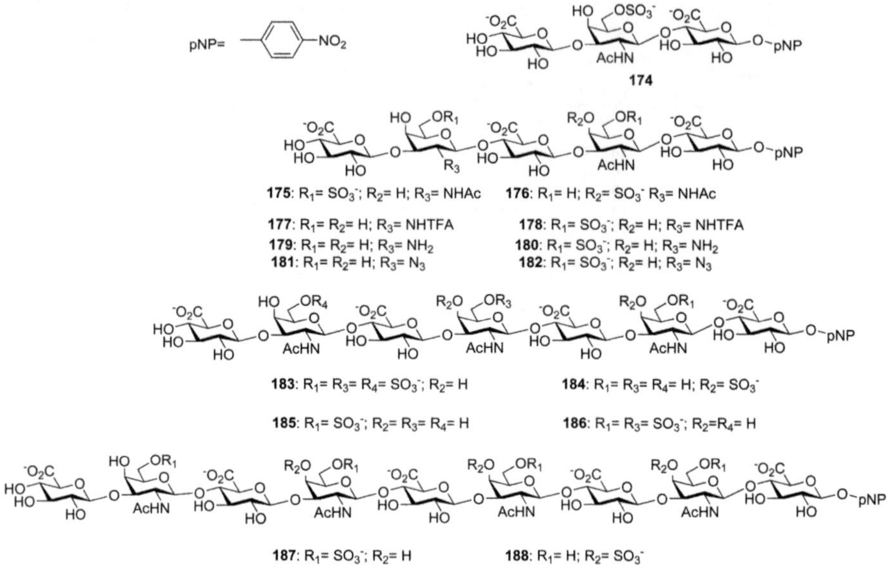

Scheme 8.18 Structures of 4-*O*-sulfated and 6-*O*-sulfated CS oligosaccharides **175–188**.

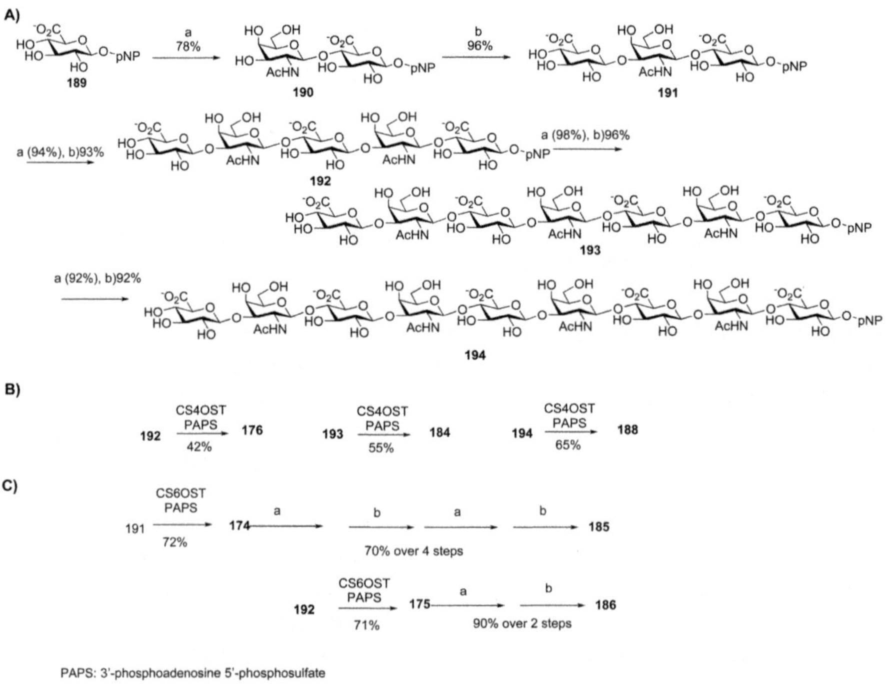

Scheme 8.19 (A) Synthesis of chondroitin backbone; Chemoenzymatic synthesis of (B) CS-A oligosaccharides and (C) CS-C oligosaccharides: (a) KfoC, UDP-GalNAc; (b) KfoC, UDP-GlcA.

GalNAc pyrophosphorylase AGX1, thus overcoming the high cost of UDP-GalNAc.[69] KfoC enzyme was found to be capable of transferring a GalNTFA moiety from UDP-GalNTFA to chondroitin backbone **177**, which could be converted to $GalNH_2$ derivative **180** or azide derivative **181**.

Using the 4-*O*-sulfotransferase (CS4OST) and 3′-phosphoadenosine-5′-phosphosulfate (PAPS) as a sulfate donor, 4-*O*-sulfated oligosaccharides **176**, **184**, and **188** were successfully synthesized (Scheme 8.19B). Structural analysis of these compounds showed the GalNAc closest to the non-reducing end was not 4-*O*-sulfated, which was attributed to substrate specificity of CS4OST. Using the 6-*O*-sulfotransferase (CS6OST), 6-*O*-sulfated oligosaccharides **174, 175, 178, 180, 182, 183, 185, 186,** and **187** were obtained. The fully 6-*O*-sulfated oligosaccharides were achieved from the chondroitin backbone (tri- to nona-saccharide) with CS6OST, while the partially 6-*O*-sulfated compounds **185** and **186** were produced from **174** and **175** respectively following chain elongation (Scheme 8.19C). The *p*-nitrophenyl (pNP) moieties at the reducing end of all these oligosaccharides were used to facilitate the detection because it has characteristic UV_{max} absorption at 310 nm and it enabled the oligosaccharide to bind to C18 beads for good purification. The pNP group could be converted to amine or azide groups for microarray or other biological studies.

8.4 Chemical Synthesis of CS Glycopolymers and Glycoconjugates

As it remains challenging to synthesize large CS oligosaccharides approaching the sizes of natural CS polysaccharides, an alternative approach is to mimic natural CS using synthetic conjugates or polymers. In 2008, the Hsieh-Wilson group reported the synthesis of CS glycopolymers (Scheme 8.20) and their biological activities.[70] The monomer **195** bearing a polymerizable cyclooctene handle was obtained by coupling disaccharide donor **196** with the cyclooctene acceptor **197**. Radical reduction of TCA and benzylidene hydrolysis gave diol **198**, which upon sulfation furnished **199**. Compounds **255** and **256** underwent ruthenium-catalyzed ring opening metathesis polymerization producing glycopolymers **200** and **201**, which were then deprotected leading to CS glyco-mimetics **202** and **203** bearing CS disaccharide units. Polymer **204** with tetrasaccharide repeating units was obtained in a similar manner from monomer **205**. The biological activities of these glycopolymers were analyzed and it was found that glycopolymers **203** and **204** mimicked the activities of natural polysaccharides enriched with CS-E towards inhibition of neuronal growth. In contrast, unsulfated glycopolymer **202** had few effects, emphasizing the importance of sulfation for biological activity. Using the same technology, norbornene-based CS glycopolymers,[71] which contained longer polymer backbone compared to cyclooctene-based ones, were synthesized.

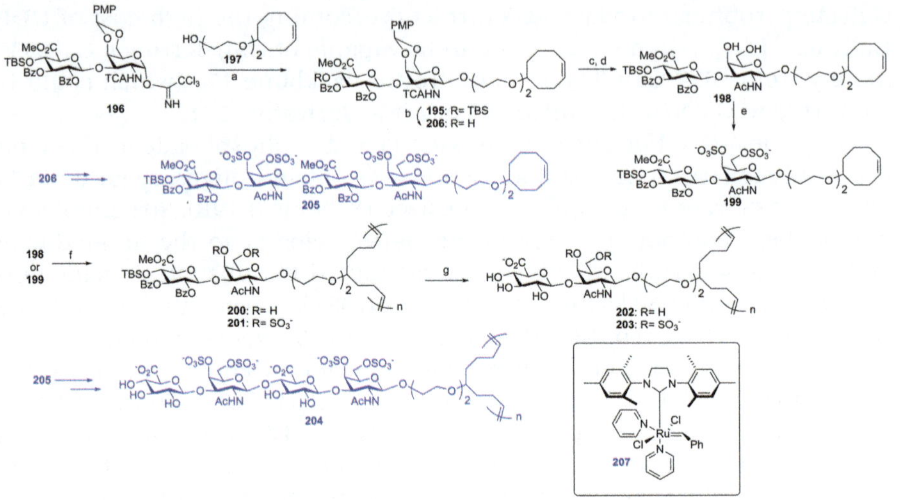

Scheme 8.20 Synthesis of CS glycopolymers: (a) TMSOTf, DCM, −20 °C, 69% for **253** and 45% for **6**. (b) HF·pyridine, THF, pyridine, 25 °C, 95%. (c) AIBN, Bu_3SnH, benzene, 50 °C, 92%. (d) DDQ, CH_3CN/H_2O, 25 °C, 89%. (e) SO_3·TMA, DMF, 80 °C, 85%. (f) **207**, 1 : 5 MeOH/DCM, 55 °C, 24 h; (g) (i) HF·pyridine, THF, pyridine, 25 °C; (ii) LiOH, H_2O_2, then NaOH, MeOH, 25 °C, 87–97% (two steps).

In nature, CS can be degraded by chondroitinase. CS analogs bearing unnatural glycosidic linkages should be more stable towards chondroitinases and inhibit these enzymes. Linhardt and his co-worker developed a new synthetic strategy for the construction of unnatural CS disaccharide and tetrasaccharide, where GlcA is linked to GalN through a triazole linkage obtained from the copper-catalyzed azide–alkyne cycloaddition reaction (CuAAC).[72] To access these molecules, GlcA azide building block **208** and 3-*O*-propargylated GalNAc **209** were coupled *via* the CuAAC reaction affording disaccharide **210** (Scheme 8.21). Disaccharide **210** was then propargylated with propargyl trichloroacetimidate to afford **211**. Alternatively, **210** was transformed to the azide bearing analog **212**. Coupling between **211** and **212** gave the protected tetrasaccharide **213**, which was deprotected to yield the target tetrasaccharide **214**. This method provided an efficient tool for preparing larger oligosaccharide analogs, which could aid in the understanding of biological functions of these molecules.

Recently, Rojo and co-workers reported the synthesis of dendrimers[73] bearing CS-E disaccharide units. They applied the CuAAC chemistry between the alkyne-terminated dendrimer core and the azido sugars. To synthesize CS-E disaccharides **216** and **217** (Scheme 8.22), glycosylation between trichloroacetimidate donor **218** and acceptor **219** furnished disaccharide **220** in 80% yield. Removal of the silylene group followed by sulfation using microwave irradiation produced 4,6-*O* disulfated product **221**. Basic hydrolysis and

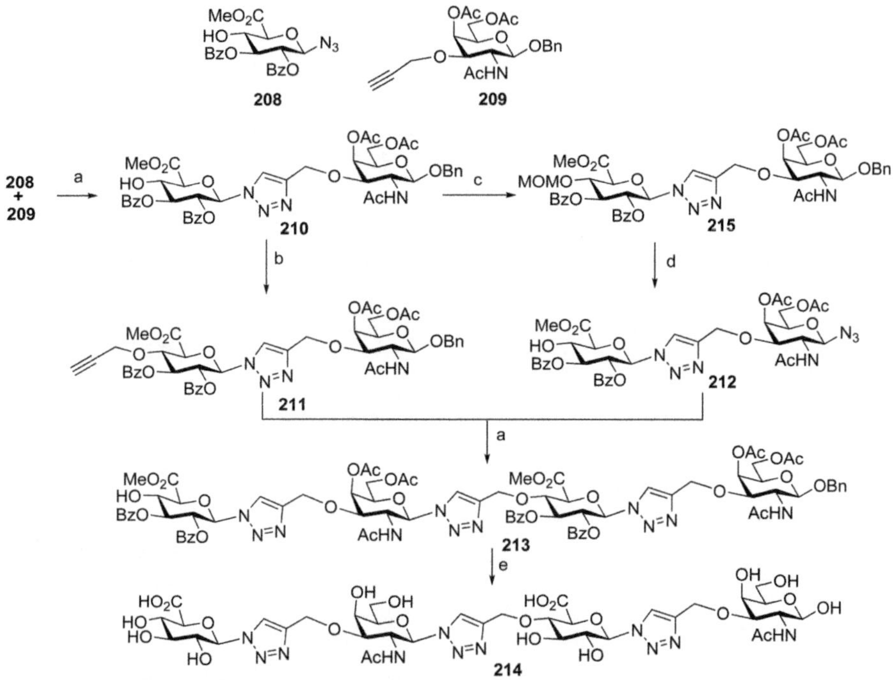

Scheme 8.21 Synthesis of chondroitin tetrasaccharide analog **214**: (a) CuI, DIPEA, CH_3CN, DCM, rt; (b) propargyl trichloroacetimidate, trifluoromethane sulfonic acid, DCM, molecular sieves 4 Å, rt; (c) MOMCl, DIPEA, DMAP, DCM, 35 °C; (d) (i) H_2, 10% Pd/C, MeOH/EtOAc (ii) $SnCl_4$, $TMSN_3$, DCM, 0 °C; (e) (i) LiOOH, THF; (ii) 2 N NaOH, MeOH; (iii) H_2, 10% Pd/C, MeOH.

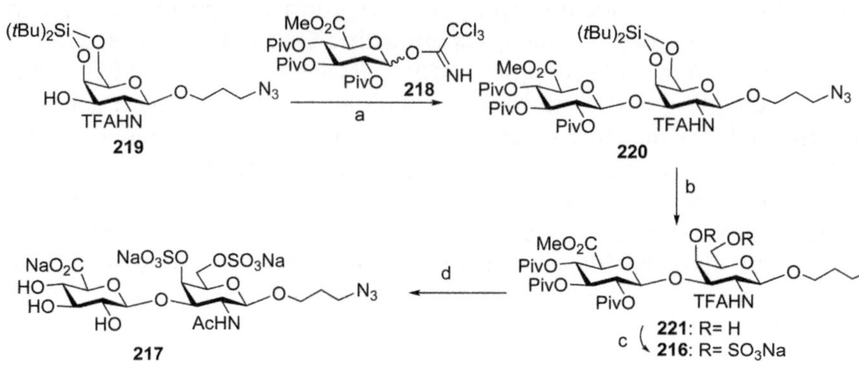

Scheme 8.22 Synthesis of CS-E disaccharide analogs: (a) TMSOTf, DCM, 0 °C, 80%; (b) HF·pyridine, THF, 0 °C, 95%; (c) $SO_3 \cdot Me_3N$, DMF, 100 °C, MW, 85%; (d) LiOH, H_2O_2, THF; NaOH, MeOH/H_2O; Ac_2O, Et_3N, MeOH/H_2O, 73%.

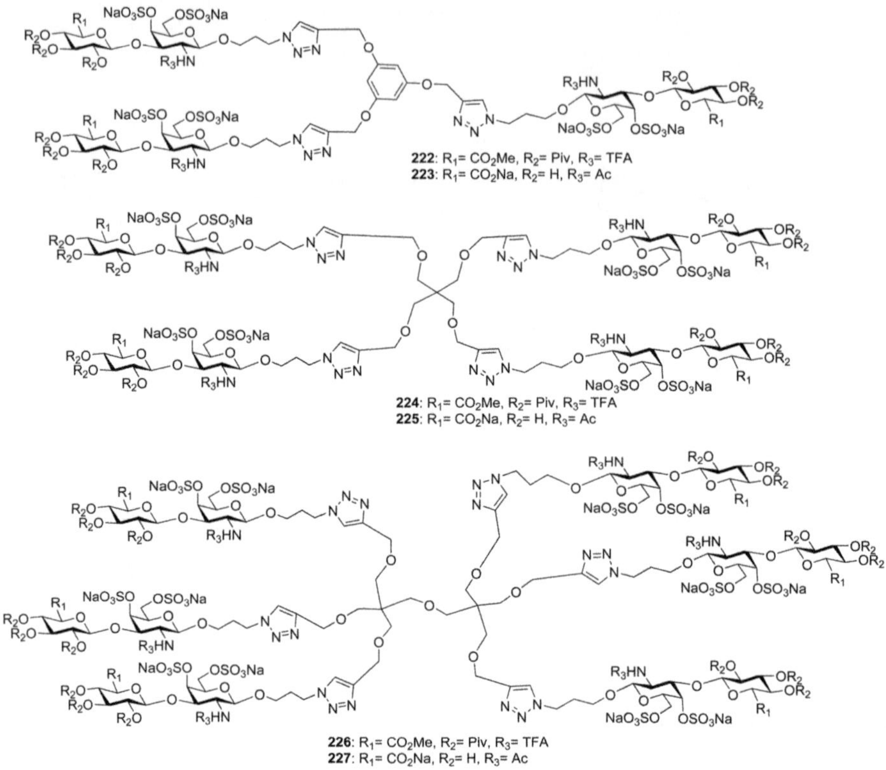

Scheme 8.23 Structures of tri-, tetra- and hexa-valent CS dendrimers.

selective *N*-acetylation of **216** gave azido disaccharide **217**. Next, CuAAC reactions between dendritic cores and sulfated disaccharides **216** and **217** (Scheme 8.23) were carried out using $CuSO_4$, sodium ascorbate and tris(benzyltriazolylmethyl)amine as the ligand affording **222, 224, 226, 223, 225** and **227** with three, four, and six copies of CS-E disaccharides respectively. They found that coupling of the fully deprotected disaccharide **217** with dendritic cores proceeded smoothly. In comparison, when the protected disaccharide **216** was used, complex mixtures were obtained during deprotection of the resulting dendrimers. With these multivalent systems, they studied the binding affinity to the growth factor midkine and found that even with a low valency of the disaccharide, a significant increase in the binding avidity was observed compared to the free disaccharide suggesting the potential of this type of polyvalent construct.

8.5 Synthesis of CS Glycopeptide

As discussed before, CS exists in nature as part of CSPG. Although many biological functions are directed by CS, clustering of CS chains on a protein backbone can potentially increase the affinity with their receptors.[74]

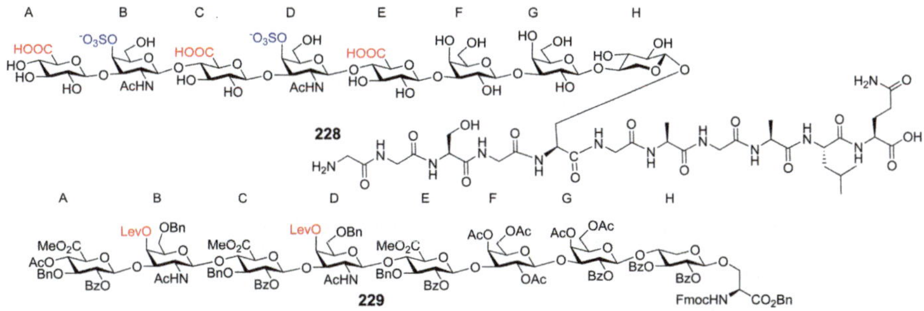

Figure 8.7 Structure of syndecan-1 CS glycopeptide **228** and CS-A octasaccharide linked serine **229**.

Furthermore, the core protein can bestow activities synergistic to the glycan chains. Thus, the availability of structurally well-defined CSPG can greatly aid in biological studies. One major challenge in CSPG synthesis is that with their structural complexity, they are unstable under many common synthetic operations. For example, due to the high sensitivity of *O*-sulfates to acid, commonly used acid cleavable amino acid side chain protective groups such as *t*-butyloxycarbonyl, *t*-butyl and trityl should be avoided for glycopeptide synthesis. Furthermore, care need to be taken as the glycoside-serine linkage is prone to base promoted β-elimination.[75–77] Thus, the reaction routes and conditions must be carefully designed and established.

Having all these obstacles in mind, Huang's group reported the first chemical synthesis of a syndecan-1 CS glycopeptide **228** (Figure 8.7) bearing CS-A tetrasaccharide attached through a tetrasaccharide linker to the core protein (corresponding to amino acid residues 45–53 of human syndecan-1) with two more glycine residues at its N-terminal for future sortase mediated ligation.[37] A cassette approach[78] was adopted to access the glycopeptide with the CS-A octasaccharide linked serine **229** designed as a cassette.

To access the octasaccharide module, they first tried 3 + 2 + 3 glycosylation strategy, which was developed for the synthesis of heparan sulfate glycopeptides.[77,79–81] However, the coupling of ABC trisaccharide unit with the DE disaccharide *en route* to **229** was very low yielding (<10%) despite many variations with reaction conditions and protecting groups. To overcome this difficulty, they switched to a new 2 + (3 + 3) synthetic route, where coupling of CDE trisaccharide donor **230** and FGH acceptor **231** was performed first (Scheme 8.24).[53] After protective group manipulations and oxidation state adjustment, hexasaccharide acceptor **232** was obtained (Scheme 8.24A), which was subsequently glycosylated by AB disaccharide donor **235** (Scheme 8.24B). Protective group adjustment and sulfation of the resulting octasaccharide module **236** produced 4,6-*O* disulfated octasaccharide

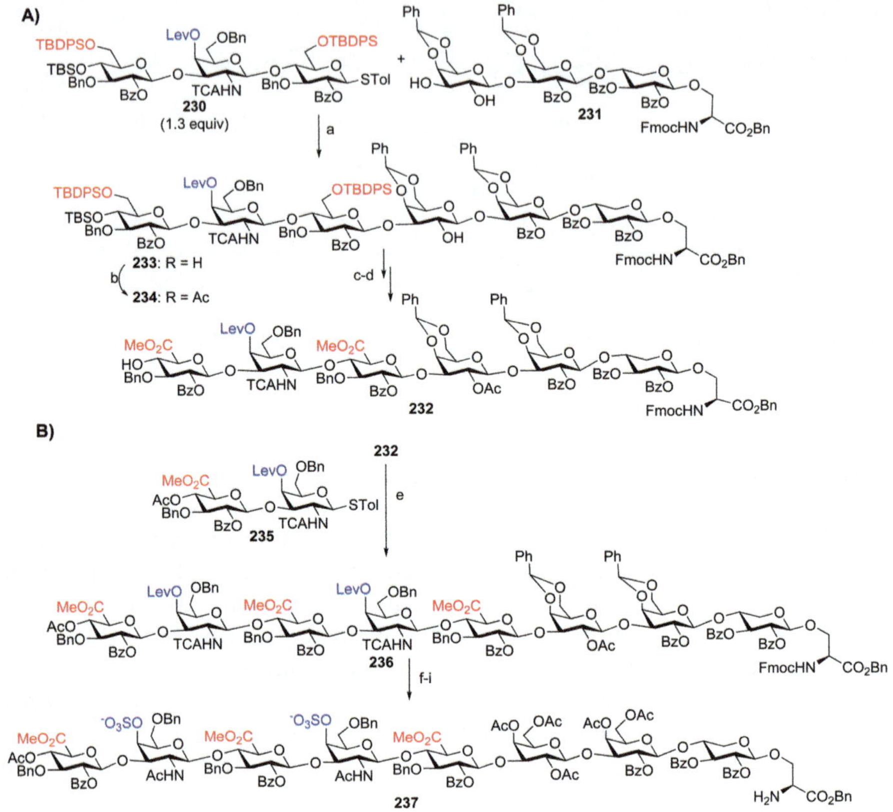

Scheme 8.24 (A) Synthesis of hexasaccharide CDEFGH; (B) Synthesis of Octasaccharide ABCDEFGH: (a) NIS, TfOH, −60 °C to rt, 3 h (95%); (b) Ac_2O, pyridine (1 : 2), rt, overnight (93%); (c) HF-pyridine, pyridine, 0 °C, overnight (92%); (d) (i) TEMPO, NaBr, *n*-Bu_4NBr, $NaHCO_3$, NaOCl, DCM/H_2O, then *t*-BuOH, 2-methylbut-2-ene, $NaClO_2$; (ii) MeI, K_2CO_3, DMF, rt, overnight; (e) NIS, TfOH, −60 °C to rt, 3 h (70%); (f) (i) 80% AcOH, DCM, 60 °C; (ii) Ac_2O, py., (1 : 2), rt, overnight (84% in 2 steps); (iii) Zn–Cu (5 equiv./Cl), AcOH, 50 °C, 24 h (70%); (g) NH_2NH_3OAc (10 equiv.), THF/MeOH, 0 °C, 2 h (90%); (h) $SO_3{\cdot}Et_3N$ (4 equiv.), DMF, 50 °C, overnight (81%); (i) piperidine, DMF, rt, 20 min (93%).

cassette **237** for peptide chain elongation. Coupling between **237** and peptide **238** using *O*-(7-azabenzotriazol-1-yl)-*N,N,N′,N′*-tetramethyluronium hexafluorophosphate (HATU), followed by catalytic hydrogenolysis afforded the carboxylic acid glycopeptide **239** in high yield (Scheme 8.25). **239** was further extended at the C-terminus producing the desired glycopeptide **228** after final deprotection.

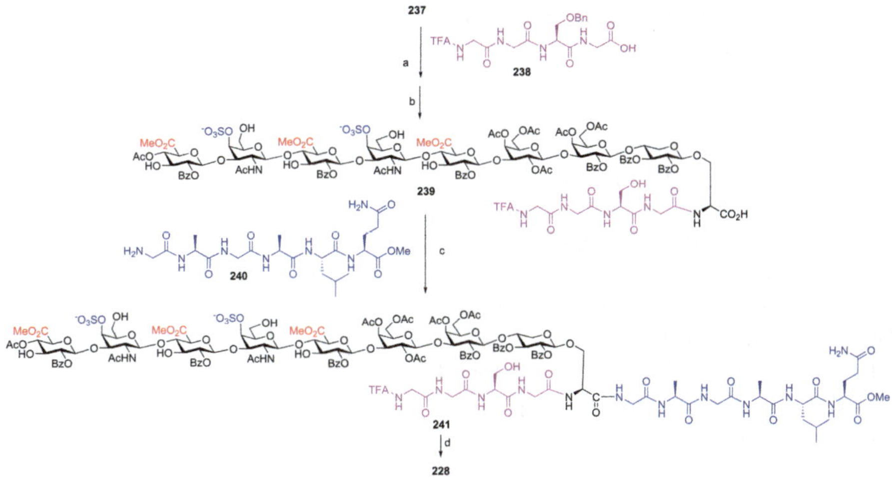

Scheme 8.25 Synthesis of syndecan-1 chondroitin sulfate glycopeptide **228**: (a) HATU, 2,6-Collidine, DMF, rt, 3 h; (b) H_2, $Pd(OH)_2$, MeOH/DCM (1 mL), overnight; (c) HATU (1.5 equiv.), DIPEA (1.5 equiv.), DMF, rt, 4 h (60%); (d) (i) LiOH, H_2O_2, THF, H_2O, pH = 9, 0 °C, 2 h; (ii) NH_2NH_2 hydrate, MeOH, rt, overnight (71%).

8.6 Conclusions

In this chapter, we summarized the efforts to access well-defined CS oligosaccharides with different sulfation patterns using either total synthesis or semi-synthesis strategies from natural CS. CS glycopolymers and glycoconjugates offer a powerful approach to explore the activities of such macromolecules and for manipulating CS functions in biological systems. In addition, the efforts for the synthesis of CS glycopeptide with well-defined sulfate patterns were reviewed. There is still a need for synthesis of full library of CS oligosaccharides longer than disaccharides that can cover most sulfation subtypes to facilitate the biological studies of these interesting molecules. Automated synthesis can play an important role in the future for the assembly of CS oligosaccharides and CS glycopeptides. Furthermore, a chemo-enzymatic approach, which integrates chemical and enzymatic synthesis, is a promising direction to pursue to prepare a wide range of CS structures.

Abbreviations

AIBN Azobisisobutyronitrile
Bn Benzyl
Bu Butyl
Bz Benzoyl

CAN	Cerium ammonium nitrate
Cbz	Carboxybenzyl
ClAc	Chloroacetyl
CS	Chondroitin sulfate
CS4OST	4-*O*-Sulfotransferase
CS6OST	6-*O*-Sulfotransferase
CSA	Camphorsulfonic acid
CSPG	Chondroitin sulfate proteoglycan
CuAAC	Copper-catalyzed azide–alkyne cycloaddition
DBU	1,8-Diazabicyclo(5.4.0)undec-7-ene
DCM	Dichloromethane
DDQ	2,3-Dichloro-5,6-dicyano-1,4-benzoquinone
DMAP	*N*,*N*-Dimethylamino pyridine
DMF	*N*,*N*-Dimethylformamide
Fmoc	Fluorenylmethyloxycarbonyl
GAG	Glycosaminoglycan
Gal	D-Galactose
GalN	2-Amino-2-deoxy-D-galactose
GalNAc	2-Acetamido-2-deoxy-D-galactose
GlcA	D-Glucuronic acid
HATU	*O*-(7-Azabenzotriazol-1-yl)-*N*,*N*,*N*′,*N*′-tetramethyluronium hexafluorophosphate
Lev	Levulinoyl
MBz	4-Methylbenzoyl
MOM	Methoxymethyl
MP	*p*-Methoxyphenyl
MS	Molecular sieves
MW	Microwave
NAHK	*N*-Acetylhexosamine kinase
PA	Pyridylamino
PAPS	3′-Phosphoadenosine-5′-phosphosulfate
pNP	*p*-Nitrophenyl
TBDPS	*tert*-Butyldiphenylsilyl
TBS	*tert*-Butyldimethylsilyl
TBTH	Tributyltin hydride
TCA	Trichloroacetamido
TCP	*N*-Tetrachlorophthaloyl
TFA	Trifluoroacetyl
THF	Tetrahydrofuran
TMA	Trimethylamine
TMSOTf	Trimethylsilyl triflate
Troc	2,2,2-Trichloroethoxycarbonyl
UDP	Uridyl diphosphate
Xyl	Xylose

Acknowledgement

We are grateful for the financial support from National Science Foundation (CHE 1507226) and the National Institute of General Medical Sciences, NIH (R01GM072667, U01GM116262).

References

1. N. Volpi, *Chondroitin Sulfate: Structure, Role and Pharmacological Activity*, Academic Press, San Diego, 2006.
2. S. Yamada and K. Sugahara, *Curr. Drug Discovery Technol.*, 2008, **5**, 289.
3. L. Kjellen and U. Lindahl, *Annu. Rev. Biochem.*, 1991, **60**, 443.
4. T. Mikami and H. Kitagawa, *Biochim. Biophys. Acta*, 2013, **1830**, 4719.
5. K. Sugahara and H. Kitagawa, *Curr. Opin. Struct. Biol.*, 2000, **10**, 518.
6. H. Kitagawa, T. Izumikawa, T. Uyama and K. Sugahara, *J. Biol. Chem.*, 2003, **278**, 23666.
7. M. Kusche-Gullberg and L. Kjellen, *Curr. Opin. Struct. Biol.*, 2003, **13**, 605.
8. H. E. Murrey and L. C. Hsieh-Wilson, *Chem. Rev.*, 2008, **108**, 1708.
9. S. Mizuguchi, T. Uyama, H. Kitagawa, K. H. Nomura, K. Dejima, K. Gengyo-Ando, S. Mitani, K. Sugahara and K. Nomura, *Nature*, 2003, **423**, 443.
10. W. W. Du, B. B. Yang, T. A. Shatseva, B. L. Yang, Z. Deng, S. W. Shan, D. Y. Lee, A. Seth and A. J. Yee, *PLoS One*, 2010, **5**, e13828.
11. R. Lin, T. W. Rosahl, P. J. Whiting, J. W. Fawcett and J. C. F. Kwok, *PLoS One*, 2011, **6**, e21499.
12. S. E. Tully, R. Mabon, C. I. Gama, S. M. Tsai, X. Liu and L. C. Hsieh-Wilson, *J. Am. Chem. Soc.*, 2004, **126**, 7736.
13. C. M. Galtrey and J. W. Fawcett, *Brain Res. Rev.*, 2007, **54**, 1.
14. M. Vallières and P. du Souich, *Osteoarthritis Cartilage*, 2010, **18**, S1.
15. N. Volpi, *Inflammopharmacology*, 2011, **19**, 299.
16. D. A. DeWitt, J. Silver, D. R. Canning and G. Perry, *Exp. Neurol.*, 1993, **121**, 149.
17. D. M. Snow, D. A. Steindler and J. Silver, *Dev. Biol.*, 1990, **138**, 359.
18. R. Bravo, M. Arimon, J. J. Valle-Delgado, R. Garcia, N. Durany, S. Castel, M. Cruz, S. Ventura and X. Fernandez-Busquets, *J. Biol. Chem.*, 2008, **283**, 32471.
19. C. I. Gama, S. E. Tully, N. Sotogaku, P. M. Clark, M. Rawat, N. Vaidehi, W. A. Goddard, A. Nishi and L. C. Hsieh-Wilson, *Nat. Chem. Biol.*, 2006, **2**, 467.
20. J. E. Turnbull and R. J. Linhardt, *Nat. Chem. Biol.*, 2006, **2**, 449.
21. M. Mende, C. Bednarek, M. Wawryszyn, P. Sauter, M. B. Biskup, U. Schepers and S. Bräse, *Chem. Rev.*, 2016, **116**, 8193.
22. Y. G. Shi, Y. C. Meng, J. R. Li, J. Chen, Y. H. Liu and X. Bai, *J. Chem. Technol. Biotechnol.*, 2014, **89**, 1445.

23. A. Vibert, J. C. Jacquinet and C. Lopin-Bon, *J. Carbohydr. Chem.*, 2011, **30**, 393.
24. J.-i. Tamura, K. W. Neumann, S. Kurono and T. Ogawa, *Carbohydr. Res.*, 1997, **305**, 43.
25. J. Tamura and M. Tokuyoshi, *Biosci., Biotechnol., Biochem.*, 2004, **68**, 2436.
26. J.-i. Tamura, Y. Nakada, K. Taniguchi and M. Yamane, *Carbohydr. Res.*, 2008, **343**, 39.
27. L. Liao and F.-I. Auzanneau, *J. Org. Chem.*, 2005, **70**, 6265.
28. C. Coutant and J. C. Jacquinet, *J. Chem. Soc., Perkin Trans. 1*, 1995, 1573.
29. J. C. Jacquinet, C. Lopin-Bon and A. Vibert, *Chem. - Eur. J.*, 2009, **15**, 9579.
30. C. Lopin and J.-C. Jacquinet, *Angew. Chem., Int. Ed.*, 2006, **45**, 2574.
31. A. Vibert, C. Lopin-Bon and J.-C. Jacquinet, *Chem. - Eur. J.*, 2009, **15**, 9561.
32. A. Vibert, C. Lopin-Bon and J. C. Jacquinet, *Eur. J. Org. Chem.*, 2011, **2011**, 4183.
33. C. I. Gama, S. E. Tully, N. Sotogaku, P. M. Clark, M. Rawat, N. Vaidehi, W. A. Goddard 3rd, A. Nishi and L. C. Hsieh-Wilson, *Nat. Chem. Biol.*, 2006, **2**, 467.
34. A. Vibert, C. Lopin-Bon and J.-C. Jacquinet, *Tetrahedron Lett.*, 2010, **51**, 1867.
35. X. Lu, M. N. Kamat, L. Huang and X. Huang, *J. Org. Chem.*, 2009, **74**, 7608.
36. P. Chassagne, L. Raibaut, C. Guerreiro and L. A. Mulard, *Tetrahedron*, 2013, **69**, 10337.
37. S. Ramadan, W. Yang, Z. Zhang and X. Huang, *Org. Lett.*, 2017, **19**, 4838.
38. H. Gold, S. Munneke, J. Dinkelaar, H. S. Overkleeft, J. M. Aerts, J. D. Codee and G. A. van der Marel, *Carbohydr. Res.*, 2011, **346**, 1467.
39. A. Vibert, C. Lopin-Bon and J. C. Jacquinet, *Tetrahedron Lett.*, 2010, **51**, 1867.
40. M. Guillemineau and F.-I. Auzanneau, *J. Org. Chem.*, 2012, **77**, 8864.
41. M. T. C. Walvoort, A. G. Volbeda, N. R. M. Reintjens, H. van den Elst, O. J. Plante, H. S. Overkleeft, G. A. van der Marel and J. D. C. Codée, *Org. Lett.*, 2012, **14**, 3776.
42. J. Dinkelaar, H. Gold, H. S. Overkleeft, J. D. C. Codée and G. A. van der Marel, *J. Org. Chem.*, 2009, **74**, 4208.
43. S. Maza, M. Mar Kayser, G. Macchione, J. Lopez-Prados, J. Angulo, J. L. de Paz and P. M. Nieto, *Org. Biomol. Chem.*, 2013, **11**, 3510.
44. A. F. G. Bongat, M. N. Kamat and A. V. Demchenko, *J. Org. Chem.*, 2007, **72**, 1480.
45. H. Weiss and C. Unverzagt, *Angew. Chem., Int. Ed.*, 2003, **42**, 4261.
46. A. Imamura, A. Kimura, H. Ando, H. Ishida and M. Kiso, *Chem. - Eur. J.*, 2006, **12**, 8862.
47. S. Maza, G. Macchione, R. Ojeda, J. Lopez-Prados, J. Angulo, J. L. de Paz and P. M. Nieto, *Org. Biomol. Chem.*, 2012, **10**, 2146.
48. S. Maza, J. L. de Paz and P. M. Nieto, *Tetrahedron Lett.*, 2011, **52**, 441.

49. G. Macchione, S. Maza, M. Mar Kayser, J. L. de Paz and P. M. Nieto, *Eur. J. Org. Chem.*, 2014, **2014**, 3868.
50. G. Despras, C. Bernard, A. Perrot, L. Cattiaux, A. Prochiantz, H. Lortat-Jacob and J.-M. Mallet, *Chem. - Eur. J.*, 2013, **19**, 531.
51. P. A. Levene, *J. Biol. Chem.*, 1941, **140**, 267.
52. E. A. Davidson and K. Meyer, *J. Am. Chem. Soc.*, 1954, **76**, 5686.
53. R. R. Schmidt, *Angew. Chem.*, 1986, **98**, 213.
54. S. Eller, M. Collot, J. Yin, H. S. Hahm and P. H. Seeberger, *Angew. Chem., Int. Ed.*, 2013, **52**, 5858.
55. N. Afratis, C. Gialeli, D. Nikitovic, T. Tsegenidis, E. Karousou, A. D. Theocharis, M. S. Pavao, G. N. Tzanakakis and N. K. Karamanos, *FEBS J.*, 2012, **279**, 1177.
56. K. J. Svensson, H. C. Christianson, P. Kucharzewska, V. Fagerstrom, L. Lundstedt, S. Borgquist, K. Jirstrom and M. Belting, *Int. J. Oncol.*, 2011, **39**, 1421.
57. R. D. Prinz, C. M. Willis, A. Viloria-Petit and M. Kluppel, *Genet. Mol. Res.*, 2011, **10**, 3901.
58. Z. Wei Poh, C. Heng Gan, E. J. Lee, S. Guo, G. W. Yip and Y. Lam, *Sci. Rep.*, 2015, **5**, 14355.
59. W. Pilgrim and P. V. Murphy, *Org. Lett.*, 2009, **11**, 939.
60. A. R. Banaag and M. A. Tius, *J. Am. Chem. Soc.*, 2007, **129**, 5328.
61. R. U. Lemieux and R. M. Ratcliffe, *Can. J. Chem.*, 1979, **57**, 1244.
62. A. Vibert, C. Lopin-Bon and J.-C. Jacquinet, *Eur. J. Org. Chem.*, 2011, **2011**, 4183.
63. K. Miyachi, M. Wakao and Y. Suda, *Bioorg. Med. Chem. Lett.*, 2015, **25**, 1552.
64. K. Ait-Mohand, A. Mirault, J.-C. Jacquinet and C. Lopin-Bon, *Org. Biomol. Chem.*, 2016, **14**, 7962.
65. K. Aït-Mohand, C. Lopin-Bon and J.-C. Jacquinet, *Carbohydr. Res.*, 2012, **353**, 33.
66. N. Sugiura, S. Shimokata, T. Minamisawa, J. Hirabayashi, K. Kimata and H. Watanabe, *Glycoconjugate J.*, 2008, **25**, 521.
67. J. Li, G. Su and J. Liu, *Angew. Chem., Int. Ed.*, 2017, **56**, 11784.
68. G. H. Zhao, W. Y. Guan, L. Cai and P. G. Wang, *Nat. Protoc.*, 2010, **5**, 636.
69. V. Bourgeaux, F. Piller and V. Piller, *Bioorg. Med. Chem. Lett.*, 2005, **15**, 5459.
70. M. Rawat, C. I. Gama, J. B. Matson and L. C. Hsieh-Wilson, *J. Am. Chem. Soc.*, 2008, **130**, 2959.
71. S. G. Lee, J. M. Brown, C. J. Rogers, J. B. Matson, C. Krishnamurthy, M. Rawat and L. C. Hsieh-Wilson, *Chem. Sci.*, 2010, **1**, 322.
72. S. Bera and R. J. Linhardt, *J. Org. Chem.*, 2011, **76**, 3181.
73. P. Domínguez-Rodríguez, J. J. Reina, S. Gil-Caballero, P. M. Nieto, J. L. de Paz and J. Rojo, *Chem. - Eur. J.*, 2017, **23**, 11338.
74. C. Y. Pumphrey, A. M. Theus, S. Li, R. S. Parrish and R. D. Sanderson, *Cancer Res.*, 2002, **62**, 3722.

75. J. A. Borgia, N. B. Malkar, H. U. Abbasi and G. B. Fields, *J. Biomol. Tech.*, 2001, **12**, 44.
76. P. Sjölin, M. Elofsson and J. Kihlberg, *J. Org. Chem.*, 1996, **61**, 560.
77. B. Yang, K. Yoshida, Z. Yin, H. Dai, H. Kavunja, M. H. El-Dakdouki, S. Sungsuwan, S. B. Dulaney and X. Huang, *Angew. Chem., Int. Ed.*, 2012, **51**, 10185.
78. X.-T. Chen, D. Sames and S. J. Danishefsky, *J. Am. Chem. Soc.*, 1998, **120**, 7760.
79. K. Yoshida, B. Yang, W. Yang, Z. Zhang, J. Zhang and X. Huang, *Angew. Chem., Int. Ed.*, 2014, **53**, 9051.
80. W. Yang, K. Yoshida, B. Yang and X. Huang, *Carbohydr. Res.*, 2016, **435**, 180.
81. S. Ramadan, W. Yang and X. Huang, in *Chemical Biology of Glycoproteins*, The Royal Society of Chemistry, 2017, p. 209.

CHAPTER 9

Chemoenzymatic Synthesis of Heparan Sulfate and Heparin

YONGMEI XU* AND JIAN LIU*

Division of Chemical Biology and Medicinal Chemistry, Eshelman School of Pharmacy, University of North Carolina, Chapel Hill, North Carolina, USA
*E-mail: yongmeix@email.unc.edu, jian_liu@unc.edu

9.1 Introduction

9.1.1 Heparan Sulfate

Heparan sulfate (HS) is ubiquitously expressed on cell surfaces and in the extracellular matrix. HS is a highly negatively charged polysaccharide. Hundreds of proteins have evolved the capacity to interact with HS, often with great specificity.[1] HS is present in the form of HS proteoglycan (HSPG) in nature, consisting of a core protein and HS polysaccharide side chains. HSPGs display essential functions in development and homeostasis, as well as playing various roles in disease processes, including embryonic development, inflammatory responses, bacterial/viral infection and blood coagulation.[2] The functions of HSPGs are mainly dependent on interactions between the HS-side chains with a variety of proteins including cytokines, growth factors, and their receptors.[2] In a given HS polysaccharide,

Chemical Biology No. 11
Synthetic Glycomes
Edited by Wanyi Guan, Lei Li and Peng George Wang

Published by the Royal Society of Chemistry, www.rsc.org

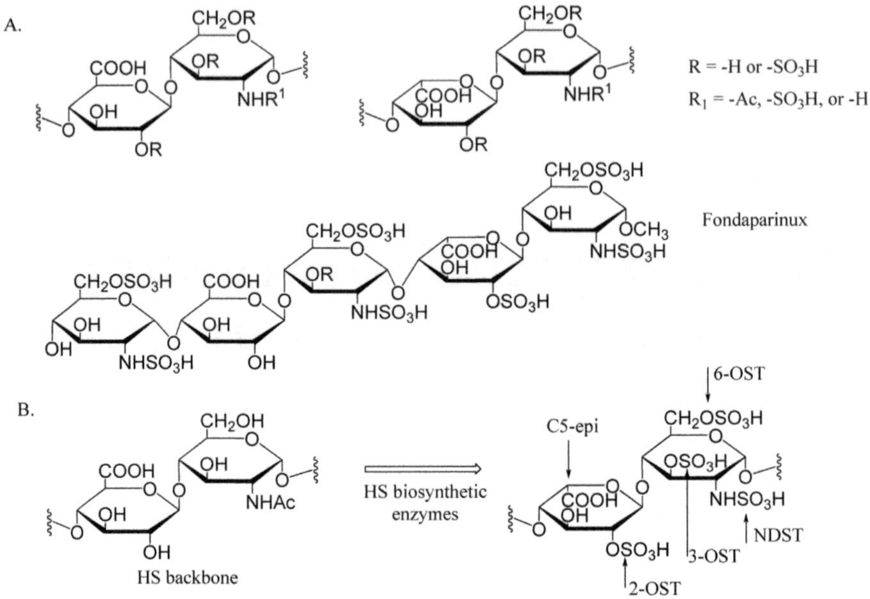

Figure 9.1 Chemical structures of HS, fondaparinux and HS biosynthetic enzymes. Panel A shows the chemical structures of HS disaccharide repeating units and fondaparinux. Panel B shows the enzymes involved to convert a HS backbone to a fully elaborated HS structure. The sites of modification by each individual enzyme are indicated by arrows. The enzymes are presented with abbreviated names: C5-epi represents C5-epimerase; 2-OST represents 2-*O*-sulfotransferase; 6-OST represents 6-*O*-sulfotransferase; 3-OST represents 3-*O*-sulfotransferase; and NDST represents *N*-deacetylase/*N*-sulfotransferase.

negatively charged sulfate and the configuration of carboxylate groups are placed on sugar residues through biosynthetic enzymes with enormous potential for structural variability. The wide range of functions have attracted considerable interest in the development of new medicines using HS as a structural scaffold.

HS consists of a disaccharide repeating unit of either iduronic acid (IdoA) or glucuronic acid (GlcA) and glucosamine (GlcN) residues, each capable of carrying sulfate groups (Figure 9.1). The presence of sulfate groups and the location of IdoA and GlcA residues dictate the functional selectivity of HS. The challenge in developing HS-based drugs centers on the synthesis of HS oligosaccharides and polysaccharides having the desired size and sulfation patterns.

9.1.2 Heparin

Heparin, a highly sulfated form of HS, is a widely used anticoagulant used to prevent and treat arterial and venous thrombosis.[2,3] Discovered in 1916, heparin has been the drug of choice to treat thrombotic disorders for nearly 90 years.[4] The discovery of heparin contributed significantly to the development of many advanced medical and surgical procedures.[5] Three forms of heparin are approved by the US Food and Drug Administration (FDA): unfractionated heparin (UFH) average molecular weight (MWavg 16 000 Da), low-molecular weight heparin (MWavg 3500–6000 Da) and fondaparinux (MW 1508 Da). UFH is a safe intravenous drug for the treatment of renal-impaired patients and its effects can be reversed using the cationic-polypeptide drug protamine;[6] however, it shows a 1–6% incidence of heparin-induced thrombocytopenia (HIT), a life-threatening complication.[7] LMWHs are subcutaneously administered and have a longer half-life than UFH, permitting use with outpatients and self-administration.[4] However, LMWH can only be used in renal-impaired patients at the reduced doses[8] and is incompletely neutralized with protamine, thereby increasing the risks of bleeding. Fondaparinux, a synthetic pentasaccharide, is subcutaneously bioavailable and has reduced risks of HIT and osteoporosis.[9] However, it is primarily excreted through the kidney and, thus, is not suitable for renal-impaired patients, and it lacks an antidote.[8]

Both UFH and LMWH are considered to be animal-sourced products, whereas fondaparinux is a synthetic.[10] UFH is isolated from porcine intestine or bovine lung or intestine, while LMWHs are a depolymerized product of heparin from porcine intestine. A worldwide contamination of heparin in 2007 affected the purity and safety of both UFH and LMWHs associated with over 200 deaths in the US.[11,12] This incident revealed the fragility of the heparin supply chain. Although fondaparinux's supply is independent of the animal sources,[13] the drug does not display all the clinical benefits that UFH and LMWH do. A cost-effective method to prepare a synthetic heparin and LMWH that can be manufactured under FDA's guidelines will eliminate the need for animal sourced materials.[12] Furthermore, the synthetic heparins could be engineered to introduce additional beneficial pharmacological properties that do not exist in UFH and LMWHs from natural sources.[4]

Heparin and HS have similar disaccharide repeating units; however, heparin has higher sulfation levels than HS. Heparin contains 2.6 sulfo groups per disaccharide compared to 0.6 sulfo groups in HS. In addition, nearly 90% of the disaccharide units in heparin contain IdoA, while only 20% of the disaccharide units in HS contain IdoA. HS can be isolated from many cell types, whereas heparin is an exclusive product of mast cells.

9.1.3 Chemoenzymatic Synthesis of HS and Heparin

HS oligosaccharides have been synthesized by a purely chemical approach; however, this synthesis is difficult, due to the required use of complex protecting/deprotecting procedures. The syntheses of diversified HS structures are further complicated because they require the preparation of a large number of precursor compounds. Therefore, HS oligosaccharide synthesis can be only completed by skilled synthetic chemists in a small number of highly specialized labs.[13–22] In recent years, a chemoenzymatic synthesis has emerged using glycosyltransferases, epimerase and sulfotransferases.[23,24] Compared to chemical synthesis, the chemoenzymatic approach offers shorter synthetic routes, excellent recovery yields, and utilizes a few common precursors to prepare diverse HS structures, which more resemble the authentic domain structure.

The chemoenzymatic synthesis of HS requires three classes of biosynthetic enzymes, some comprised of multiple isoforms.[12] Glycosyltransferases are used to build the HS backbone structure consisting of a disaccharide repeating unit of GlcA and *N*-acetylglucosamine (GlcNAc).[25] An epimerase (C5-epi) is responsible for converting a GlcA residue to an IdoA residue (Figure 9.1B).[26] Sulfo groups are transferred from 3′-phosphoadenosine-5′-phosphosulfate (PAPS) to oligosaccharide by HS sulfotransferases. For example, 2-*O*-sulfotransferase (2-OST) transfers a sulfo group to the 2-position of an IdoA residue and, to a lesser extent, to the 2-position of a GlcA residue (Figure 9.1B).[27,28] The 6-*O*-sulfotransferase (6-OST) and the 3-*O*-sulfotransferase (3-OST) transfer sulfo groups to the 6-position and 3-position of glucosamine residues.[28,29] *N*-deacetylase/*N*-sulfotransferase (NDST) converts a GlcNAc residue to a *N*-sulfoglucosamine (GlcNS) residue.[30] NDST is unique in displaying two activities, *N*-deacetylase activity removing *N*-acetyl group from a GlcNAc residue, and *N*-sulfotransferase activity transferring a sulfo group to a GlcN residue, forming a GlcNS residue (Figure 9.1B).[31,32]

The enzymes used for the chemoenzymatic synthesis of heparin are predominantly mammalian proteins, presenting challenges for making these recombinant proteins accessible in large quantities. Currently, most of these HS biosynthetic enzymes, with the exception of NDST, EXT-1 and EXT-2, have been expressed in *Escherichia coli* with sufficient efficiency for high milligram-scale synthesis.[33] EXT-1 and EXT-2 are HS polymerase involving in the biosynthesis of HS backbones. Only the *N*-sulfotransferase (NST) domain of the NDST can be effectively expressed in *E. coli* and, thus, this protein lacks *N*-deacetylase activity. Two bacterial glycosyltransferases known as KfiA (an *N*-acetyl glucosaminyl transferase from the *E. coli* K5 strain) and pmHS2 (heparosan synthase 2 from *Pasteurella multocida*) are used as substitutes of EXT1 and EXT2, to build the HS backbone. Both KfiA and pmHS2 enzymes can be readily expressed in *E. coli*.[34,35]

9.2 Key Techniques of Synthesis

9.2.1 Design of Sugar Nucleotides for the Chemoenzymatic Synthesis of HS

Chemoenzymatic synthesis of HS requires a cascade of enzymatic modifications. The *N*-sulfation of glucosamine is essential for the downstream enzymatic modifications to occur.[36] The synthesis of the HS

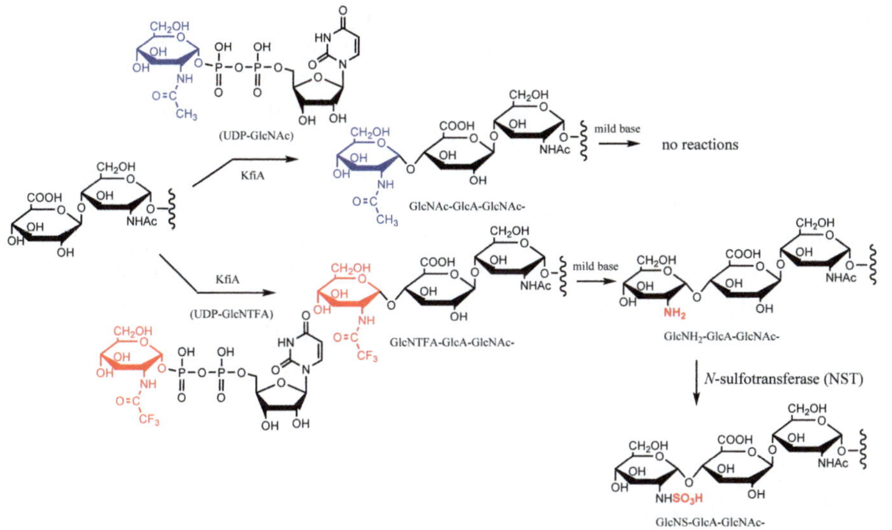

Figure 9.2 Synthesis of oligosaccharides containing GlcNS residues using a non-natural sugar nucleotide.[72] KfiA transfers a GlcNAc residue to the primer substrate, where the sugar nucleotide donor is UDP-GlcNAc, a natural substrate for KfiA. KfiA also transfers a GlcNTFA residue to the primer substrate, where the sugar nucleotide donor is UDP-GlcNTFA. The trifluoroacetyl group can be removed under mild base conditions to form an N-unsubstituted glucosamine (GlcN) residue. The GlcN residue is then converted to GlcNS by *N*-sulfotransferase (NST).

backbone containing GlcNS residues represented a major roadblock due to the lack of efficient recombinant NDST. An unnatural sugar nucleotide, UDP-GlcNTFA (uridine diphosphate *N*-trifluoroacetylated glucosamine), was introduced to overcome this problem.[37,38] UDP-GlcNTFA was first reportedly used in the synthesis of sugars of *O*-linked glycoprotein by "core-2" GlcNAc transferase with a modest yield of 56%.[39] For HS backbone synthesis, UDP-GlcNTFA is an excellent substrate for KfiA and pmHS2.[40] The GlcNTFA residue is then converted to a GlcNS residue by removing the trifluoroacetyl group under mild alkaline conditions followed by *N*-sulfotransferase modification (Figure 9.2). The use of UDP-GlcNTFA provides a means to introduce a GlcNS at precisely the desired position within an oligosaccharide. This strategy is now used to synthesize structurally homogeneous ultra-low molecular weight heparins, low-molecular weight heparins[23,24] and heterogeneous heparin-like polysaccharides.[41,42]

A novel strategy to utilize additional unnatural sugar nucleotides to synthesize heparin-like oligosaccharide was also proposed by Chen and coworkers.[43] In this study, the 6-hydroxyl group of GlcN was substituted by an azido group to synthesize sugar nucleotides, including UDP-[6-azido]GlcNAc or UDP-[6-azio]GlcNTFA. Chen and coworkers demonstrated the synthesis of oligosaccharides, which carry the azido groups at the 6-position of the GlcN residue, in excellent yields using pmHS2 enzyme. The azido group was then converted to an amino group followed by *N*-sulfation to generate 6-*N*-sulfo GlcN residues. Ultimately, HS-analog oligosaccharides, containing *N*-sulfate groups at both the 2- and the 6-positions of GlcN were synthesized, mimicking the 2-*N*-sulfo, 6-*O*-sulfo GlcN residues in natural HS and heparins. Because this synthesis only involves the use of glycosyltransferase and chemoselective *N*-sulfation to produce sulfated saccharides, the approach eliminates the potential complexity in introducing a sulfate group at a desired site by relying on the substrate specificity of the *O*-sulfotransferases. Because of this unnatural 6-*N*-sulfation these products might be resistant to heparanase and lysosomal degradation, increasing the biological half-life of oligosaccharides.

9.2.2 Sequences of Enzymatic Modifications for the Synthesis of Different Oligosaccharides

The chemoenzymatic synthesis of heparin requires a series of enzymatic modification steps. The design of an appropriate modification sequence is closely related to the substrate specificities of the biosynthetic enzymes. *In vivo*, a HS biosynthetic pathway is responsible for installing the multiple layers of sulfation on the polysaccharide backbone by following a specific order. The cells probably utilize the modification sequence to partially control HS structures. Here are three specific examples:

Example 1. Control of the enzymatic modification sequences by HS sulfotransferases

The 2-*O*-sulfation step catalyzed by 2-OST occurs only after the *N*-sulfation step, but prior to the 6-*O*-sulfation step. This modification sequence is attributed to the substrate specificity of 2-OST. The unique structures and the spatial orientations of amino acid side chains in 2-OST contribute to the substrate recognition. Three amino acid residues of 2-OST, Arg-80, Lys-350 and Arg-190, directly interact with the *N*-sulfo group of GlcNS residue flanked on both ends of the IdoA residue. Mutation any of three residues that interact with the *N*-sulfo group result in the loss of sulfotransferase activity. Likewise, substituting the *N*-sulfo group with an *N*-acetyl group or hydrogen on the GlcN residue of the substrate abolishes its reactivity to 2-OST modification.[44] In contrast, two amino acid residues, Tyr-173 and Pro-82 near that active site of 2-OST are probably used by the enzyme to exclude the 6-*O*-sulfo group, suggesting that 6-*O*-sulfation occurs after 2-*O*-sulfation in the process of biosynthesizing highly sulfated domains consisting of -IdoA2S-GlcNS6S-. The understanding of the contribution of individual amino acid residues to the substrate recognition offers a road map to engineer unique 2-OST enzymes exhibiting distinct substrate specificity of the wild type enzyme.

Example 2. Preparation of 3-*O*-sulfated oligosaccharides

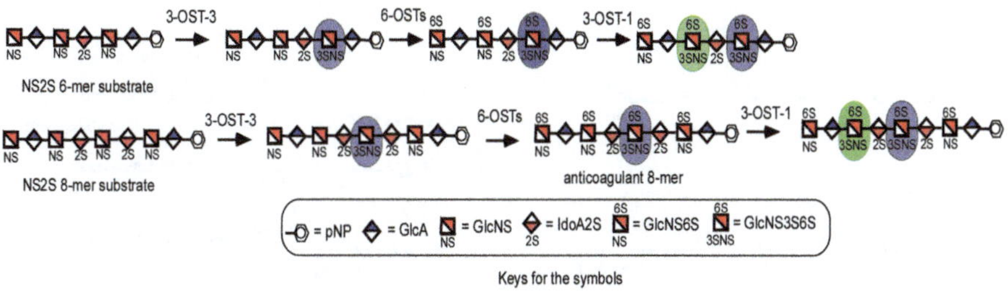

Figure 9.3 Structures and synthetic schemes for 3OST1/3different HS oligosaccharides. The 3-*O*-sulfation by 3-OST-3 enzyme is performed before the 6-*O*-sulfation. But, 3-*O*-sulfation by 3-OST-1 enzyme is performed after the 6-*O*-sulfation.

The 3-*O*-sulfation carried out by different 3-OST isoforms is regulated by preexisting 6-*O*-sulfation. The 3-*O*-sulfation is a rare modification found in HS isolated from natural sources. But, the 3-*O*-sulfation has been attributed to many different biological functions, *i.e.* binding to antithrombin to serve as an anticoagulant, to glycoprotein D of the herpes simplex virus to serve as an entry receptor on the host cells, and to neuropilin to guide neuron growth.[45,46] Recently, 3-*O*-sulfation is reportedly bound to FGF-10 to regulate epithelial differentiation.[47] There are seven different isoforms of 3-OST, and each isoform may generate the 3-*O*-sulfated glucosamine residue in different sulfated saccharide sequences. A recent study found that 3-OST-3 modification, to synthesize oligosaccharides comprising the -IdoA2S-GlcNS3S- or -IdoA2S-GlcNS3S6S- disaccharide unit, must precede the 6-*O*-sulfation step, whereas 3-OST-1 modification can occur only after 6-*O*-sulfation in order to generate the -GlcA-GlcNS3S6S disaccharide unit in Figure 9.3.[48] The distinct and unique substrate requirements between 3-OST-1 and 3-OST-3 raise a possibility that 3-Osulfated HS modified by different isoforms of 3-OST are biosynthesized through different pathways.

Example 3. Synthesis of oligosaccharides with repeating -IdoA2S-GlcNS- units

The synthesis of a domain with repeating -IdoA2S-GlcNS- disaccharide units is critically important for preparing heparin and the high sulfation domains of HS. In heparin, ~90% of the disaccharides are composed of -IdoA2S-GlcNS6S-.[47] Attempts to synthesize a domain with more than one -IdoA2S-GlcNS- motif results in a complex mixture.[49] The difficulty in installing repeating -IdoA2S-GlcNS6S units stems from a unique property of C5-epi. Unlike other HS biosynthetic enzymes, C5-epi is a two-way catalyst: it transforms a GlcA to an IdoA residue, and also converts an IdoA to a GlcA residue (Figure 9.4A).[50] Therefore, the final product is a mixture of GlcA and IdoA after C5-epi modification. However, the IdoA2S residue cannot be converted to GlcA2S residue by C5-epi. A recent study revealed that C5-epi is a one-way catalyst towards substrates having a unique distribution of GlcNAc and GlcNS residues.[51] The results indicate that C5-epi

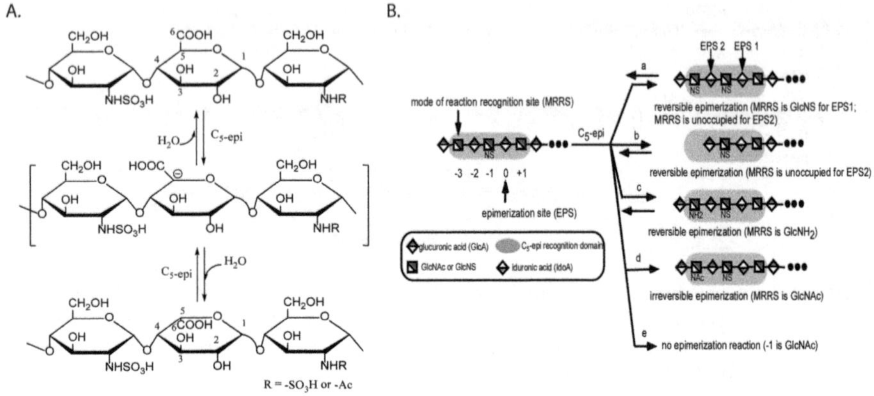

Figure 9.4 The reaction catalyzed by C5-epi and its substrate specificity. Panel A shows the reaction catalyzed by C5-epi. A trisaccharide segment is shown. C5-epi removes the proton from C5 of the GlcA residue to form a putative carbanion intermediate. Conversely, C5-epi can catalyze the reverse reaction, converting an IdoA residue to a GlcA residue. Panel B shows the substrate specificity of C5-epi. The designated epimerization site (EPS) is at the residue 0. The residue −1 must be a GlcNS residue to serve as an EPS. If residue −1 is a GlcNAc residue, the site is not reactive to C5-epi (see reaction e). The N-substitution status at the mode of reaction recognition site (MRRS) determines the mode of reaction of C5-epi. If the MRRS site (residue −3) is a GlcNS or GlcN residue or unoccupied, C5-epi displays a reversible reaction mode (see reaction a, c and b). If a GlcNAc residue is at the MRRS site, C5-epi displays an irreversible reaction mode (see reaction d). The grey shaded box indicates the pentasaccharide domain recognized by C5-epi.

can display both two-way and one-way catalytic modes by recognizing a sequence motif designated as the C5-epi recognition pentasaccharide domain (Figure 9.4B).[51] This pentasaccharide domain consists of an epimerization site (EPS) and a mode of reaction recognition site (MRRS). The MRRS is located three residues away from the non-reducing end side of the EPS (Figure 9.4B). The N-substitution status of the MRRS residue, controlled in biosynthesis by NDST,[52] dictates the mode of action of C5-epi. A GlcNAc residue directs C5-epi to display an irreversible reaction mode, while a GlcNS or GlcN residue (or the MRRS site being unoccupied) drives C5-epi to display a reversible reaction mode. The presence of GlcNS at the −1 residue is essential for rendering the susceptibility to C5-epi modification.

A synthetic pathway was employed for synthesizing IdoA2S-GlcNS repeating units to avoid by-products from the action of C5-epi (Figure 9.5).[51] Seven steps (Figure 9.5, Steps a, b, a, b, c, a, and d, transformation of 1 to 8) are required to synthesize the first IdoA2S residue from a monosaccharide starting material. Each repetition of these steps adds

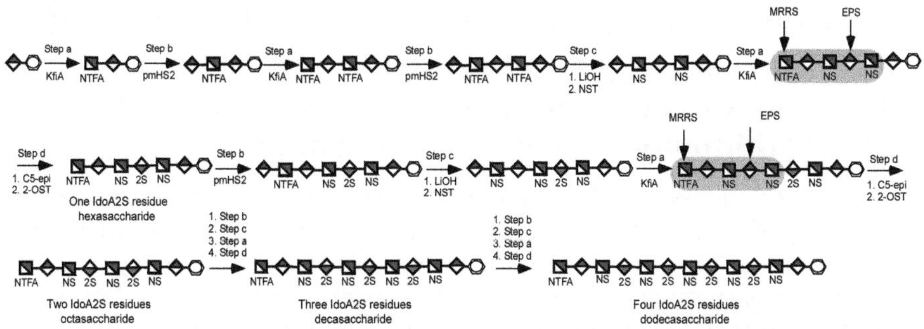

Figure 9.5 Route to synthesize oligosaccharides with IdoA2S-GlcNS repeats. The key in this strategy is to construct a C5-epi recognition pentasaccharide module to utilize the irreversible reaction. The C5-epi recognition pentasaccharide module is highlighted in a grey box. The synthesis initiates from GlcA-pnp (1, *p*-nitrophenyl glucuronide). The monosaccharide is converted to a pentasaccharide in four enzymatic steps. Hexasaccharide consists of the C5-epi recognition pentasaccharide module. The synthesis of "One IdoA2S residue hexasaccharide" is achieved by two enzymatic steps (step d) using both C5-epi and 2-OST. The construction of another C5-epi recognition pentasaccharide module involves four enzymatic steps, b, c, a and d octasaccharide. Repeating those steps can reach a decasaccharide and dodecasaccharide with 3 and 4 IdoA2S residues, respectively.

an additional IdoA2S residue, transforming the oligosaccharides containing two, three and four IdoA2S-GlcNS repeats, respectively. The construction of a C5-epi recognition pentasaccharide module with the structure of GlcNTFA-GlcA-GlcNS-GlcA-GlcNS- is the centerpiece of this strategy (Figure 9.5). This pentasaccharide module allows C5-epi to exhibit the irreversible reaction mode, leading to a single product. The resultant IdoA residue in the pentasaccharide domain is immediately 2-*O*-sulfated to form an IdoA2S residue (Figure 9.5, one IdoA2S residue hexasaccharide). Subsequent construction of the C5-epi recognition pentasaccharide module involves three steps (Figure 9.5, Step c, b and a) to introduce additional -IdoA2S-GlcNS- motifs. For example, the hexasaccharide, one IdoA2S residue hexasaccharide, is elongated to an octasaccharide with a structure of GlcNTFA-GlcA-GlcNS-GlcA-GlcNS-IdoA2S-GlcNS-GlcApnp, containing the C5-epi recognition pentasaccharide domain (highlighted in bold). The modifications by C5-epi and 2-OST (Step d and e) result in an octasaccharide of GlcNTFA-GlcA-GlcNS-IdoA2S-GlcNS-IdoA2S-GlcNS-GlcApnp (two IdoA2S residues octasaccharide). The 2-*O*-sulfation to form an IdoA2S residue is essential because it prevents further action of C5-epi on this residue and accelerates the C5-epimerization reaction.[50] These steps can be further repeated to form decasaccharide and dodecasaccharide to introduce three and four repeats of IdoA2S-GlcNS disaccharides as described in a previous publication.[24]

9.3 Chemoenzymatic Synthesis of Different HS and Heparin Oligosaccharides

9.3.1 Synthesis of HS Oligosaccharide Library and a New Antithrombin-binding Octasaccharide

Chemoenzymatic synthesis has opened up a new avenue to synthesize structurally defined HS oligosaccharides for different biological studies. The main advantage of the chemoenzymatic method is that it is relatively easier to prepare much longer oligosaccharides and the method only uses a limited number of precursors, involving only four cofactors and a single starting monosaccharide. The synthesis is readily repeatable at milligram scales for further studies, *i.e.* cocrystalization between oligosaccharides and the proteins and the intervention of biological systems using soluble oligosaccharides. This method is particularly suitable for the synthesis of a HS oligosaccharide library. A recent paper reports the synthesis of a library of HS oligosaccharides covering 65 different compounds using the chemoenzymatic method.[53] The availability of a HS library opens the opportunity to investigate the relationship between saccharide sulfation/conformation and biological functions, a major step forward in dissecting the structure and function.

One notable finding is that we discovered a new antithrombin(AT)-binding saccharide sequence -GlcNS3S6S-IdoA2S- that can be used as an anticoagulant drug. The previously known AT-binding sequence comprises a disaccharide -GlcA-GlcNS3S ± 6S- unit.[54] The substitution of the -GlcA-GlcNS3S6S- disaccharide unit with the -IdoA2SGlcNS3S6S- disaccharide unit in HS was perceived to abolish AT-binding affinity.[29,55] Our recent study challenged this assertion. One octasaccharide has a structure of GlcNS6S-GlcA-GlcNS6S-IdoA2S-GlcNS3S6S-IdoA2S-GlcNS6S-GlcA-pNP.[48] This octasaccharide contains the -IdoA2S-GlcNS3S6S- (but not -GlcA-GlcNS3S6S-) disaccharide unit, displays strong anti-FXa activity. The symbolic structure of the 8-mer named as "anticoagulant 8-mer" is shown in Figure 9.3. Furthermore, this octasaccharide displays faster clearance in a rat model than that of fondaparinux and a fondaparinux-like octasaccharide. Given the fact that a fast-clearing anticoagulant drug likely reduces the bleeding side effect, the new octasaccharide has the potential to become an anticoagulant drug for those patients with special needs.

9.3.2 HS Microarray to Target HS and Protein Interaction

Glycan microarray technology is a powerful method to investigate glycan–protein interactions.[56–59] A number of defined neutral and sulfated saccharides have been made for glycan microarrays. New classes of HS will expand the structural diversity in glycan microarray.[60–63] Recently, an HS microarray library containing 14 HS oligosaccharides was synthesized

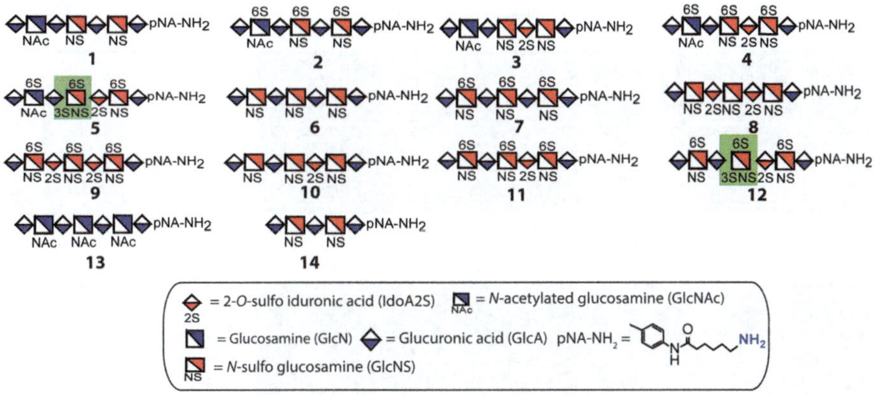

Figure 9.6 Structures of HS oligosaccharides on the array.

using a chemoenzymatic approach (Figure 9.6).[23,24] The array was probed with two glycan-binding proteins, HS 3-*O*-sulfotransferase 1 (3-OST-1) and antithrombin, demonstrating the binding selectivity between HS and proteins.

Thirteen different heptasaccharides and one hexasaccharide were used in this study. The array slides were hybridized with fluorescently labeled proteins. 3-OST-1 is an enzyme in the HS biosynthetic pathway that is responsible for introducing a 3-*O*-sulfo group to the HS.[64] Fluorescently labeled 3-OST-1 binds strongly with compounds **5**, **9**, and **12** (Figure 9.7A), and has lower binding with **4** and **11**. The results are consistent with the anticipated selectivity to the saccharide sequence based on the co-crystal structures of 3-OST-1.[65] Antithrombin (AT) is an inhibitor of factor Xa (FXa) and other proteases in the coagulation cascade; heparin exhibits anti-FXa activity by binding to AT. AT primarily binds to two oligosaccharides, compounds **5** and **12** (Figure 9.7B). Among the oligosaccharides on the array, only compounds **5** and **12** contain 3-*O*-sulfo groups, consistent with other studies, showing that the interaction between HS and AT requires 3-*O*-sulfation.[66,67] Expanding the coverage of the HS microarray to larger oligosaccharides and more diversified sulfated saccharide is underway. The HS microarray technique will provide a reliable tool for the investigation of the structure and biological function relationship.

9.3.3 Design of Heparin Drugs

Two specific examples for preparing heparin oligosaccharides with anticoagulant activity have been recently demonstrated. In the first example, it was used to synthesize a homogenous heptasaccharide (Figure 9.8). The synthesis, initiated from a disaccharide (GlcA-AnMan, AnMan represents 2,5-anhydromannitol), was completed in nine enzymatic steps with 43% recovery yield.[23] The synthesis includes the elongation of the starting

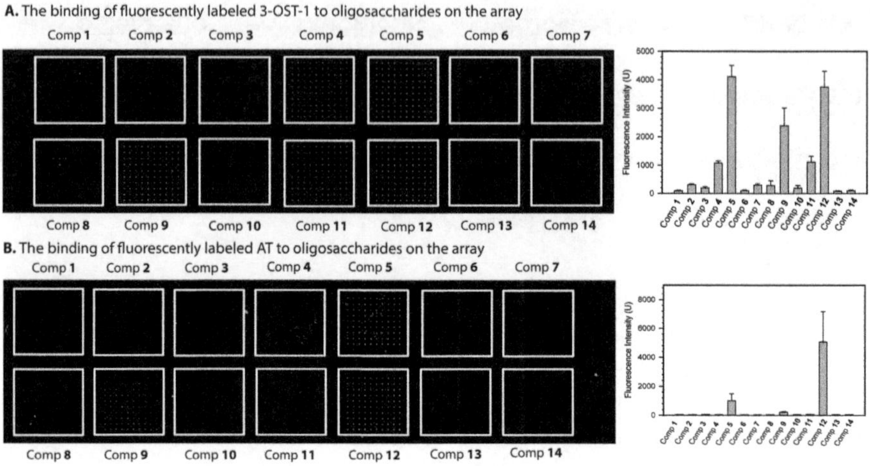

Figure 9.7 Images of HS-array analysis. Panel A shows the image of the array slide hybridized with fluorescently labeled 3-OST-1. A total of 14 oligosaccharides were affixed on the slide, and each compound was spotted up to 100 spots. The histogram of fluorescence intensity analysis is shown on the right. Panel B shows the image of the array slide hybridized with fluorescently labeled AT. The histogram of fluorescence intensity analysis is shown on the right. Both 3-OST-1 and AT were labeled with Alexa Fluo 488.

disaccharide to a heptasaccharide using KfiA and pmHS2. The heptasaccharide backbone is then converted to the final product by a series of reactions, including conversion of the GlcNTFA residue to GlcNS (Figure 9.8, step c), epimerization and 2-*O*-sulfation (Figure 9.8, step e and f), 6-*O*-sulfation (Figure 9.8, step g) and 3-*O*-sulfation (Figure 9.8, step h). The product has very similar anticoagulant potency to that of fondaparinux as measured by *in vitro* and *ex vivo* experiments. Synthesis of fondaparinux using a purely chemical approach requires a much longer route to complete.[13]

The second example is to design the structures of LMWHs with the desired pharmacological properties (Figure 9.9). Desirable properties of LMWHs include: a homogeneous structure prepared from non-animal ingredients, safe for kidney-impaired patients, and has reversible anticoagulant activity. Toward these goals, five different oligosaccharides were synthesized.[24] Among these were a 6-mer ULMW heparin target compound, very similar to fondaparinux, as well as a 10-mer and 12-mer that were in the size range of LMWHs. Because of the size and structural complexity of LMWHs, a more efficient chemoenzymatic method was required. A commercially available aglycone, 1-*O*-(*para*-nitrophenyl) glucuronide (GlcA-pNP), was used as a starting material. The presence of a pNP motif provides a UV chromophore for monitoring the product during the synthesis and purification. Nonsulfated oligosaccharides carrying this pNP motif bind to conventional C18-columns, allowing easy purification of intermediates from proteins and unreacted cofactors. Moreover, this pNP group can be removed at the end of the

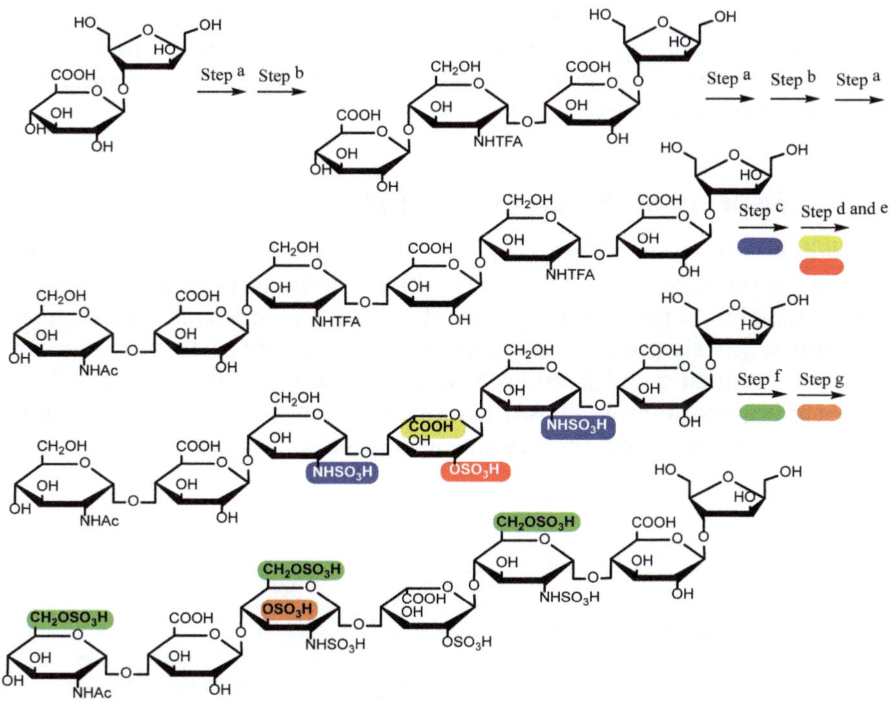

Figure 9.8 Route for the chemoenzymatic synthesis of ultra-low molecular weight heparin.

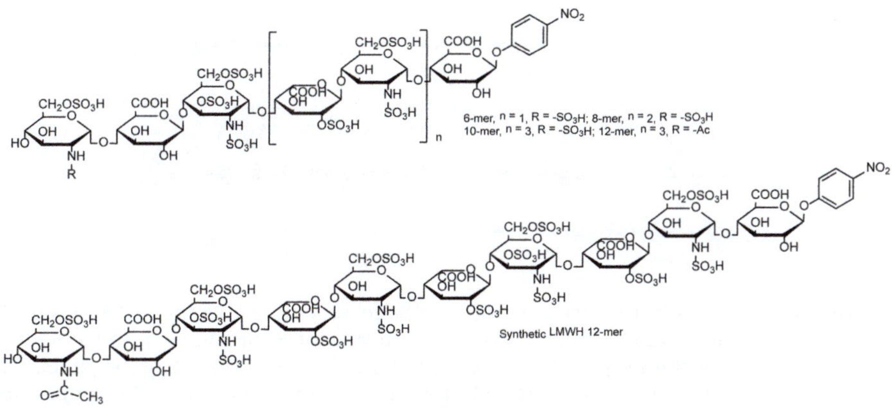

Figure 9.9 Structure of synthetic low-molecular weight heparins.

synthesis.[68] The final products were purified by anion exchange chromatography to obtain a purity of >98%. In a mouse model, oligosaccharides larger than 10-mer were metabolized in the liver, suggesting that these synthetic LMWHs will be safe for use in renal-impaired patients. Most interestingly, the anticoagulant activity of the synthetic LMWH 12-mer was reversed by protamine, an FDA approved drug to neutralize unfractionated heparin. The protamine reversibility of this 12-mer could potentially reduce the risk of bleeding in its clinical use.

9.3.4 Scale-up Synthesis of HS 12-mer

In addition to demonstrating that the chemoenzymatic method is able to synthesize HS oligosaccharide library with diversified structures, the method is also scalable to prepare 1.3 grams of 12-mer-1 compound.[69] To make the scale-up synthesis possible, a slight modification of the 12-mer was made at the nonreducing end from an *N*-acetyl glucosamine (GlcNAc) residue to an *N*-sulfo glucosamine (GlcNS) residue in Figure 9.10. This modification, suggested by analyzing the crystal structures of the 6-*O*-sulfotransferase

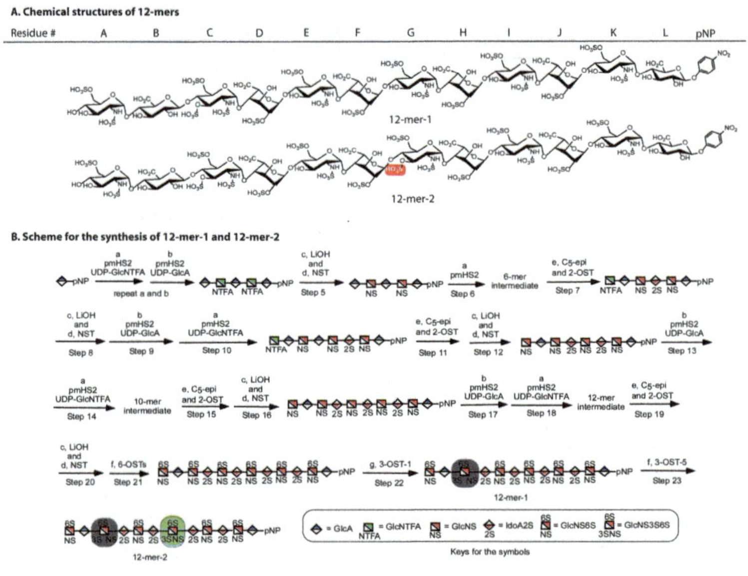

Figure 9.10 Chemical structures and synthetic scheme of synthetic LMWHs, 12-mer-1 and 12-mer-2. (A) Chemical structures of 12-mer-1 and 12-mer-2. The name of each residue (A to L) is indicated at the top of the panel. The differences between two 12-mers are the number of 3-*O*-sulfo groups present: one in 12-mer-1 in residue C and two in 12-mer-2 in residues C and G (highlighted). pNP represents *p*-nitrophenyl group. (B) Synthetic schemes for 12-mer-1 and 12-mer-2 using shorthand symbols. Each reaction step uses different enzymes and chemicals: a, PmHS2 (heparosan synthase 2 from *P. multocida*) and UDP-GlcNTFA; b, PmHS2 and UDP-GlcA; c, LiOH; d, *N*-sulfotransferase (NST) and 3′-phosphoadenosine 5′-phosphosulfate (PAPS); e, C5-epimerase (C5-epi), 2-*O*-sulfotransferase (2-OST), and PAPS; f, 6-OST-1, 6-OST-3, and PAPS; g, 3-*O*-sulfotransferase isoform 1 (3-OST-1) and PAPS; h, 3-OST-5 and PAPS.

(6-OST),[70] increased the reactivity of the substrate, making complete 6-*O*-sulfation considerably easier (reaction f, Figure 9.10B). Six 6-*O*-sulfo groups were installed into each dodecasaccharide in a single reaction step. Complete 6-*O*-sulfation reaction avoided the necessity of purifying the product from a mixture of partial 6-*O*-sulfated 12-mers, increasing product purity and yield. The 12-mer-1 maintains the anti-FXa activity demonstrated in both *in vitro* and *ex vivo* experiments. In addition, the anticoagulant activity of this 12-mer can be reversed by protamine.

Improvements in the enzyme and cofactor production also contributed to the success of the chemoenzymatic synthesis. Both C5-epi and 2-OST were expressed in insect cells using the baculovirus expression system. The activity levels of recombinant C5-epi and 2-OST from insect cells were 225- and 75-fold higher,[69] respectively, than the same enzymes expressed in *E. coli*. The cost for insect cells (Sf9 cells)-conditioned medium was ~$60 per liter, and the cost for *E. coli* culture was $3 per liter. The excellent expression of C5-epi and 2-OST offsets the higher costs of culturing insect cells.

12-mer-1 was evaluated in disease-related mouse models, used for pharmacodynamic analysis in nonhuman primates, and examined by toxicological analysis in rats. 12-mer-1 reduced the size of blood clots in the mouse model of deep vein thrombosis and attenuated circulating procoagulant markers in the mouse model of sickle cell disease. 12-mer-1 was also examined in a nonhuman primate model to determine its pharmacodynamics parameters. A 7-day toxicity study in a rat model showed no toxic effects. The data suggest that a synthetic homogeneous oligosaccharide can replace animal-sourced LMWHs.

9.4 Conclusion

There has been steady improvement in chemoenzymatic synthesis, focused on making synthesis more flexible for the preparation of different targets with increased diversity, purity and synthetic scale. The scale of synthesis has increased from 10 μg as reported in 2003[10] to 10 g, or a 1 million-fold increase.[69] This improvement in synthesis is attributable to improved enzyme expression, higher efficiency synthesis for enzyme cofactors and an improved understanding of the substrate specificities of HS biosynthetic enzymes. Recently, over 100 HS oligosaccharides have been synthesized, and this number is continuously growing. Researchers can now purchase most of these compounds from a commercial source (http://www.glycantherapeutics.com). Efforts to further expand the coverage of the HS library to larger oligosaccharides are underway. The oligosaccharides will have more diversified sulfated saccharide sequences resembling the structures of HS polysaccharides isolated from natural sources. The availability of a structurally defined HS library will accelerate dissecting the structure and function relationship.

The next goal for the chemoenzymatic method is to synthesize cost-effective homogeneous heparins to replace animal sourced heparin. The current synthesis of the 12-mer requires 22 synthetic steps with an overall yield of about 10%, which is a very efficient synthesis compared with chemical synthesis.[71] However, the cost for the 12-mer synthesis is still considerably higher than the production cost of animal-sourced low-molecular weight heparins. The favorable pharmacological benefits of synthetic 12-mer may offset its high cost to penetrate the anticoagulant drug market. Another example is the synthesis of short-acting anticoagulant 8-mer, which can be cleared from the circulation quickly. The 8-mer compound would be particularly beneficial to those patients with high bleeding risk. These efforts should accelerate the modernization of heparin drugs.

References

1. S. Sarrazin, W. C. Lamanna and J. D. Esko, *Cold Spring Harbor Perspect. Biol.*, 2011, **3**, a004952.
2. J. R. Bishop, M. Schuksz and J. D. Esko, *Nature*, 2007, **446**, 1030.
3. S. R. Kahn, W. Lim, A. S. Dunn, M. Cushman, F. Dentali, E. A. Akl, D. J. Cook, A. A. Balekian, R. C. Klein, H. Le, S. Schulman and M. H. Murad, *Chest*, 2012, **141**, e195S.
4. R. J. Linhardt and J. Liu, *Curr. Opin. Pharmacol.*, 2012, **12**, 217.
5. J. A. Marcum, *J. Hist. Med. Allied Sci.*, 2000, **55**, 37.
6. J. Hirsh, M. O'Donnell and J. W. Eikelboom, *Circulation*, 2007, **116**, 552.
7. G. M. Arepally and T. L. Ortel, *N. Engl. J. Med.*, 2006, **355**, 809.
8. S. Harder, *J. Clin. Pharmacol.*, 2012, **52**, 964.
9. J. I. Weitz and L. A. Linkins, *Expert Opin. Invest. Drugs*, 2007, **16**, 271.
10. U. Bhaskar, E. Sterner, A. M. Hickey, A. Onishi, F. Zhang, J. S. Dordick and R. J. Linhardt, *Appl. Microbiol. Biotechnol.*, 2012, **93**, 1.
11. J. E. Turnbull, *Science*, 2011, **334**, 462.
12. H. Liu, Z. Zhang and R. J. Linhardt, *Nat. Prod. Rep.*, 2009, **26**, 313.
13. M. Petitou, P. Duchaussoy, I. Lederman, J. Choay, J. C. Jacquinet, P. Sinay and G. Torri, *Carbohydr. Res.*, 1987, **167**, 67.
14. S. Arungundram, K. Al-Mafraji, J. Asong, F. E. Leach 3rd, I. J. Amster, A. Venot, J. E. Turnbull and G. J. Boons, *J. Am. Chem. Soc.*, 2009, **131**, 17394.
15. Y. P. Hu, S. Y. Lin, C. Y. Huang, M. M. Zulueta, J. Y. Liu, W. Chang and S. C. Hung, *Nat. Chem.*, 2011, **3**, 557.
16. J. L. de Paz, C. Noti and P. H. Seeberger, *J. Am. Chem. Soc.*, 2006, **128**, 2766.
17. T. Polat and C. H. Wong, *J. Am. Chem. Soc.*, 2007, **129**, 12795.
18. Z. Wang, Y. Xu, B. Yang, G. Tiruchinapally, B. Sun, R. Liu, S. Dulaney, J. Liu and X. Huang, *Chemistry*, 2010, **16**, 8365.
19. S. U. Hansen, G. J. Miller, C. Cole, G. Rushton, E. Avizienyte, G. C. Jayson and J. M. Gardiner, *Nat. Commun.*, 2013, **4**, 2016.
20. P. Sinay, *Nature*, 1999, **398**, 377.

21. A. Canales, J. Angulo, R. Ojeda, M. Bruix, R. Fayos, R. Lozano, G. Gimenez-Gallego, M. Martin-Lomas, P. M. Nieto and J. Jimenez-Barbero, *J. Am. Chem. Soc.*, 2005, **127**, 5778.
22. R. Schworer, O. V. Zubkova, J. E. Turnbull and P. C. Tyler, *Chemistry*, 2013, **19**, 6817.
23. Y. Xu, S. Masuko, M. Takieddin, H. Xu, R. Liu, J. Jing, S. A. Mousa, R. J. Linhardt and J. Liu, *Science*, 2011, **334**, 498.
24. Y. Xu, C. Cai, K. Chandarajoti, P. H. Hsieh, L. Li, T. Q. Pham, E. M. Sparkenbaugh, J. Sheng, N. S. Key, R. Pawlinski, E. N. Harris, R. J. Linhardt and J. Liu, *Nat. Chem. Biol.*, 2014, **10**, 248.
25. C. McCormick, Y. Leduc, D. Martindale, K. Mattison, L. E. Esford, A. P. Dyer and F. Tufaro, *Nat. Genet.*, 1998, **19**, 158.
26. J. Li, A. Hagner-McWhirter, L. Kjellen, J. Palgi, M. Jalkanen and U. Lindahl, *J. Biol. Chem.*, 1997, **272**, 28158.
27. M. Kobayashi, H. Habuchi, O. Habuchi, M. Saito and K. Kimata, *J. Biol. Chem.*, 1996, **271**, 7645.
28. H. Habuchi, M. Kobayashi and K. Kimata, *J. Biol. Chem.*, 1998, **273**, 9208.
29. N. W. Shworak, J. Liu, L. M. Fritze, J. J. Schwartz, L. Zhang, D. Logeart and R. D. Rosenberg, *J. Biol. Chem.*, 1997, **272**, 28008.
30. A. Orellana, C. B. Hirschberg, Z. Wei, S. J. Swiedler and M. Ishihara, *J. Biol. Chem.*, 1994, **269**, 2270.
31. Y. Kakuta, T. Sueyoshi, M. Negishi and L. C. Pedersen, *J. Biol. Chem.*, 1999, **274**, 10673.
32. M. B. Duncan, M. Liu, C. Fox and J. Liu, *Biochem. Biophys. Res. Commun.*, 2006, **339**, 1232.
33. S. Peterson, A. Frick and J. Liu, *Nat. Prod. Rep.*, 2009, **26**, 610.
34. M. Chen, A. Bridges and J. Liu, *Biochemistry*, 2006, **45**, 12358.
35. A. E. Sismey-Ragatz, D. E. Green, N. J. Otto, M. Rejzek, R. A. Field and P. L. DeAngelis, *J. Biol. Chem.*, 2007, **282**, 28321.
36. J. D. Esko and S. B. Selleck, *Annu. Rev. Biochem.*, 2002, **71**, 435.
37. M. Weiwer, T. Sherwood, D. E. Green, M. Chen, P. L. DeAngelis, J. Liu and R. J. Linhardt, *J. Org. Chem.*, 2008, **73**, 7631.
38. S. Masuko, S. Bera, D. E. Green, M. Weiwer, J. Liu, P. L. DeAngelis and R. J. Linhardt, *J. Org. Chem.*, 2012, **77**, 1449.
39. R. F. Sala, S. L. MacKinnon, M. M. Palcic and M. E. Tanner, *Carbohydr. Res.*, 1998, **306**, 127.
40. R. Liu, Y. Xu, M. Chen, M. Weiwer, X. Zhou, A. S. Bridges, P. L. DeAngelis, Q. Zhang, R. J. Linhardt and J. Liu, *J. Biol. Chem.*, 2010, **285**, 34240.
41. E. Sterner, S. Masuko, G. Li, L. Li, D. E. Green, N. J. Otto, Y. Xu, P. L. DeAngelis, J. Liu, J. S. Dordick and R. J. Linhardt, *J. Biol. Chem.*, 2014, **289**, 9754.
42. K. Chandarajoti, Y. Xu, E. Sparkenbaugh, N. S. Key, R. Pawlinski and J. Liu, *Glycobiology*, 2014, **24**, 476.
43. Y. Chen, Y. Li, H. Yu, G. Sugiarto, V. Thon, J. Hwang, L. Ding, L. Hie and X. Chen, *Angew. Chem., Int. Ed. Engl.*, 2013, **52**, 11852.

44. C. Liu, J. Sheng, J. M. Krahn, L. Perera, Y. Xu, P. H. Hsieh, W. Dou, J. Liu and L. C. Pedersen, *J. Biol. Chem.*, 2014, **289**, 13407.
45. J. Liu and L. C. Pedersen, *Appl. Microbiol. Biotechnol.*, 2007, **74**, 263.
46. B. E. Thacker, E. Seamen, R. Lawrence, M. W. Parker, Y. Xu, J. Liu, C. W. Vander Kooi and J. D. Esko, *ACS Chem. Biol.*, 2016, **11**, 971.
47. V. N. Patel, I. M. Lombaert, S. N. Cowherd, N. W. Shworak, Y. Xu, J. Liu and M. P. Hoffman, *Dev. Cell*, 2014, **29**, 662.
48. Z. Wang, P. H. Hsieh, Y. Xu, D. Thieker, E. J. Chai, S. Xie, B. Cooley, R. J. Woods, L. Chi and J. Liu, *J. Am. Chem. Soc.*, 2017, **139**, 5249.
49. Y. Xu, E. H. Pempe and J. Liu, *J. Biol. Chem.*, 2012, **287**, 29054.
50. J. P. Li, F. Gong, K. El Darwish, M. Jalkanen and U. Lindahl, *J. Biol. Chem.*, 2001, **276**, 20069.
51. J. Sheng, Y. Xu, S. B. Dulaney, X. Huang and J. Liu, *J. Biol. Chem.*, 2012, **287**, 20996.
52. J. Sheng, R. Liu, Y. Xu and J. Liu, *J. Biol. Chem.*, 2011, **286**, 19768.
53. X. Zhang, V. Pagadala, H. M. Jester, A. M. Lim, T. Q. Pham, A. M. P. Goulas, J. Liu and R. J. Linhardt, *Chem. Sci.*, 2017, **8**, 7932.
54. M. Guerrini, P. A. Mourier, G. Torri and C. Viskov, *Glycoconjugate J.*, 2014, **31**, 409.
55. D. Xu, A. F. Moon, D. Song, L. C. Pedersen and J. Liu, *Nat. Chem. Biol.*, 2008, **4**, 200.
56. S. Park, J. C. Gildersleeve, O. Blixt and I. Shin, *Chem. Soc. Rev.*, 2013, **42**, 4310.
57. Y. Liu, A. S. Palma and T. Feizi, *Biol. Chem.*, 2009, **390**, 647.
58. D. F. Smith and R. D. Cummings, *Curr. Opin. Virol.*, 2014, **7**, 79.
59. S. R. Stowell, C. M. Arthur, R. McBride, O. Berger, N. Razi, J. Heimburg-Molinaro, L. C. Rodrigues, J. P. Gourdine, A. J. Noll, S. von Gunten, D. F. Smith, Y. A. Knirel, J. C. Paulson and R. D. Cummings, *Nat. Chem. Biol.*, 2014, **10**, 470.
60. X. Song, H. Ju, Y. Lasanajak, M. R. Kudelka, D. F. Smith and R. D. Cummings, *Nat. Methods*, 2016, **13**, 528.
61. X. Song, Y. Lasanajak, B. Xia, J. Heimburg-Molinaro, J. M. Rhea, H. Ju, C. Zhao, R. J. Molinaro, R. D. Cummings and D. F. Smith, *Nat. Methods*, 2011, **8**, 85.
62. Z. Wang, Z. S. Chinoy, S. G. Ambre, W. Peng, R. McBride, R. P. de Vries, J. Glushka, J. C. Paulson and G. J. Boons, *Science*, 2013, **341**, 379.
63. Z. Xiao, Y. Guo, Y. Liu, L. Li, Q. Zhang, L. Wen, X. Wang, S. M. Kondengaden, Z. Wu, J. Zhou, X. Cao, X. Li, C. Ma and P. G. Wang, *J. Org. Chem.*, 2016, **81**, 5851.
64. S. C. Edavettal, K. Carrick, R. R. Shah, L. C. Pedersen, A. Tropsha, R. M. Pope and J. Liu, *Biochemistry*, 2004, **43**, 4680.
65. A. F. Moon, Y. Xu, S. M. Woody, J. M. Krahn, R. J. Linhardt, J. Liu and L. C. Pedersen, *Proc. Natl. Acad. Sci. U. S. A.*, 2012, **109**, 5265.
66. D. H. Atha, J. C. Lormeau, M. Petitou, R. D. Rosenberg and J. Choay, *Biochemistry*, 1985, **24**, 6723.
67. U. Lindahl, G. Backstrom, M. Hook, L. Thunberg, L. A. Fransson and A. Linker, *Proc. Natl. Acad. Sci. U. S. A.*, 1979, **76**, 3198.

68. C. Cai, L. Li, C. Harvey, J. Liu and R. J. Linhardt, *Tetrahedron Lett.*, 2013, **54**, 4471.
69. Y. Xu, K. Chandarajoti, X. Zhang, V. Pagadala, W. Dou, D. M. Hoppensteadt, E. M. Sparkenbaugh, B. Cooley, S. Daily, N. S. Key, D. Severynse-Stevens, J. Fareed, R. J. Linhardt, R. Pawlinski and J. Liu, *Sci. Transl. Med.*, 2017, **9**, eaan5954.
70. Y. Xu, A. F. Moon, S. Xu, J. M. Krahn, J. Liu and L. C. Pedersen, *ACS Chem. Biol.*, 2017, **12**, 73.
71. P. A. Driguez, P. Potier and P. Trouilleux, *Nat. Prod. Rep.*, 2014, **31**, 980.
72. J. Liu and R. J. Linhardt, *Nat. Prod. Rep.*, 2014, **31**, 1676.

CHAPTER 10

Synthesis of Glycosphingolipids (GSLs)

ABHISHEK SANTRA, HAI YU* AND XI CHEN*

Department of Chemistry, University of California, One Shields Avenue, Davis, CA 95616, USA
*E-mail: hyu@ucdavis.edu, xiichen@ucdavis.edu

10.1 Introduction

Glycosphingolipids (GSLs) are found in cell membranes of diverse organisms ranging from bacteria to human. GSLs exhibit a large structural heterogeneity, with structural variations on both the lipid part and the glycan component. GSLs with more than 300 different glycan structures have been identified,[1,2] and more than 500 GSLs including synthetic analogs are included in the LipidBank (http://lipidbank.jp/index.html). The structural diversity of glycan is due to the presence of diverse sugar units with or without further post-glycosylational modification (PGMs)[3] such as sulfation and acetylation,[2] as well as various types of glycosidic linkage between the sugar units with different regio- and stereo-chemistry. Based on the glycan core structures, GSLs are classified into several series including ganglio-, globo-, isoglobo-, lacto-, neolacto-series that are abundant in vertebrates as well as arthro-, gala-, neogala-, muco-, mollu-, schisto- and spirometo-series that have been found in

Chemical Biology No. 11
Synthetic Glycomes
Edited by Wanyi Guan, Lei Li and Peng George Wang

Published by the Royal Society of Chemistry, www.rsc.org

Table 10.1 Core structures of glycosphingolipid subfamilies.[1,2,4–6] Symbols and abbreviations: ▢ *N*-acetylgalactosamine (GalNAc), ○ galactose (Gal), ■ *N*-acetylglucosamine (GlcNAc), ● mannose (Man), ● glucose (Glc), ceramide (Cer), sphingosine (Sph).

Series	Abbreviations	Core Structure
Ganglio	Gg	
Globo	Gb	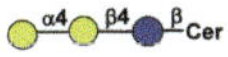
Isoglobo	iGb	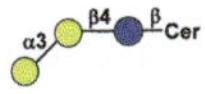
Lacto	Lc	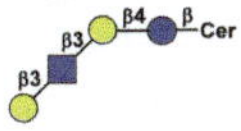
Neolacto	nLc	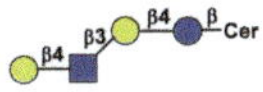
Arthro	At/Ar	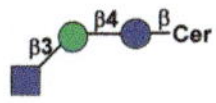
Gala	Ga	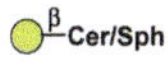
Neogala	nGa	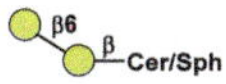
Muco	Mc	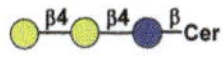
Mollu	Mu/Ml	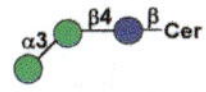
Schisto		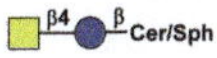
Spirometo	Sp	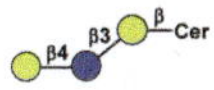

invertebrates and other organisms (Table 10.1).[1,2,4–6] GSLs can also be divided into neutral, charged, and sulfated categories.[2] On the cell surface of vertebrates, GSLs participate in various biological processes including influencing raft formation and membrane protein function, viral and bacterial infection, and cell–cell recognition and interaction.[1,2,6] In addition to being key components in cell recognition, structurally diverse glycosphingolipids are involved in cell signaling.[7] Therefore, obtaining structurally defined GSLs will facilitate structure–activity relationship studies of these important compounds at the molecular level.

A traditional approach to obtaining GSLs for functional studies is extraction, mainly from animal tissues.[8,9] The purification processes can result in heterogeneity in the products and generate complications in data analysis including challenges in identifying ligands that are responsible for binding to specific proteins, antibody, or cells. Chemical synthesis of GSLs is an alternative approach and the strategies for doing so have been advanced significantly during the past decades. A number of GSLs, especially gangliosides, have been synthesized by sophisticated chemical methods.[10] Recently, several enzyme-involved strategies have been developed.[11–15] Despite the progress in chemical and enzymatic synthesis, most glycosphingolipids found in nature are still not readily available. So far, only about 50 naturally occurring GSLs (Table 10.2) have been synthesized. Most of them have been

Table 10.2 List of glycosphingolipids have been chemical or chemoenzymatically synthesized. Symbols and abbreviations: *N*-acetylneuraminic acid (Neu5Ac), *N*-glycolylneuraminic acid (Neu5Gc), fucose (Fuc), galactose (Gal), *N*-acetylglucosamine (GlcNAc), glucose (Glc), galactopyranose (Gal*f*), arabinofuranose (Ara*f*), ceramide (Cer), sphingosine (Sph).

		Reference	
Name and structure	Abbreviation	Chemical synthesis	Chemoenzymatic synthesis
Ganglio-series			
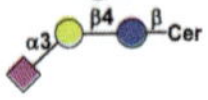	GM3	16–18	11–13,19–21
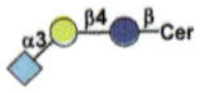	(Neu5Gc)GM3	22	
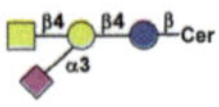	GM2	23–25	13,21
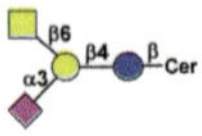	6′-GM2	26	
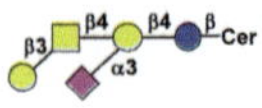	GM1	23,27,28	13
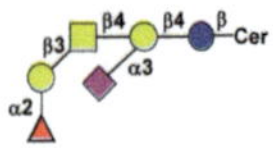	Fucosyl GM1		13
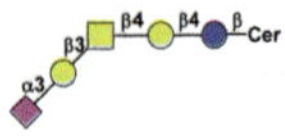	GM1b	29–31	
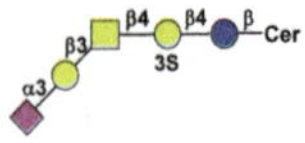	3′-*O*-sulfo-GM1b	32	
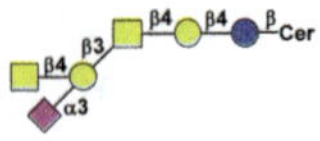	GalNAc-GM1b	33,34	
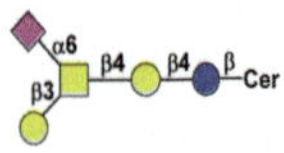	GM1α	35,36	
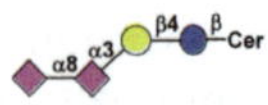	GD3	37–40	13,21
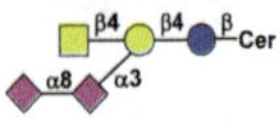	GD2	41–43	13
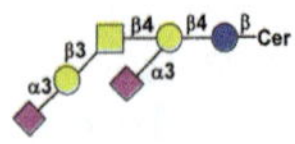	GD1a	27,28	
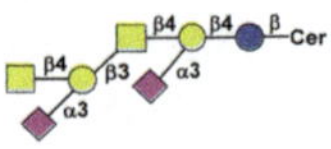	GalNAc-GD1a	44,45	
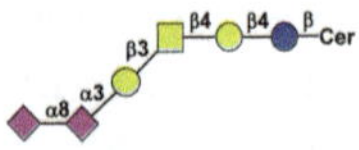	GD1c	46	
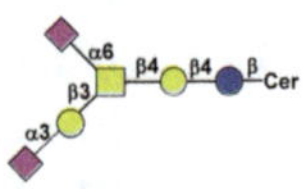	GD1α	47,48	

Table 10.2 (*continued*)

		Reference	
Name and structure	Abbreviation	Chemical synthesis	Chemoenzymatic synthesis
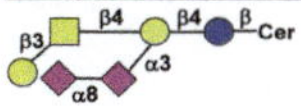	GD1b	32,49	
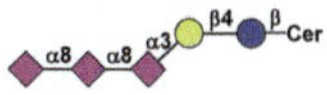	GT3	40	21
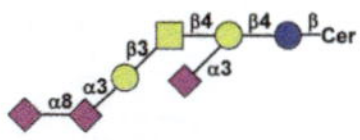	GT1a	46	
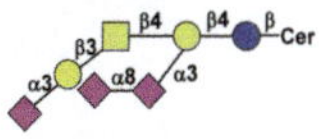	GT1b	49	
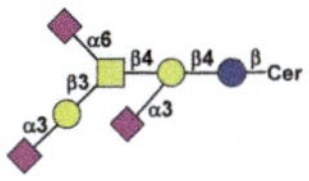	GT1aα	50,51	
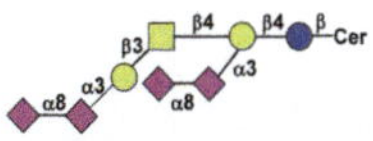	GQ1b	49,52,53	
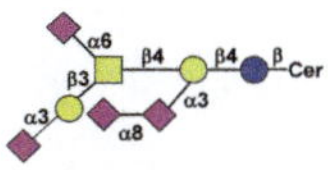	GQ1bα	54	
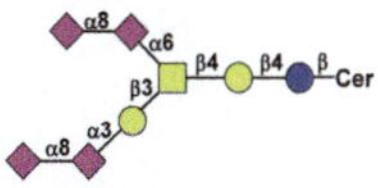	GQ1β	55	
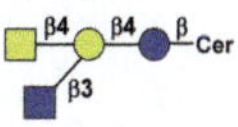	M1-xGL-1 (LcGg4)	56	
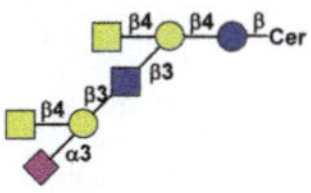	X1	57	
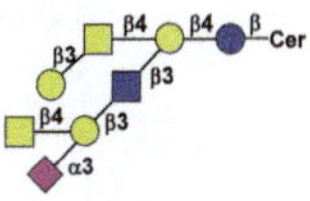	X2	58	
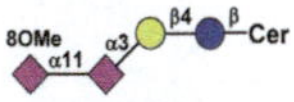	LLG-3	59	
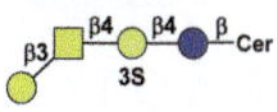	SM1a	60,61	
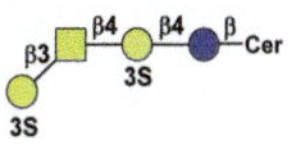	SB1a	61	
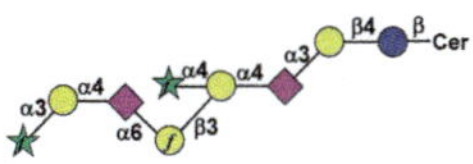	GP3	62	
Lacto-series			
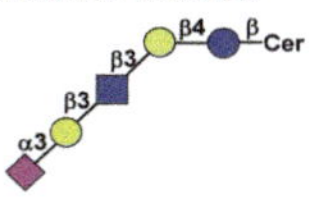	3′-isoLM1 (Sialyl Lc4)	63	

(*continued*)

Table 10.2 (*continued*)

Name and structure	Abbreviation	Reference: Chemical synthesis	Reference: Chemoenzymatic synthesis
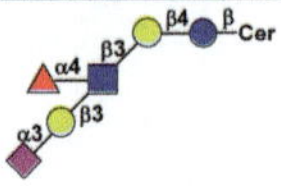	Fuc-3′-isoLM1 (Sialyl Lewis a)	64	
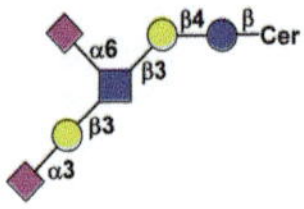	DSLc4	65	
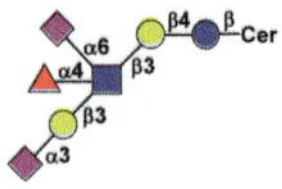	DSLe^a	65	
Neolacto-series			
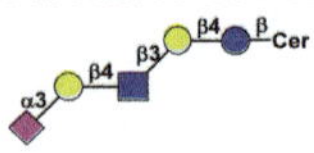	LM1 (Sialyl nLc4)	63,66	15
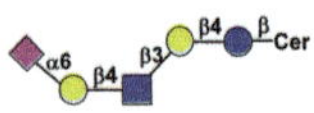	6′-LM1		15
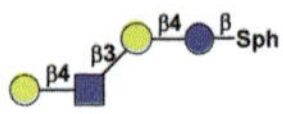	nLc$_4$		14
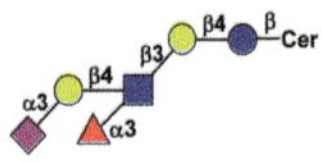	Fuc-3′-LM1 (Sialyl Lewis x)	67–69	
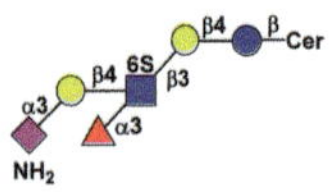	6-Sulfo sLe^x	70	
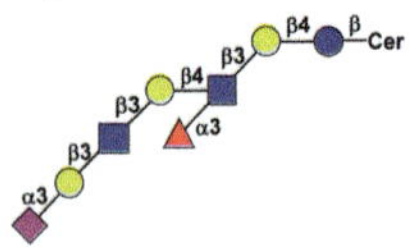	VIM-2 (Sialyl Le^a type)	71	
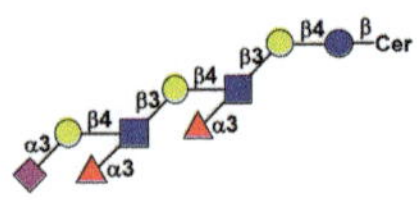	Sialyl dimeric Lewis x	72	
Globo-series			
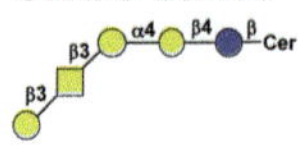	Gb5	73	
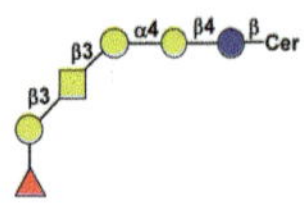	Globo-H	74	
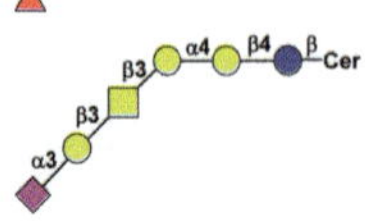	SGG (V^{3}Neu5 AcGb$_5$Cer)	75–77	
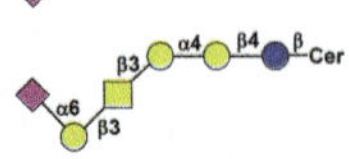	V^{6}Neu5AcGb$_5$Cer	76	

Table 10.2 (*continued*)

		Reference	
Name and structure	Abbreviation	Chemical synthesis	Chemoenzymatic synthesis
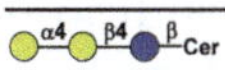	Gb_3	78–83	84
Isoglobo-series			
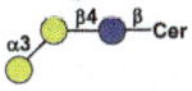	iGb_3	79,80,83,85	84
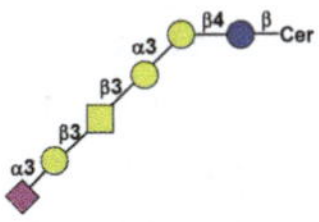	$V^3Neu5AciGb_5Cer$	86	
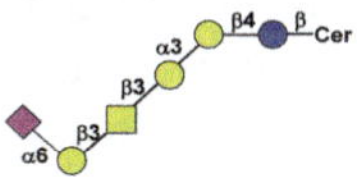	$V^6Neu5AciGb_5Cer$	86	
Gala-series			
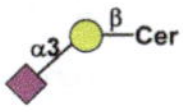	GM4	87,88	
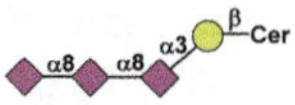	GT4	40	
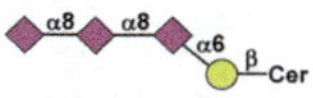	α2–6GT4	40	
Other types of GSLs			
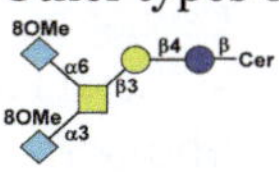	GAA-7	89	
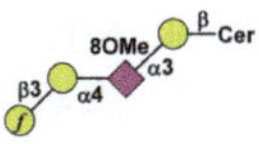	PNG-2A	90	
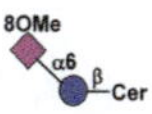	DSG-A	91	
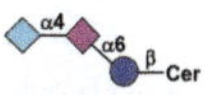	HLG-2	92	

obtained by chemical synthetic approaches while an increasing number of GSLs are being synthesized chemoenzymatically with recent developed efficient sequential one-pot multienzyme (OPME) methods plus facile purification schemes for preparing glycosyl sphingosines, followed by a one-step highly efficient acylation process.[13,14]

10.2 Chemical Strategies for Synthesizing Glycosphingolipids

Chemical methods developed for the synthesis of GSLs differ mainly on the types of the glycosyl donors used, the types of the lipids for initial glycosylation, and the order of forming glycosidic bonds with the lipid or with other glycan components. The discussion below centers on the types of glycosyl donors used.

10.2.1 Synthesis Using Glycosyl Trichloroacetimidate Donors

Glycosyl trichloroacetimidates (Schmidt donors) are the most popular glycosyl donors for chemical synthesis of complex glycosides. Most chemically synthesized GSLs were obtained by glycosylation using this type of glycosyl donors. This is mainly contributed to by the mild glycosylation conditions required and the need for only a catalytic amount of Lewis acid such as TMSOTf or $BF_3{\cdot}Et_2O$.

A generally adopted strategy for chemical synthesis of GSLs starts with chemical construction of the protected glycan component with an easily removable protecting group at the anomeric carbon of the reducing end. After removal of the reducing end anomeric carbon protecting group, the glycan is converted to the corresponding glycosyl trichloroacetimidate donor. A partially protected 2-azido-sphingosine, an analog of the ceramide precursor, is typically used as a glycosyl acceptor. Followed by glycosylation, the azido group of the glycosylation product is reduced to form an amino group which is conjugated with an acyl chain to form the desired ceramide structure. Deprotection of the resulting glycoconjugates produces the target ganglioside.[93] This method was used successfully for the synthesis of complex GSLs, including ganglio-series GM3,[17] GM2,[24,25] 6′-GM2,[26] GM1,[27,28] GM1b,[29–31] 3′-*O*-sulfo-GM1b,[32] GM1α,[35,36] GD3,[38] GD2,[42,43] GD1c,[46] GD1a,[27,28] GD1α,[47,48] GD1b,[49] GT3,[40] GT1a,[46] GT1b,[49] GT1aα,[50,51] GQ1bα,[54] GQ1b,[49,52] GQ1β;[55] lacto-series 3′-isoLM1 (sialyl LC4),[63] sialyl Lewis a,[64] DSLc4,[65] DSLe^a;[65] neolacto-series LM1(sialyl nLC4),[63,66] VIM-2 (sialyl Lewis a type),[71] sialyl Lewis x,[67,69] sialyl dimeric Lewis x,[72] 6-sulfo sLe^x;[70] globo-series Gb3,[83,94] SGG (V^{3}Neu5AcGb5Cer),[75–77] V^{6}Neu5AcGb5Cer;[76] isoglobo-series iGb3,[83,94] V^{3}Neu5AciGb5Cer,[86] V^{6}Neu5AciGb5Cer;[86] gala-series GM4,[88] GT4,[40] α2–6GT4;[40] and ganglioside PNG-2A,[90] HLG-2.[92]

As shown in Figure 10.1, taking the synthesis of VIM-2 as an example,[71] the protected octasaccharide was synthesized through multiple chemical glycosylation steps. It was then converted into the key glycosyl trichloroacetimidate donor and coupled with an azido-sphingosine acceptor (47% glycosylation yield). Converting the sphingosine to ceramide followed by deprotection led to the formation of the desired VIM-2.[71]

In comparison, glycosylation of protected oligosaccharide trichloroacetimidate donors directly with the ceramide moiety led to very low yields (*e.g.* 13% yield for synthesizing Gb3 [79]) which was caused by poor nucleophilicity of the ceramide due to intramolecular hydrogen bonding. Nevertheless, a number of GSLs were synthesized by coupling glycosyl trichloroacetimidates with the ceramide acceptors, including ganglio-series GM_3,[16,18] (Neu5Gc)GM_3,[22] GD_3,[37] GM2,[23] GM1,[23] GalNAc-GM1b,[33] GD2,[41] X1;[57] gala-series GM4;[87] globo-series Gb3,[79] Gb5;[73] and isoglobo-series iGb3.[79,85]

To overcome the poor yields resulting from direct glycosylation of a ceramide acceptor with the glycan donor, a glucosyl ceramide (GlcβCer) cassette approach was developed.[95] A number of ganglio-series GSLs have been synthesized using this strategy, including GalNAc-GM1b,[34] GalNAc-GD1a,[44,45]

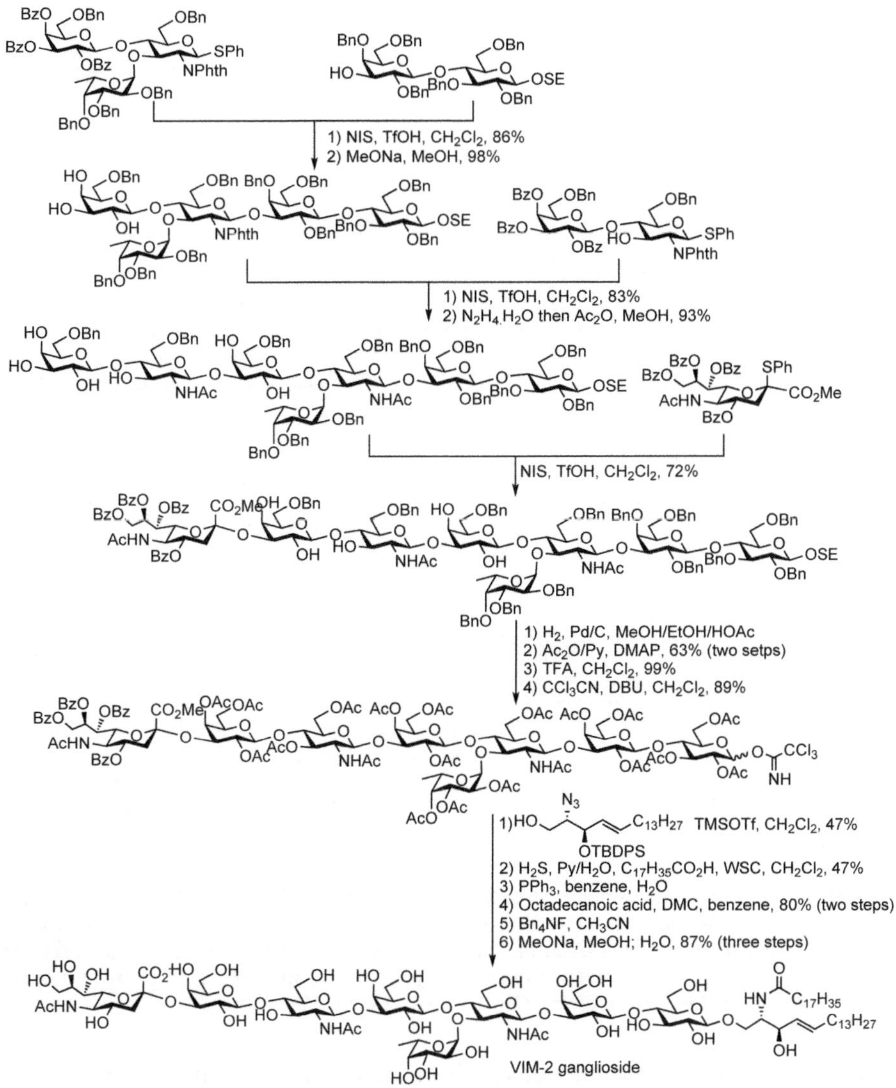

Figure 10.1 Chemical synthesis of VIM-2 ganglioside by coupling a glycosyl trichloroacetimidate donor with an azido-sphingosine acceptor followed by acylation and deprotection. Adapted from ref. 71 with permission from Elsevier, Copyright 1996.

GQ1b,[53] LLG-3,[59] X2,[58] SM1a,[60,61] SB1a,[61] GAA-7,[89] GP3.[62] An example is shown in Figure 10.2 for the synthesis of ganglioside GalNAc-GD1a.[44] The hexasaccharide component carrying two sialic acid residues was synthesized efficiently using a GM2-trisaccharide unit as a common building block. The potential difficulty in coupling the oligosaccharide and the ceramide moieties was overcome by carrying out the reactions using a cyclic glucosyl

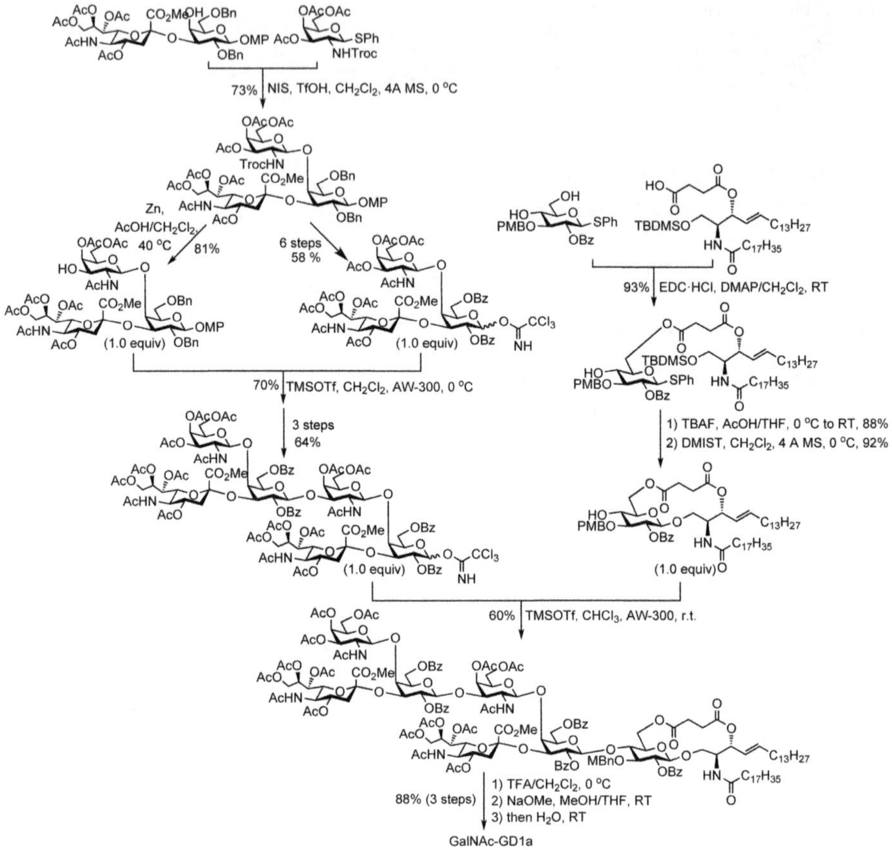

Figure 10.2 Chemical synthesis of GalNAc-GD1a ganglioside using a glycosyl trichloroacetimidate donor with a GlcβCer cassette approach. Abbreviations: PMB, *para*-methoxybenzyl, MP, *p*-methoxyphenyl. Adapted from ref. 44 with permission from John Wiley and Sons, Copyright © 2011 WILEY-VCH Verlag GmbH & Co. KGaA, Weinheim.

ceramide as a coupling partner for the hexasaccharyl trichloroacetimidate. The strategy successfully provided the framework of the target compound (60% yield for glycosylation). Global deprotection delivered the homogenous ganglioside GalNAc-GD1a.[44]

10.2.2 Synthesis with Glycosyl *N*-Phenyl Trifluoroacetimidates

Glycosyl *N*-phenyl trifluoroacetimidates developed by the Yu group[96] were utilized extensively for the synthesis of glycans and complex glycosides as alternative imidate donors.[97] It was applied by van der Marel *et al.* for the synthesis of Gb3 sphingosine.[98] As shown in Figure 10.3, condensation of azido-sphingosine with a trisaccharide *N*-phenyl trifluoroacetimidate

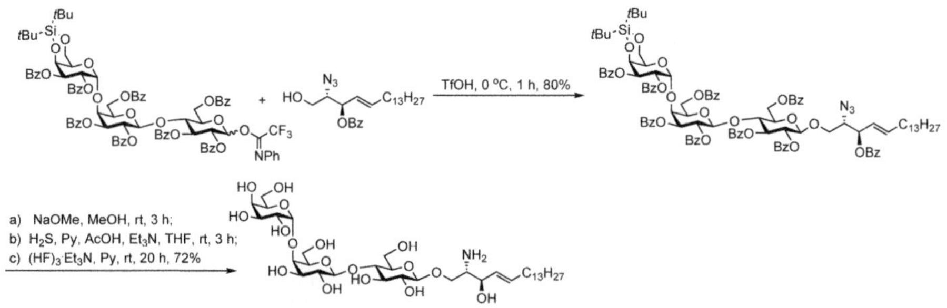

Figure 10.3 Chemically synthesis of Gb3 sphingosine using a glycosyl *N*-phenyl trifluoroacetimidate donor. Adapted from ref. 98 with permission from John Wiley and Sons, Copyright © 2011 WILEY-VCH Verlag GmbH & Co. KGaA, Weinheim.

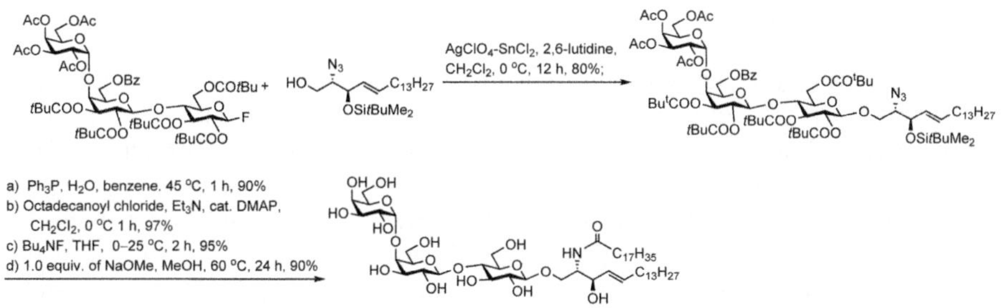

Figure 10.4 Chemically synthesis of Gb3 using a glycosyl fluoride donor. Adapted from ref. 80 with permission from Elsevier, Copyright 1990.

in the presence of triflic acid afforded the protected glycolipid with a yield of 80%. Reduction of the azide and deprotection produced a Gb3 sphingosine.

10.2.3 Synthesis Using Glycosyl Fluoride Donors

The strategy of activating glycosyl fluoride using $AgClO_4$-$SnCl_2$ for stereoselective glycosylation was introduced to carbohydrate synthesis by Mukaiyama *et al.*[99] This method gained attention due to the higher stability of the glycosyl fluorides compared to glycosyl bromides or glycosyl iodides and their milder activation conditions. The strategy was used for the total syntheses of globotriaosylceramide (Gb3),[79–81] lysoglobotriaosylceramide (lysoGb3),[80] and lacto-ganglio-series glycosphingolipid Ml-xGL-1 (LcGg4).[56] For example, for the synthesis of Gb3 (Figure 10.4), $AgClO_4$-$SnCl_2$ was used as a promoter, glycosylation of the azido sphingosine acceptor using a fully protected trisaccharide fluoride proceeded smoothly to produce the desired glycosyl product with a yield of 80%. After deprotection and acylation, Gb3 was obtained successfully.[80]

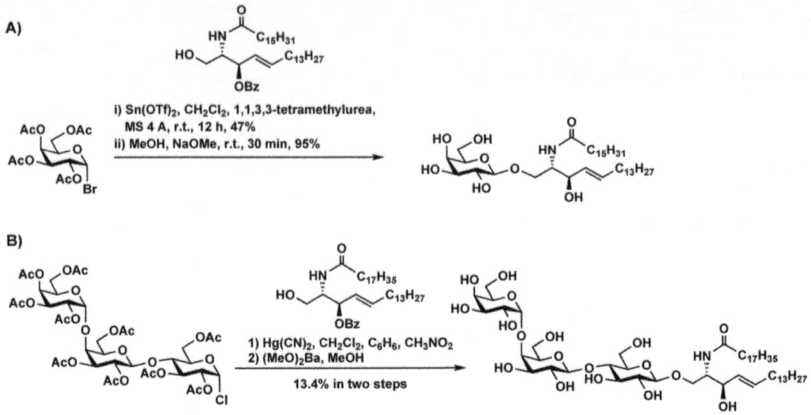

Figure 10.5 Chemically synthesis of glycosyl ceramides such as A) GalβCer and B) Gb3βCer by Koenigs–Knorr glycosylation using glycosyl halide donors. Adapted from ref. 101 with permission from Elsevier, Copyright 1988.

10.2.4 Synthesis Using Koenigs–Knorr Glycosylation Reactions

Koenigs–Knorr glycosylation is achieved by activation of glycosyl halides (bromide and chloride) in the presence of $AgCO_3$ or Ag_2O for glycosylation of the glycosyl acceptor.[100] Early attempts using the method of synthesizing galactosyl ceramide using α-D-bromo-tetraacetylgalactose and sphingoceramide with Thornton's procedure [Hg(CN)$_2$, CH_3NO_2, 80 °C, 1.5 h] led to the formation of a mixture. The method was modified to achieve a higher glycosylation yield by changing the catalyst to stannous triflate and using 1,1,3,3-tetramethylurea as a base (Figure 10.5A).[101] The classical Koenigs–Knorr method was later used for the first total synthesis of Gb3 by coupling the trisaccharide chloride with a ceramide acceptor in the presence of Hg(CN)$_2$ (18 equivalents) which led to the formation of the desired Gb3 yield of 13.4% after deprotection (Figure 10.5B).[82]

10.2.5 Synthesis Using Thioglycoside Glycosyl Donors

Thioglycosides are easy to prepare and much more stable than other glycosyl donors. A perbenzoylated SBox galactosyl donor activated by AgOTf was used to glycosylate azidosphingosine to produce galactosylsphingosine (Figure 10.6).[102] Thioglycosides were also used for the synthesis of α-linked galactosyl ceramide, a GSL that was originally found in marine sponge but which can stimulate human invariant natural killer T (iNKT) cells and has potent immunostimulating activity.[103] This achieved a yield of 39% by coupling perbenzylated thiogalactose with sphingosine using dimethyl(methylthio) sulfonium triflate (DMTST) as a promoter. NIS/AgOTf was found to be an unsuitable promotor system for the glycosylation as it led to the formation of several byproducts resulting from the addition of electrophiles to the double bond of the sphingosine acceptor.[103]

Kondo *et al.*[39] synthesized GD3 using a thioglycoside donor. As shown in Figure 10.7, glycosylation of a disaccharide acceptor with an α2–8-linked disialyl fluoride donor in the presence of AgOTf and $SnCl_2$ gave the

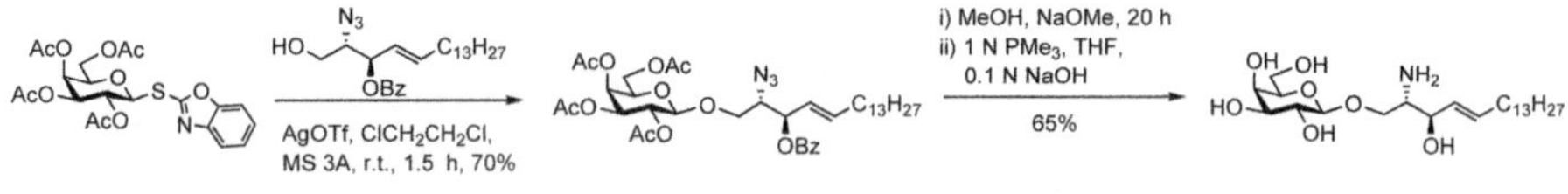

Figure 10.6 Chemically synthesis of galactosyl sphingosine using a perbenzoylated SBox thioglycoside donor. Adapted from ref. 102 with permission from John Wiley and Sons, Copyright © 2010 WILEY-VCH Verlag GmbH & Co. KGaA, Weinheim.

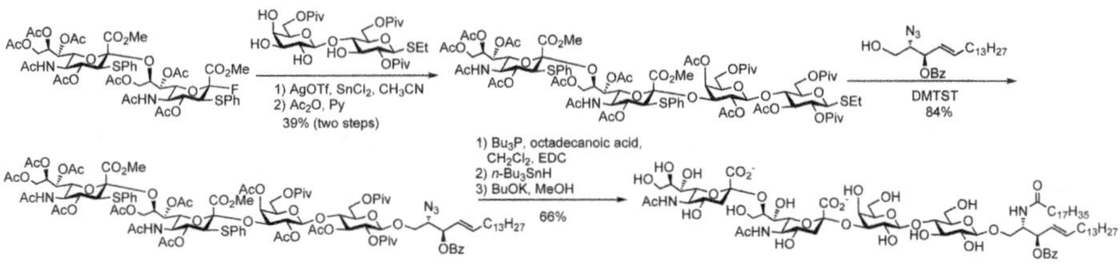

Figure 10.7 Chemically synthesis of GD3 using a tetrasaccharide thioglycoside donor.

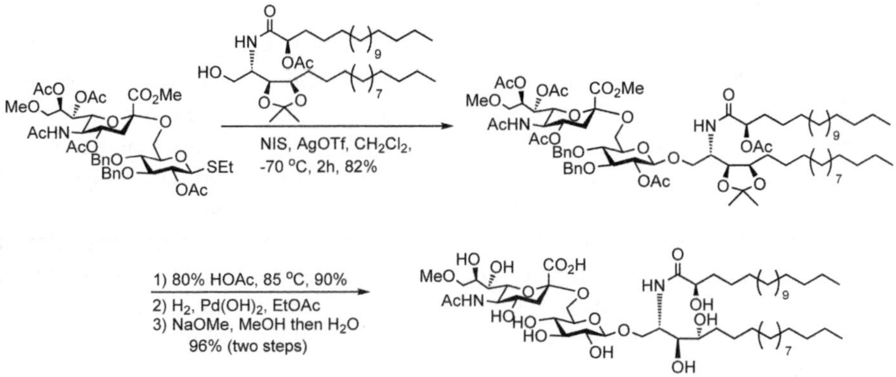

Figure 10.8 Chemically synthesis of DSG-A by coupling a thioglycoside with phytoceramide.

tetrasaccharide thioglycoside after acetylation. Condensation of the thiotetrasaccharide with azdiosphingosine activated by DMTST produced the corresponding glycosylazidosphingosine. After transformation of the glycosyl sphingosine to glycosyl ceramide and deprotection, the desired GD3 ganglioside was obtained.

Recently, the total synthesis of ganglioside DSG-A was achieved *via* coupling a disaccharide thioglycoside with a protected phytoceramide acceptor.[91] As shown in Figure 10.8, the protected phytoceramide was glycosylated with the

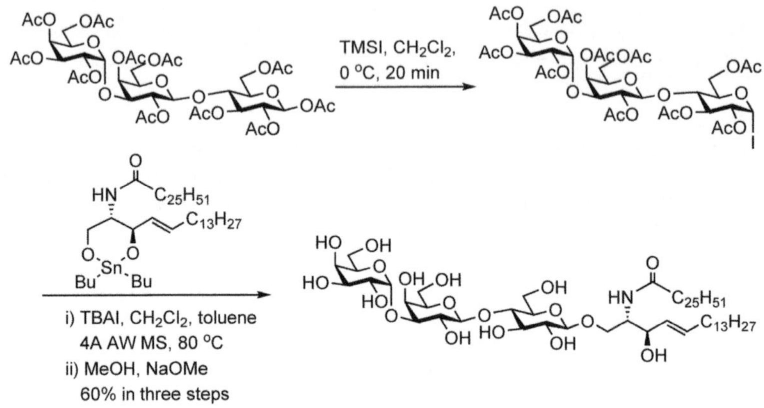

Figure 10.9 Chemically synthesis of Gb3 ganglioside using a glycosyl iodide donor. Adapted from ref. 106 with permission from John Wiley and Sons, Copyright © 2009 WILEY-VCH Verlag GmbH & Co. KGaA, Weinheim.

thioglycoside in the presence of promoter NIS/AgOTf to produce the protected DSG-A analog yield of 82%. After deprotection, the desired DSG-A was obtained.

10.2.6 Synthesis Using α-Glycosyl Iodide Donors

α-Glycosyl iodides have also been used as efficient donors for synthesizing glycolipids. For example, Gervay-Hague *et al.* reported the use of glycosyl iodides for the synthesis of α-linked glycolipids.[104,105] Reaction of per-*O*-benzylated galactosyl iodide with an partially protected sphingosine in the presence of tetrabutylammonium iodide (TBAI)/DIPEA produced a glycolipid product yield of 77%.[104] Recently, α-glycosyl iodide was used with a stannyl-ceramide in the presence of TBAI as a promoter and AW 300 molecular sieves as a drying agent for efficient stereoselective glycosylation. The strategy was used successfully to form galactosyl ceramide, lactosyl ceramide, and iGb3 by directly coupling oligosaccharyl iodides and the stannylceramide acceptor (Figure 10.9).[106]

10.2.7 Synthesis Using Glycosyl Mesylate Donors

Recently, *p*-methoxybenzyl (PMB)-protected glycosyl mesylate donors were used in a diarylborinic acid-catalyzed direct glycosylation of unprotected ceramides for the formation of lactosyl ceramide and GM3 at room temperature under mild homogeneous conditions (Figure 10.10).[107] This method allowed the formation of the beta linkage (a characteristic of major mammalian glycosphingolipids) between the carbohydrate component and the ceramide without the use of armed glycosyl donors or the protection of the ceramide moiety.

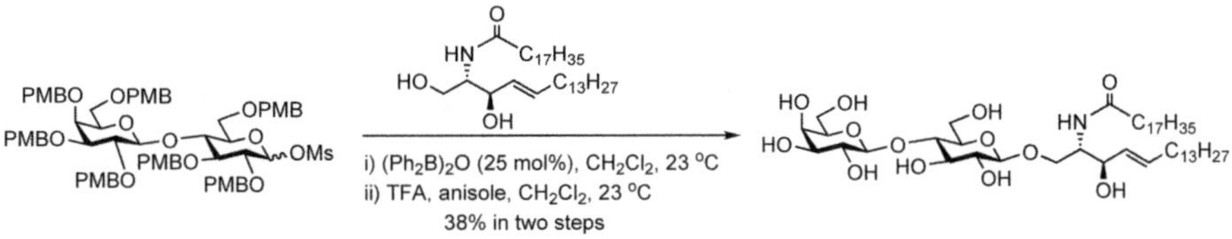

Figure 10.10 Chemically synthesis of lactosyl ceramide using glycosyl mesylate as donor.

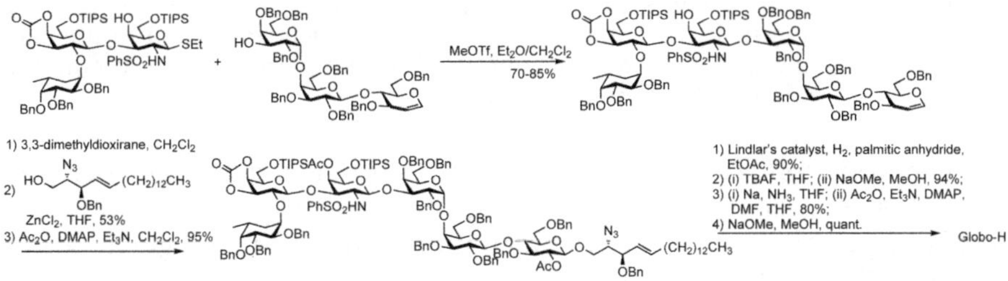

Figure 10.11 Chemical synthesis of globo-H ceramide using a glycal donor. Adapted from ref. 74 with permission from American Chemical Society, Copyright 1996.

10.2.8 Synthesis Using Glycal Donors

Danishefsky *et al.*[74] reported the synthesis of globo-H glycosphingolipid using an oligosaccharide glycal donor. As shown in Figure 10.11, the hexasaccharide glycal was assembled by coupling two trisaccharide building blocks. The glycal reacted with 3,3-dimethyldioxirane to produce a 1,2-epoxide derivative, which reacted with a sphingosine derivative using $ZnCl_2$ as a promoter to form glycolipid. Reduction of the azide group followed by palmitoylation and deprotection resulted in the desired globo-H glycosphingolipid.

10.2.9 Synthesis of α-Galactosylceramide Analogs from Naturally Configured α-Galactosides

The α-glycosidic linkage in α-galactosylceramide is critical for its biological functions. However, chemical glycosylation reactions usually produce mixtures of α and β isomers which are difficult to separate.

To overcome the difficulties of producing a mixture of α/β isomers in the synthesis of galactosylceramide by chemical glycosylation, Wang *et al.*[108] reported the synthesis of α-galactosyl lipids from raffinose, a naturally configured α-galactoside. The advantage of this approach was to avoid chemical glycosylation for the glycosyl lipid. From raffinose, the key azido galactoside intermediated was obtained in five steps. Deacylation by a base-catalyzed

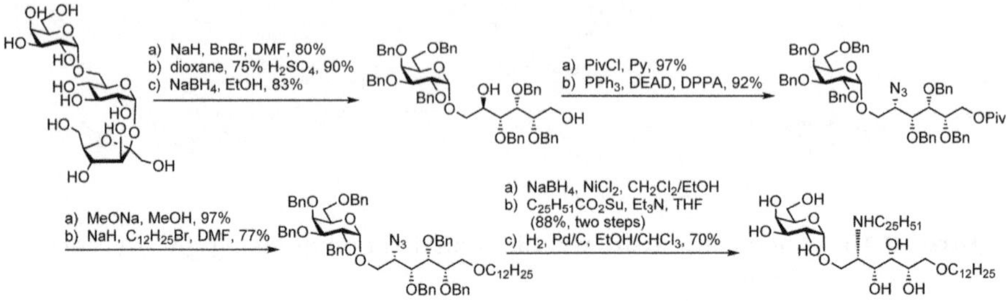

Figure 10.12 Chemically synthesis of an α-galactosylceramide analog from raffinose by retaining the original α-galactosidic bond in the starting material. Adapted from ref. 108 with permission from American Chemical Society, Copyright 2011.

methanolysis followed by alkylation with 1-bromododecane in the presence of NaH in DMF afforded an ether, which was then converted into an α-galactosylceramide analog by reduction, esterification, and global debenzylation (Figure 10.12).

10.3 Chemoenzymatic Synthesis of GSLs

In recent years, an increasing number of chemoenzymatic strategies have been developed for efficient synthesis of GSLs. Enzymatic glycosylation has the advantages of high stereo- and regio-selectivity along with high yields and being environmentally friendly. The processes have been greatly supported by recent advances in the identification and characterization of an increasing number of highly efficient and stable glycosyltransferases and other carbohydrate-active enzymes in large amounts.

10.3.1 Endoglycoceramidase-derived Glycosynthase-catalyzed Synthesis

Mutants of *Rhodococcus* sp. endoglycoceramidase II (EGCase) were successfully used as glycosynthases for conjugating sphingosine or derivatives with chemically or chemoenzymatically synthesized non-protected glycosyl fluorides.[11,109] As shown in Figure 10.13, chemically synthesized lactosyl fluoride was coupled to sphingosine by EGCase mutants to form lactosyl sphingosine which was further sialylated by CjCst-I, a recombinant *Campylobacter jejuni* α2–3 sialyltransferase, to produce GM3 sphingosine with a yield of 51%. EGCase mutants were also able to catalyze the transfer of chemoenzymatically synthesized larger complex glycosyl fluorides to sphingosine for the formation of more complex glycosphingosines such as lyso-GM3, lyso-GM1, lyso-Gb3, lyso-sialyl LNnT, and lyso-Galα1–4LNnT, which can be acylated for the formation of the desired GSLs.[11,109]

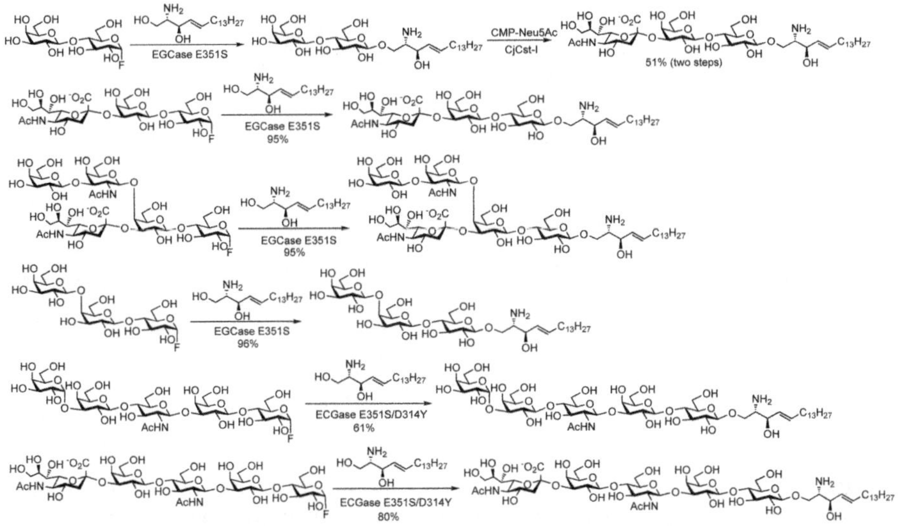

Figure 10.13 Endoglycoceramidase II (EGCase) mutant-catalyzed enzymatic synthesis of glycosyl sphingosines using glycosyl fluorides as donors and sphingosine as the acceptor substrate. Adapted from ref. 109 with permission from American Chemical Society, Copyright 2006.

10.3.2 Enzymatic Synthesis of Gb3 and iGb3 Using Lactosyl Ceramide as Acceptor Substrate

In general, naturally occurring lactosyl ceramides have very low solubility in water and are not suitable glycosyltransferase acceptors for *in vitro* enzymatic reactions. A lactosyl ceramide containing a short acyl chain (C8) was used for enzymatic synthesis of iGb3 and Gb3.[84] It was found that the activity of bovine α1–3-galactosyltransferase for the synthesis of iGb3 with a short acyl chain in the ceramide was very poor and the activity of *Neisseria meningitidis* α1–4-galactosyltransferase for the synthesis of Gb3 with a short acyl chain in the ceramide was better. Nevertheless, both reactions were carried out in a limited amount using 66.5 nmol of lactosyl ceramide and the actual yields of Gb3 and iGb3 obtained were not reported.

10.3.3 Enzymatic Assembly of Oligosaccharides Followed by Chemical Glycosylation

GSLs were also successfully synthesized by chemical protection of enzymatically assembled oligosaccharides followed by activation to generate glycosyl donors for chemical glycosylation with a partially protected azido-analog of sphingosine.[12,110]

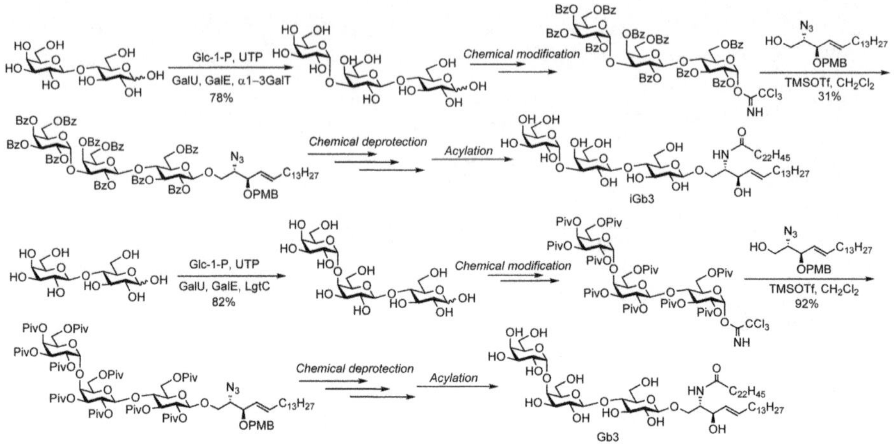

Figure 10.14 Chemoenzymatic synthesis of iGb3 and Gb3 by enzymatic assembly of oligosaccharides followed by chemical glycosylation using glycosyl trichloroacetimidate donors. Adapted from ref. 110 with permission from American Chemical Society, Copyright 2006.

As shown in Figure 10.14, isoglobotrihexose and globotrihexose were successfully synthesized by glycosyltransferase-catalyzed reactions with the *in situ* generation of sugar nucleotide donor uridine-5′-diphosphate galactose (UDP-Gal) from glucose-1-phosphate (Glc-1-P), uridine 5′-triphosphate (UTP), and lactose by a one-pot three-enzyme system containing a galactosyltransferase, uridine-5′-diphosphate glucose (UDP-Glc) pyrophosphorylase (GalU), and a UDP-Gal 4-epimerase (GalE). A yield of 82% was achieved for globotrihexose using a recombinant *C. jejuni* α1–4-galactosyltransferse and a yield of 78% was obtained for isoglobotrihexose using a recombinant bovine α1–3-galactosyltransferase. These two enzymatically synthesized trisaccharides were chemically converted into glycosyl trichloroacetimidates as glycosyl donors, and coupled with partial protected 2-azido-shingosine acceptor. Gb3 and iGb3 were produced after deprotection and acylation. Between benzoyl and pivaloyl group tested as protecting groups for the glycosyl donors for glycosylation, the pivaloyl group was proved to be superior.[110]

Paulson *et al.*[20] synthesized ganglioside GM3 using an enzymatically produced sialoside glycosyl donor. As shown in Figure 10.15, a 2-*O*-pivaloyl 2-(trimethylsilyl)ethyl lactoside was used as a substrate. The chosen bulky pivaloyl (Piv) group at C-2 helped the late-stage 1,2-*trans* glycosylation with ceramide. Preparative-scale sialylation of this lactoside was carried out using a CMP-Neu5Ac recycling protocol to produce the trisaccharide yield of 78%. The product was converted to a trichloroacetimidate donor of 78% in three steps. Coupling the glycosyl trichloroacetimidate with ceramide by the action of trimethylsilyl triflate formed ganglioside GM3 after deprotection.

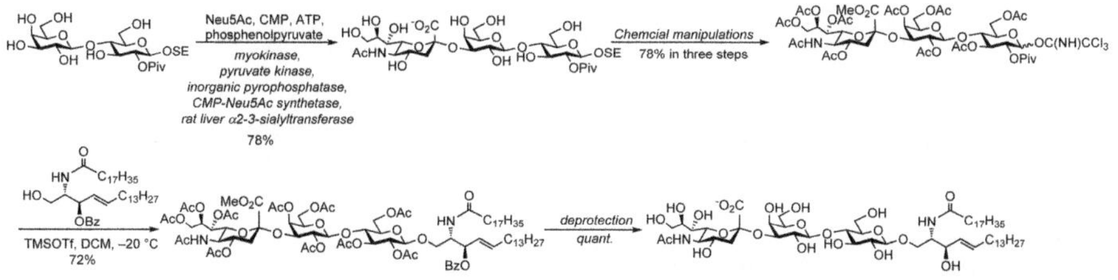

Figure 10.15 Chemoenzymatic synthesis of GM3 using a glycosyl trichloroacetimidate donor derived from enzymatically synthesized GM3 oligosaccharide. Adapted from ref. 20 with permission from American Chemical Society, Copyright 1993.

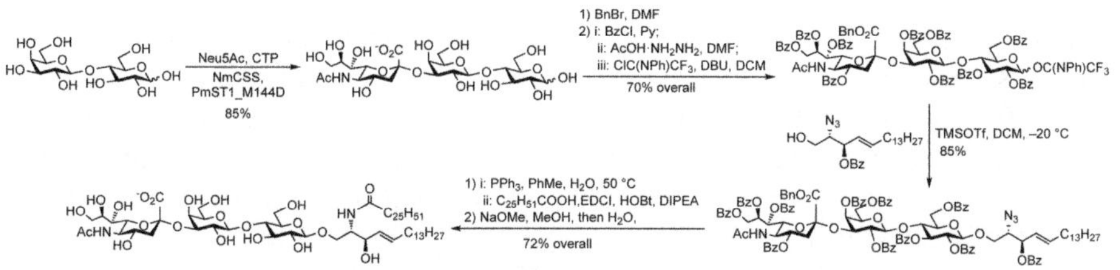

Figure 10.16 Chemoenzymatic synthesis of GM3 using glycosyl *N*-phenyltrifluoroacetimidate donor derived from enzymatically synthesized GM3 oligosaccharide. Adapted from ref. 12 with permission from John Wiley and Sons, Copyright © 2016 WILEY-VCH Verlag GmbH & Co. KGaA, Weinheim.

More recently, it was found that perbenzoylated glycosyl *N*-phenyltrifluoroacetimidate was a better glycosyl donor than the perbenzoylated glycosyl trichloroacetimidates. In addition, benzoyl is a preferred protecting group compared to PMB for the azido-derivative of sphingosine acceptor.[12] As shown in Figure 10.16, taking the synthesis of ganglioside GM3 as an example, the enzymatically synthesized GM3 oligosaccharide was converted to glycosyl *N*-phenyltrifluoroacetimidate donor, which was coupled to 2-azido-3-*O*-benzoylsphingosine in the presence of TMSOTf to form glycosylation product (60% over five steps). Reduction, *N*-acylation, and global deprotection led to the formation of the desired ganglioside GM3 (72% over three steps).[12]

10.3.4 Enzymatic Sialylation of Glycolipids

Danieshefsky *et al.*[19] reported a concise chemoenzymatic approach for synthesizing GM3. Lactal was used as a starting material to form a per-TES 1,2-anhydrolactal derivative by reacting with triethylsilyl triflate in pyridine followed by treating with 2,2-dimethyldioxirane in acetone. Coupling the 1,2-anhydrosugar with a partially protected sphingosine

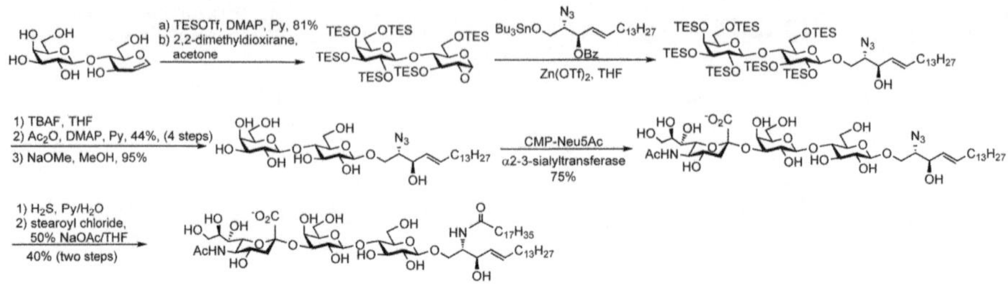

Figure 10.17 Chemoenzymatic synthesis of GM3 from lactosyl sphingosine. Adapted from ref. 19 with permission from John Wiley and Sons, Copyright © 1996 WILEY-VCH Verlag GmbH & Co. KGaA, Weinheim.

acceptor produced lactosyl sphingosine after deprotection. Enzymatic sialylation of the lactosyl sphingosine was achieved by an α2–3-sialyltransferase (E.C.2.4.99.4) in the presence of CMP-Neu5Ac with a yield of 75%. After reduction of the azide group, the crude amine was stearoylated in a biphasic system (THF/50% NaOAc) to produce GM3 with a yield of 40% (Figure 10.17).

DeFrees *et al.* developed enzymatic approaches for preparing gangliosides.[21] Sialylation of lactosyl sphingosine was carried out by an α2–3-*trans*-sialidase-catalyzed reaction using 3′-sialyllactose as a sialyl donor to form lyso-GM3 (12.0 g) with an 82% yield. Lyso-GM2 (9.7 g) was synthesized from lyso-GM3 and UDP-GalNAc in the presence of a GM2-synthetase with a 79% yield. The synthesis of lyso-GM1 (4.8 g) was carried out by GM1-synthetase-catalyzed galactosylation from lyso-GM2 and UDP-Gal which resulted in 84% yield. GM3 ceramide (6.1 g, 90%) was produced from lactosyl ceramide (5.12 g) and 3′-sialyllactose (253 g) in the presence of Zwittergent 3-14 (9.4 g) using a cell homogenate containing the α2–3-*trans*-sialidase. The purification involved the addition of a large amount of potassium chloride (64 g) to the reaction mixture and extraction procedures with large volumes (29 L and 10 L) of chloroform and methanol (2 : 1 by volume) mixed solvent. GM2 (7.6 g, 77%) was synthesized from GM3 and UDP-GalNAc/UDP-GlcNAc by GM2 synthetase and UDP-GalNAc. Treatment of GM3 and CMP-Neu5Ac by an α2–8-sialyltransferase (CST-68) produced a mixture of GD3 (38%), GT3 (15%), and GQ3 (7%).

10.3.5 One-pot Multienzyme (OPME) Chemoenzymatic Strategies

The chemoenzymatic methods described above for the synthesis of GLSs involve the assembly of non-protected oligosaccharide using enzymatic glycosylation reactions before coupling with sphingosine or derivatives by chemical or glycosynthase-catalyzed reactions. The purification processes for oligosaccharides are not trivial. In order to develop an efficient approach to allow high-yield synthesis and easy purification, glycolipids can be used directly as the acceptor substrates for glycosyltransferase-catalyzed reactions. In this case, the lipid component can be used for facile purification of product by simple C18-cartridge solid-phase extraction. One of the requirements is that the glycolipid acceptors need to be soluble in aqueous solution. Most complex vertebrate GLSs share a common lactosyl ceramide core. Lactosyl ceramide (LacβCer) has low water solubility and is not (or very poor) an acceptor for *in vitro* enzymatic reactions using glycosyltransferases in aqueous solutions.[111–113] It was reported that enzyme-catalyzed sialylation of glycosphingolipid paragloboside failed to produce the desired sialylated product even in the presence of non-ionic detergents.[15] In comparison, the solubility of lactosylsphingosine (LacβSph) in aqueous solution is much better than its counterpart with the ceramide as the lipid component and is a well suited acceptor for glycosyltransferase-catalyzed extension of the glycan chain. Once the desired glycosphingosine is achieved, it can be readily converted to the corresponding glycosphingolipids by simple acylation. This strategy also allows easy installation of different acyl chains in the last step to provide the diversity desired for downstream studies.

To decrease the cost, a sugar nucleotide *in situ* generation system can be used together with a suitable glycosyltransferase for the synthesis of the desired glycosphingosine. The strategy was initially demonstrated for the synthesis of α-Gal pentasaccharyl ceramide Galα3nLc$_4$βCer, a neolacto-series glycosphingolipid.[14] As shown in Figure 10.18, α-Gal pentasaccharyl sphingosine Galα3nLc$_4$βSph was synthesized by enzymatic extension of chemically synthesized LacβSph using a sequential OPME strategy involving three OPME reactions. Formation of trisaccharyl sphingosine Lc$_3$βSph from LacβSph was achieved by a one-pot four-enzyme GlcNAc activation and transfer system containing *Bifidobacterium longum* strain

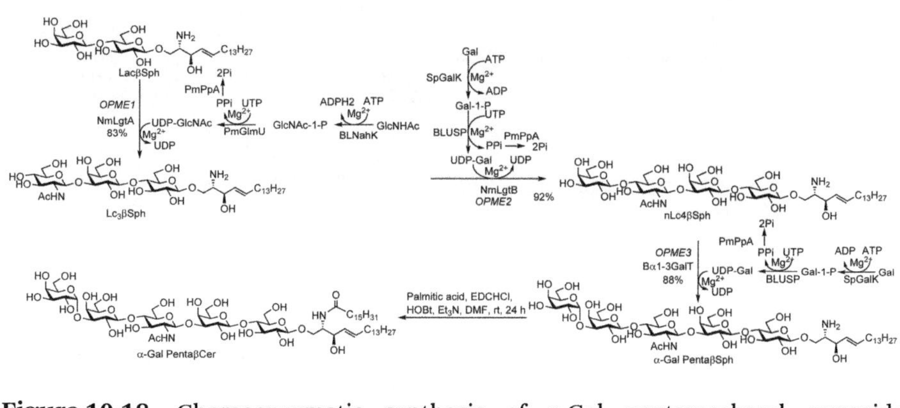

Figure 10.18 Chemoenzymatic synthesis of α-Gal pentasaccharyl ceramide (Galα3nLc$_4$βCer) from LacβSph.

ATCC55813 *N*-acetylhexosamine-1-kinase (BLNahK), *Pasteurella multocida* *N*-acetylglucosamine uridylyltransferase (PmGlmU), *P. multocida* inorganic pyrophosphatase (PmPpA), and *N. meningitidis* β1–3-*N* acetylglucosaminyltransferase (NmLgtA) with an 83% yield. nLc$_4$βSph was then synthesized from Lc$_3$βSph using an OPME galactose activation and transfer system containing *Streptococcus pneumoniae TIGR4* galactokinase (SpGalK), *B. longum* UDP-sugar pyrophosphorylase (BLUSP), PmPpA, and *N. meningitidis* β1–4-galactosyltransferase (NmLgtB) with a 92% yield. The last OPME reaction was carried out to convert the obtained nLc$_4$βSph to Galα3nLc$_4$βSph using a galactose activation and transfer system containing SpGalK, BLUSP, PmPpA, and bovine α1–3-galactosyltransferase (Bα1–3GalT) with an 88% yield. The target Galα3nLc$_4$βCer was readily obtained with an 85% yield by *N*-acylation of Galα3nLc$_4$βSph with palmitic acid in the presence of EDC·HCl, HOBt, and Et$_3$N.[14]

More recently, the strategy was demonstrated to be highly efficient for the synthesis of gangliosides including GM3, GM2, GM1, fucosyl GM1, GD2, GD3.[13] As shown in Figure 10.19, the key intermediate GM3 sphingosine (GM3βSph) lacking the fatty acyl chain was readily obtained from LacβSph and Neu5Ac using a one-pot two-enzyme sialylation system containing *N. meningitidis* CMP-sialic acid synthetase (NmCSS) and *P. multocida*

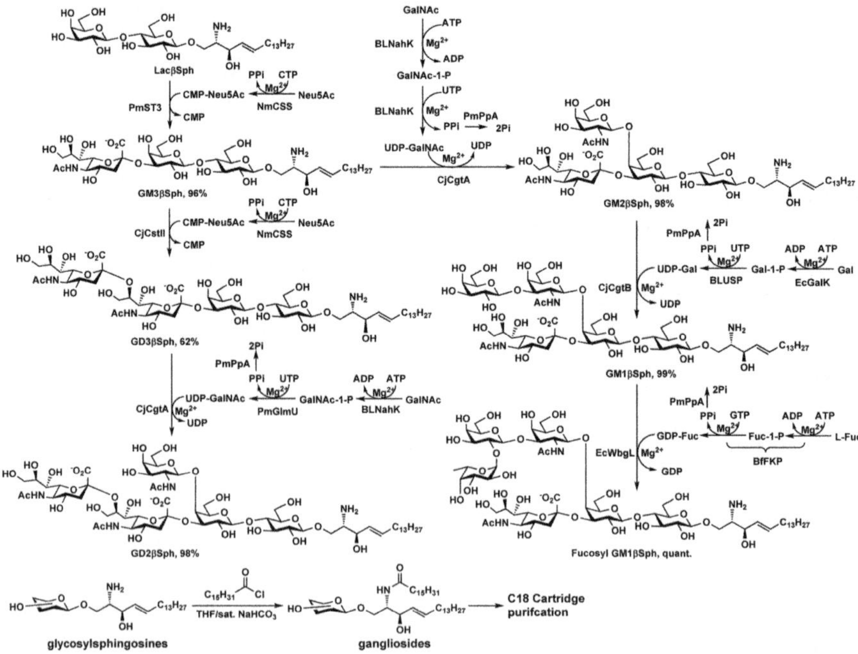

Figure 10.19 Sequential one-pot multienzyme (OPME) synthesis of GM3, GM2, GM1, fucosyl GM1, GD3, and GD2 glycosyl sphingosines followed by chemical acylation to form the corresponding gangliosides.

α2–3-sialyltransferase 3 (PmST3). GM3βSph was obtained with an excellent 96% yield in a gram-scale (1.44 g). With GM3βSph in hand, a one-pot four-enzyme *N*-acetylgalactosamine (GalNAc)-activation and transfer system[114] containing BlNahK, PmGlmU, PmPpA, and *C. jejuni* β1–4-*N*–acetylgalactosaminyltransferase (CjCgtA), was applied in glycosylating GM3βSph to form GM2βSph with a 98% yield (120 mg). GM1βSph was then synthesized from GM2βSph using a one-pot four-enzyme galactose-activation and transfer system containing EcGalK, BLUSP, PmPpA, and *C. jejuni* β1–3-galactosyltransferase (CjCgtB). GM1βSph (57 mg) was synthesized with a 99% yield. Finally, fucosyl GM1βSph (55 mg) was synthesized with a quantitative yield from GM1βSph using a one-pot three-enzyme fucose-activation and transfer system containing EcWbgL and guanosine 5′-diphosphate fucose (GDP-Fuc) biosynthetic enzymes including BfFKP and PmPpA. Overall, the formation of complex glycolipid Fuc-GM1βSph from chemically synthesized LacβSph involved four sequential OPME reaction with a total yield of 93%.[13]

Furthermore, GD3βSph was successfully synthesized from GM3βSph and Neu5Ac using a one-pot two-enzyme sialylation reaction containing NmCSS and *C. jejuni* α2–3/8-sialyltransferase (CjCstII) with a 62% yield (910 mg). Similar to the synthesis of GM2βPh from GM3βSph, GD2βSph was readily synthesized from GD3βSph using the one-pot multienzyme GalNAc-activation and glycosylation system containing BLNahK, PmGlmU, PmPpA, and CjCgtA. GD2βSph (34 mg) was obtained with an excellent 98% yield. Finally, the obtained glycosylsphingosines were converted into their corresponding gangliosides under improved acylation conditions by reacting with palmitoyl chloride in THF and saturated $NaHCO_3$ solution (1 : 1, v/v). Gangliosides GM3, GM2, GM1, fucosyl GM1, GD3, and GD2 were produced with 98–99% yields (Figure 10.19).[13]

10.4 Conclusion

Considerable advances have been made in the last two decades for the synthesis of GSLs using either chemical synthetic strategies or chemoenzymatic methods. Nevertheless, only a limited number of GSLs have been synthesized. While chemical synthesis will continue to be indispensable for accessing structurally diverse GSLs, sequential one-pot multienzyme (OPME) systems combined with facile product purification by C18-cartridge-based solid-phase extraction have become an effective strategy to allow highly efficient production of glycosphingosines from water soluble lactosyl sphingosine. Using one-step chemical acylation with C18-cartridge purification, target GSLs can be obtained readily. This chemoenzymatic strategy can be adapted by non-specialists for synthesizing the desired GSLs given the availability of lactosyl sphingosine and enzymes involved in the extension of the glycan chains in glycosylsphingosines. Future efforts in identifying additional glycosyltransferases that can be used for the synthesis of complex GSLs and expanding the available GSL pools will assist in the exploring their biological functions and therapeutic potentials.

Acknowledgement

This work was supported by the National Institute of General Medical Sciences (NIGMS) of the National Institutes of Health (NIH) under Award Number U01GM120419. The content is solely the responsibility of the authors and does not necessarily represent the official views of the National Institutes of Health. H.Y. and X.C. are co-founders of Glycohub, Inc., a company focused on the development of carbohydrate-based reagents, diagnostics, and therapeutics. Glycohub, Inc., played no role in the design, execution, interpretation, or publication of this work.

References

1. R. L. Schnaar and T. Kinoshita, in *Essentials of Glycobiology*, ed. A. Varki, R. D. Cummings, J. D. Esko, P. Stanley, G. W. Hart, M. Aebi, A. G. Darvill, T. Kinoshita, N. H. Packer, J. H. Prestegard, R. L. Schnaar and P. H. Seeberger, Cold Spring Harbor Laboratory Press, Cold Spring Harbor, NY, 3rd edn, 2015, pp. 125–135.
2. R. K. Yu, M. Yanagisawa and T. Ariga, in *Comprehensive Glycoscience*, ed. H. Kamerling, Elsevier, Oxford, 2007, pp. 73–122.
3. H. Yu and X. Chen, *Org. Biomol. Chem.*, 2007, **5**, 865.
4. M. Wuhrer, R. D. Dennis, M. J. Doenhoff, G. Lochnit and R. Geyer, *Glycobiology*, 2000, **10**, 89.
5. Y. Kawakami, K. Nakamura, H. Kojima, M. Suzuki, F. Inagaki, A. Suzuki, J. Ikuta, A. Uchida, Y. Murata and Y. Tamai, *Eur. J. Biochem.*, 1996, **239**, 905.
6. J. Kopitz, *Histochem. Cell Biol.*, 2017, **147**, 175.
7. S. Hakomori, *Proc. Natl. Acad. Sci. U. S. A.*, 2002, **99**, 225.
8. X. Song, Y. Lasanajak, B. Xia, J. Heimburg-Molinaro, J. M. Rhea, H. Ju, C. Zhao, R. J. Molinaro, R. D. Cummings and D. F. Smith, *Nat. Methods*, 2011, **8**, 85.
9. R. L. Schnaar, *Methods Enzymol.*, 1994, **230**, 348.
10. H. Ando, H. Ishida and M. Kiso, *Methods Enzymol.*, 2010, **478**, 521.
11. J. R. Rich, A.-M. Cunningham, M. Gilbert and S. G. Withers, *Chem. Commun.*, 2011, **47**, 10806.
12. Y. Liu, L. Wen, L. Li, M. R. Gadi, W. Guan, K. Huang, Z. Xiao, M. Wei, C. Ma, Q. Zhang, H. Yu, X. Chen, P. G. Wang and J. Fang, *Eur. J. Org. Chem.*, 2016, **2016**, 4315.
13. H. Yu, A. Santra, Y. Li, J. B. McArthur, T. Ghosh, X. Yang, P. G. Wang and X. Chen, *Org. Biomol. Chem.*, 2018, **16**, 4076.

14. A. Santra, Y. Li, H. Yu, T. J. Slack, P. G. Wang and X. Chen, *Chem. Commun.*, 2017, **53**, 8280.
15. J. J. Gaudino and J. C. Paulson, *J. Am. Chem. Soc.*, 1994, **116**, 1149.
16. M. Sugimoto and T. Ogawa, *Glycoconjugate J.*, 1985, **2**, 5.
17. T. Murase, H. Ishida, M. Kiso and A. Hasegawa, *Carbohydr. Res.*, 1989, **188**, 71.
18. R. I. Duclos, *Carbohydr. Res.*, 2000, **328**, 489.
19. K. K.-C. Liu and S. J. Danieshefsky, *Chem. - Eur. J.*, 1996, **2**, 1359.
20. Y. Ito and J. C. Paulson, *J. Am. Chem. Soc.*, 1993, **115**, 1603.
21. S. Defrees and E. J. McGuire, Chemo-enzymatic synthesis of sialylated oligosaccharides, *WO Pat.* 2003016469 A2, PCT Int. Appl., 2003.
22. M. Numata, M. Sugimoto, S. Shibayama and T. Ogawa, *Carbohydr. Res.*, 1988, **174**, 73.
23. M. Sugimoto, M. Numata, K. Koike, Y. Nakahara and T. Ogawa, *Carbohydr. Res.*, 1986, **156**, C1.
24. A. Hasegawa, T. Nagahama, H. Ohki and M. Kiso, *J. Carbohydr. Chem.*, 1992, **11**, 699.
25. J. C. Castro-Palomino, G. Ritter, S. R. Fortunato, S. Reinhardt, L. J. Old and R. R. Schmidt, *Angew. Chem., Int. Ed. Engl.*, 1997, **36**, 1998.
26. H. Ishida, Y. Ito, E. Tanahashi, Y.-T. Li, M. Kiso and A. Hasegawa, *Carbohydr. Res.*, 1997, **302**, 223.
27. A. Hasegawa, T. Nagahama and M. Kiso, *Carbohydr. Res.*, 1992, **235**, C13.
28. A. Hasegawa, H. Ishida, T. Nagahama and M. Kiso, *J. Carbohydr. Chem.*, 1993, **12**, 703.
29. M. Sugimoto, K. Fujikura, S. Nunomura, T. Horisaki, Y. Ito and T. Ogawa, *Tetrahedron Lett.*, 1990, **31**, 385.
30. H. Prabhanjan, H. Ishida, M. Kiso and A. Hasegawa, *Carbohydr. Res.*, 1991, **211**, C1.
31. H. Prabhanjan, A. Kameyama, H. Ishida, M. Kiso and A. Hasegawa, *Carbohydr. Res.*, 1991, **220**, 127.
32. T. Komori, S. Kondo, H. Ando, H. Ishida and M. Kiso, *Carbohydr. Res.*, 2002, **337**, 1679.
33. M. Sugimoto, K. Fujikura, S. Nunomura, Y. Ito and T. Ogawa, *Tetrahedron Lett.*, 1990, **31**, 1435.
34. M. Konishi, A. Imamura, K. Fujikawa, H. Ando, H. Ishida and M. Kiso, *Molecules*, 2013, **18**, 15153.
35. K. Hotta, S. Komba, H. Ishida, M. Kiso and A. Hasegawa, *J. Carbohydr. Chem.*, 1994, **13**, 665.
36. Y. Takeda and S. Horito, *Carbohydr. Res.*, 2005, **340**, 211.
37. Y. Ito, M. Numata, M. Sugimoto and T. Ogawa, *J. Am. Chem. Soc.*, 1989, **111**, 8508.
38. H. Ishida, Y. Ohta, Y. Tsukada, M. Kiso and A. Hasegawa, *Carbohydr. Res.*, 1993, **246**, 75.

39. K. Tadao, T. Toshiyuki, A. Hiroyuki, I. Minoru and G. Toshio, *Chem. Lett.*, 1996, **25**, 337.
40. H. Ando, H. Ishida, M. Kiso and A. Hasegawa, *Carbohydr. Res.*, 1997, **300**, 207.
41. Y. Matsuzaki, S. Nunomura, Y. Ito, M. Sugimoto, Y. Nakahara and T. Ogawa, *Carbohydr. Res.*, 1993, **242**, C1.
42. A. Hasegawa, H. Ishida, Y. Isogai, H. Ishida and M. Kiso, *J. Carbohydr. Chem.*, 1993, **12**, 1217.
43. H. Ishida, Y. Ohta, Y. Tsukada, Y. Isogai, H. Ishida, M. Kiso and A. Hasegawa, *Carbohydr. Res.*, 1994, **252**, 283.
44. K. Fujikawa, S. Nakashima, M. Konishi, T. Fuse, N. Komura, T. Ando, H. Ando, N. Yuki, H. Ishida and M. Kiso, *Chem. - Eur. J.*, 2011, **17**, 5641.
45. K. Fujikawa, T. Nohara, A. Imamura, H. Ando, H. Ishida and M. Kiso, *Tetrahedron Lett.*, 2010, **51**, 1126.
46. H.-K. Ishida, H. Ando, H. Ito, H. Ishida, M. Kiso and A. Hasegawa, *J. Carbohydr. Chem.*, 1997, **16**, 413.
47. H. Prabhanjan, M. Kiso and A. Hasegawa, *Carbohydr. Res.*, 1991, **222**, C1.
48. H. Prabhanjan, K. Aoyama, M. Kiso and A. Hasegawa, *Carbohydr. Res.*, 1992, **233**, 87.
49. H.-K. Ishida, H. Ishida, M. Kiso and A. Hasegawa, *Tetrahedron: Asymmetry*, 1994, **5**, 2493.
50. H. Ito, H. Ishida, M. Kiso and A. Hasegawa, *Carbohydr. Res.*, 1997, **304**, 187.
51. H. Ito, H. Ishida, H. Waki, S. Ando and M. Kiso, *Glycoconjugate J.*, 1999, **16**, 585.
52. H.-K. Ishida, H. Ishida, M. Kiso and A. Hasegawa, *Carbohydr. Res.*, 1994, **260**, C1.
53. A. Imamura, H. Ando, H. Ishida and M. Kiso, *J. Org. Chem.*, 2009, **74**, 3009.
54. K. Hotta, H. Ishida, M. Kiso and A. Hasegawa, *J. Carbohydr. Chem.*, 1995, **14**, 491.
55. K. Hotta, T. Kawase, H. Ishida, M. Kiso and A. Hasegawa, *J. Carbohydr. Chem.*, 1995, **14**, 961.
56. Y. Ito, M. Sugimoto, S. Sato and T. Ogawa, *Tetrahedron Lett.*, 1986, **27**, 4753.
57. S. Nakashima, H. Ando, A. Imamura, N. Yuki, H. Ishida and M. Kiso, *Chem. - Eur. J.*, 2011, **17**, 588.
58. S. Nakashima, H. Ando, R. Saito, H. Tamai, H. Ishida and M. Kiso, *Chem. - Asian J.*, 2012, **7**, 1041.
59. H. Tamai, H. Ando, H.-N. Tanaka, R. Hosoda-Yabe, T. Yabe, H. Ishida and M. Kiso, *Angew. Chem., Int. Ed. Engl.*, 2011, **50**, 2330.
60. P. Zhang, K. Wang, J. Zhang, C. Li and H. Guan, *Eur. J. Org. Chem.*, 2015, **2015**, 570.

61. H. Hirose, H. Tamai, C. Gao, A. Imamura, H. Ando, H. Ishida, T. Feizi and M. Kiso, *Org. Biomol. Chem.*, 2015, **13**, 11105.
62. K. Goto, M. Sawa, H. Tamai, A. Imamura, H. Ando, H. Ishida and M. Kiso, *Chem. - Eur. J.*, 2016, **22**, 8323.
63. A. Kameyama, H. Ishida, M. Kiso and A. Hasegawa, *Carbohydr. Res.*, 1990, **200**, 269.
64. A. Kameyama, H. Ishida, M. Kiso and A. Hasegawa, *J. Carbohydr. Chem.*, 1994, **13**, 641.
65. T. Ando, H. Ishida and M. Kiso, *Carbohydr. Res.*, 2003, **338**, 503.
66. L. F. Tietze, C. O. Janssen and J.-A. Gewert, *Eur. J. Org. Chem.*, 1998, **1998**, 1887.
67. A. Kameyama, H. Ishida, M. Kiso and A. Hasegawa, *Carbohydr. Res.*, 1991, **209**, C1.
68. A. Kameyama, H. Ishida, M. Kiso and A. Hasegawa, *J. Carbohydr. Chem.*, 1991, **10**, 549.
69. M. Yamaguchi, H. Ishida, C. Galustian, T. Feizi and M. Kiso, *Carbohydr. Res.*, 2002, **337**, 2111.
70. S. Komba, C. Galustian, H. Ishida, T. Feizi, R. Kannagi and M. Kiso, *Angew. Chem., Int. Ed. Engl.*, 1999, **38**, 1131.
71. T. Ehara, A. Kameyama, Y. Yamada, H. Ishida, M. Kiso and A. Hasegawa, *Carbohydr. Res.*, 1996, **281**, 237.
72. A. Kameyama, T. Ehara, Y. Yamada, H. Ishida, M. Kiso and A. Hasegawa, *J. Carbohydr. Chem.*, 1995, **14**, 507.
73. S. Nunomura and T. Ogawa, *Tetrahedron Lett.*, 1988, **29**, 5681.
74. T. K. Park, I. J. Kim, S. Hu, M. T. Bilodeau, J. T. Randolph, O. Kwon and S. J. Danishefsky, *J. Am. Chem. Soc.*, 1996, **118**, 11488.
75. J. Lassaletta and R. R. Schmidt, *Tetrahedron Lett.*, 1995, **36**, 4209.
76. H. Ishida, R. Miyawaki, M. Kiso and A. Hasegawa, *J. Carbohydr. Chem.*, 1996, **15**, 163.
77. J. M. Lassaletta, K. Carlsson, P. J. Garegg and R. R. Schmidt, *J. Org. Chem.*, 1996, **61**, 6873.
78. O. M. Schutte, A. Ries, A. Orth, L. J. Patalag, W. Romer, C. Steinem and D. B. Werz, *Chem. Sci.*, 2014, **5**, 3104.
79. K. Koike, M. Sugimoto, S. Sato, Y. Ito, Y. Nakahara and T. Ogawa, *Carbohydr. Res.*, 1987, **163**, 189.
80. K. C. Nicolaou, T. J. Caulfield and H. Katoaka, *Carbohydr. Res.*, 1990, **202**, 177.
81. K. C. Nicolaou, T. Caulfield, H. Kataoka and T. Kumazawa, *J. Am. Chem. Soc.*, 1988, **110**, 7910.
82. D. Shapiro and A. J. Acher, *Chem. Phys. Lipids*, 1978, **22**, 197.
83. D. Qiu and R. R. Schmidt, *Liebigs Ann. Chem.*, 1992, **1992**, 217.
84. D. Adlercreutz, J. T. Weadge, B. O. Petersen, J. Ø. Duus, N. J. Dovichi and M. M. Palcic, *Carbohydr. Res.*, 2010, **345**, 1384.
85. A. Kimura, A. Imamura, H. Ando, H. Ishida and M. Kiso, *Synlett*, 2006, **2006**, 2379.

86. H. Ishida, R. Miyawaki, M. Kiso and A. Hasegawa, *Carbohydr. Res.*, 1996, **284**, 179.
87. M. Numata, M. Sugimoto, K. Koike and T. Ogawa, *Carbohydr. Res.*, 1987, **163**, 209.
88. T. Murase, A. Kameyama, K. P. R. Kartha, H. Ishida, M. Kiso and A. Hasegawa, *J. Carbohydr. Chem.*, 1989, **8**, 265.
89. H. Tamai, A. Imamura, J. Ogawa, H. Ando, H. Ishida and M. Kiso, *Eur. J. Org. Chem.*, 2015, **2015**, 5199.
90. K. Goto, T. Suzuki, H. Tamai, J. Ogawa, A. Imamura, H. Ando, H. Ishida and M. Kiso, *Asian J. Org. Chem.*, 2015, **4**, 1160.
91. Y.-F. Wu, Y.-F. Tsai, J.-R. Guo, C.-P. Yu, H.-M. Yu and C.-C. Liao, *Org. Biomol. Chem.*, 2014, **12**, 9345.
92. Y. Iwayama, H. Ando, H. Ishida and M. Kiso, *Chem. - Eur. J.*, 2009, **15**, 4637.
93. R. R. Schmidt and P. Zimmermann, *Angew. Chem., Int. Ed. Engl.*, 1986, **25**, 725.
94. O. M. Schütte, A. Ries, A. Orth, L. J. Patalag, W. Römer, C. Steinem and D. B. Werz, *Chem. Sci.*, 2014, **5**, 3104.
95. H. Sakamoto, S. Nakamura, T. Tsuda and S. Hashimoto, *Tetrahedron Lett.*, 2000, **41**, 7691.
96. B. Yu and H. Tao, *Tetrahedron Lett.*, 2001, **42**, 2405.
97. B. Yu and J. Sun, *Chem. Commun.*, 2010, **46**, 4668.
98. H. Gold, R. G. Boot, J. M. F. G. Aerts, H. S. Overkleeft, J. D. C. Codée and G. A. van der Marel, *Eur. J. Org. Chem.*, 2011, **2011**, 1652.
99. M. Teruaki, M. Yoshiyuki, S. Shin-ichiro, *Chem. Lett.*, 1981, **10**, 431.
100. K. Wilhelm and K. Eduard, *Ber. Dtsch. Chem. Ges.*, 1901, **34**, 957.
101. K. Ohashi, S. Kosai, M. Arizuka, T. Watanabe, M. Fukunaga, K. Monden, T. Uchikoda, Y. Yamagiwa and T. Kamikawa, *Tetrahedron Lett.*, 1988, **29**, 1189.
102. A. R. Parameswar, J. A. Hawkins, L. K. Mydock, S. S. Mark and A. V. Demchenko, *Eur. J. Org. Chem.*, 2010, **2010**, 3269.
103. O. Plettenburg, V. Bodmer-Narkevitch and C.-H. Wong, *J. Org. Chem.*, 2002, **67**, 4559.
104. W. Du, S. S. Kulkarni and J. Gervay-Hague, *Chem. Commun.*, 2007, **0**, 2336.
105. W. Du and J. Gervay-Hague, *Org. Lett.*, 2005, **7**, 2063.
106. J. A. Morales-Serna, Y. Díaz, M. I. Matheu and S. Castillón, *Eur. J. Org. Chem.*, 2009, **2009**, 3849, Corrections in *Eur. J. Org. Chem.*, 2010, **2010**, 371.
107. K. A. D'Angelo and M. S. Taylor, *Chem. Commun.*, 2017, **53**, 5978.
108. Z. Zhang, W. Zhao, B. Wang, C. Xia, W. Zhang and P. G. Wang, *Org. Lett.*, 2011, **13**, 4530.
109. M. D. Vaughan, K. Johnson, S. DeFrees, X. Tang, R. A. J. Warren and S. G. Withers, *J. Am. Chem. Soc.*, 2006, **128**, 6300.
110. Q. Yao, J. Song, C. Xia, W. Zhang and P. G. Wang, *Org. Lett.*, 2006, **8**, 911.

111. J. I. Rearick, J. E. Sadler, J. C. Paulson and R. L. Hill, *J. Biol. Chem.*, 1979, **254**, 4444.
112. S. C. Basu, *Glycobiology*, 1991, **1**, 469.
113. B. Guilbert and S. L. Flitsch, *J. Chem. Soc., Perkin Trans. 1*, 1994, 1181.
114. Z. Xiao, Y. Guo, Y. Liu, L. Li, Q. Zhang, L. Wen, X. Wang, S. M. Kondengaden, Z. Wu, J. Zhou, X. Cao, X. Li, C. Ma and P. G. Wang, *J. Org. Chem.*, 2016, **81**, 5851.

CHAPTER 11

Enzymatic and Chemoenzymatic Synthesis of Human Milk Oligosaccharides (HMOS)

HAI YU AND XI CHEN*

Department of Chemistry, University of California, One Shields Avenue, Davis, CA 95616, USA
*E-mail: xiichen@ucdavis.edu

11.1 Introduction

Human milk oligosaccharides (HMOS) are a complex mixture of structurally diverse oligosaccharides that constitute the third major component of human milk after lactose and lipids. The concentration of HMOS varies in lactation stages and individual mothers with a broad range of 5–24 g L^{-1} being reported.[1,2] HMOS are either linear or branched free oligosaccharides, most of which are extended from lactose by glycosyltransferase-catalyzed reactions. More than 200 different HMOS structures have been identified with over 100 structures being elucidated.[3–7] While 60–77% of HMOS are fucosylated, the percentage of sialylated HMOS vary from less than 28% to 61% for different samples.[3,6,8] The most abundant HMOS are trisaccharides including monofucosylated lactose such as 2′-fucosyllactose (2′-FL) and 3-fucosyllactose (3-FL), and monosialylated lactose such as

Chemical Biology No. 11
Synthetic Glycomes
Edited by Wanyi Guan, Lei Li and Peng George Wang

Published by the Royal Society of Chemistry, www.rsc.org

3′-sialylactose (3′-SL) and 6′-sialylactose (6′-SL); tetrasaccharides lacto-*N*-tetraose (LNT) and lacto-*N*-neotetraose (LNnT) as well as their mono- or di-fucosylated or sialylated structures. In bovine milk, only trace amounts of oligosaccharides are detectable, with sialylated lactose being the major component.[1,9]

HMOS have received increasing attention for their roles in selectively stimulating the growth of beneficial bacteria in the intestine,[10,11] acting as decoy receptors to inhibit the binding of some pathogens,[12] and immune modulation.[13] These functions help to create protective intestinal microbiota dominated by bifidobacterial species that contribute to the health of breast-fed infants which have fewer incidents of diarrhea, allergy, and other health problems.[12–15] For example, different fractions of HMOS have been shown to inhibit the binding of *Escherichia coli*, *Vibrio cholerae*, and *Salmonella fyris* to intestinal cells.[16] Fucosylated HMOS are known to inhibit the binding of pathogenic bacteria and some such as Le^b, Le^x, and sialyl Le^x have been shown to possess anti-inflammatory activities.[17] One of the two simplest fucosylated human milk oligosaccharides, 2′-FL, was shown to protect against necrotizing enterocolitis in a neonatal mouse model.[18] The addition of 2′-FL was shown to help strengthen the immune system of formula-fed infants.[19] 2′-FL was recently approved for marketing as a supplement to infant formula in the European Union in 2016. A double-blind randomized clinical trial for the effects of adding 2′-FL to infant formula started on July 15, 2017 in the US (ClinicalTrials.gov Identifier: NCT03109223).

All structurally known HMOS are comprised of only five types of monosaccharides: D-glucose (Glc), D-galactose (Gal), L-fucose, *N*-acetyl-D-glucosamine (GlcNAc), and *N*-acetylneuraminic (Neu5Ac). Among the five monosaccharide building blocks of HMOS, Glc resides only at the reducing end and is part of the lactose core structure. HMOS share some common backbones (20 identified so far, Figure 11.1) including lactose and more complex structures extended from lactose.[5,20] The more complex backbones have lactose at the reducing end which is extended with Galβ1–4GlcNAc (*N*-acetyllactoamine or LacNAc) repeating unit β1–3- or β1–6-linked to the Gal residue in the lactose with or without a Galβ1–3GlcNAc (lacto-*N*-biose or LNB) cap. These backbones can be fucosylated with α1–2, α1–3, and/or α1–4-linkages and/or sialylated with α2–3 and/or α2–6-linkages, leading to 157 structures that have been elucidated so far (Figure 11.2). In addition, 12 deviant structures missing reducing end lactose or containing less common GalNAcα1–3Gal or Galβ1–3/4/6Gal have also been identified.[5] HMOS core structures as well as detailed information about HMOS structures that have been characterized and the related references can be found in a recent review and two articles.[5–7]

The important roles of HMOS in promoting the growth of beneficial bacteria, inhibiting pathogenic bacteria binding, regulating immune functions, providing nutrients for brain development, and preventing diseases in infants, as well as the lack of access to complex HMOS in a pure form and in sufficient amounts for various biological studies and clinical testing

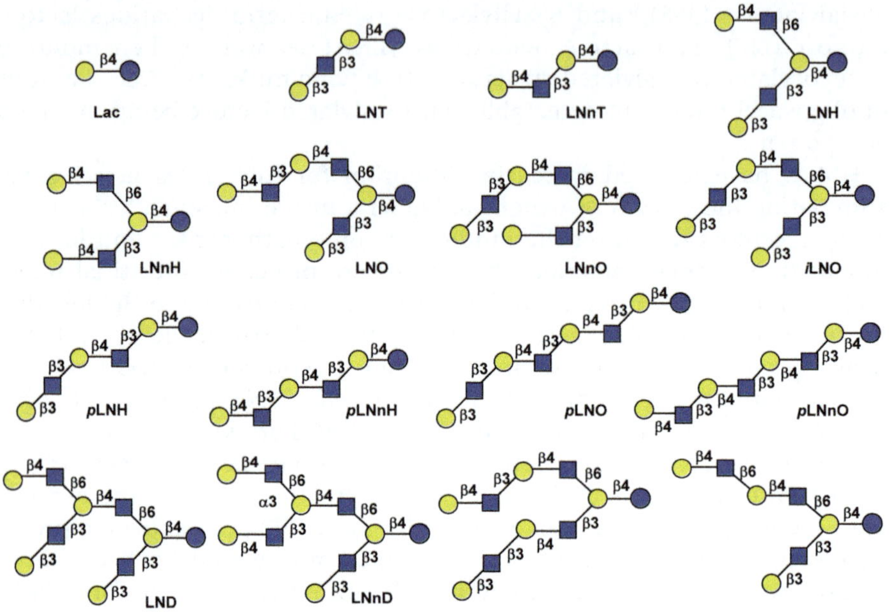

Figure 11.1 Core structures of diverse human milk oligosaccharides (HMOS) that have been reported.[5] Except for lactose (Lac) which itself is not a human oligosaccharide, all other HMOS core structures are part of HMOS themselves. Symbols and abbreviations: galactose (Gal), *N*-acetylglucosamine (GlcNAc), glucose (Glc).

provide an excellent opportunity for developing synthetic methods for preparative and large-scale production of HMOS. While derivatization of oligosaccharides from human milk followed by column purification in a shot-gun glycomics strategy[6,8] provided HMOS probes for glycan microarray studies, the amount of available HMOS is limited. Obtaining HMOS using chemical synthesis alone is time consuming due to the complex multistep routes involving protecting/deprotecting operations which often result in relatively low yields. A limited number of HMOS with sizes up to octasaccharides have been synthesized by chemical methods, including core structures LNnT,[21] LNT,[21–23] lacto-*N*-neohexaose (*p*LNnH), and lacto-*N*-neooctaose (*p*LNnO);[24] fucosylated compounds 3-FL,[25] 2′-FL,[25] 3,2′-difucosyl-lactose (LDFP),[25] lacto-*N*-fucopentaose I (LNFP I),[26] lacto-*N*-fucopentaose III (LNFP III);[22,26,27] and sialylated compounds 3′-SL,[28–30] 6′-SL,[28,31] sialyllacto-*N*-tetraose (LST a).[32–34]

Compared to chemical synthesis of HMOS, enzymatic and chemoenzymatic approaches have great advantages and have been developed rapidly in recent years.[5,35–37] Here we summarize our recent efforts on synthesizing structurally defined HMOS using sequential one-pot multienzyme (OPME)

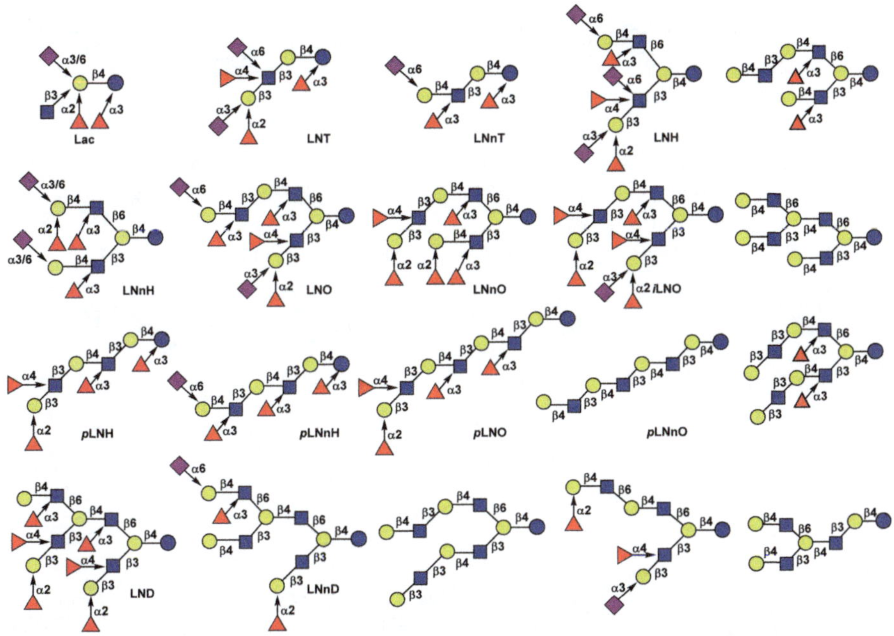

Figure 11.2 Possible fucosylation and/or sialylation sites for the formation of diverse human milk oligosaccharides (HMOS) that have been reported.[5] Symbols and abbreviations: ◆ *N*-acetylneuraminic acid (Neu5Ac), ▲ fucose (Fuc), ○ galactose (Gal), ■ *N*-acetylglucosamine (GlcNAc), ● glucose (Glc).

glycosylation methods. Other enzyme-dependent approaches for producing HMOS are also reviewed including those catalyzed by glycosyltransferases, transglycosidases, glycosidases and mutants, enzymatic synthesis, chemoenzymatic synthesis, whole-cell production and fermentation of engineered *E. coli* cells. So far less than 30 reported HMOS structures (Figure 11.3) have been synthesized, some are in sub-milligram amounts. It is expected that this list will grow rapidly with the ongoing efforts on synthetic method development and mining glycosyltransferases with the desired substrate specificity and expression level.

11.2 Bacterial and Mammalian Glycosyltransferases (GTs) that Have Been Used for the Synthesis of HMOS

The key requirement to produce HMOS by enzymatic and chemoenzymatic methods is to access suitable glycosyltransferases with the desired substrate specificity. So far, several bacterial glycosyltransferases have been successfully

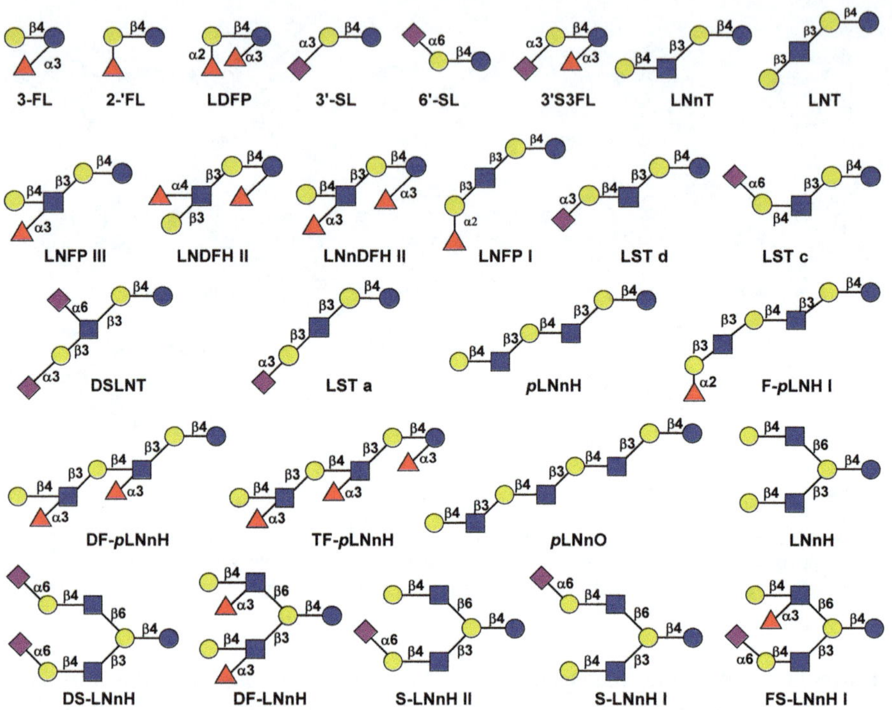

Figure 11.3 Structures of HMOS that have been synthesized so far. Symbols and abbreviations: *N*-acetylneuraminic acid (Neu5Ac), fucose (Fuc), galactose (Gal), *N*-acetylglucosamine (GlcNAc), glucose (Glc).

identified and applied in preparative and/or gram-scale synthesis of HMOS. A limited number of HMOS have been synthesized at industrial scale.[38–42] In addition, mammalian glycosyltransferase cloning and expression effort supported by the National Institutes of Health (NIH)[43] has also facilitated the access to complex HMOS by chemoenzymatic approaches. Bacterial and mammalian glycosyltransferases (GTs) that have been used for synthesizing HMOS are listed in Table 11.1.

11.3 Synthesis of HMOS *via* One-pot Multienzyme (OPME) Approaches

11.3.1 One-pot Multienzyme (OPME) Glycosylation Systems

In the past several years, we have developed highly efficient one-pot multienzyme (OPME) glycosylation methods for the synthesis of complex glycans and glycoconjugates by combining the sugar nucleotide biosynthetic processes with glycosyltransferase-catalyzed reactions. The methods can be readily

Table 11.1 Bacterial and mammalian glycosyltransferases (GTs) that have been used for synthesizing HMOS.

Glycosidic linkages	Abbreviations	Symbol	Bacterial GTs	Mammalian GTs
Galactosidic bonds	Galβ1–3GlcNAc		*E. coli* β1–4-GalT (WbgO)[44,45]	B3GALT5 [46]
	Galβ1–4GlcNAc		*N. meningitidis* β1–4-GalT[47] *N. gonorrhoeae* β1–4-GalT[38]	GalT1 [46]
N-acetyl-glucosaminidic bond	GlcNAcβ1–3Gal		*N. meningitidis* β1–3-*N*-GlcNAcT[48] *N. gonorrhoeae* β1–3-*N*-GlcNAcT[38]	B3GNT2 [46]
	GlcNAcβ1–6Gal			GCNT2 [46]
Fucosidic bond	Fucα1–2Gal		*T. elongatus* BP-1 α1–2-FucT[49] *H. pylori* α1–2-FucT[50,51] *H. mustelae* α1–2-FucT[47] *E. coli* α1,2-FucT (WbgL)[52]	FUT1 [46,53]
	Fucα1–3Glc		*H. pylori* α1–3-FucT[54]	FUT3 [46] FUT5 [46]
	Fucα1–3GlcNAc		*H. pylori* α1–3/4-FucT[55] *H. pylori* α1–3FucT[47]	FUT3 [46,56] FUT5 [46] FUT6 [56] FUT9 [46]
	Fucα1–4GlcNAc	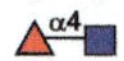	*H. pylori* α1–3/4-FucT[55]	FUT3 [53,56]
Sialosidic bond	Neu5Acα2–3Gal		*P. multocida* α2–3-SiaT[47,57] *P. dagmatis* α2–3-SiaT[58,59] *N. meningitides* α2–3-SiaT[60]	ST3GAL4 [46]
	Neu5Acα2–6Gal	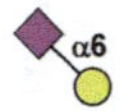	*P. damselae* α2–6-SiaT[47,48,61,62] PmST1_P34H/M144L[63] *P. dagmatis* SiaT_P7H/M117A[58]	ST6GAL1 [46]
	Neu5Acα2–6GlcNAc	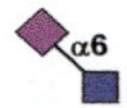		ST6GALNAC5 [46]

applied for the production of HMOS. Despite their structural complexity for some members, all HMOS in principal can be enzymatically synthesized from lactose (Lac) using the desired glycosyltransferases with four sugar nucleotide activation systems for the formation of uridine 5′-diphosphate-*N*-ace tylglucosamine (UDP-GlcNAc), uridine 5′-diphosphate-galactose (UDP-Gal), guanidine 5′-diphosphate-fucose (GDP-Fuc), and cytidine monophosphate-*N*-acetylneuraminic acid (CMP-Neu5Ac). Nucleoside diphosphate (NDP)-activated monosaccharides including UDP-GlcNAc, UDP-Gal, and GDP-Fuc can be formed from their corresponding simple free monosaccharides using a suitable glycokinase (GlyK) and a suitable nucleotidyltransferase or a sugar nucleotide pyrophosphorylase. An inorganic pyrophosphatase can also be added to drive the reaction towards the *in situ* formation of sugar nucleotides. CMP-Neu5Ac can be formed from Neu5Ac and cytidine 5′-triphosphate (CTP) using a CMP-sialic acid synthetase (CSS)-catalyzed reaction or from *N*-acetylmannosamine (ManNAc), pyruvate, and CTP using CSS and a sialic acid aldolase (NanA). The sugar nucleotide biosynthetic enzymes and the desired glycosyltransferases can be used in one-pot for the formation of target glycans with the desired glycosidic linkages from simple free monosaccharides and suitable acceptors. The desired glycosidic linkage is determined by the glycosyltransferase used. Multiple OPME reactions can be carried out to build up more complex glycan targets.[64] The advantage of OPME glycosylation approaches is that they allow the production of the desired glycans from readily available, low cost monosaccharides and nucleoside triphosphates. Each OPME process adds one or more monosaccharide units to the linear or branched glycan acceptors. With a good understanding of the acceptor substrate specificity of glycosyltransferase and altered OPME reaction sequences, different glycan targets can be produced even with the same set of enzymes.

Various OPME systems (**OPME1–OPME10**) that have been used in the synthesis of HMOS[48,57,65–67] are shown in Figure 11.4.

11.3.2 OPME Synthesis of Core Glycans LNTri II (Lc_3) and LNnT

Human milk trisaccharide GlcNAcβ1–3Galβ1–4Glc (LNTri II or Lc_3) was synthesized from inexpensive disaccharide lactose and monosaccharide *N*-acetylglucosamine (GlcNAc) using a one-pot four-enzyme GlcNAc activation and transfer system (OPME1, Figure 11.4) containing *Bifidobacterium longum* (strain ATCC55813) *N*-acetylhexosamine-1-kinase (BLNahK),[68] *Pasteurella multocida* *N*-acetylglucosamine uridylyltransferase (PmGlmU),[69] *P. multocida* inorganic pyrophosphatase (PmPpA),[70] and *Neisseria meningitidis* β1–3-*N*-acetylglucosaminyltransferase (NmLgtA).[71,72] In this system, adenosine 5′-triphosphate (ATP) and GlcNAc were used by BLNahK-catalyzed reaction to form GlcNAc-1-P, which was used with uridine 5′-triphosphate (UTP) by PmGlmU to form UDP-GlcNAc, the sugar nucleotide donor for NmLgtA for the production of Lc_3 from lactose. All four enzymes were quite active in the

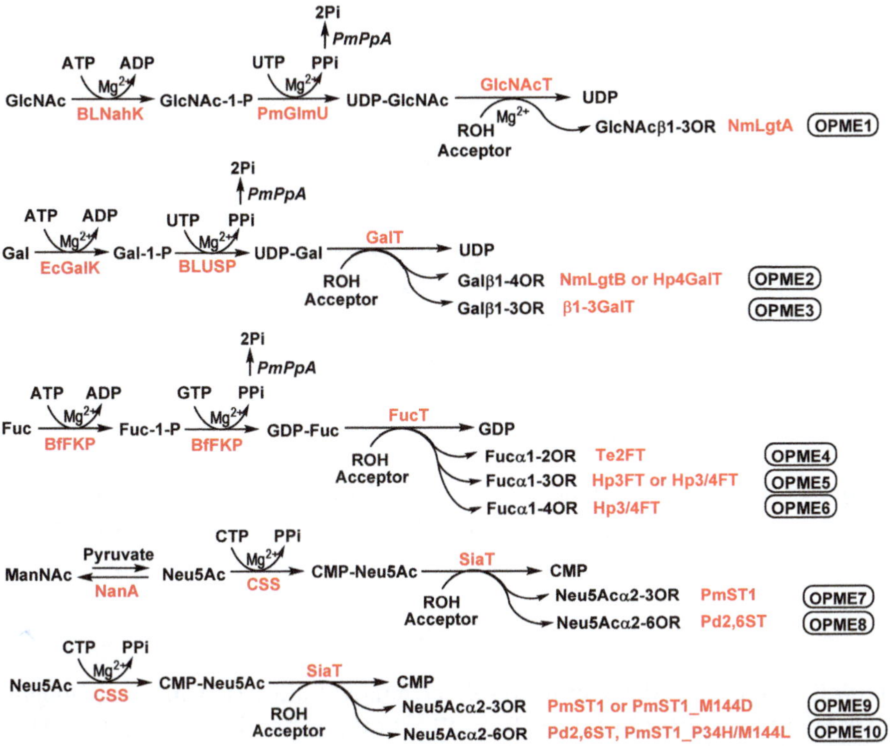

Figure 11.4 One-pot multienzyme (OPME) GlcNAc, Gal, Fuc, and Neu5Ac activation and transfer systems. Enzyme abbreviations: BLNahK, *B. longum* *N*-acetylhexosamine-1-kinase; PmGlmU, *P. multocida* *N*-acetylglucosamine 1-phosphate uridylyltransferase; PmPpA, *P. multocida* inorganic pyrophosphatase; GlcNAcT, *N*-acetylglucosaminyltransferase; NmLgtA, *N. meningitidis* β1–3-*N*-acetylglucosaminyltransferase; EcGalK, *E. coli* K-12 galactose kinase; BLUSP, *B. longum* UDP-sugar synthase; GalT, galactosyltransferase; NmLgtB, *N. meningitidis* β1–4-galactosyltransferase; BfFKP, a bifunctional enzyme from *B. fragilis* that has both L-fucokinase and GDP-fucose pyrophosphorylase activities; FucT, fucosyltransferase; Te2FT, α1–2-fucosyltransferase from *T. elongatus* BP-1; Hp3FT, α1–3-fucosyltransferase from *H. pylori*; Hp3/4FT, *H. pylori* UA948 α1–3/4-fucosyltransferase; NanA, sialic acid aldolase; NmCSS, *N. meningitidis* CMP-sialic acid synthetase; SiaT, sialyltransferases; PmST1, *P. multocida* α2–3-sialyltransferase; PmST1_M144D, *P. multocida* α2–3-sialyltransferase 1 M144D mutant; Pd2,6ST, *P. damselae* α2–6-sialyltransferase.

Tris-HCl buffer at pH 8.0. LNTri II (1.36 g) was obtained with an excellent 95% yield (Figure 11.5).[48]

Gram-scale synthesis of lacto-*N*-neotetraose (LNnT) tetrasaccharide Galβ1–4GlcNAcβ1–3Galβ1–4Glc was achieved from LNTri II and monosaccharide galactose (Gal) using a OPME galactose activation and transfer system

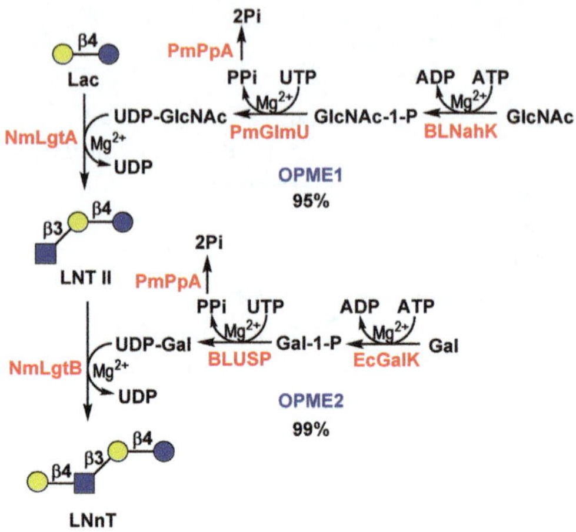

Figure 11.5 Gram-scale OPME synthesis of LNTri II and LNnT. Symbols and abbreviations: galactose (Gal), *N*-acetylglucosamine (GlcNAc), glucose (Glc).

(OPME2, Figure 11.4)[48] containing *E. coli* galactokinase (EcGalK),[73] *B. longum* UDP-sugar pyrophosphorylase (BLUSP),[74] PmPpA,[70] and *N. meningitidis* β1–4-galactosyltransferase (NmLgtB).[70] Sugar nucleotide donor UDP-Gal was generated *in situ* from Gal, ATP, and UTP using EcGalK, BLUSP, and PmPpA, and was used by NmLgtB with LNTri II to produce LNnT (1.28 g) with an excellent 99% yield (Figure 11.5).[48]

11.3.3 OPME Enzymatic Synthesis of Fucose-containing HMOS

11.3.3.1 OPME Synthesis of LNFP I

α1–2-Linked fucosides are major structural components of neutral HMOS in the breast milk of individuals with an α1–2-fucosyltransferase (FUT2) encoded by the Secretor (*Se*) gene (*Se*⁺ individuals). We recently reported the discovery and characterization of a novel α1–2-fucosyltransferase from thermophilic cyanobacterium *Thermosynechococcus elongatus* BP-1 (Te2FT) with high fucosyltransferase activity and low donor hydrolysis activity.[49] Te2FT was applied in an efficient OPME fucosylation system (OPME4, Figure 11.4) for high-yield gram-scale synthesis of an important HMOS lacto-*N*-fucopentaose I (LNFP I, Fucα1–2Galβ1–3GlcNAcβ1–3Galβ1–4Glc). In this system, GDP-Fuc was formed from L-fucose, ATP, and guanosine 5′-triphosphate (GTP) using a bifunctional *Bacteroides fragilis* enzyme (BfFKP) with both L-fucokinase and

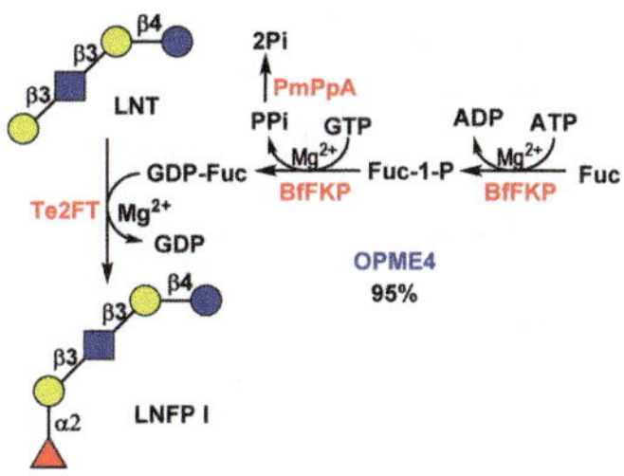

Figure 11.6 OPME gram-scale synthesis of LNFP I. Symbols and abbreviations: ▲ fucose (Fuc), ● galactose (Gal), ■ *N*-acetylglucosamine (GlcNAc), ● glucose (Glc).

GDP-Fuc pyrophosphorylase activities.[75] PmPpA was used to break down the pyrophosphate byproduct to drive the formation of GDP-Fuc which was used as the donor substrate for fucosyltransferases for the formation of fucosides. As shown in Figure 11.6, starting from L-fucose, GTP and ATP, GDP-Fuc produced *in situ* using BfFKP[75] was used by Te2FT for the production of pentasaccharide LNFP I (1.1 grams) from commercially available tetrasaccharide lacto-*N*-tetraose (LNT) with an excellent 95% yield.[49] The ability of production of LNFP I at gram-scale allowed the performance of a bacterial growth study which showed that LNFP I was selectively consumed as a carbon source by *Bifidobacterium longum* subsp. *infantis* (*B. infantis*, known to predominate in healthy breast-fed infants) but not *Bifidobacterium animalis* subsp. *lactis* (*B. lactis*, predominant in adults).[49]

11.3.3.2 OPME Synthesis of LNFP III

α1–3-Fucosyltransferase from *Helicobacter pylori* (Hp3FT) is highly efficient in adding α1–3-linked fucose to the GlcNAc residue in LNnT.[76,77] The OPME fucosylation system (OPME5, Figure 11.4)[78] containing BfFKP, PmPpA, and Hp3FT was applied for the synthesis of HMOS LNFP III Galβ1–4(Fucα1–3)GlcNAcβ1–3Galβ1–4Glc (181 mg) with a 94% yield from LNnT (Figure 11.7).[47]

11.3.3.3 OPME Synthesis of 3-FL, LNFP III, LNDFH II, and LNDFH III

H. pylori UA948 α1–3/4-fucosyltransferase (Hp3/4FT) can use both type I and type II glycans as acceptors,[79,80] which makes it a very attractive catalyst in the synthesis of fucose-containing HMOS. It was used in a highly efficient

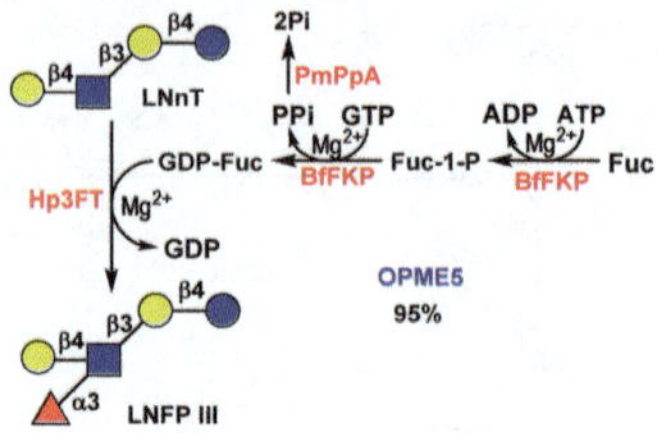

Figure 11.7 OPME synthesis of HMOS LNFP III. Symbols and abbreviations: ▲ fucose (Fuc), ● galactose (Gal), ■ *N*-acetylglucosamine (GlcNAc), ● glucose (Glc).

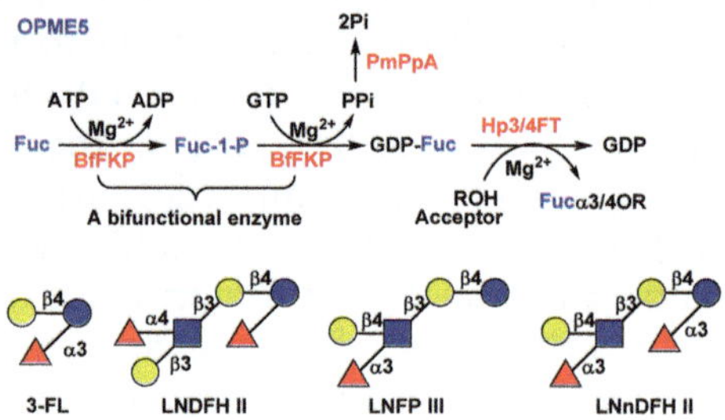

Figure 11.8 Hp3/4FT-catalyzed one-pot three-enzyme (OPME5) synthesis of human milk oligosaccharides 3-FL, LNFP III, LNDFH II, and LNDFH III. Symbols and abbreviations: ▲ fucose (Fuc), ● galactose (Gal), ■ *N*-acetylglucosamine (GlcNAc), ● glucose (Glc).

one-pot three-enzyme fucosylation system (OPME5 and OPME6, Figure 11.4) for the synthesis of diverse α1–3- and α1–4-linked fucosylated HMOS, including 3-fucosyllactose (3-FL), lacto-*N*-fucopentaose (LNFP) III, as well as lacto-*N*-difuco-hexaose (LNDFH) II and III (Figure 11.8).[55]

As shown in Figure 11.8, mono-fucosylation of lactose led to the formation of 3-fucosyllactose (3-FL) Galβ1–4(Fucα1–3)Glc with a 90% yield. 3-FL is one of the most abundant fucosylated HMOS in the milk of individuals with an α1–3/4-fucosyltransferase (FUT3) encoded by the Lewis (*Le*) gene (Le^+ individuals). It could inhibit bacterial or viral adhesion to human epithelial cells.[81]

Both GlcNAc and Glc in Galβ1–3/4GlcNAc and Galβ1–4Glc components in LNT and LNnT[63,82] are suitable fucosylation sites for Hp3/4FT although the enzyme prefers to fucosylate Galβ1–3/4GlcNAc over Galβ1–4Glc. Fucosylation of LNT with an excess amount of L-fucose, ATP, and GTP (3 equivalents each) led to the formation of difucosylated hexasaccharides lacto-*N*-difuco-hexaose II (LNDFH II) Galβ1–3(Fucα1–4)GlcNAcβ1–3Galβ1–4(Fucα1–3)Glc with a 98% yield. While human milk hexasaccharide lacto-*N*-neodifucohexaose II (LNnDFH II or LNDFH III) Galβ1–4(Fucα1–3)GlcNAcβ1–3Galβ1–4(Fucα1–3) Glcβ$ProN_3$ was obtained with a 99% yield in a similar procedure when LNnT was used as the fucosylation acceptor. When the ratio of GDP-Fuc to LNnT was set to 1.2 : 1, Hp3/4FT selectively catalyzed the transfer of fucose to the LacNAc (instead of lactose) component in LNnT and led to the formation of mono-fucosylated product LNFP III Galβ1–4(Fucα1–3)GlcNAcβ1–3Galβ1–4Glc with an 88% yield.

11.3.4 OPME Synthesis of Sialylated HMOS

Efficient OPME sialylation approaches for the synthesis of α2–3/6-linked sialosides[61,66,67,83] have been successfully applied for the synthesis of sialic acid-containing HMOS. In this system, commercially available Neu5Ac was used together with CTP and a suitable acceptor for the synthesis of target sialosides using a one-pot two-enzyme system containing a cytidine 5′-monophosphate sialic acid (CMP-Sia) biosynthetic enzyme *N. meningitidis* CMP-sialic acid synthetase (NmCSS)[84] and a suitable sialyltransferase. The six-carbon precursor of Neu5Ac, *N*-acetylmannosamine (ManNAc), and pyruvate can also be used as the starting materials to replace Neu5Ac, in this case a one-pot three-enzyme system is needed with the addition of a sialic acid aldolase, such as *P. multocida* sialic acid aldolase (PmNanA)[85] or *E. coli* sialic acid aldolase (EcNanA),[84] in addition to NmCSS and a sialyltransferase.

11.3.4.1 OPME Synthesis of 3′-SL and 6′-SL

3′-Sialyllactose (3′-SL) Neu5Acα2–3Lac and 6′-sialyllactose (6′-SL) Neu5Acα2–6Lac are the simplest and predominant sialylated HMOS. As shown in Figure 11.9A, 3′-SL was synthesized from lactose and ManNAc with a 79% yield using a one-pot three-enzyme system (OPME7, Figure 11.4) containing EcNanA, NmCSS, and *P. multocida* α2–3-sialyltransferase 1 (PmST1).[83] It was necessary to control the reaction time, pH value, and the amount of PmST1 as well as to monitor the reaction progress to reach optimal yields due to the presence of the α2–3-sialidase activity of PmST1.[83] PmST1 M144D mutant (PmST1_M144D)[57] with decreased α2–3-sialidase and donor hydrolysis activity was a better catalyst than the wild-type PmST1 in synthesizing 3′-SL. It was synthesized in a 2.1-gram scale from lactose as the acceptor substrate and Neu5Ac as the donor precursor using a one-pot two-enzyme system (OPME9, Figure 11.4) containing NmCSS and PmST1_M144D with an excellent 98% yield.[47]

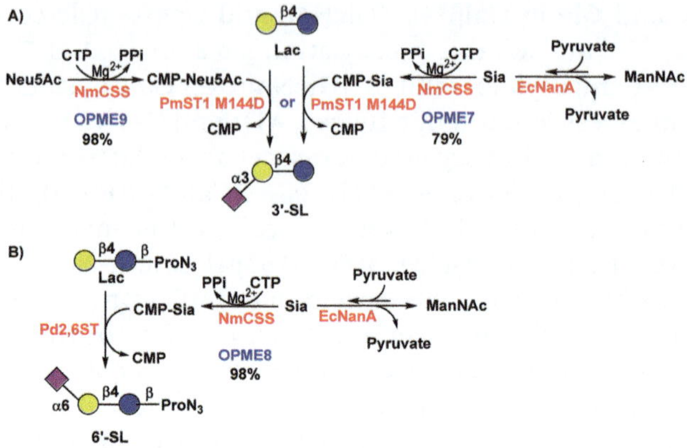

Figure 11.9 OPME enzymatic synthesis of HMOS 3′-SL (A) and 6′-SL (B). Symbols and abbreviations: ◆ *N*-acetylneuraminic acid (Neu5Ac), ○ galactose (Gal), ● glucose (Glc).

6′-SL with a propyl azide linker was chemoenzymatically synthesized *via* OPME α2–6-sialylation system (OPME8, Figure 11.4) which contained EcNanA, NmCSS, and *Photobacterium damselae* α2–6-sialyltransferase (Pd2,6ST) from a lactoside and ManNAc. This OPME was very efficient and a 98% sialylation yield was achieved (Figure 11.9B).[61]

11.3.4.2 OPME Synthesis of LST a, LST d, Sialylated LNFP III

PmST1_M144D-catalyzed OPME sialylation system was also successfully applied in the synthesis of more complex sialic acid-containing HMOS including LST a, LST d, and sialylated LNFP III.[47] As shown in Figure 11.10, both sialyllacto-*N*-tetraose a (LST a) and sialyllacto-*N*-neotetraose d (LST d) were successfully synthesized from Neu5Ac as the donor precursor using OPME9 (Figure 11.4). When LNT used as acceptor, LST a (107 mg) was obtained with 94% yield using PmST1_M144D-catalyzed OPME system. While LST d (126 mg) was obtained with 93% yield when LNnT was used as the sialyltransferase acceptor substrate. Similarly, sialylated LNFP III (51 mg) was synthesized at 86% when LNFP III was used as PmST1_M144D acceptor substrate.

The OPME sialyation system was also applied in the synthesis of 3′-sialyl LNT (LST a) with a propyl azide linker using chemically or chemoenzymatically synthesized LNTβProN$_3$ as the acceptor substrate.[86,87]

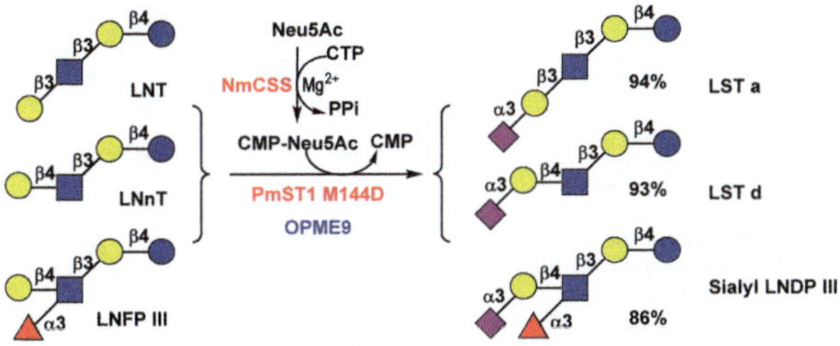

Figure 11.10 OPME enzymatic synthesis of HMOS LST a, LST d, sialylated LNFP III. Symbols and abbreviations: ◆ *N*-acetylneuraminic acid (Neu5Ac), ▲ fucose (Fuc), ● galactose (Gal), ■ *N*-acetylglucosamine (GlcNAc), ● glucose (Glc).

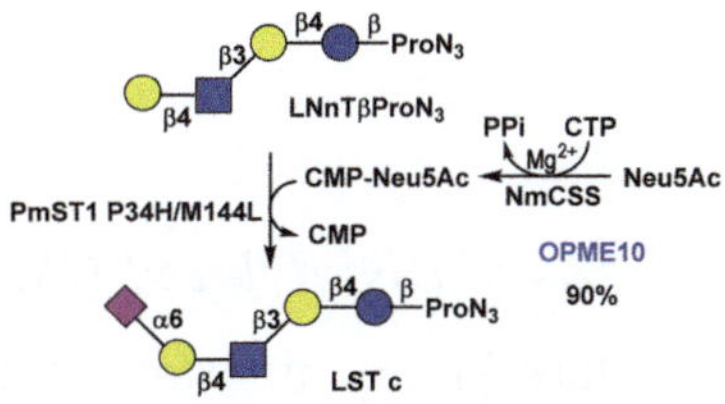

Figure 11.11 OPME enzymatic synthesis of HMOS LST c. Symbols and abbreviations: ◆ *N*-acetylneuraminic acid (Neu5Ac), ● galactose (Gal), ■ *N*-acetylglucosamine (GlcNAc), ● glucose (Glc).

11.3.4.3 OPME Synthesis of LSTc

Recently, a highly active PmST1 mutant P34H/M144L with α2–6-sialyltransferase activity was generated by protein crystal-structure-based saturation mutagenesis. This mutant was found to be the most effective α2–6-sialyltransferase and displayed 50% reduced donor hydrolysis and 50-fold reduced sialidase activity compared to the wild-type PmST1.[63] It displayed 8.7-fold α2–6ST activity compared to Pd2,6ST. PmST1 P34H/M144L mutant preferred to sialylated the terminal Gal residue in a multi-galactose-containing tetrasaccharide LNnT. HMOS sialyllacto-*N*-neotetraose c (LST c) (38 mg) was successfully synthesized with a 90% yield from LNnT and Neu5Ac using a PmST1 P34H/M144L mutant-catalyzed OPME sialylation system (Figure 11.11).[63]

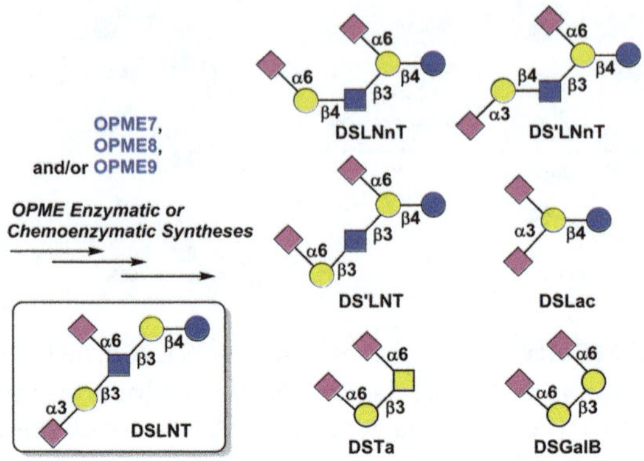

Figure 11.12 OPME enzymatic synthesis of disialyl oligosaccharides as analogs of DSLNT. Symbols and abbreviations: ◆ *N*-acetylneuraminic acid (Neu5Ac), ○ galactose (Gal), □ *N*-acetylgalactosamine (GalNAc), ■ *N*-acetylglucosamine (GlcNAc), ● glucose (Glc).

11.3.4.4 OPME Synthesis of Disialylated HMOS

Disialyl lacto-*N*-tetraose (DSLNT) is an important HMOS. As shown in Figure 11.12, a number of disialyl analogs of DSLNT were successfully synthesized using sequential OPME systems. Multiple OPME systems were used sequentially for building up complex HMOS structures. The combination of several OPME systems was used for the synthesis of α2–3/6-linked disialyl oligosaccharides including DSLNnT (1.70 g, 96% yield), DS'LNT (1.79 g, 95% yield), DS'LNnT (121 mg, 93% yield), DSTa (249 mg, 93% yield), DSGalB (38 mg, 84% yield), DSLac (112 mg, 93% yield).[48,62] Similar to natural HMOS DSLNT and the HMOS pool, both synthetic DSLNnT and DSLNT were shown to protect neonatal rats from necrotizing enterocolitis.[48,62]

11.4 Glycosyltransferase-catalyzed Enzymatic Synthesis of HMOS

Other than sequential OPME systems for synthesizing diverse HMOS from lactose and simple free monosaccharides, glycosyltransferases can be used directly for the synthesis of HMOS from a suitable acceptor and a desired sugar nucleotide donor. Bacterial glycosyltransferases are commonly used. For example, trisaccharide LNTri II was synthesized in 250 grams from lactose, UDP-GlcNAc, and a bacterial β1–3-*N*-acetyl-glucosaminyltransferase with an 85% yield.[38] LNnT (300 g) was then synthesized from LNTri II and UDP-Gal with a yield of over 85% using a bacterial β1–4-galactosyltransferase

(LgtB).[38] LNnT was further converted to yield sialyllacto-*N*-neotetraose D (LST d) using *Trypanosoma cruzi* α2–3-*trans*-sialidase.[38] HMOS derivatives such as benzyl β-lacto-*N*-tetraoside (LNTβOBn, 8.7 mg) was also successfully synthesized from benzyl β-lacto-*N*-triose with a yield of 87% using a purified *E. coli* β1–3-galactosyltransferase (EcWbgO).[44] 2′-FL and 3-FL were synthesized enzymatically from lactose and GDP-L-fucose using *H. pylori* NCTC 364 α1–2-fucosyltransferase and an α1–3-fucosyltransferase, respectively.[50,52] Recently, a *H. pylori* α1–3-fucosyltransferase (Δ52FutA) quadruple mutant (A128N/H129E/Y132I/S46F) was successfully applied in the synthesis of 3-FL with an improved (>96%) yield using 5 mM of GDP-Fuc with a one hour reaction time frame, which was a 40-fold increase of yield compared to the reactions using the wild-type FutA (2.4% yield, 14 h).[54] To produce 3′-SL in large amounts with the use of expensive CMP-Neu5Ac, a fusion protein of *N. meningitidis* CMP-Neu5Ac synthetase and α2–3-sialyltransferase was constructed and used with mykokinase and pyruvate kinase for synthesizing 3′-SL in 77 g with a 88% yield from lactose, sialic acid, phosphoenolpyruvate, and catalytic amounts of ATP and CMP.[60] More recently, *Pasteurella dagmatis* α2–3-sialyltransferase was cloned and used in synthesizing 3′-SL from lactose and CMP-Neu5Ac.[59]

Glycosyltransferases from humans and other animals have also been used in enzymatic synthesis of HMOS. For example, a variety of fucosylated LNT and LNnT were produced using recombinant human α1–3-and α1–4-fucosyltransferases.[56] LNnT was converted into LNFP-III by an α1–3-fucosyltransferase from chicken serum.[88] LNDFH-I was also enzymatically synthesized *via* fucosylations of LNT using a combination of human fucosyltransferase I (FUT1) and a commercially available human fucosyltransferase III (FUT3).[53]

11.5 Enzymatic Synthesis of HMOS Using Transglycosidases, Glycosidases and Mutants

In addition to glycosyltransferases, transglycosidases, glycosidases and mutants (*e.g.* glycosynthases) have been used in the synthesis of HMOS.[35,36,89–92] For example, an inverting α1–2-L-fucosidase from *B. bifidum* was converted into an α1–2-fucosynthase by Asp766Gly mutation and was used to synthesize 2′-FL from lactose and a chemically synthesized β-L-fucosyl fluoride donor with a 6% yield.[89] The use of α-L-fucosidase variants from *B. bifidum* for enzymatic preparation of 3-FL (using lactose as the acceptor), LDFT (using 2′-FL as the acceptor), and LNFP II (using LNT as the acceptor) with β-L-fucopyranosylfluoride as donor were also demonstrated, with a yield of 13, 5.5, and 41% respectively.[90] These methods require the use of chemically synthesized donors and are commonly accompanied with challenges of low yields and poor regioselectivity. Therefore, they are not ideal choices for large-scale production of HMOS.

11.6 Chemoenzymatic Synthesis of HMOS

Combining the flexibility of chemical synthesis and the regio- and stereo-selectivity of glycosyltransferase-catalyzed reactions, chemoenzymatic syntheses have proven to be powerful strategies to construct libraries of complex HMOS with different chain lengths, various branching patterns, and diverse degree of fucosylation or sialylation. One such approach is called the core synthesis/enzymatic extension (CSEE) strategy.[47] As shown in Figure 11.13, three branched symmetric and asymmetric core structures (HMO1, HMO2, and HMO3) were chemically synthesized using glycosyl thioether and bromide as glycosylation donors. Each core structure was extended by glycosyltransferase-catalyzed reactions with or without *in situ* generation of a sugar nucleotide donor in OPME systems to a diverse collection of up to 11 different HMOS. It was impressive that only four glycosyltransferases including *N. meningitidis* β1–3-*N*-acetylglucosaminyltransferase (NmLgtB), *P. damselae* α2–6-sialyltransferase (Pd2,6ST), and *H. pylori* α1–3-fucosyltransferase (Hp3FT), and *Helicobacter mustelae* α1–3-fucosyltransferase (Hm2FT), were needed to extend three core structures to generate a library of 31 branched oligosaccharides sharing the same core structure of HMOS lacto-*N*-neohexaose (LNnH). Among these oligosaccharides,[47] five and possibly an additional three were oligosaccharides identified previously in human milk[5] (Figure 11.13). Some are HMOS analogs having *N*-glycolylneuraminic acid (Neu5Gc) as the sialic acid form.

More recently, chemoenzymatic strategy was used to synthesize a library of 60 tagged HMOS and HMOS-liked oligosaccharides (Figure 11.14).[46] The tag introduced at the reducing end of lactose contained a hydrophobic chromophore to facilitate reaction monitoring and product purification, a primary amine group to allow conjugation to *N*-hydroxysuccinamide or epoxide-activated microarray slide for microarray printing, and an acid labile linkage to allow the release of the tag to produce free oligosaccharides had also been described.[46] Extension of the tagged lactose was achieved by a collection of ten mammalian glycosyltransferases including β1–3-galactosyltransferase 5 (B3GALT5), β1–4-galactosyltransferase (GalT1), β1–3-*N*-acetylglucosaminyltransferase 2 (B3GNT2), *N*-acetyllactosaminide β1–6-*N*-acetylglucosaminyltransferase (GCNT2), galactoside α1–2-fucosyltransferase 1 (FUT1), lactosamine α1–3/4 fucosyltransferase 3 and 5 (FUT3 and FUT5), lactosamine α1–3-fucosyltransferase 9 (FUT9), β-galactoside α2–3-sialyltransferase 4 (ST3GAL4), β-galactoside α2–6-sialyltransferase 1 (ST6GAL1), *N*-acetylgalactosaminide α2–6-sialyltransferase 5 (ST6GALNAC5). Noticeably, a Neu5Acα2–6-GlcNAc linkage was formed by ST6GALNAC5-catalyzed reaction. Structure branching was achieved using mammalian *N*-acetyllactosaminide β1–6-*N*-acetylglucosaminyltransferase (GCNT2) that selectively added β1–6-linked GlcNAc at the penultimate galactose residue in the acceptor substrate. Many complex asymmetric structures were obtained by clever design of the enzymatic synthetic route considering the acceptor

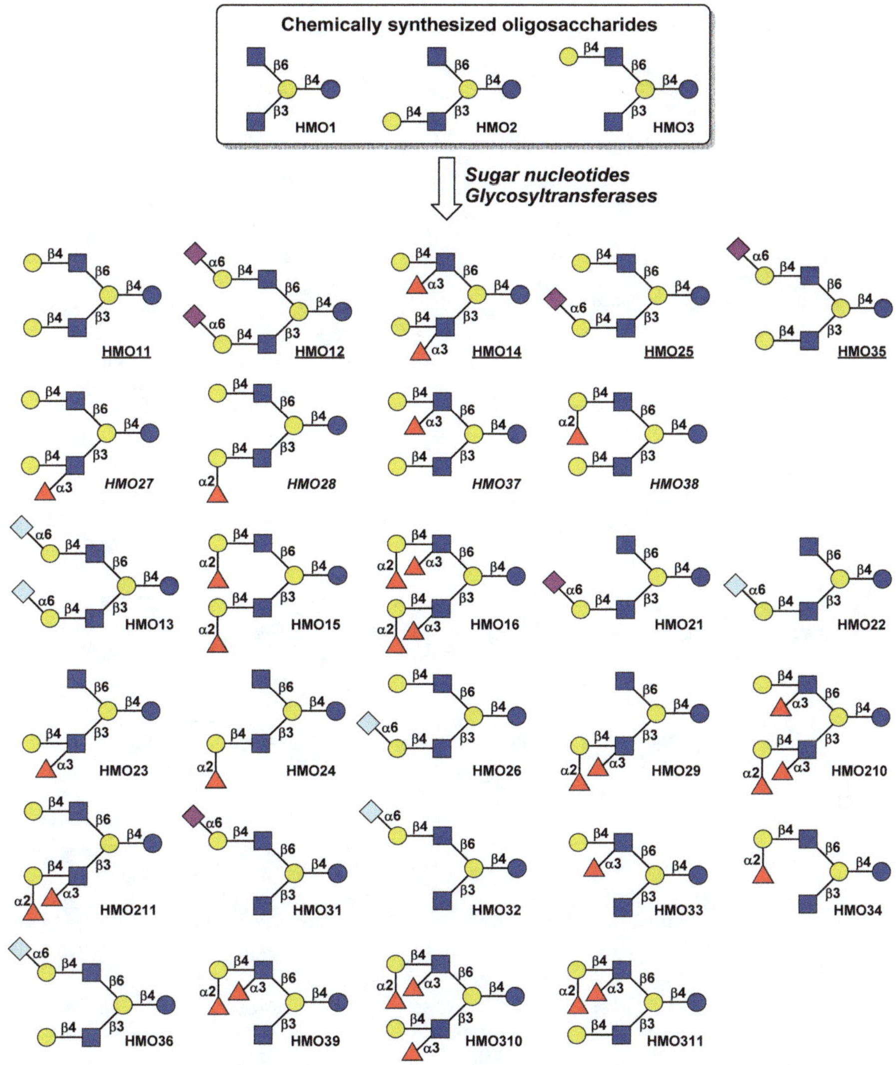

Figure 11.13 Structures of HMOS-like oligosaccharides that were synthesized using chemical core synthesis/enzymatic extension (CSEE) strategy.[47] Among these, HMO11, HMO12, HMO14, HMO25, HMO35 (underlined) were previously identified HMOS structures. HMO27, HMO28, HMO37, and HMO38 (italicized) were possible HMOS structures. Symbols and abbreviations: ◆ *N*-acetylneuraminic acid (Neu5Ac), ◇ *N*-glycolylneuraminic acid (Neu5Gc), ▲ fucose (Fuc), ● galactose (Gal), ■ *N*-acetylglucosamine (GlcNAc), ● glucose (Glc). Adapted from ref. 47 with permission from American Chemical Society, Copyright 2016.

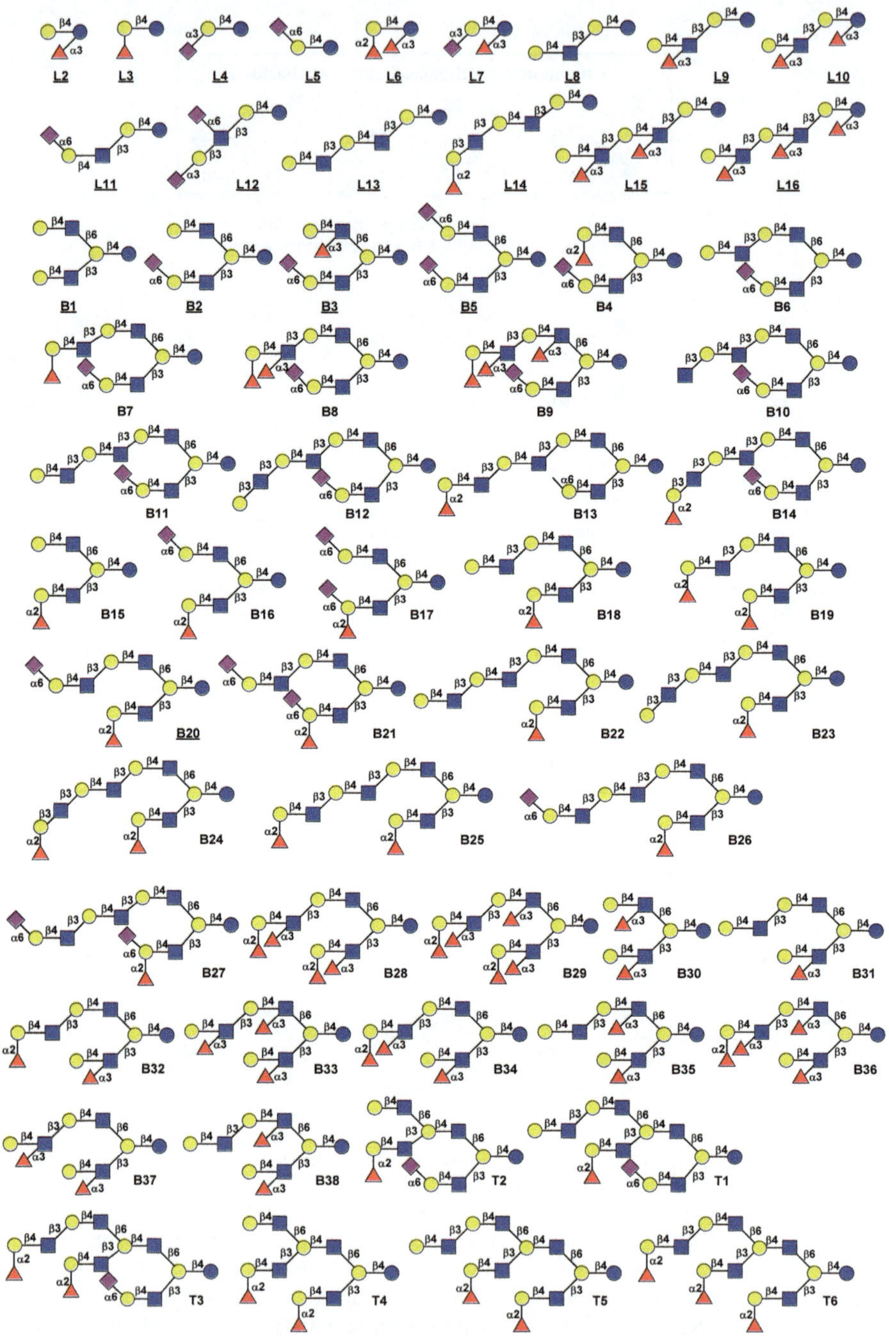

Figure 11.14 Chemoenzymatic synthesized chromophore-tagged asymmetric multiantennary HMOS (underlined numbers) and HMOS-like oligosaccharides.[46] Symbols and abbreviations: ◆ *N*-acetylneuraminic acid (Neu5Ac), ▲ fucose (Fuc), ● galactose (Gal), ■ *N*-acetylglucosamine (GlcNAc), ● glucose (Glc). Reproduced from ref. 46 with permission from National Academy of Sciences.

substrate specificity of the glycosyltransferases used. Glycosidases (*e.g.* sialyl linkage-unspecific *Arthrobacter ureafaciens* sialidase and a microbial α1–2-fucosidase) were used to generate additional HMOS analogs. Among the 60 compounds synthesized, 20 (whose numbers are underlined in Figure 11.14) have been previously identified in human milk.

11.7 Whole-cell Production and Fermentation of Engineered *E. coli* Cells

HMOS have also been synthesized by whole-cell production and fermentation of engineered *E. coli* cells.[93] These methods are practical and economically sound for producing short chain HMOS. Both use genetically engineered bacterial strains and take advantage of some endogenous genes involved in the synthesis of nucleotides and/or sugar nucleotides that are related to the glycosyltransferase-catalyzed production of HMOS. Between these two methods, the fermentation method requires the manipulation of a lower number of genes and allows the possibility of eliminating the use of some expensive starting materials.

The whole cell production of HMOS was demonstrated for 3′-SL using the combination of engineered *Corynebacterium ammoniagenes* strain overexpressing UTP-producing enzymes from orotic acid, an inexpensive starting material and three *E. coli* strains engineering to overexpress CTP synthetase, CMP-Neu5Ac synthetase, and *Neisseria gonorrhoeae* α2–3-sialyltransferase, respectively. In this system, engineered cells were permeabilized and used directly as catalysts for reactions. Starting from orotic acid (10 g L^{-1}), fructose (100 g L^{-1}), *N*-acetylneuraminic acid (40–50 g L^{-1}), and lactose (50 g L^{-1}), multiple grams of 3′-SL (33 g L^{-1} in 30 mL-scale and 36 g L^{-1} in 2 L-scale reactions, respectively) were synthesized after 11–12 h.[42]

3′-SL (2.6 g L^{-1}) was also produced from lactose and Neu5Ac with a purified yield of 49% by fermenting *E. coli* JM107 (*lacY*$^+$, *lacZ*−, *nanT*$^+$, *nanA*−) cells engineered to express *N. meningitidis* CMP-Neu5Ac synthetase (NmCSS) and *N. meningitidis* L3 strain MC58 α2–3-sialyltransferase (Nm2–3ST).[39] Additional genetic modification of the strain by knocking out the ManNAc kinase *nanK* gene and incorporating plasmids for the expression of *Campylobacter jejuni* strain ATCC43438 *neuABC* genes encoding GlcNAc-6-phosphate 2-epimerse, sialic acid synthase, and CMP-Neu5Ac synthetase eliminated the need to feed the culture with expensive Neu5Ac to lower the cost for 3′-SL production. A higher concentration (25 g L^{-1}) of 3′-SL was able to be obtained with the improved *E. coli* strain under an optimized condition.[94] A similar strategy was applied for the synthesis of 6′-SL using *E. coli* strain (*lacY*$^+$, *lacZ*$^-$, *nanA*$^-$, *nanK*$^-$) engineered to express *Photobacterium* sp. JT-ISH-224 (Psp2,6ST) and *Campylobacter jejuni* strain ATCC43438 *neuABC* genes.[95] Due to the donor substrate promiscuity of Psp2,6ST, a 2-keto-3-deoxy-manno-octonyllacto se (KDO)-containing trisaccharide KDOα2–6Lac was a major product when the *neuABC* genes were under a promoter weaker than the one for Psp2,6ST.

When both *neuABC* genes and the gene for Psp2,6ST were under the control of a strong promoter, a disialyl compound 6′,6-disialyllactose was a side product when the lactose was not supplied in a sufficient amount. The side product was minimized by culturing cells with a continuous supply of lactose and 6′-SL was accumulated to 34 g L^{-1}.[95]

More recently, 2′-FL (1.23 g L^{-1}) was produced from lactose (14.5 g L^{-1}) by batch fermentation of *lacZ*− *E. coli* JM109(DE3) strains engineered to overproduce GDP-fucose and overexpress *H. pylori* 26695 strain (ATCC 700392) α1–2-fucosyltransferase.[96] The production of 2′-FL (6.4 g L^{-1}, with the yield of 0.225 g 2′-FL per g Lac) was improved using an engineered *E. coli* BL21star (DE3) strain through deletion of the whole endogenous lactose operon and introduction of the modified lactose operon containing *lacZ*Δ*M15* from *E. coli* K-12.[97] To avoid the use of antibiotics in 2′-FL production, the *E. coli* JM109 (*lacY*$^+$, *lacZ*−) strain was chromosomally integrated with GDP-L-fucose *de novo* biosynthetic genes and *H. pylori* α1–2-fucosyltranferase *futC* gene. Fed-batch fermentation (13.5 L) of the engineered *E. coli* led to a large-scale production of 2′-FL (20 g L^{-1}) from lactose with glycerol as the carbon and energy source.[41]

Multigram-scale production of LNTri II (6 g L^{-1}, 73% yield) was achieved by fermenting *E. coli* JM109 cells (*lacY*$^+$ *lacZ*−) engineered to overexpress *N. meningitidis* β1–3-*N*-acetylglucosaminyltransferase (NmLgtA) with glycerol as the carbon and energy source.[39] When fermenting the strain further engineered with an additional glycosyltransferase β1–4GalT (NmLgtB) with glucose as the carbon and energy source, LNnT (>5 g L^{-1}) was obtained. Lower concentrations of longer HMOS including *p*LNnH, *p*LNnO, and lacto-*N*-neodecaose were also identified, especially when the concentration of the lactose added was lowered.[39] The production of LNTri was also achieved using *E. coli* strain LJ110 (*lacY*$^+$ *lacZ*−) chromosomally integrated with genes encoding NmLgtA LNT II. LNTri II was successfully obtained and its yield was higher (~2.5 g L^{-1}) when glycerol, instead of glucose (~1.9 g L^{-1} yield of LNTri II), was used as the carbon source.[45] When the gene for *E. coli* O55:H7 β1–3-galactosyltransferase (EcWbgO) was further chromosomally integrated, both LNT (219 mg L^{-1} with glucose and 162 mg L^{-1} with glycerol) and LNTri II (1.9 g L^{-1} with glucose and 2.1 g L^{-1} with glycerol) were produced.[45] The yield of LNT production (811 mg L^{-1}) was improved by 3.6-fold by substituting the glucose with galactose as the sole energy and carbon source. Large-scale production of LNT (12.7 g L^{-1}) and LNTri II (13.7 g L^{-1}) was achieved using fed-batch cultivation, leading to the formation of 173 g of LNT.[40]

11.8 Conclusion

Compared to chemical synthesis which requires multiple protection and deprotection steps and the use of environmentally unfriendly reagents, enzyme-catalyzed methods for producing HMOS have numerous advantages. Transglycosidase, glycosidase and glycosynthase-catalyzed reactions can be limited by the lack of access to glycosyl donors in large amounts and lower regio- and stereo-selectivity and/or lower yields in some cases. Glycosyltransferase-catalyzed reactions without *in situ* generation of sugar nucleotide donors can lead to high cost of HMOS production. In comparison, one-pot multienzyme (OPME) and sequential OPME systems using purified glycosyltransferases and sugar nucleotide biosynthesis enzymes can allow the production of HMOS from lactose, simple monosaccharides, and nucleoside 5′-triphosphates (*e.g.* ATP, CTP, UTP, and/or GTP). The OPME approaches significantly lower the cost of HMOS production and make the large-scale production affordable and practical. The core synthesis/enzymatic extension (CSEE) chemoenzymatic strategy bypasses the need for some glycosyltransferases by chemical construction of some common scaffolds of HMOS, including those asymmetric structures. It is extremely useful when some glycosyltransferases with the desired acceptor substrate selectivity are not readily available. Chemoenzymatic synthesis using lactosides tagged with a cleavable chromophore or fluorophore as starting glycosyltransferase acceptors allows easy monitoring of the reaction processes and the production of both HMOS as standards and their immobilizable analogs for conjugation. Innovation of enzymatic and/or chemoenzymatic syntheses including clever design of glycosylation sequences[48,64,98] as well as chemical-assisted[99] and glycosyltransferase/glycosidase-catalyzed[100] protection strategies, identifying glycosyltransferases with the desired acceptor specificities by gene mining and mutagenesis will be key aspects to accessing the structure diversity represented by natural HMOS.

The living cell fermentation approaches will be the method of choice for industrial scale production of HMOS with simpler structures. More challenging targets will rely on additional innovations especially in identifying glycosyltransferases with the desired acceptor substrate specificity. Innovation in product purification is also a high demand for industrial scale production of HMOS.

Combining efforts to identify suitable glycosyltransferases with the desired acceptor specificity, clever design of enzymatic/chemoenzymatic synthetic routes, and innovation of product purification, pure HMOS of desired structures in sufficient amounts will be made available for various research efforts to identify HMOS-based therapeutics and/or nutraceuticals. Industrial-scale production of the desired structures, including more complex ones, will also become a reality.

Acknowledgement

The authors would like to acknowledge the financial support from National Institutes of Health (NIH) Grants R01HD065122 (to X. C.) and R43HD087077 (to H.Y.) for projects related to human milk oligosaccharides and analogs. H.Y. and X.C. are co-founders of Glycohub, Inc., a company focused on the development of carbohydrate-based reagents, diagnostics, and therapeutics. Glycohub, Inc., played no role in the design, execution, interpretation, or publication of this work.

References

1. C. Kunz, S. Rudloff, W. Baier, N. Klein and S. Strobel, *Annu. Rev. Nutr.*, 2000, **20**, 699.
2. G. V. Coppa, O. Gabrielli, P. Pierani, C. Catassi, A. Carlucci and P. L. Giorgi, *Pediatrics*, 1993, **91**, 637.
3. M. R. Ninonuevo, Y. Park, H. Yin, J. Zhang, R. E. Ward, B. H. Clowers, J. B. German, S. L. Freeman, K. Killeen, R. Grimm and C. B. Lebrilla, *J. Agric. Food Chem.*, 2006, **54**, 7471.
4. D. Dai, N. N. Nanthkumar, D. S. Newburg and W. A. Walker, *J. Pediatr. Gastroenterol. Nutr.*, 2000, **30**(Suppl 2), S23.
5. X. Chen, in *Advances in Carbohydrate Chemistry and Biochemistry*, ed. D. C. Baker and D. Horton, Academic Press, 2015, vol. 72, pp. 113–190.
6. Y. Yu, Y. Lasanajak, X. Song, L. Hu, S. Ramani, M. L. Mickum, D. J. Ashline, B. V. V. Prasad, M. K. Estes, V. N. Reinhold, R. D. Cummings and D. F. Smith, *Mol. Cell. Proteomics*, 2014, **13**, 2944.
7. D. J. Ashline, Y. Yu, Y. Lasanajak, X. Song, L. Hu, S. Ramani, V. Prasad, M. K. Estes, R. D. Cummings, D. F. Smith and V. N. Reinhold, *Mol. Cell. Proteomics*, 2014, **13**, 2961.
8. Y. Yu, S. Mishra, X. Song, Y. Lasanajak, K. C. Bradley, M. M. Tappert, G. M. Air, D. A. Steinhauer, S. Halder, S. Cotmore, P. Tattersall, M. Agbandje-McKenna, R. D. Cummings and D. F. Smith, *J. Biol. Chem.*, 2012, **287**, 44784.
9. P. Chaturvedi, C. D. Warren, M. Altaye, A. L. Morrow, G. Ruiz-Palacios, L. K. Pickering and D. S. Newburg, *Glycobiology*, 2001, **11**, 365.
10. R. E. Ward, M. Niñonuevo, D. A. Mills, C. B. Lebrilla and J. B. German, *Mol. Nutr. Food Res.*, 2007, **51**, 1398.
11. J. B. German, S. L. Freeman, C. B. Lebrilla and D. A. Mills, *Nestle Nutr. Inst. Workshop Ser., Pediatr. Program*, 2008, **62**, 205.
12. D. S. Newburg, G. M. Ruiz-Palacios and A. L. Morrow, *Annu. Rev. Nutr.*, 2005, **25**, 37.
13. J. T. Smilowitz, C. B. Lebrilla, D. A. Mills, J. B. German and S. L. Freeman, *Annu. Rev. Nutr.*, 2014, **34**, 143.
14. A. L. Morrow, G. M. Ruiz-Palacios, X. Jiang and D. S. Newburg, *J. Nutr.*, 2005, **135**, 1304.
15. M. B. Stepans, S. L. Wilhelm, M. Hertzog, T. K. Rodehorst, S. Blaney, B. Clemens, J. J. Polak and D. S. Newburg, *Breastfeed. Med.*, 2006, **1**, 207.
16. G. V. Coppa, L. Zampini, T. Galeazzi, B. Facinelli, L. Ferrante, R. Capretti and G. Orazio, *Pediatr. Res.*, 2006, **59**, 377.
17. L. Bode, C. Kunz, M. Muhly-Reinholz, K. Mayer, W. Seeger and S. Rudloff, *Thromb. Haemostasis*, 2004, **92**, 1402.

18. M. Good, C. P. Sodhi, Y. Yamaguchi, H. P. Jia, P. Lu, W. B. Fulton, L. Y. Martin, T. Prindle, D. F. Nino, Q. J. Zhou, C. R. Ma, J. A. Ozolek, R. H. Buck, K. C. Goehring and D. J. Hackam, *Br. J. Nutr.*, 2016, **116**, 1175.
19. K. C. Goehring, B. J. Marriage, J. S. Oliver, J. A. Wilder, E. G. Barrett and R. H. Buck, *J. Nutr.*, 2016, **146**, 2559.
20. A. Kobata, *Proc. Jpn. Acad., Ser. B*, 2010, **86**, 731.
21. M. R. E. Aly, E. S. I. Ibrahim, E. S. H. El Ashry and R. R. Schmidt, *Carbohydr. Res.*, 1999, **316**, 121.
22. L. Manzoni, L. Lay and R. R. Schmidt, *J. Carbohydr. Chem.*, 1998, **17**, 739.
23. K. M. Craft and S. D. Townsend, *Carbohydr. Res.*, 2017, **440–441**, 43.
24. M. R. E. Aly, E.-S. I. Ibrahim, E.-S. H. E. El-Ashry and R. R. Schmidt, *Eur. J. Org. Chem.*, 2000, **2000**, 319.
25. A. Fernandez-Mayoralas and M. Martin-Lomas, *Carbohydr. Res.*, 1986, **154**, 93.
26. C. Arboe Jennum, T. Hauch Fenger, L. M. Bruun and R. Madsen, *Eur. J. Org. Chem.*, 2014, **2014**, 3232.
27. Y.-M. Zhang, J. Esnault, J.-M. Mallet and P. Sinaÿ, *J. Carbohydr. Chem.*, 1999, **18**, 419.
28. Y. Liu, T. Feizi, M. A. Campanero-Rhodes, R. A. Childs, Y. Zhang, B. Mulloy, P. G. Evans, H. M. Osborn, D. Otto, P. R. Crocker and W. Chai, *Chem. Biol.*, 2007, **14**, 847.
29. T. Ogawa and M. Sugimoto, *Carbohydr. Res.*, 1985, **135**, C5.
30. D. Esposito, M. Hurevich, B. Castagner, C.-C. Wang and P. H. Seeberger, *Beilstein J. Org. Chem.*, 2012, **8**, 1601.
31. K. Furuhata, K. Anazawa, M. Itoh, Y. Shitori and H. Ogura, *Chem. Pharm. Bull.*, 1986, **34**, 2725.
32. D. Schmidt and J. Thiem, *Beilstein J. Org. Chem.*, 2010, **6**, 18.
33. A. A. Sherman, O. N. Yudina, Y. V. Mironov, E. V. Sukhova, A. S. Shashkov, V. M. Menshov and N. E. Nifantiev, *Carbohydr. Res.*, 2001, **336**, 13.
34. R. J. Fair, H. S. Hahm and P. H. Seeberger, *Chem. Commun.*, 2015, **51**, 6183.
35. B. Petschacher and B. Nidetzky, *J. Biotechnol.*, 2016, **235**, 61.
36. B. Zeuner, C. Jers, J. D. Mikkelsen and A. S. Meyer, *J. Agric. Food Chem.*, 2014, **62**, 9615.
37. G. A. Sprenger, F. Baumgärtner and C. Albermann, *J. Biotechnol.*, 2017, **258**, 79.
38. K. F. Johnson, *Glycoconjugate J.*, 1999, **16**, 141.
39. B. Priem, M. Gilbert, W. W. Wakarchuk, A. Heyraud and E. Samain, *Glycobiology*, 2002, **12**, 235.
40. F. Baumgartner, G. A. Sprenger and C. Albermann, *Enzyme Microb. Technol.*, 2015, **75–76**, 37.
41. F. Baumgartner, L. Seitz, G. A. Sprenger and C. Albermann, *Microb. Cell Fact.*, 2013, **12**, 40.
42. T. Endo, S. Koizumi, K. Tabata and A. Ozaki, *Appl. Microbiol. Biotechnol.*, 2000, **53**, 257.

43. K. W. Moremen, A. Ramiah, M. Stuart, J. Steel, L. Meng, F. Forouhar, H. A. Moniz, G. Gahlay, Z. Gao, D. Chapla, S. Wang, J. Y. Yang, P. K. Prabhakar, R. Johnson, M. D. Rosa, C. Geisler, A. V. Nairn, J. Seetharaman, S. C. Wu, L. Tong, H. J. Gilbert, J. LaBaer and D. L. Jarvis, *Nat. Chem. Biol.*, 2018, **14**, 156.
44. X. W. Liu, C. Xia, L. Li, W. Y. Guan, N. Pettit, H. C. Zhang, M. Chen and P. G. Wang, *Bioorg. Med. Chem.*, 2009, **17**, 4910.
45. F. Baumgartner, J. Conrad, G. A. Sprenger and C. Albermann, *ChemBioChem*, 2014, **15**, 1896.
46. A. R. Prudden, L. Liu, C. J. Capicciotti, M. A. Wolfert, S. Wang, Z. Gao, L. Meng, K. W. Moremen and G. J. Boons, *Proc. Natl. Acad. Sci. U. S. A.*, 2017, **114**, 6954.
47. Z. Xiao, Y. Guo, Y. Liu, L. Li, Q. Zhang, L. Wen, X. Wang, S. M. Kondengaden, Z. Wu, J. Zhou, X. Cao, X. Li, C. Ma and P. G. Wang, *J. Org. Chem.*, 2016, **81**, 5851.
48. H. Yu, K. Lau, V. Thon, C. A. Autran, E. Jantscher-Krenn, M. Xue, Y. Li, G. Sugiarto, J. Qu, S. Mu, L. Ding, L. Bode and X. Chen, *Angew. Chem., Int. Ed. Engl.*, 2014, **53**, 6687.
49. C. Zhao, Y. Wu, H. Yu, I. M. Shah, Y. Li, J. Zeng, B. Liu, D. A. Mills and X. Chen, *Chem. Commun.*, 2016, **52**, 3899.
50. C. Albermann, W. Piepersberg and U. F. Wehmeier, *Carbohydr. Res.*, 2001, **334**, 97.
51. D. B. Stein, Y.-N. Lin and C.-H. Lin, *Adv. Synth. Catal.*, 2008, **350**, 2313.
52. L. Engels and L. Elling, *Glycobiology*, 2014, **24**, 170.
53. T. Miyazaki, T. Sato, K. Furukawa and K. Ajisaka, *Methods Enzymol.*, 2010, **480**, 511.
54. Y. H. Choi, J. H. Kim, B. S. Park and B. G. Kim, *Biotechnol. Bioeng.*, 2016, **113**, 1666.
55. H. Yu, Y. Li, Z. Wu, L. Li, J. Zeng, C. Zhao, Y. Wu, N. Tasnima, J. Wang, H. Liu, M. R. Gadi, W. Guan, P. G. Wang and X. Chen, *Chem. Commun.*, 2017, **53**, 11012.
56. M. Nimtz, E. Grabenhorst, U. Gambert, J. Costa, V. Wray, M. Morr, J. Thiem and H. S. Conradt, *Glycoconjugate J.*, 1998, **15**, 873.
57. H. Yu, K. Lau, Y. Li, G. Sugiarto and X. Chen, *Curr. Protoc. Chem. Biol.*, 2012, **4**, 233.
58. K. Schmolzer, T. Czabany, C. Luley-Goedl, T. Pavkov-Keller, D. Ribitsch, H. Schwab, K. Gruber, H. Weber and B. Nidetzky, *Chem. Commun.*, 2015, **51**, 3083.
59. K. Schmolzer, D. Ribitsch, T. Czabany, C. Luley-Goedl, D. Kokot, A. Lyskowski, S. Zitzenbacher, H. Schwab and B. Nidetzky, *Glycobiology*, 2013, **23**, 1293.
60. M. Gilbert, R. Bayer, A. M. Cunningham, S. DeFrees, Y. Gao, D. C. Watson, N. M. Young and W. W. Wakarchuk, *Nat. Biotechnol.*, 1998, **16**, 769.
61. H. Yu, S. Huang, H. Chokhawala, M. Sun, H. Zheng and X. Chen, *Angew. Chem., Int. Ed. Engl.*, 2006, **45**, 3938.

62. H. Yu, X. Yan, C. A. Autran, Y. Li, S. Etzold, J. Latasiewicz, B. M. Robertson, J. Li, L. Bode and X. Chen, *J. Org. Chem.*, 2017, **82**, 13152.
63. J. B. McArthur, H. Yu, J. Zeng and X. Chen, *Org. Biomol. Chem.*, 2017, **15**, 1700.
64. H. Yu and X. Chen, *Org. Biomol. Chem.*, 2016, **14**, 2809.
65. L. Zhang, K. Lau, J. Cheng, H. Yu, Y. Li, G. Sugiarto, S. Huang, L. Ding, V. Thon, P. G. Wang and X. Chen, *Glycobiology*, 2010, **20**, 1077.
66. H. Yu, H. A. Chokhawala, S. Huang and X. Chen, *Nat. Protoc.*, 2006, **1**, 2485.
67. H. Yu, J. Cheng, L. Ding, Z. Khedri, Y. Chen, S. Chin, K. Lau, V. K. Tiwari and X. Chen, *J. Am. Chem. Soc.*, 2009, **131**, 18467.
68. Y. Li, H. Yu, Y. Chen, K. Lau, L. Cai, H. Cao, V. K. Tiwari, J. Qu, V. Thon, P. G. Wang and X. Chen, *Molecules*, 2011, **16**, 6396.
69. Y. Chen, V. Thon, Y. Li, H. Yu, L. Ding, K. Lau, J. Qu, L. Hie and X. Chen, *Chem. Commun.*, 2011, **47**, 10815.
70. K. Lau, V. Thon, H. Yu, L. Ding, Y. Chen, M. M. Muthana, D. Wong, R. Huang and X. Chen, *Chem. Commun.*, 2010, **46**, 6066.
71. L. Cai, L. Ban, W. Guan, M. Mrksich and P. G. Wang, *Carbohydr. Res.*, 2011, **346**, 1576.
72. O. Blixt, I. van Die, T. Norberg and D. H. van den Eijnden, *Glycobiology*, 1999, **9**, 1061.
73. X. Chen, Z. Liu, J. Zhang, W. Zhang, P. Kowal and P. G. Wang, *ChemBioChem*, 2002, **3**, 47.
74. M. M. Muthana, J. Qu, Y. Li, L. Zhang, H. Yu, L. Ding, H. Malekan and X. Chen, *Chem. Commun.*, 2012, **48**, 2728.
75. W. Yi, X. Liu, Y. Li, J. Li, C. Xia, G. Zhou, W. Zhang, W. Zhao, X. Chen and P. G. Wang, *Proc. Natl. Acad. Sci. U. S. A.*, 2009, **106**, 4207.
76. G. Sugiarto, K. Lau, H. Yu, S. Vuong, V. Thon, Y. Li, S. Huang and X. Chen, *Glycobiology*, 2011, **21**, 387.
77. Z. Khedri, Y. Li, H. Cao, J. Qu, H. Yu, M. M. Muthana and X. Chen, *Org. Biomol. Chem.*, 2012, **10**, 6112.
78. L. Zhang, K. Lau, J. Cheng, H. Yu, Y. Li, G. Sugiarto, S. Huang, L. Ding, V. Thon, P. G. Wang and X. Chen, *Glycobiology*, 2010, **20**, 1077.
79. D. A. Rasko, G. Wang, M. M. Palcic and D. E. Taylor, *J. Biol. Chem.*, 2000, **275**, 4988.
80. B. Ma, G. F. Audette, S. Lin, M. M. Palcic, B. Hazes and D. E. Taylor, *J. Biol. Chem.*, 2006, **281**, 6385.
81. J. Montreuil, *C. R. Hebd. Seances Acad. Sci.*, 1956, **242**, 192.
82. H. Yu, J. Zeng, Y. Li, V. Thon, B. Shi and X. Chen, *Org. Biomol. Chem.*, 2016, **14**, 8586.
83. H. Yu, H. Chokhawala, R. Karpel, H. Yu, B. Wu, J. Zhang, Y. Zhang, Q. Jia and X. Chen, *J. Am. Chem. Soc.*, 2005, **127**, 17618.
84. H. Yu, H. Yu, R. Karpel and X. Chen, *Bioorg. Med. Chem.*, 2004, **12**, 6427.
85. Y. Li, H. Yu, H. Cao, K. Lau, S. Muthana, V. K. Tiwari, B. Son and X. Chen, *Appl. Microbiol. Biotechnol.*, 2008, **79**, 963.

86. W. Yao, J. Yan, X. Chen, F. Wang and H. Cao, *Carbohydr. Res.*, 2015, **401**, 5.
87. H. Shengshu, Y. Hai and C. Xi, *Sci. China: Chem.*, 2011, **54**, 117.
88. K. Totani, K. Shimizu, Y. Harada, T. Murata and T. Usui, *Biosci., Biotechnol., Biochem.*, 2002, **66**, 636.
89. J. Wada, Y. Honda, M. Nagae, R. Kato, S. Wakatsuki, T. Katayama, H. Taniguchi, H. Kumagai, M. Kitaoka and K. Yamamoto, *FEBS Lett.*, 2008, **582**, 3739.
90. H. Sakurama, S. Fushinobu, M. Hidaka, E. Yoshida, Y. Honda, H. Ashida, M. Kitaoka, H. Kumagai, K. Yamamoto and T. Katayama, *J. Biol. Chem.*, 2012, **287**, 16709.
91. P. M. Danby and S. G. Withers, *ACS Chem. Biol.*, 2016, **11**, 1784.
92. A. Saumonneau, E. Champion, P. Peltier-Pain, D. Molnar-Gabor, J. Hendrickx, V. Tran, M. Hederos, G. Dekany and C. Tellier, *Glycobiology*, 2016, **26**, 261.
93. N. S. Han, T. J. Kim, Y. C. Park, J. Kim and J. H. Seo, *Biotechnol. Adv.*, 2012, **30**, 1268.
94. N. Fierfort and E. Samain, *J. Biotechnol.*, 2008, **134**, 261.
95. S. Drouillard, T. Mine, H. Kajiwara, T. Yamamoto and E. Samain, *Carbohydr. Res.*, 2010, **345**, 1394.
96. W. H. Lee, P. Pathanibul, J. Quarterman, J. H. Jo, N. S. Han, M. J. Miller, Y. S. Jin and J. H. Seo, *Microb. Cell Fact.*, 2012, **11**, 48.
97. Y. W. Chin, J. Y. Kim, W. H. Lee and J. H. Seo, *J. Biotechnol.*, 2015, **210**, 107.
98. C. Chen, Y. Zhang, M. Xue, X. W. Liu, Y. Li, X. Chen, P. G. Wang, F. Wang and H. Cao, *Chem. Commun.*, 2015, **51**, 7689.
99. X. Meng, W. Yao, J. Cheng, X. Zhang, L. Jin, H. Yu, X. Chen, F. Wang and H. Cao, *J. Am. Chem. Soc.*, 2014, **136**, 5205.
100. I. A. Gagarinov, T. Li, J. S. Torano, T. Caval, A. D. Srivastava, J. A. Kruijtzer, A. J. Heck and G. J. Boons, *J. Am. Chem. Soc.*, 2017, **139**, 1011.

CHAPTER 12

Synthesis of Marine Polysaccharides/ Oligosaccharides and Their Derivatives

CHAO CAI[a,b], JUN LI[a], FEI FAN[a], GUOYUN LI[a,b], CHUNXIA LI[a,b] AND GUANGLI YU*[a,b]

[a]Ocean University of China, Key Laboratory of Marine Drugs of Ministry of Education, Shandong Provincial Key Laboratory of Glycoscience and Glycotechnology, School of Medicine and Pharmacy, No. 5 Yushan Road, Qingdao, 266003, China; [b]Qingdao National Laboratory for Marine Science and Technology, Laboratory for Marine Drugs and Bioproducts, No.1 Wenhai Road, Qingdao, 266237, China
*E-mail: glyu@ouc.edu.cn

12.1 Introduction

The marine ecosystem offers an abundant source of valuable products such as carbohydrates, proteins, lipids, *etc.* Bioactive compounds isolated from marine animals, plants, and microorganisms have been considered as a major source of novel marine drugs due to their biodiversity.[1–4] Marine carbohydrate is one of the primarily developed bioactive substances and has been used infood additives, biomedical materials, and pharmaceutics.[5–7]

Chemical Biology No. 11
Synthetic Glycomes
Edited by Wanyi Guan, Lei Li and Peng George Wang

Published by the Royal Society of Chemistry, www.rsc.org

Therefore, they have received much more attention in this century, and the evolution of new technology and accumulated knowledge in this field have enhanced the comprehensive understanding of natural carbohydrates. The carbohydrates from the marine organism are mainly in the form of polysaccharides and oligosaccharides,[5] which are classified into three groups in terms of their original species, including marine plant-derived, animal-derived, and microorganism-derived carbohydrates.

As depicted in Table 12.1, marine plant-derived polysaccharides are commonly isolated from the cell wall of different types of seaweeds with heterogeneous structures corresponding to the versatile sources.[7,8] Polysaccharides obtained from brown algae mainly consist of alginate, fucoidan, and laminaran. Alginate, composed of linear α-L-glucuronic acid/β-D-mannuronic acid (G/M) polymers, is the most abundant polysaccharide in brown algae.[9] The use of alginate in the food and beverage industry has been allowed by major regulatory agencies, including the United States Food & Drug Administration (FDA) and the European Commission (EC), and emerging pharmaceutical applications are expected to drive their future development. Fucoidan, a linear acidic polysaccharide composed of sulfated α-L-fucopyranose residues, is divided into two categories in accordance with their linkage patterns: α1–3 linkages (type I) and α1–3/4 alternative linkages (type II).[10] Due to the complex structure and versatile biological functions of fucoidan, several groups are devoted to obtaining homogeneous fucan poly-/oligosaccharides. Laminaran is a storage glucan composed of β1–3-linked glucose as the backbone and β1–6-linked branches, which is similar to polysaccharides extracted from traditional Chinese herbs such as *Ganoderma lucidum*, *Lentinula edodes*, and *Poria cocos*.[11] Agarose and κ/λ/ι-carrageenan collected from red seaweeds are both jelly-like substances and have been widely applied in the food additive, and chemical industries as well as in medicinal scientific research. Sulfated galactans composed of α1–4-linked 3,6-anhydro-L-galactose alternating with β1–3-linked D-galactose are classed as agarose, whereas the D-series of 3,6-anhydro-galactose are termed carrageenans.[12] Furthermore, various types of seaweed such as green algae, blue algae, and microalgae also yield complex polysaccharides, as listed in Table 12.1.

Marine animal-derived polysaccharides are structurally distinct with respect to different species. Chitin, a cationic polysaccharide with β1–4-linked *N*-acetyl-D-glucosamine (GlcNAc), is commercially extracted from the shells of crustaceans such as crabs and shrimps and generally collected from seafood processing by-products.[30,31] Chitosan, a deacetylated derivative of chitin, is currently used in cosmetics, biofertilizers, food additives, biomedical material and dietary supplements.[32–34] The global market for chitosan is projected to exceed 118 000 tons by 2018.[35] Fucosylated chondroitin sulfate (fCS), a glycosaminoglycan-like polysaccharide composed of glucuronic acid (GlcA) and *N*-acetyl-galactosamine (GalNAc) as the linear backbone and sulfated fucose as the branch, is extracted from the body wall of sea cucumbers such as *Holothuria mexicana*, *Ludwigothurea grisea*, *Thelenota ananas etc.*[36] Several reports have demonstrated that fCS possess potent anticoagulant

Table 12.1 Representative structures and references of marine organism derived polysaccharides.

Structure and Name	Source/ Reference	Structure and Name	Source/Reference
Marine plant-derived polysaccharides			
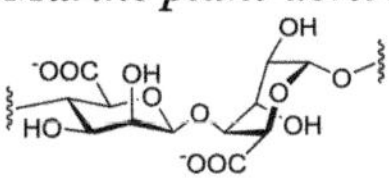 Alginate	Brown algae[13]	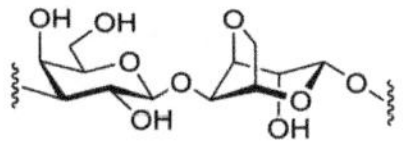 Agarose	Red algae *Gracilaria* spp. and *Gelidium* spp.[14,15]
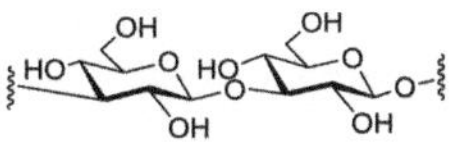 Laminaran	Brown algae *Laminaria* spp.[16]	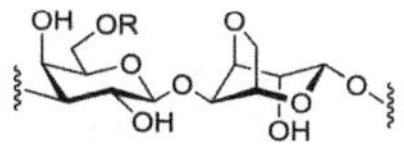 Porphyran $R = H, SO_3^-, Me$	Red algae *Porphyra umbilicalis*[17]
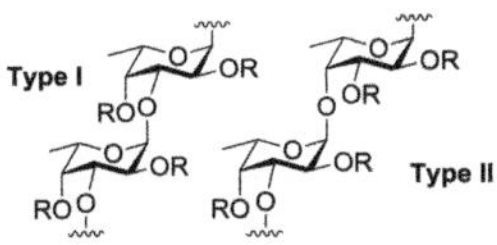 Fucoidan $R = H$ or SO_3^-	Brown algae[10]	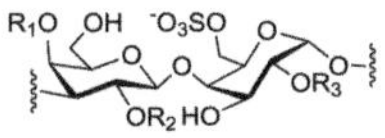 λ: $R_1 = H, R_2 = R_3 = SO_3^-$ 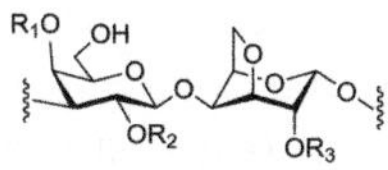 κ: $R_1 = SO_3^-$, $R_2 = R_3 = H$ τ: $R_1 = R_3 = SO_3^-$, $R_2 = H$ Carrageenan	Red algae[18]
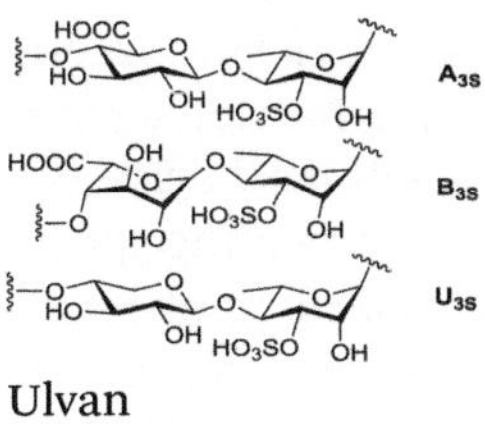 Ulvan	Green algae[19]	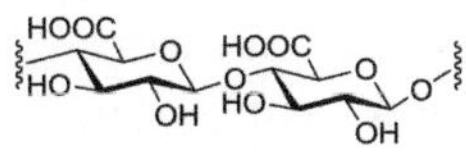 Glucuronan	Green algae *Enteromorpha compressa*[20]
Marine animal-derived polysaccharides			
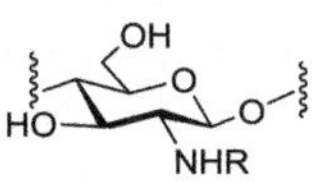 Chitin: R = Ac Chitosan: R = H	Shell of crustaceans[21,22]	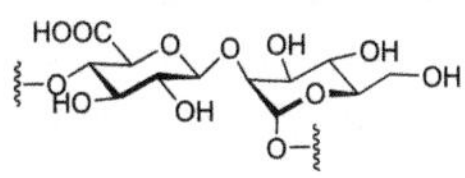 AGSP (Polysaccharide from abalone gonad)	*Haliotis discus hannai* Ino[23]

(*continued*)

Table 12.1 (*continued*)

Structure and Name	Source/ Reference	Structure and Name	Source/Reference
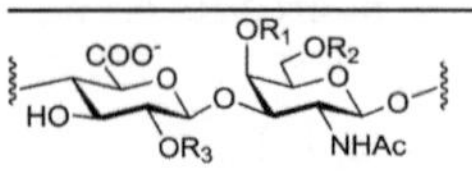 Chondroitin sulfate $R_1 = R_2 = R_3 = H$ or SO_3^-	Shark fins and fish cartilage[24,25]	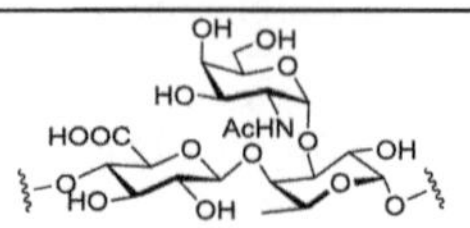 SIP (Squid ink polysaccharide)	Ink of the squid *Ommastrephes bartrami*[26]
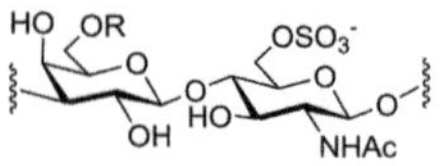 Keratan sulfate $R = H$ or SO_3^-	Shark cartilage[27]	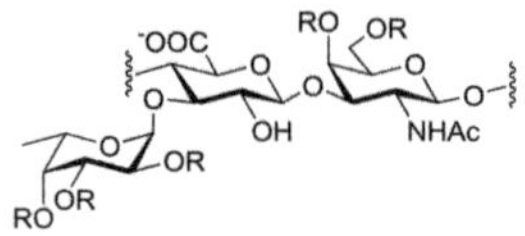 Fucosylated chondroitin sulfate $R = H$ or SO_3^-	Body wall of *Cucumaria frondosa*[28] *Holothuria mexicana*[29]

activity comparable to heparin.[37] Keratan sulfate (KS) obtained from the skin and cartilage of marine animals plays a significant role in diverse physiological and biological processes. KS contains linear disaccharide repeating units with β1–3-linked galactose and β1–4-linked *N*-acetyl glucosamine.[38] In addition, a series of chondroitin sulfate with diverse sulfation patterns were also discovered in marine animals.[39]

Marine microbes and single-celled organisms, including microalgae, bacteria, fungi, and viruses, live in the ocean and account for more than 98% of the marine biomass. Marine microbial poly-/oligosaccharides are mostly obtained from microbial exopolysaccharides, which are one of the most important secondary metabolites secreted externally to the cell surface of marine microorganisms for growth promotion, functioning as protective barrier against predators, and adhering to solid surfaces.[40] For example, a polysaccharide isolated from the marine bacterium *Vibrio diabolicus* contains repeating glucuronic acid and *N*-acetylhexosamine units as a hyaluronic acid-like polymer.[41] Meanwhile, the fermentation processes of marine exopolysaccharides are being greatly expanded due to their promising industrial applications.[42]

Marine carbohydrates are highly abundant bioactive compounds and closely related to our daily life. However, many functional marine carbohydrates also suffer from certain drawbacks such as batch-to-batch structural variability, limited supply, and heavy metal contamination. In addition, factors such as species, collecting locations, harvest times and processing methods could result in structural diversity with heterogeneous carbohydrates.[7] The structurally defined substances are essential for the study of marine biological process and functions. Until now, the attainment of marine poly-/oligosaccharides with well-defined structures has yet to be fully accomplished. This chapter aims to provide a summary of current efforts in preparing structurally defined marine carbohydrates.

12.2 Synthesis of Marine Polysaccharides

Marine polysaccharides are abundantly found in marine organisms, such as algae, fish, and shellfish. For commercial applications, some of them are produced by direct extraction from field harvest, but most structures of marine polysaccharides cannot be controlled using the conventional approach. This makes it very challenging to address their biological functions especially for highly complex and variable glycans. The access to marine polysaccharides with relatively defined structures plays an important role in better understanding their essential characteristics. Thus, novel approaches and methodologies are emerging in the fields of biosynthesis, semi-synthesis, mimic synthesis, and modification of marine polysaccharides. Recent advances in the development of marine polysaccharides with well-defined structures are summarized in the following sections.

12.2.1 Biosynthesis of Marine Polysaccharides

Biosynthetic motifs and mechanisms discovered on marine polysaccharides have attracted widespread attention because of their potential industrial applications. In general, the bioengineering approach is advantageous compared to conventional extraction due to its full operation in laboratory, standard quality control, and free design of novel compounds through genetic modification.[43,44] Therefore, biotechnological methods have been extensively applied to the efficient production of marine polysaccharides with improved properties and enhanced yields. In addition, the potential integration of molecular and bioengineering techniques is under development as a competitive bioprocess for creating novel marine polysaccharides.[45–48]

Through bioengineering, numerous studies have reported the biosynthesis of marine polysaccharides with a wide range of molecular weights and relative monomer contents. Taking alginate as an example, bacterial alginate biosynthesis is of great importance due to its role in the virulence of the human and plant pathogenic *Pseudomonas* species, as well as its potential industrial applications.[49] Two types of bacteria, namely *Azotobacter* and *Pseudomonas*,[50–53] are capable of producing alginate and are promising candidates for alginate production by fermentation. Howell and co-workers[54] recently found that the binding of dimeric c-di-GMP was required for alginate biosynthesis. Their results explained why Alg44 regulated the alginate production in *Pseudomonas aeruginosa*. For accumulating alginate overproduction, Yu's group[55] reported that alginate induction in PAO1 was primarily due to the increased MucA degradation, which resulted in the activation of AlgU. Furthermore, the growth on PIA-AMV induced the synthesis of palmitoylated lipid A. Altogether, these findings established a potent model to regulate alginate production by overproducing *Pseudomonas* strains. Meanwhile, other characteristic marine polysaccharides such as chitin,[56,57] agar,[58] fucoidan,[59] chondroitin sulfate,[44,60,61] and fCS[62,63] have also been investigated for the biosynthesis.

12.2.2 Semi-synthesis of Marine-derived Polysaccharides

Semi-synthesis is a robust alternative to achieve the rare and complex natural products. To better decipher the biological functions, the complicated marine polysaccharides have been continuously supplied by semi-synthesis from homogeneous substances. Chondroitin sulfate E (CS-E) was originally isolated from squid cartilage with the repeating units of –4GlcAβ1–3-4,6-disulfo-GalNAcβ1–.[64,65] Linhardt and co-workers[25] (Scheme 12.1) reported the semi-synthesis of CS-E from commercially available chondroitin sulfate A (CS-A). A facile synthetic route using trimethylamine sulfur trioxide was employed to achieve the regioselective sulfation and prepare the relatively expensive CS-E. The interaction between semi-synthetic CS-E and fibroblast growth factors (FGF) displayed very similar binding affinities and kinetics to natural CS-E from squid cartilage. The current effort on the semi-synthesis of marine polysaccharides offers an economical approach for accessing rare carbohydrates for commercial development.

The group of Bedini developed a concise and practical semi-synthesis of fCS polysaccharides by using bioengineered chondroitin from a microbial source as the starting material (Scheme 12.2). The manipulation of multiple protection-deprotection steps was handled to regioselectively release free

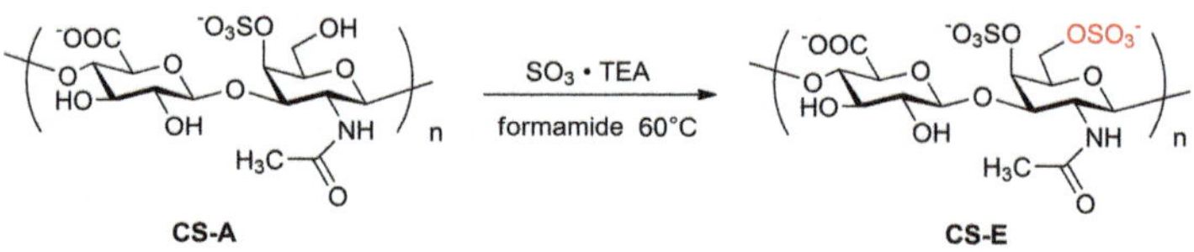

Scheme 12.1 Chemical semi-synthesis of CS-E from CS-A. Reproduced from ref. 25 with permission from Elsevier, Copyright 2011.

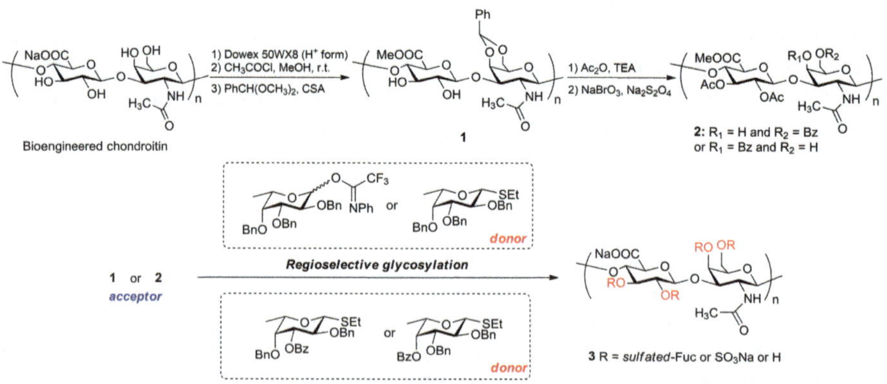

Scheme 12.2 Representative route for the semi-synthesis of fCS polysaccharides. Adapted from ref. 67 with permission from John Wiley and Sons, Copyright © 2016 Wiley-VCH Verlag GmbH & Co. KGaA, Weinheim.

hydroxyl groups.[66] The challenging step of this protocol is the chemical fucosylation on the backbone of the CS polysaccharide. A detailed 2D nuclear magnetic resonance (NMR) analysis on semi-synthetic fCSs was then conducted to determine the regio- and stereochemistry of the fucose branches as well as the sulfation patterns. The efficacy of their anticoagulant activities against FXa was preliminarily estimated to be similar to the low molecular weight (LMW) fCS obtained from the natural source. Subsequently, Bedini and co-workers further improved this modular approach for accessing to the semi-synthesis of fCS with specific sulfation and fucosylation patterns in 2016.[67] The well-defined sulfation patterns on the Fuc and/or GalNAc sites were established and confirmed by 2D NMR spectroscopic analysis. A small library of six fCS polysaccharides with well-defined structures was established to decipher their biological functions.

12.2.3 Synthesis of Glycopolymers and Glycoclusters to Mimic Natural Marine Polysaccharides

Glyco-macroligands, such as glycopolymers, glycoclusters, and glycodendrimers with pendant saccharides in linear or multi-branching structures, possess an intrinsic "cluster effect". The "cluster effect" has been well noted in biological recognition as it takes advantage of the amplification of carbohydrate-receptor binding affinity.[68] Versatile synthetic approaches have been developed to mimic marine bioactive polysaccharides for deciphering their biological functions. Chitosan as a cationic polymer offers a wide range of antimicrobial characteristics, which is highly correlated with its molecular weight (M_w), polydispersity, pH, and pattern of deacetylation. Kasko's group[69] reported the synthesis of poly(methyl 6-acryloyl-β-D-glucosaminoside) exhibiting the M_w-dependant bacterial inhibitory property of natural chitosan (Scheme 12.3). The monomer **5**

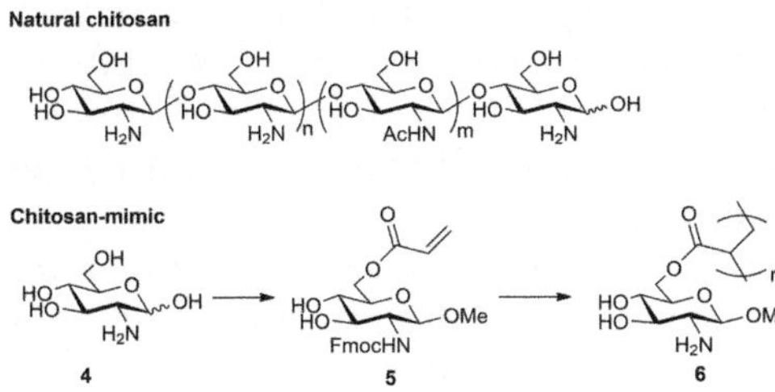

Scheme 12.3 Synthesis of chitosan-mimetic glycopolymers. Reproduced from ref. 69 with permission from American Chemical Society, Copyright 2017.

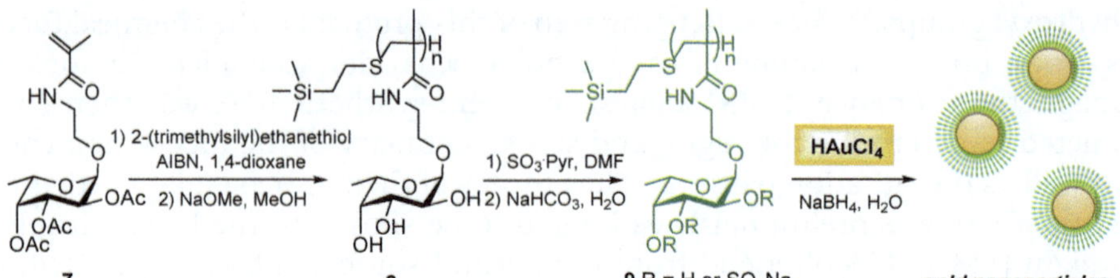

Scheme 12.4 Synthesis of fucoidan-mimetic glycopolymers and GNPs. Reproduced from ref. 72 with permission from The Royal Society of Chemistry.

containing 6-acryloyl converted from glucosamine **4** was polymerized under free radical polymerization. Their findings demonstrated that poly(methyl 6-acryloyl-β-D-glucosaminoside) maintained its antibacterial activity with no cytotoxic effects under biological pH levels.

The synthesis of fucoidan-mimetic glycopolymers by cyanoxyl-mediated free-radical polymerization was reported by Tengdelius and co-workers (Scheme 12.4). (Meth)acrylamide fucoside monomers were synthesized through coupling NHS-methacrylate and free amino groups on fucose derivatives. Global *O*-sulfation was conducted following reversible addition-fragmentation chain transfer (RAFT) polymerization instead of on the monomer due to the obstacle that emerged during the polymerization of the sulfated monomer. These synthetic glycopolymers possessed significant effects against HSV-1 infection presumably during the virus-cell adsorption step.[70] Meanwhile, the synthetic glycopolymers showed aggregation in the cellular responses and platelet surface changes, which were comparable to natural fucoidan.[71] The glycopolymers induced platelet activation, which significantly depended on their sulfate contents. Afterwards, Tengdelius also developed gold nanoparticles (GNPs) coated with fucoidan-mimetic glycopolymers (Scheme 12.4).[72] GNPs with an average size of 40 nm were formed using $NaBH_4$ as the reducing agent and glycopolymers as the capping agent. The biological evaluation of GNPs displayed selective cytotoxicity to human colon cancer cell lines (HCT116) but no toxicity to mouse fibroblast cells (NIH3T3). Furthermore, a similar approach was employed to develop the novel lipoglycopolymers composed of octadecyl, dioctadecyl, and cholesteryl linked with fucoidan glycomimetics.[73]

Recently, Yu's group[74] developed a robust approach to fabricate glycopolymers under microwave-assisted ring-opening metathesis polymerization (ROMP) in an emulsion system. Six sulfated fucoidan glycopolymers were synthesized with well-defined sulfation patterns (0*S*, 2*S*, 3*S*, 2,3*S*, 3,4*S*, and 2,3,4*S*), narrow polydispersity index (PDI: 1.06–1.19), and high molecular weights (64.8–127.3 kDa) (Figure 12.1).[75] 2-*O*-sulfation fucoidan glycopolymers exhibited good inhibitory activities against both the H1N1 and H3N2 viruses, while 3-*O*-sulfation glycopolymers only possessed inhibition on

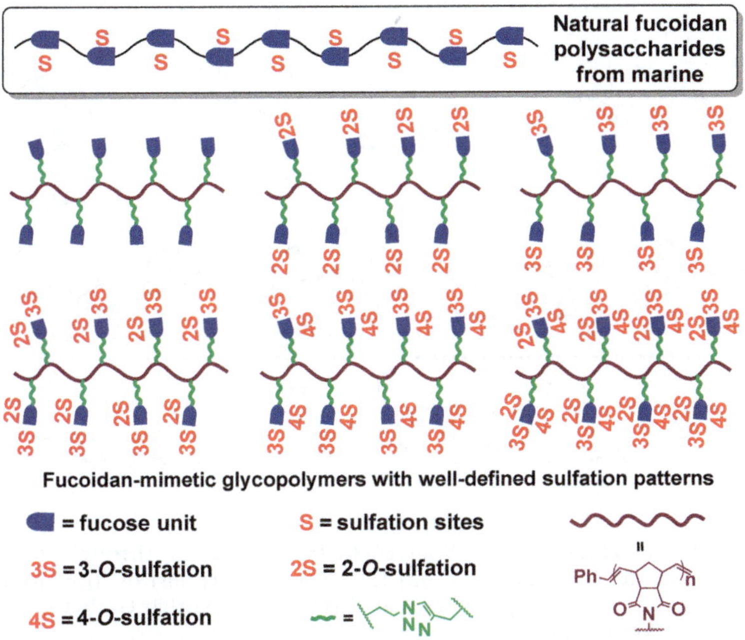

Figure 12.1 Representative structures of fucoidan-mimetic glycopolymers with well-defined sulfation patterns. Reproduced from ref. 75 with permission from American Chemical Society, Copyright 2018.

the H3N2 virus. Therefore, it was speculated that the antiviral activities of fucoidan-mimetic glycopolymers depended more on their sulfation patterns instead of their sulfation degree. In addition, further increases in the content of sulfation did not improve their antiviral activities.

In addition, the chemical synthesis of glycoclusters that mimic the natural fCS polysaccharide was reported by Li's group[76] (Figure 12.2). The protecting groups and regioselective glycosylation were manipulated to synthesize fCS trisaccharides with well-defined sulfate groups from degraded chondroitin sulfate. The assembly of glycoclusters were finally achieved using the copper(I)-catalyzed azide alkyne cycloaddition reaction with different types of scaffolds. The glycocluster bearing eight trisaccharide epitopes possessed the most potent anticoagulant activity, which presumably corresponded to the relatively high molecular weight.

12.2.4 Synthesis and Modification of Marine Polysaccharide Derivatives

Since structural diversity is a distinguishing feature of marine polysaccharides, their biological activities are highly variable depending on the natural sources. The derivatives obtained from the modification of marine

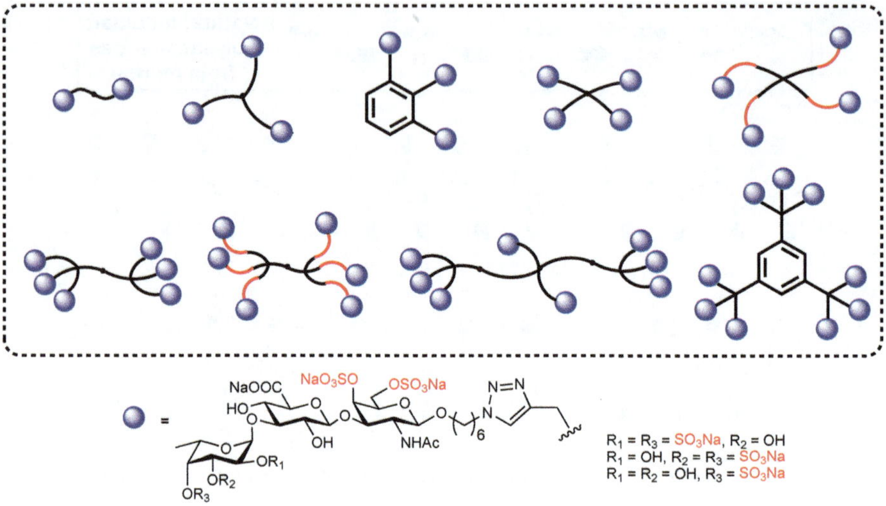

Figure 12.2 Representative fCS glycoclusters obtained through the "click" reaction. Adapted from ref. 76 with permission from John Wiley and Sons, Copyright © 2018 Wiley-VCH Verlag GmbH & Co. KGaA, Weinheim.

polysaccharides have been extensively studied in regulating biological activity, mitigating toxicity, and improving bioavailability.[77] The chemical approaches, including carboxymethylation, phosphorylation, sulfonation, acylation, and alkylation, provide a classical pathway to rapidly achieve functional derivatives from marine polysaccharides.

Chitosan decorated with carboxymethyl has emerged as a promising candidate for biomedical applications, particularly in the field of wound repair and regeneration.[78–81] Water-soluble *N*-carboxymethyl chitosan has been prepared under a mild condition,[80] which was also designed to be a degradable membrane *in vivo*. Interestingly, the results indicated that degradation was faster in the musculature than in sub-cutaneous tissue. Jiang *et al.*[82] reported the formation of a thermo-sensitive injectable hydrogel with carboxymethyl chitin. The hydrogel was developed for three-dimensional (3D) cell culture by the *in situ* formation of the hydrogel at body temperature. More importantly, an *in vivo* mouse study on the hydrogel showed the promotion of cell survival and proliferation as well as *in situ* gel formation and biocompatibility.

Grøndahl *et al.*[83] reported the phosphorylation of alginates using a urea/phosphate system. The degree of substitution under various conditions was predominantly determined by the amount of urea presented in the system. The hydrogel formed by ionized alginate with phosphate groups exhibited an enhanced resistance to the degradation by chelating agents. In addition, a high content of sulfation is a characteristic feature of numerous marine polysaccharides, though the extraction process may lead to some loss of sulfate groups at specific sites.

Moreover, the chemical sulfonation of marine polysaccharides is a concise protocol to obtain glycomimetics of natural glycosaminoglycans (GAGs). Usoltseva's group reported the chemical sulfonation of laminarans obtained from the brown alga *Dictyota dichotoma*.[84] The sulfated laminaran presented potent inhibition to DLD-1 cells but exhibited weak inhibition to HCT-116 and HT-29 cells. Li[85,86] reported a series polyanionic derivatives of polymannuronic acid (PM) and polyguluronate (PG) prepared by chemical modification employing phosphate ($H_3PO_4/(NH_2)_2CO$), H-phosphonate ($H_3PO_3/(NH_2)_2CO$) and sulfate ($ClSO_3H$) reagents. The anticoagulant and antioxidant activities of these derivatives were evaluated *in vitro*. Their results indicated that both phosphate and H-phosphonate derivatives improved the hydroxyl radical scavenging activity, while sulfate derivative enhanced the superoxide radical scavenging activity. Moreover, it was also found that sulfate groups were the best substituents to improve anticoagulant activities and activate FGF/FGFR1c signaling pathway as evaluated *in vitro*.

12.3 Synthesis of Marine Oligosaccharides

12.3.1 Chemical Synthesis of Marine Oligosaccharides

Marine oligosaccharides can be easily obtained by enzymatic/chemical degradation, though the resulting residues are mostly heterogeneous mixtures. Therefore, the attainment of versatile types of marine oligosaccharides with well-defined structures has remained a challenge. Chemical synthesis is a fundamental tool for accessing to homogeneous and structurally defined oligosaccharides.[87] With the recent development of carbohydrate chemistry, numerous marine oligosaccharides have been synthesized by the chemical approach, thereby significantly facilitating the structure–activity relationship study of marine-derived oligosaccharides.

12.3.1.1 Synthesis of Chitin/Chitosan Oligosaccharides

The chemical synthesis of chitin/chitosan oligosaccharides was initially difficult due to the inherently low reactivity of hexosamine and the lability of β-bromoglycoside donors. Since 1990, the synthesis of chitin/chitosan oligosaccharides has been extensively explored with the emergence of numerous glycosylation donors and strategies.[88–92] Yamago *et al.*[93] developed *N*-phthaloyl (Phth)-protected thioglycosides for both of glycosyl donors and acceptors, which was deemed a robust approach for the fabrication of chitosan oligosaccharides (Scheme 12.5). The coupling conditions of glycosylation were screened with 1-benzenesulfinyl piperidine (BSP) and triflic anhydride (Tf_2O), benzenesulfenyl triflate (PhSOTf), benzenesulfenyl chloride, and silver bis[(trifluoromethyl)sulfonyl]amide (triflimide, NTf_2). The glycosyl triflate intermediate was observed by NMR analysis to promote glycosylation. Their findings offered an iterative glycosylation strategy for the rapid and efficient

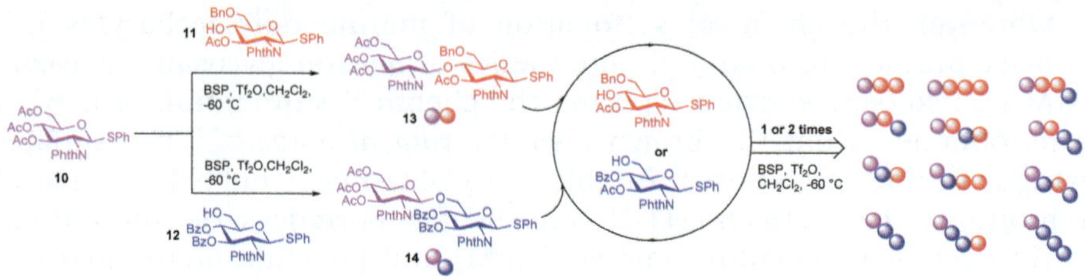

Scheme 12.5 Concise assembly of chitosan oligosaccharides. Adapted from ref. 93 with permission from John Wiley and Sons, Copyright © 2004 WILEY-VCH Verlag GmbH & Co. KGaA, Weinheim.

assembly of the β-glycosidic bond of glucosamine. Meanwhile, the chitin tetrasaccharide was sequentially assembled with thioglycosides in a one-pot process by Huang.[94] Afterwards, the synthesis of chitin/chitosan oligosaccharides with a fully-automated oligosaccharide synthesizer was reported by Seeberger's group.[95] The thioglycoside of *N*-trichloroacetyl glucosamine performed significantly better during the automated synthesis as compared to the trifluoroacetimidate donor. The target hexasaccharide was furnished by a preparative reverse-phase high-performance liquid chromatography (HPLC) and hydrogenolysis followed by gel filtration.

12.3.1.2 Synthesis of Alginate Oligosaccharides

Alginates, which are abundant cell-wall polysaccharides from brown algae, are linear polymers bearing β1–4-mannuronic acid (M) and α1–4-L-guluronic acid (G) residues. It is composed of different types of repeating units, including homogeneous polymers (MMMMM and GGGGG) as well as block (MMMGGG), alternative (MGMGMG), and random (GMMGMGGM) polymers. The ratio of M/G from natural alginate polysaccharides varies depending on the diverse sources and species.[9] Due to the barrier, which forms a 1,2-*cis* glycosidic bond for β-D-mannoside, a 4,6-benzylidene protecting group was developed and introduced as an elegant solution. The construction of alginate oligomers was initially achieved by the preglycosylation oxidation[96] or post-glycosylation oxidation[97] approach. Recently, several groups reported that the presence of carboxylic ester on C5 of the mannuronic acid promoted the generation of excellent β configuration because of the electronic effects. This facilitated the chemical synthesis of alginate oligosaccharides.[98,99]

van der Marel *et al.*[98] first reported the chemical synthesis of homogeneous β-D-mannuronic acid alginate oligosaccharides and α-L-guluronic acid alginate oligosaccharides. A trisaccharide of β-linked mannuronic acids was synthesized from the thioglycoside in Ph_2SO/Tf_2O or NIS/TMSOTf mediated glycosylation.[100] In 2008, they further reported the synthesis of a dimer and trimer of α-L-guluronic alginates.[97] In contrast to the above mentioned β-linked mannuronic acid, 4,6-*O*-di-*tert*-butylsilylidene protection on gulose promoted the most efficient α-selectivity. The primary hydroxyls were finally oxidized to achieve carboxylic acid on the oligosaccharides. In 2009, Hung's group[96] also reported a chemical synthesis of alginate tetrasaccharide containing repeated guluronic acids. Commercially available L-ascorbic acid was converted to the key synthon 1,6-anhydro-L-gulose, which was then applied to the assembly of alginate oligosaccharides with α1–4-linked guluronate. Furthermore, the synthesis of hybrid alginate oligosaccharides was more challenging. Recently, Codée *et al.*[101] developed an excellent synthetic route towards the assembly of sequence-mixed alginate oligosaccharides (Scheme 12.6). The silylidene-protected gulose imidate donor was condensed with α-*S*-tolyl mannuronic acid acceptor to furnish the GM dimer in excellent α-configuration with a yield of 91%. The alginate oligosaccharides with alternative G/M from di- to hexasaccharide (GMGMGM, **19**) were successfully synthesized by using GM disaccharide building blocks. Their findings indicated that the conformational flexibility of the GM acceptor was a significant factor in the efficiency of the glycosylation process. Thus, the steric transition state during glycosylation could be potentially overcome by driving the acceptor towards a more favorable shape. Furthermore, a hybrid hexasaccharide (GMGGMG, **23**) was obtained by a similar approach.

The automated solid-phase oligosaccharide synthesizer was first developed by Seeberger,[102] which has been widely used in carbohydrate research. Codée *et al.*[103] reported the automated solid-phase synthesis of β-mannuronic acid alginates (Scheme 12.7A). Merrifield resin with a butene-1,4-diol linker was designed to attach and release alginate oligosaccharides. Tetra-, octa-, and dodecamannuronates were accomplished with an automatic synthesizer, of which the final products were isolated by preparative HPLC in multimilligram quantities. Subsequently, Pohl's group[99] reported an alternative approach for the automated solution-phase synthesis of hexamannuronate

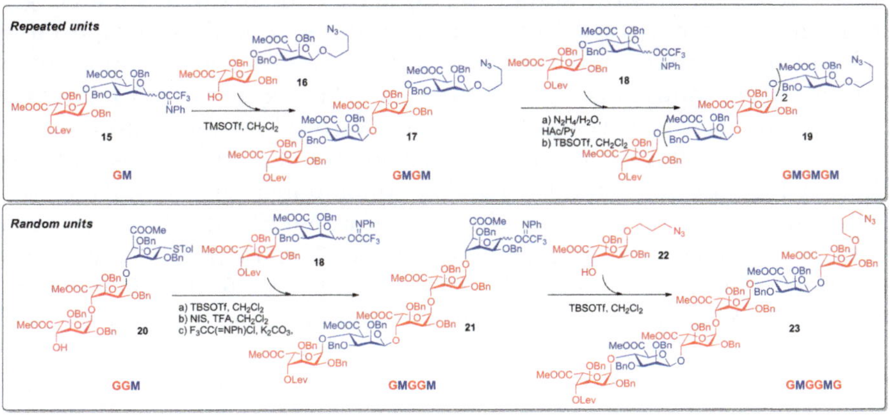

Scheme 12.6 Modular assembly of alginate oligosaccharides. Adapted from ref. 101 with permission from John Wiley and Sons, Copyright © 2015 WILEY-VCH Verlag GmbH & Co. KGaA, Weinheim.

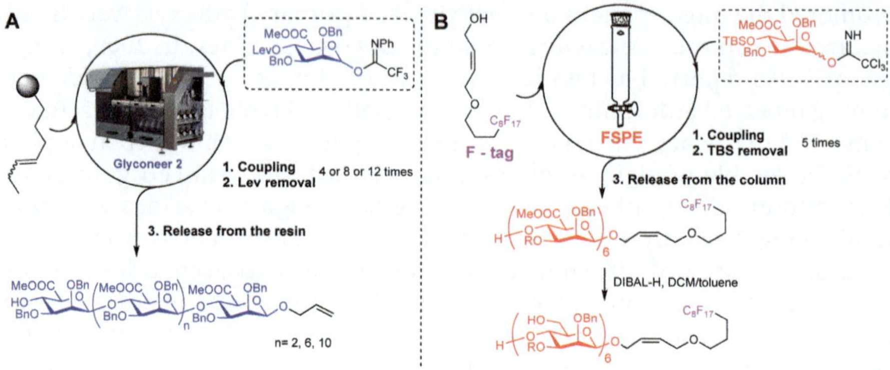

Scheme 12.7 (A) Solid phase[103] and (B) fluorous-tag assisted synthesis of iterative alginate oligosaccharides. (A) Adapted from ref. 103 with permission from John Wiley and Sons, Copyright © 2012 WILEY-VCH Verlag GmbH & Co. KGaA, Weinheim. (B) Adapted from ref. 99 with permission from American Chemical Society, Copyright 2015.

by using an acceptor attached with a fluorous tag and rapid purification with fluorous solid-phase extraction (FSPE, Scheme 12.7B). The mannuronate hexamer was successfully achieved through cyclic glycosylation, fluorous-tag assisted purification, oligosaccharide release from the tag, and global deprotection steps. The advantage of this approach is the presence of the homogeneous reaction solution as compared to the solid phase synthesis during glycosylated coupling.

12.3.1.3 Synthesis of Fucosylated Chondroitin Sulfate Oligosaccharides

Fucosylated chondroitin sulfate (fCS), a class of distinct sulfated glycosaminglycans extracted from sea cucumbers, is mainly composed of a chondroitin sulfate (CS) backbone and sulfated fucose as the branch. The repeating motifs on the backbone of CS are alternative β1–4-linked D-glucuronic acid (GlcA) and β1–3-linked *N*-acetyl-D-galactosamine (GalNAc). To obtain fCS with potent anticoagulant activity and less adverse effects which are comparable to unfractionated heparin and LMWH, several studies have reported the synthesis of well-defined fCS oligosaccharides to decipher their biological activities. Tamura *et al.*[104] reported a concise approach to access an fCS trisaccharide with four sulfate groups. Afterwards, Nifantiev *et al.* developed a modular assembly of a series of fCS oligosaccharides.[105–107] The total chemical synthesis of the fCS repeating trisaccharides was first reported by Li's group[76] in 2017 (Scheme 12.8). A concise "2 + 1" synthetic route from a CS disaccharide building block which was degraded and prepared from the natural CS polysaccharide was designed. Fucose thioglycoside was employed to furnish the newly formed glycosidic bond at C3 of the glucuronic acid under NIS/AgOTf.

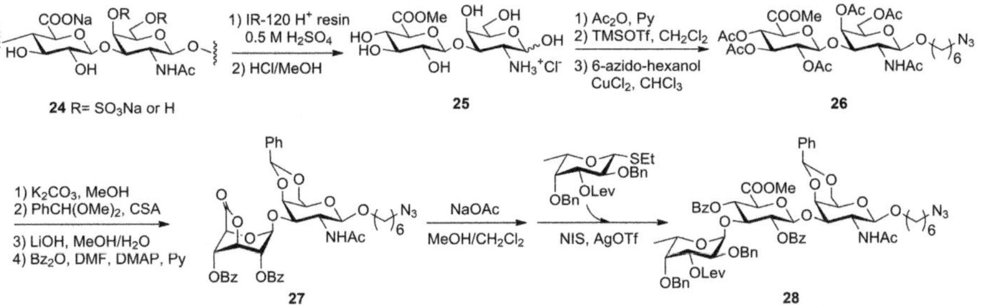

Scheme 12.8 Synthesis of fCS repeating trisaccharide unit. Adapted from ref. 76 with permission from John Wiley and Sons, Copyright © 2018 Wiley-VCH Verlag GmbH & Co. KGaA, Weinheim.

Three trisaccharides with defined sulfation patterns were finally achieved through sulfation with $SO_3{\cdot}NMe_3$ and global deprotection steps. As mentioned above, these trisaccharides serve as the building blocks of glycoclusters to mimic natural fCS polysaccharides.

12.3.1.4 Synthesis of Fucoidan Oligosaccharides

Fucoidan as a major component in the cell wall of algae possesses versatile biological activities. Due to its structural complexity, chemists have explored a series of approaches to access to fucoidan oligosaccharides for a better understanding of their biological activities since 2000. Nifant'ev *et al.*[108] first reported the chemical synthesis of αl–2- and αl–3-linked fucoidan di- and trisaccharides. Fucosyl bromide donors with a benzoyl group on C4 supply excellent α-selectivity as compared to benzylated donors. This phenomenon was potentially attributed to the participation of the remote neighboring group. The conformational properties of oligosaccharides are highly correlated to the chemical shift, especially for the carbon NMR spectra.[108–112] Du *et al.*[113] reported the synthesis of alternative αl–3/4-linked pentasaccharide with a specific sulfate pattern using the convergent "3 + 2" strategy (Scheme 12.9). Thioglycoside donor catalyzed by NIS/TMSOTf exhibited better yield and α-selectivity as compared to bromide and phosphate donors. The synthetic fucoidan pentasaccharide with a β-configured octyl at the reducing end showed antitumor activity *in vivo* and anticoagulant activity *in vitro*. Li *et al.*[114] developed an efficient approach to fabricate fucoidan tetra- to octasaccharides to probe the structure–activity relationship (SAR). Both thioglycosides and trichloroacetimidates were selected as glycosyl donors for glycosylated coupling with stepwise elongation. The tetrasaccharides were carried out according to the "2 + 2" condensation. In addition, the subsequent "*n*+1" chain-assembly strategy was employed to achieve hexa- to octasaccharides.

Until recently, fucoidan oligosaccharides with type I and II linkages were systemically synthesized with structurally defined sulfation patterns

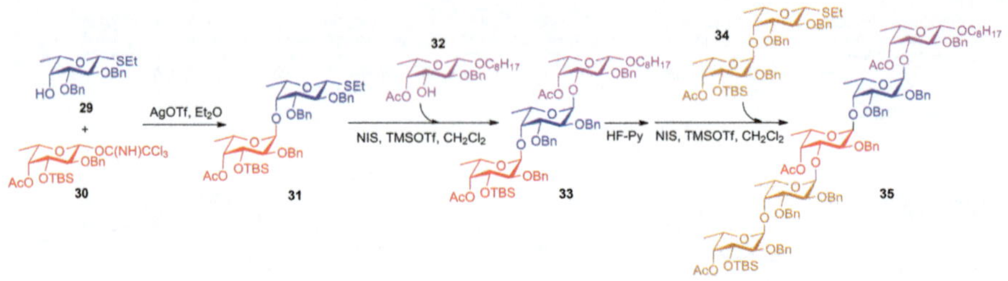

Scheme 12.9 Representative synthetic routes of fucoidan oligosaccharides. Reproduced from ref. 113 with permission from Elsevier, Copyright 2004.

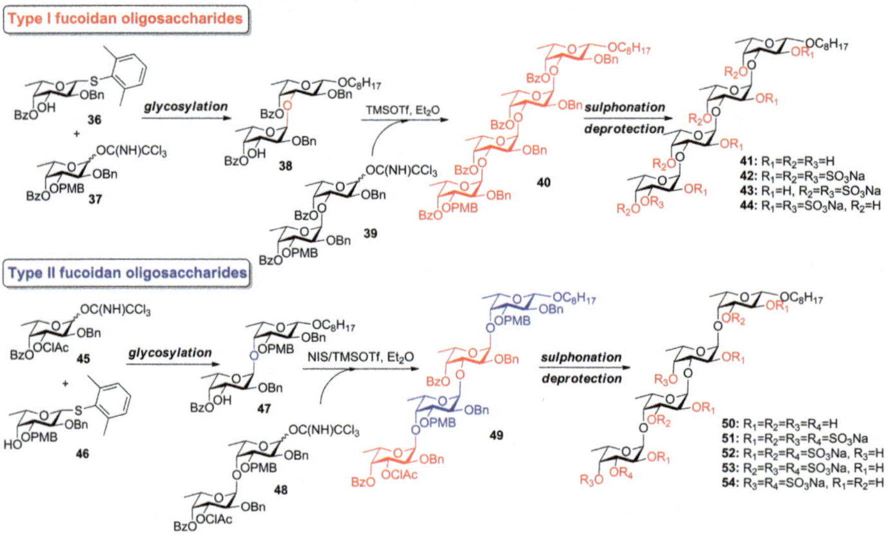

Scheme 12.10 Modular assembly of type I and type II fucoidan oligosaccharides. Reproduced from ref. 115 and 116 with permission from The Royal Society of Chemistry.

(Scheme 12.10).[115,116] The disaccharide with α-fucoside was smoothly obtained through the glycosylation of **47** and **48** using a catalytic amount of $Yb(OTf)_3$. After the regioselective removal of the protecting groups and *O*-sulfation steps, five terafucosides were readily furnished with specific sulfation patterns. Their inhibitory effects on growth of breast cancer MCF-7 cells were evaluated in a dose-dependent manner, of which 3,4-*O*-sulfated tetrafucoside showed the highest anti-proliferative activity, which is comparable to natural fucoidan isolated from *Fucus vesiculosus*. The tetrafucoside epitope was determined to be the minimum structural motifs for presenting anti-tumor activity.

The total synthesis of a sulfated galactofucan (Scheme 12.11) was reported by Hu *et al.*[117] with a rare sequence isolated from the brown alga *Sargassum polycystum*. The synthetic route started from the stereoselective construction

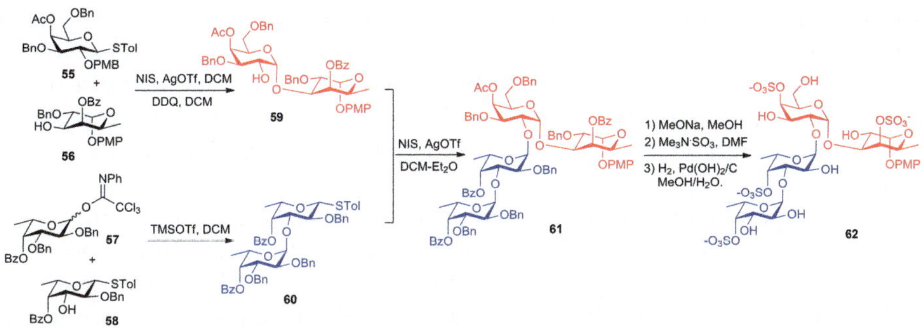

Scheme 12.11 Total synthesis of sulfated galactofucan tetrasaccharide. Adapted from ref. 117 with permission from American Chemical Society, Copyright 2014.

of four 1,2-*cis* glycosidic bonds on both fucose and galactose, and the convergent "2 + 2" assembly was employed to condense the major building motif **61**. The final tetrasaccharide **62** was accomplished by regioselective *O*-sulfation with $SO_3{\cdot}NMe_3$ in DMF at 50 °C and global deprotection with hydrogenolysis over $Pd(OH)_2/C$. Moreover, the NMR analysis was once again confirmed to be a helpful technique to clearly resolve the detailed structure of complex carbohydrates.

12.3.2 Accessing Marine Oligosaccharides Through Enzymatic/Chemical Degradation

12.3.2.1 Enzymatic Depolymerization

Glycosidases (*O*-glycoside hydrolases)[118] and lyases[119,120] are various groups of enzymes for carbohydrate metabolism. Glycosidases display the cleavage of *O*-glycoside bonds in carbohydrates and carbohydrate-containing biopolymers. Although they are involved in complicated biological functions, marine glycosidases and lyases can be utilized as tools for the preparation of structurally sequenced oligosaccharides. For the development of potential biological activities and industrial applications of marine oligosaccharides, numerous studies have explored specific enzymes for both glycosidases and lyases.

Yu *et al.*[121] cloned an endo-type β-agarase gene, AgaB, from marine *Pseudoalteromonas* sp. CY24. AgaB hydrolyzed agarose furnished DP8 and DP10 as the main products, whereas most β-agarases A and B yielded shorter DP4 and DP6 residues. AgaB showed higher activities on the degradation of longer oligosaccharides DP ≥ 12 as compared to the extremely low activity observed on DP10. Meanwhile, endolytic alginate lyases have also been widely explored from *Vibrio splendidus* 12B01 and *Pseudomonas* sp. HZJ 216 in the production of unsaturated oligosaccharides (di-, tri-, and tetrasaccharides). Ulvans, which are cell wall matrix polysaccharides in green algae, are mainly composed of a major repeating unit of α-L-rhamnose (Rha) and β-D-glucuronic acid (GlcA) or its C5 epimer iduronic acid (IduA) as well as small amounts of xylose (Xyl) residues. Helbert *et al.*[122] reported a novel unsaturated β-glucuronyl hydrolase from *N. ulvanivorans*. This glycosyl hydrolase is capable of degrading unsaturated oligosaccharides produced by ulvan lyase.

Marine-derived glycosaminoglycans, usually including high sulphated chondroitin sulfate (CS)[123,124] and keratan sulfate (KS), are linear polysaccharides consisting of repeating disaccharide units. There are commercial bacterial lyases chondroitinase ABC, AC, B and C as tools to generate the CS-derived oligosaccharides with a 4,5-unsaturated uronic acid residue at their non-reducing ends. KS degrading enzymes are glycosidases classified based on their cleavage sites and substrate specificity. Keratanase I is an endo-β-galactosidase from *Pseudomonas* sp. that cleaves at the galactosyl linkages in the KS chain. Keratanase II, identified as endo-β-*N*-acetylglucosaminidase, cleaves *N*-acetylglucosaminyl linkages of the KS chain, releasing Galβ1–4GlcNAc disaccharides with one or two sulfo groups, or tetrasaccharides of Galβ1–4GlcNAcβ1–3Galβ1–4GlcNAc.[125] Linhardt *et al.*[126] used a published sequence of a highly thermostable keratanase II identified in *Bacillus circulans* to clone and express a series of truncation mutants in *Escherichia coli* BL21. The resulting truncated forms of keratanase II exhibit storage and thermal stability. Based on control enzymatic digestion of keranase II, a series of highly sulphated KS oligosaccharides also including the sialylated oligomers was prepared from the KS isolated from blue shark (*Prionace glauca*) cartilage.

Human gut microbes have been extensively explored in recent years and potentially considered as a useful tool for the degradation of marine derived carbohydrates that cannot be digested by human enzymes. Michel *et al.*[127] first reported the transfer of carbohydrate-active enzymes from marine bacteria to Japanese gut microbiota in 2010. Boraston *et al.*[128] found that *Bacteroides plebeius*, a member of the human gut microbe, possessed degradation activity on marine red algal polysaccharides. Recently, Yu *et al.*[129] systemically studied the product characteristics of agarose and alginate degraded by *Bacteroides ovatus* and *Bacteroides uniforms*, which were isolated from human gut microbes (Figure 12.3). The results indicated that *B. ovatus* G19 could effectively digest alginates due to the presence of both α1–4-guluronanlyase and β1–4-mannuronanlyase. A specific agarose-hydrolyzing bacterium was also identified, namely *B. uniformis* L8, which degrades agarose into di- to octasaccharide after 48 h.

12.3.2.2 Chemical Depolymerization

As mentioned above, enzymatic degradation is a powerful tool to access marine oligosaccharides with narrow molecular weight distribution. However, this process has a high cost and exhibits strict substrate specificity.

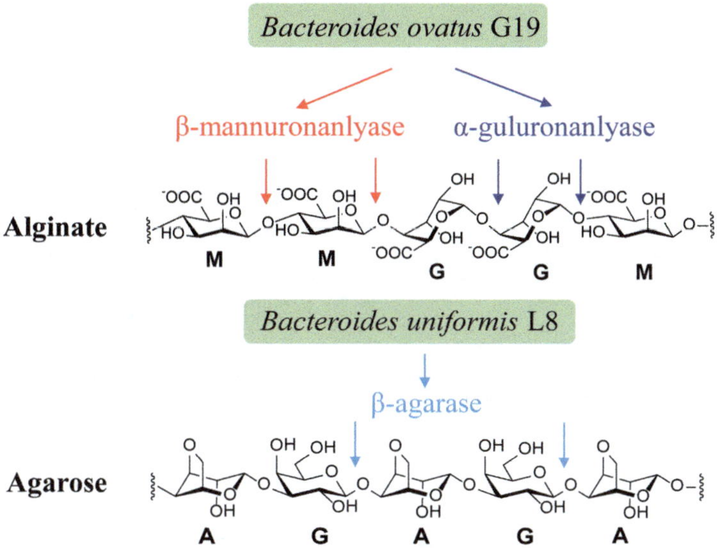

Figure 12.3 Proposed mechanism of gut microbial degradation of marine polysaccharides. Adapted from ref. 129 [DOI: https://doi.org/10.3390/md15040092] under the terms of a CC BY 4.0 license [https://creativecommons.org/licenses/by/4.0/].

To overcome the limitation of the enzymatic method, chemical depolymerization has been considered as the easiest way to obtain marine oligosaccharides. These approaches mainly include acidic hydrolysis,[130] free radical degradation,[29] β-elimination,[131] and deaminative degradation.[132]

Acidic hydrolysis, the most widely used approach for chemical depolymerization, primarily relies on the use of hydrochloric acid (HCl), sulfonic acid (H_2SO_4), and hydrogen fluoride (HF). Yu *et al.*[130] reported the highly ordered degradation of sulfated galactan (κ-carrageenan) under mild acidic conditions (Scheme 12.12). The hydrolysis of κ-carrageenan was carried out with 0.1 M HCl at 60 °C for 3 h, which was shown to be a concise approach to achieve exclusively odd-numbered oligosaccharides. The proposed mechanism was deduced to involve a two-step cleavage: initial cleavage of the α1–3 glycosidic bond followed by immediate removal of β1–4-linked 3,6-anhydro-galactose. The labile reducing terminal 3,6-anhydro-galactose could also be stabilized *via* reduction to alditol, which had a crucial effect on the specificity of the acidic hydrolysis.[133] Meanwhile, 2-*O*-sulfation was determined to stabilize the glycosidic bond at the reducing end, and the associated substrate generally required a stronger condition for complete hydrolysis. Moreover, acidic hydrolysis has been extensively reported for the degradation of versatile marine polysaccharides such as chitosan, agarose,[14,134] fCS,[135] and uronic acid-containing polysaccharides.[23,136] Emerging approaches were also developed by using solid state acid[137] or

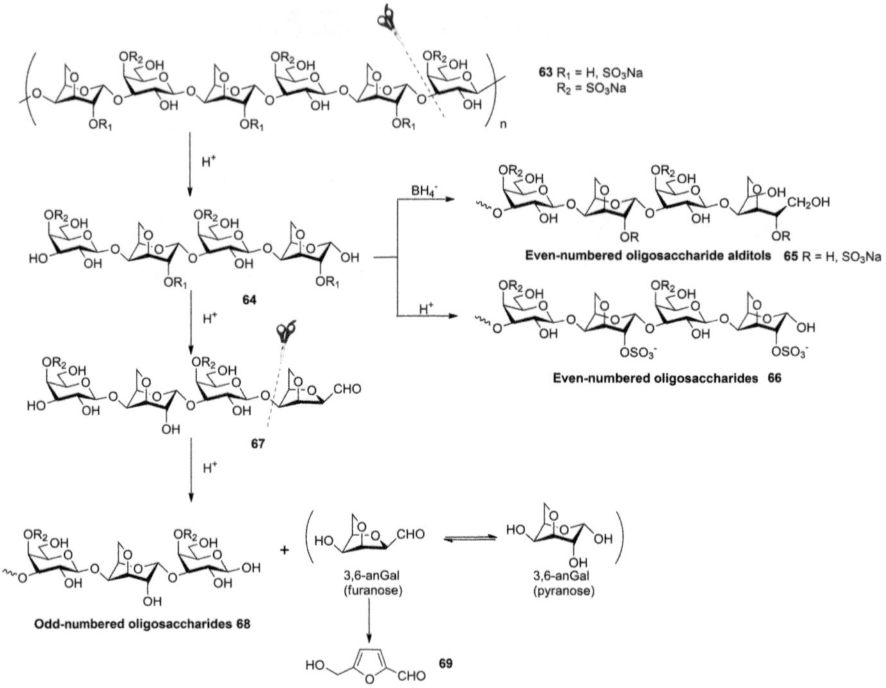

Scheme 12.12 Mild acid hydrolysis of κ-carrageenan. Reproduced from ref. 130 with permission from John Wiley and Sons, Copyright © 2009 The Authors Journal compilation © 2009 FEBS.

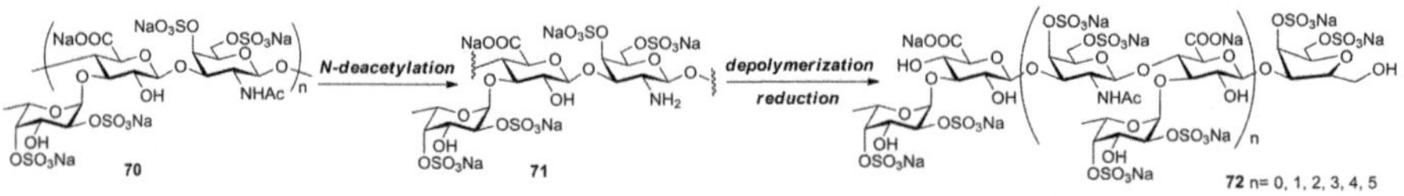

Scheme 12.13 Preparation of fCS oligosaccharides. Reproduced from ref. 132 with permission from National Academy of Sciences.

microwave-irradiated sulfonic acid-functionalized ionic liquids (SFILs) as the catalyst.[138]

Due to the limitation of the specific enzyme for the degradation of fucosylated chondroitin sulfate, Zhao *et al.*[132] first reported the deaminative degradation to obtain structurally homogeneous fCS oligomers (Scheme 12.13). The fCS polysaccharide extracted from the sea cucumber *Stichopus variegatus* was *N*-deacetylated with hydrazine hydrate, which was subjected to the cleavage with nitrous acid and subsequent reduction with $NaBH_4$ to furnish the oligomers **72** from DP3 to DP18. The fCS oligomers DP ≥ 9 exhibited strong anticoagulant activity by directly inhibiting the

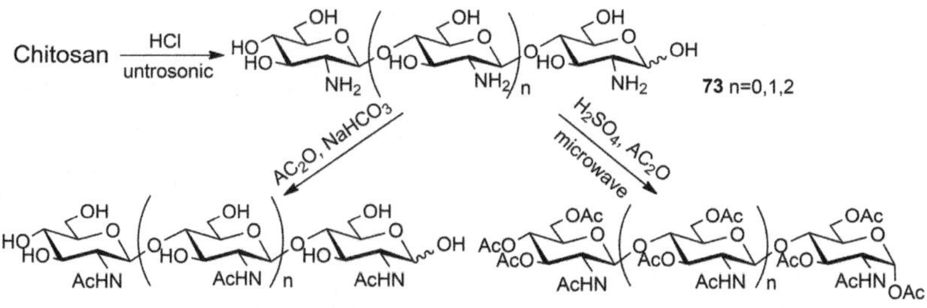

Scheme 12.14 Synthesis and modification of chitosan oligosaccharides.

human intrinsic tenase while avoiding side effects such as FXII activation and platelet aggregation. Moreover, unsaturated fCS oligomers from *Thelenota ananas* were achieved by β-eliminative depolymerization on the fCS benzyl ester in an alkaline anhydrous solution.[131] Recently, Yu *et al.*[29] isolated a novel fCS from the sea cucumber *Holothuria mexicana* and obtained fCS oligomers by free radical degradation while retaining their specific structural properties.

Chitosan oligosaccharides (COSs), depolymerized products of chitosan composed of β1–4-glucosamine units, have a broad range of biological functions such as antitumor, antifungal, and antioxidant activities. The *N*-acetylated chitosan oligosaccharides (**74**, NACOs) and peracetylated chitosan oligosaccharides (**75**, PACOs) were prepared from the COSs by chemical modification (Scheme 12.14).[139,140] Their antagonist effects against glutamate-induced PC12 cell death were investigated. The results showed that pretreatment of PC12 cells with the PACOs markedly inhibited glutamate-induced cell death in a concentration-dependent manner, suggesting the peracetylation is essential for the neuroprotective effects of chitosan oligosaccharides.

12.4 Development of Novel Drugs from Marine Poly-/Oligosaccharides and Their Derivatives

Propylene glycol alginate sodium sulfate (PSS), a sulphated polysaccharide, is a representative marine carbohydrate-based drug developed by Guan in China.[141] It has been in clinical application for the treatment of ischemic cerebrovascular disease (ICVD) and cardiovascular disease in China since 1986. PSS is prepared from sodium alginate through hydrolysis, esterification and sulfation (Scheme 12.15).[141] The alginate polysaccharide was originally degraded into oligomers with M/G ratio of 7 : 3 and a PDI values of PDI < 1.80. The propylene glycol ester and sulfate groups were subsequently introduced by 1,2-epoxypropane and HSO_3Cl/formamide, respectively.[142] PSS in the sodium sulfate form was finally achieved following sodium exchange

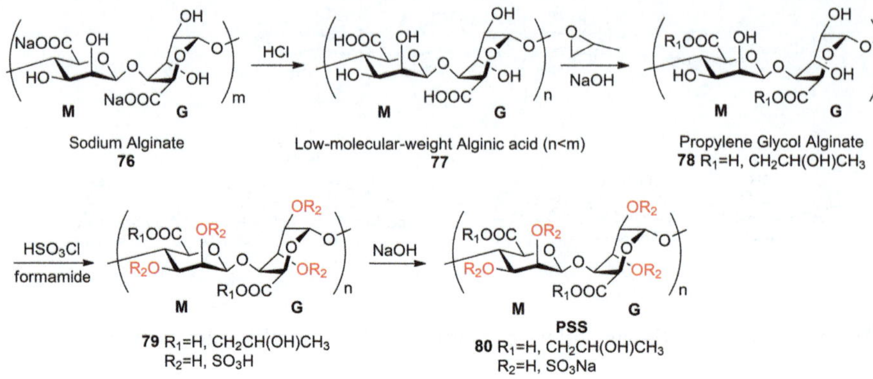

Scheme 12.15 Chemical synthesis of PSS. Reproduced from ref. 142 with permission from Elsevier, Copyright 2016.

and the purification step. Usually the relative molecular weight of PSS ranges from 10 kDa to 20 kDa, and the organic sulfur content is 9.0~14%. The emerging indications of PSS were expanded for the treatment of cancers, diabetes, hepatitis, chronic glomerulonephritis, and restless leg syndrome (RLS).[143] In recent years, re-development of PSS was carried out to improve its effectiveness and safety, expand its new clinical application and find new potential therapy agents. The combination of biological and chemical methods was developed to improve the quality control of PSS.[142] A PSS fragment with a molecular weight of 6 kDa, is an optimal antithrombotic candidate for further study.[144]

Alzheimer's disease is a chronic neurodegenerative disease that remains a pandemic disease given the lack of effective medicinal cures.[145] Oligomannurarate 971 (HS971),[146] which was originally developed by the school of medicine and pharmacy at Ocean University of China, contains a dimer to octamer in the form of homogeneous β1–4-linked D-mannuronic acid. HS971 was confirmed to prevent the aggregation and deposition of amyloid peptide for the treatment of Alzheimer's disease. Now, HS971 is currently in phase III trials to investigate its efficacy and safety for patients with moderate Alzheimer's disease symptoms. Three marine polysaccharide-derived drugs (PSS, PGMS, and FPS) have been under clinical use approved by CFDA, and five marine drug candidates (HS971, HS911, HS203, PGS, PS916) are at various clinical development stages (Scheme 12.16).

Oligomannurarate sulfate (JG3), a sulfated derivative of oligomannurarate mixtures, was initially developed as a heparanase inhibitor to combat tumor angiogenesis and metastasis targeting the basic fibroblast growth factor.[147] The sodium alginate was degraded with hydrogen peroxide at 90 °C for 2 h, and oligomannurarate mixtures were purified by ethanol precipitation and the removal of sodium salt. JG3 was obtained by using the traditional sulfating reagents $ClSO_3H$/formamide. Their findings revealed that JG3 abrogated the bFGF-induced angiogenesis *via* binding with bFGF and significantly inhibited tumor angiogenesis, metastasis, and tumor growth *in vivo*.[147]

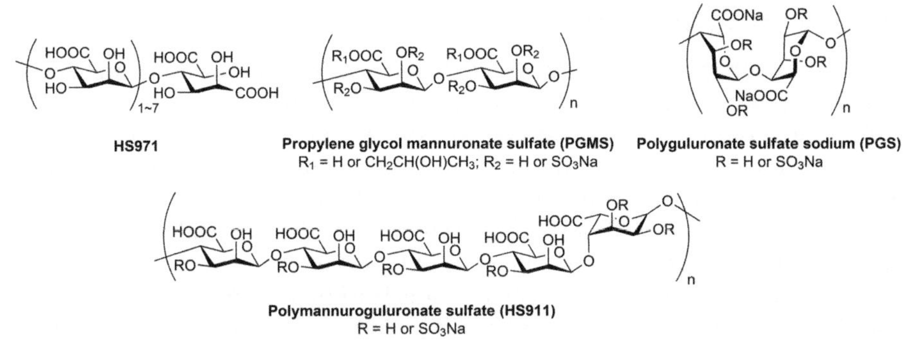

Scheme 12.16 Representative carbohydrate structures currently under clinical use and evaluation.

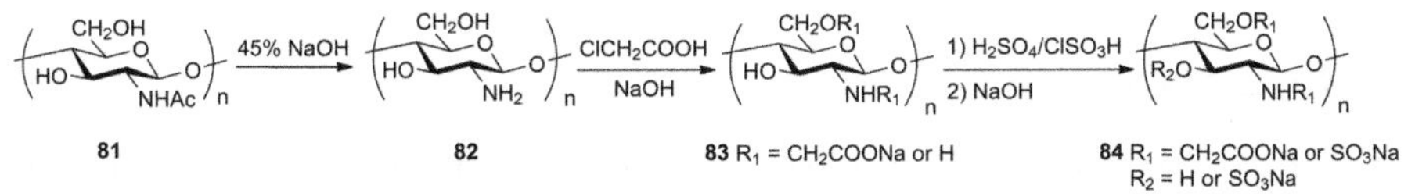

Scheme 12.17 Synthetic route of PS916.

Sulfated carboxymethylated chitosan (PS916), is an anti-atherosclerotic drug developed by Guan *et al.* in China.[148] After the removal of acetyl groups with 45% sodium hydroxide, chitin oligosaccharide was decorated with carboxymethylated groups and was subjected to the general sulfation protocol with $ClSO_3H/H_2SO_4$ to achieve chitosan ester oligosaccharides PS916 (Scheme 12.17). Lv *et al.*[149] recently reported that fluorescein-labeled PS916 was widely distributed in rats with relatively high levels in kidney and liver.

12.5 Conclusions and Prospects

As one of the bioactive compounds, carbohydrates obtained from the marine ecosystem are mainly in the form of complex polysaccharides, oligosaccharides, and glycoconjugates. Marine poly-/oligosaccharides have been extensively developed for industrial applications such as in food additives, pharmaceuticals, textiles, cosmetics, adhesives, *etc.* Highlighted from this chapter, the current approaches developed for the synthesis of marine carbohydrates with well-defined structures have inspired us to exploit the promising advanced applications. Marine polysaccharides are particularly designed as biodegradable drug carriers, self-healing materials, and thermal or pH-sensitive hydrogels, furthermore, synthetic marine oligosaccharides were developed as novel drug candidates for more specific treatments. The widening access of marine poly-/oligosaccharides offers great potentials to better understand and utilize the mysterious ocean.

Acknowledgements

The support from National Natural Science Foundation of China and NSFC-Shandong Joint Fund for Marine Science Research Centers (31670811, 21602212, U1606403), National Science and Technology Major Project for Significant New Drugs Development (2018ZX09735004), National Key Research and Development Program of China (2018YFC0310900), Major Science and Technology Projects in Shandong Province (2015ZDJS04002), Fundamental Research Funds for the Central Universities (201762002), Natural Science Foundation of Shandong Province (ZR2016BB02), Primary Research and Development Plan of Shandong Province (2017GSF221002), Basic Research Funds for Application of Qingdao (17-1-1-63-jch), Shandong Provincial Key Laboratory of Glycoscience Industry Alliance and Taishan Scholar Project Special Funds.

References

1. I. Hamed, F. Özogul, Y. Özogul and J. M. Regenstein, *Compr. Rev. Food Sci. Food Saf.*, 2015, **14**, 446.
2. L. R. Tchokouaha Yamthe, R. Appiah-Opong, P. V. Tsouh Fokou, N. Tsabang, F. Fekam Boyom, A. K. Nyarko and M. D. Wilson, *Mar. Drugs*, 2017, **15**, 323.
3. L. K. Shubina, T. N. Makarieva, D. V. Yashunsky, N. E. Nifantiev, V. A. Denisenko, P. S. Dmitrenok, S. A. Dyshlovoy, S. N. Fedorov, V. B. Krasokhin, S. H. Jeong, J. Han and V. A. Stonik, *J. Nat. Prod.*, 2015, **78**, 1383.
4. X. Wang, H. Yu, R. Xing and P. Li, *BioMed Res. Int.*, 2017, **2017**, 9746720.
5. H. K. Kang, C. H. Seo and Y. Park, *Int. J. Mol. Sci.*, 2015, **16**, 6018.
6. V. H. Pomin, *Front. Cell. Infect. Microbiol.*, 2014, **4**, 5.
7. G. Jiao, G. Yu, J. Zhang and H. S. Ewart, *Mar. Drugs*, 2011, **9**, 196.
8. S. Y. Xu, X. Huang and K. L. Cheong, *Mar. Drugs*, 2017, **15**, 388.
9. K. Y. Lee and D. J. Mooney, *Prog. Polym. Sci.*, 2012, **37**, 106.
10. M. T. Ale, J. D. Mikkelsen and A. S. Meyer, *Mar. Drugs*, 2011, **9**, 2106.
11. S. U. Kadam, B. K. Tiwari and C. P. O'Donnell, *Int. J. Food Sci. Technol.*, 2015, **50**, 24.
12. T. T. Stevenson and R. H. Furneaux, *Carbohydr. Res.*, 1991, **210**, 277.
13. M. Falkeborg, L. Z. Cheong, C. Gianfico, K. M. Sztukiel, K. Kristensen, M. Glasius, X. Xu and Z. Guo, *Food Chem.*, 2014, **164**, 185.
14. M. Li, G. Li, L. Zhu, Y. Yin, X. Zhao, C. Xiang, G. Yu and X. Wang, *PLoS One*, 2014, **9**, e91106.
15. X. T. Fu and S. M. Kim, *Mar. Drugs*, 2010, **8**, 200.
16. C. Devillé, J. Damas, P. Forget, G. Dandrifosse and O. Peulen, *J. Sci. Food Agric.*, 2004, **84**, 1030.
17. D. A. Rees, *J. Chem. Soc.*, 1961, **0**, 5168.
18. J. K. Tobacman, *Environ. Health Perspect.*, 2001, **109**, 983.
19. M. Lahaye and A. Robic, *Biomacromolecules*, 2007, **8**, 1765.
20. B. Ray, *Carbohydr. Polym.*, 2006, **66**, 408.

21. R. C. Cheung, T. B. Ng, J. H. Wong and W. Y. Chan, *Mar. Drugs*, 2015, **13**, 5156.
22. M.-T. Yen, J.-H. Yang and J.-L. Mau, *Carbohydr. Polym.*, 2009, **75**, 15.
23. H. X. Wang, J. Zhao, D. M. Li, S. Song, L. Song, Y. H. Fu and L. P. Zhang, *Carbohydr. Res.*, 2015, **402**, 95.
24. C. Schiraldi, D. Cimini and M. De Rosa, *Appl. Microbiol. Biotechnol.*, 2010, **87**, 1209.
25. C. Cai, K. Solakyildirim, B. Yang, J. M. Beaudet, A. Weyer, R. J. Linhardt and F. Zhang, *Carbohydr. Polym.*, 2012, **87**, 822.
26. S. Chen, J. Xu, C. Xue, P. Dong, W. Sheng, G. Yu and W. Chai, *Glycoconjugate J.*, 2008, **25**, 481.
27. Q. Shang, Q. Li, M. Zhang, G. Song, J. Shi, H. Jiang, C. Cai, J. Hao, G. Li and G. Yu, *Mar. Drugs*, 2016, **14**, 224.
28. X. Liu, J. Hao, X. Shan, X. Zhang, X. Zhao, Q. Li, X. Wang, C. Cai, G. Li and G. Yu, *Carbohydr. Polym.*, 2016, **152**, 343.
29. Q. Li, C. Cai, Y. Chang, F. Zhang, R. J. Linhardt, C. Xue, G. Li and G. Yu, *Carbohydr. Polym.*, 2018, **181**, 1160.
30. M. Mahata, S. Shinya, E. Masaki, T. Yamamoto, T. Ohnuma, R. Brzezinski, T. K. Mazumder, K. Yamashita, K. Narihiro and T. Fukamizo, *Carbohydr. Res.*, 2014, **383**, 27.
31. B. B. Aam, E. B. Heggset, A. L. Norberg, M. Sorlie, K. M. Varum and V. G. Eijsink, *Mar. Drugs*, 2010, **8**, 1482.
32. M. Kaya, F. Dudakli, M. Asan-Ozusaglam, Y. S. Cakmak, T. Baran, A. Mentes and S. Erdogan, *LWT–Food Sci. Technol.*, 2016, **65**, 1109.
33. W. Tan, Q. Li, F. Dong, L. Wei and Z. Guo, *Int. J. Biol. Macromol.*, 2016, **92**, 293.
34. Q. Li, W. Tan, C. Zhang, G. Gu and Z. Guo, *Int. J. Biol. Macromol.*, 2016, **91**, 623.
35. D. Gómez-Ríos, R. Barrera-Zapata and R. Ríos-Estepa, *Food Bioprod. Process.*, 2017, **103**, 49.
36. P. Myron, S. Siddiquee and S. Al Azad, *Carbohydr. Polym.*, 2014, **112**, 173.
37. S. Chen, C. Xue, L. A. Yin, Q. Tang, G. Yu and W. Chai, *Carbohydr. Polym.*, 2011, **83**, 688.
38. V. H. Pomin, *Int. J. Biol. Macromol.*, 2015, **72**, 282.
39. W. Garnjanagoonchorn, L. Wongekalak and A. Engkagul, *Chem. Eng. Process.*, 2007, **46**, 465.
40. F. Donot, A. Fontana, J. C. Baccou and S. Schorr-Galindo, *Carbohydr. Polym.*, 2012, **87**, 951.
41. J. P. Shukla, M. N. Nathani and P. B. Dave, *Int. J. Res. Biosci.*, 2017, **6**, 20.
42. I. Finore, P. Di Donato, V. Mastascusa, B. Nicolaus and A. Poli, *Mar. Drugs*, 2014, **12**, 3005.
43. P. L. DeAngelis, *Appl. Microbiol. Biotechnol.*, 2012, **94**, 295.
44. W. He, L. Fu, G. Li, J. Andrew Jones, R. J. Linhardt and M. Koffas, *Metab. Eng.*, 2015, **27**, 92.
45. Z. Zhou, Q. Li, H. Huang, H. Wang, Y. Wang, G. Du, J. Chen and Z. Kang, *Biotechnol. Bioeng.*, 2018, **115**, 1561.

46. E. Zhou, A. B. Seminara, S. K. Kim, C. L. Hall, Y. Wang and V. T. Lee, *ACS Chem. Biol.*, 2017, **12**, 3076.
47. H. Hoshino, T. Foyez, S. Ohtake-Niimi, Y. Takeda-Uchimura, M. Michikawa, K. Kadomatsu and K. Uchimura, *J. Histochem. Cytochem.*, 2014, **62**, 145.
48. D. Cimini, M. De Rosa, E. Carlino, A. Ruggiero and C. Schiraldi, *Microb. Cell Fact.*, 2013, **12**, 46.
49. S. Maleki, M. Maerk, R. Hrudikova, S. Valla and H. Ertesvag, *New Biotechnol.*, 2017, **37**, 2.
50. J. R. Wiens, A. I. Vasil, M. J. Schurr and M. L. Vasil, *mBio*, 2014, **5**, e01010.
51. G. A. Bonartseva, E. A. Akulina, V. L. Myshkina, V. V. Voinova, T. K. Makhina and A. P. Bonartsev, *Appl. Biochem. Microbiol.*, 2017, **53**, 52.
52. I. Pacheco-Leyva, F. Guevara Pezoa and A. Díaz-Barrera, *Int. J. Polym. Sci.*, 2016, **2016**, 1.
53. A. Schmidt, A. S. Hammerbacher, M. Bastian, K. J. Nieken, J. Klockgether, M. Merighi, K. Lapouge, C. Poschgan, J. Kolle, K. R. Acharya, M. Ulrich, B. Tummler, G. Unden, V. Kaever, S. Lory, D. Haas, S. Schwarz and G. Doring, *Environ. Microbiol.*, 2016, **18**, 3390.
54. J. C. Whitney, G. B. Whitfield, L. S. Marmont, P. Yip, A. M. Neculai, Y. D. Lobsanov, H. Robinson, D. E. Ohman and P. L. Howell, *J. Biol. Chem.*, 2015, **290**, 12451.
55. F. H. Damron, M. R. Davis Jr, T. R. Withers, R. K. Ernst, J. B. Goldberg, G. Yu and H. D. Yu, *Mol. Microbiol.*, 2011, **81**, 554.
56. F. J. Toledo Marante, I. Herrera Bravo de Laguna, N. V. Torres and R. Mioso, *Appl. Biochem. Biotechnol.*, 2014, **174**, 2426.
57. W.-J. Jung and R.-D. Park, *Mar. Drugs*, 2014, **12**, 5328.
58. W. K. Lee, Y. Y. Lim, A. T. Leow, P. Namasivayam, J. Ong Abdullah and C. L. Ho, *Carbohydr. Polym.*, 2017, **164**, 23.
59. A. V. Skriptsova, *Russ. J. Mar. Biol.*, 2015, **41**, 145.
60. D. Cimini, S. Fantaccione, F. Volpe, M. De Rosa, O. F. Restaino, G. Aquino and C. Schiraldi, *Appl. Microbiol. Biotechnol.*, 2014, **98**, 3955.
61. Y.-g. Shi, Y.-c. Meng, J.-r. Li, J. Chen, Y.-h. Liu and X. Bai, *J. Chem. Technol. Biotechnol.*, 2014, **89**, 1445.
62. V. H. Pomin, *Mar. Drugs*, 2014, **12**, 232.
63. E. Bedini, C. De Castro, M. De Rosa, A. Di Nola, A. Iadonisi, O. F. Restaino, C. Schiraldi and M. Parrilli, *Angew. Chem.*, 2011, **50**, 6160.
64. S. Suzuki, H. Saito, T. Yamagata, K. Anno, N. Seno, Y. Kawai and T. Furuhashi, *J. Biol. Chem.*, 1968, **243**, 1543.
65. Y. Kawai, N. Seno and K. Anno, *J. Biochem.*, 1966, **60**, 317.
66. A. Laezza, A. Iadonisi, C. D. Castro, M. De Rosa, C. Schiraldi, M. Parrilli and E. Bedini, *Biomacromolecules*, 2015, **16**, 2237.
67. A. Laezza, A. Iadonisi, A. V. Pirozzi, P. Diana, M. De Rosa, C. Schiraldi, M. Parrilli and E. Bedini, *Chem. - Eur. J.*, 2016, **22**, 18215.
68. L. L. Kiessling, J. E. Gestwicki and L. E. Strong, *Angew. Chem.*, 2006, **45**, 2348.
69. W. T. Liau and A. M. Kasko, *Biomacromolecules*, 2017, **18**, 4133.

70. M. Tengdelius, C. J. Lee, M. Grenegard, M. Griffith, P. Pahlsson and P. Konradsson, *Biomacromolecules*, 2014, **15**, 2359.
71. M. Tengdelius, C. Kardeby, K. Falker, M. Griffith, P. Pahlsson, P. Konradsson and M. Grenegard, *Macromol. Biosci.*, 2017, **17**, 1600257.
72. M. Tengdelius, D. Gurav, P. Konradsson, P. Pahlsson, M. Griffith and O. P. Oommen, *Chem. Commun.*, 2015, **51**, 8532.
73. M. Tengdelius, K. Y. Cheung, M. Griffith, P. Påhlsson and P. Konradsson, *Eur. Polym. J.*, 2018, **98**, 285.
74. F. Fan, C. Cai, L. Gao, J. Li, P. Zhang, G. Li, C. Li and G. Yu, *Polym. Chem.*, 2017, **8**, 6709.
75. F. Fan, C. Cai, W. Wang, L. Gao, J. Li, J. Li, F. Gu, T. Sun, J. Li, C. Li and G. Yu, *ACS Macro Lett.*, 2018, **7**, 330.
76. X. Zhang, W. Yao, X. Xu, H. Sun, J. Zhao, X. Meng, M. Wu and Z. Li, *Chem. - Eur. J.*, 2018, **24**, 1694.
77. G. D'Ayala, M. Malinconico and P. Laurienzo, *Molecules*, 2008, **13**, 2069.
78. Z. Jiang, J. Chi, B. Han and W. Liu, *Carbohydr. Polym.*, 2017, **174**, 282.
79. M. He, B. Han, Z. Jiang, Y. Yang, Y. Peng and W. Liu, *Carbohydr. Polym.*, 2017, **166**, 228.
80. J. Chang, W. Liu, B. Han, S. Peng, B. He and Z. Gu, *Wound Repair Regener.*, 2013, **21**, 113.
81. Z. Jiang, B. Han, H. Li, Y. Yang and W. Liu, *Carbohydr. Polym.*, 2015, **129**, 1.
82. H. Liu, J. Liu, C. Qi, Y. Fang, L. Zhang, R. Zhuo and X. Jiang, *Acta Biomater.*, 2016, **35**, 228.
83. R. J. Coleman, G. Lawrie, L. K. Lambert, M. Whittaker, K. S. Jack and L. Grøndahl, *Biomacromolecules*, 2011, **12**, 889.
84. R. V. Usoltseva, N. M. Shevchenko, O. S. Malyarenko, I. A. Ishina, S. I. Ivannikova and S. P. Ermakova, *Carbohydr. Polym.*, 2018, **180**, 21.
85. Q. Li, C. Li, C. Yang, C. Liu, G. Yu and H. Guan, *Int. J. Biol. Macromol.*, 2013, **62**, 281.
86. Q. Li, Y. Zeng, L. Wang, H. Guan, C. Li and L. Zhang, *Carbohydr. Polym.*, 2017, **155**, 313.
87. C. Kinnaert, M. Daugaard, F. Nami and M. H. Clausen, *Chem. Rev.*, 2017, **117**, 11337.
88. B. Yu, Q. Ouyang, C. Li and Y. Hui, *J. Carbohydr. Chem.*, 1996, **15**, 297.
89. C. Lamberth, J. O. Nagy, C. Kasper and M. D. Bednarski, *J. Carbohydr. Chem.*, 1994, **13**, 819.
90. H. Kuyama, Y. Nakahara, T. Nukada, Y. Ito, Y. Nakahara and T. Ogawa, *Carbohydr. Res.*, 1993, **243**, C1.
91. G. Blatter, J.-M. Beau and J.-C. Jacquinet, *Carbohydr. Res.*, 1994, **260**, 189.
92. O. Kanie, Y. Ito and T. Ogawa, *J. Am. Chem. Soc.*, 1994, **116**, 12073.
93. S. Yamago, T. Yamada, T. Maruyama and J. Yoshida, *Angew. Chem.*, 2004, **43**, 2145.
94. L. Huang, Z. Wang and X. Huang, *Chem. Commun.*, 2004, 1960.
95. L. Krock, D. Esposito, B. Castagner, C.-C. Wang, P. Bindschadler and P. H. Seeberger, *Chem. Sci.*, 2012, **3**, 1617.

96. F. C. Chi, S. S. Kulkarni, M. M. Zulueta and S. C. Hung, *Chem. - Asian J.*, 2009, **4**, 386.
97. J. Dinkelaar, L. J. van den Bos, W. F. Hogendorf, G. Lodder, H. S. Overkleeft, J. D. Codee and G. A. van der Marel, *Chem. - Eur. J.*, 2008, **14**, 9400.
98. L. J. van den Bos, J. Dinkelaar, H. S. Overkleeft and G. A. van der Marel, *J. Am. Chem. Soc.*, 2006, **128**, 13066.
99. S. L. Tang and N. L. Pohl, *Org. Lett.*, 2015, **17**, 2642.
100. J. D. C. Codée, L. J. van den Bos, A.-R. de Jong, J. Dinkelaar, G. Lodder, H. S. Overkleeft and G. A. van der Marel, *J. Org. Chem.*, 2009, **74**, 38.
101. Q. Zhang, E. R. van Rijssel, M. T. Walvoort, H. S. Overkleeft, G. A. van der Marel and J. D. Codée, *Angew. Chem.*, 2015, **54**, 7670.
102. O. J. Plante, E. R. Palmacci and P. H. Seeberger, *Science*, 2001, **291**, 1523.
103. M. T. Walvoort, H. van den Elst, O. J. Plante, L. Krock, P. H. Seeberger, H. S. Overkleeft, G. A. van der Marel and J. D. Codée, *Angew. Chem.*, 2012, **51**, 4393.
104. J.-i. Tamura, H. Tanaka, A. Nakamura and N. Takeda, *Tetrahedron Lett.*, 2013, **54**, 3940.
105. N. E. Ustyuzhanina, P. A. Fomitskaya, A. G. Gerbst, A. S. Dmitrenok and N. E. Nifantiev, *Mar. Drugs*, 2015, **13**, 770.
106. A. G. Gerbst, A. S. Dmitrenok, N. E. Ustyuzhanina and N. E. Nifantiev, *Mar. Drugs*, 2015, **13**, 936.
107. D. Z. Vinnitskiy, N. E. Ustyuzhanina, A. S. Dmitrenok, A. S. Shashkov and N. E. Nifantiev, *Carbohydr. Res.*, 2017, **438**, 9.
108. E. A. Khatuntseva, N. E. Ustuzhanina, G. V. Zatonskii, A. S. Shashkov, A. I. Usov and N. E. Nifant'ev, *J. Carbohydr. Chem.*, 2000, **19**, 1151.
109. A. Gerbst, N. Ustuzhanina, A. Grachev, E. Khatuntseva, D. Tsvetkov, D. Whitfield, A. Berces and N. Nifantiev, *J. Carbohydr. Chem.*, 2001, **20**, 821.
110. A. G. Gerbst, A. A. Grachev, N. E. Ustyuzhanina, E. A. Khatuntseva, D. E. Tsvetkov, A. I. Usov, A. S. Shashkov, M. E. Preobrazhenskaya, N. A. Ushakova and N. E. Nifantiev, *Russ. J. Bioorg. Chem.*, 2004, **30**, 137.
111. A. A. Grachev, A. G. Gerbst, N. E. Ustyuzhanina, A. S. Shashkov, A. I. Usov and N. E. Nifantiev, *J. Carbohydr. Chem.*, 2006, **25**, 315.
112. N. Nifantiev, N. Ustyuzhanina, V. Krylov, A. Grachev and A. Gerbst, *Synthesis*, 2006, **2006**, 4017.
113. Y. Hua, Y. Du, G. Yu and S. Chu, *Carbohydr. Res.*, 2004, **339**, 2083.
114. C. Zong, Z. Li, T. Sun, P. Wang, N. Ding and Y. Li, *Carbohydr. Res.*, 2010, **345**, 1522.
115. S. Arafuka, N. Koshiba, D. Takahashi and K. Toshima, *Chem. Commun.*, 2014, **50**, 9831.
116. A. Kasai, S. Arafuka, N. Koshiba, D. Takahashi and K. Toshima, *Org. Biomol. Chem.*, 2015, **13**, 10556.
117. J. Zhou, L. Yang and W. Hu, *J. Org. Chem.*, 2014, **79**, 4718.
118. V. V. Sova, M. S. Pesentseva, A. M. Zakharenko, S. N. Kovalchuk and T. N. Zvyagintseva, *Biochemistry*, 2013, **78**, 746.

119. B. Zhu, Y. Sun, F. Ni, L. Ning and Z. Yao, *Int. J. Biol. Macromol.*, 2018, **108**, 1140.
120. X. Zhu, X. Li, H. Shi, J. Zhou, Z. Tan, M. Yuan, P. Yao and X. Liu, *Mar. Drugs*, 2018, **16**, 30.
121. C. Ma, X. Lu, C. Shi, J. Li, Y. Gu, Y. Ma, Y. Chu, F. Han, Q. Gong and W. Yu, *J. Biol. Chem.*, 2007, **282**, 3747.
122. P. N. Collen, A. Jeudy, J. F. Sassi, A. Groisillier, M. Czjzek, P. M. Coutinho and W. Helbert, *J. Biol. Chem.*, 2014, **289**, 6199.
123. R. Ramachandra, R. B. Namburi, O. Ortega-Martinez, X. Shi, J. Zaia, S. T. Dupont, M. C. Thorndyke, U. Lindahl and D. Spillmann, *Glycobiology*, 2014, **24**, 195.
124. S. S. Deepa, Y. Umehara, S. Higashiyama, N. Itoh and K. Sugahara, *J. Biol. Chem.*, 2002, **277**, 43707.
125. K. Nakazawa and S. Suzuki, *J. Biol. Chem.*, 1975, **250**, 912.
126. H. Wang, W. He, P. Jiang, Y. Yu, L. Lin, X. Sun, M. Koffas, F. Zhang and R. J. Linhardt, *Glycoconjugate J.*, 2017, **34**, 643.
127. J. H. Hehemann, G. Correc, T. Barbeyron, W. Helbert, M. Czjzek and G. Michel, *Nature*, 2010, **464**, 908.
128. J. H. Hehemann, A. G. Kelly, N. A. Pudlo, E. C. Martens and A. B. Boraston, *Proc. Natl. Acad. Sci. U. S. A.*, 2012, **109**, 19786.
129. M. Li, Q. Shang, G. Li, X. Wang and G. Yu, *Mar. Drugs*, 2017, **15**, 92.
130. B. Yang, G. Yu, X. Zhao, G. Jiao, S. Ren and W. Chai, *FEBS J.*, 2009, **276**, 2125.
131. N. Gao, F. Lu, C. Xiao, L. Yang, J. Chen, K. Zhou, D. Wen, Z. Li, M. Wu, J. Jiang, G. Liu and J. Zhao, *Carbohydr. Polym.*, 2015, **127**, 427.
132. L. Zhao, M. Wu, C. Xiao, L. Yang, L. Zhou, N. Gao, Z. Li, J. Chen, J. Chen, J. Liu, H. Qin and J. Zhao, *Proc. Natl. Acad. Sci. U. S. A.*, 2015, **112**, 8284.
133. G. Yu, X. Zhao, B. Yang, S. Ren, H. Guan, Y. Zhang, A. M. Lawson and W. Chai, *Anal. Chem.*, 2006, **78**, 8499.
134. X. Zhao, G. Jiao, Y. Yang, M. Li, Q. Li, X. Wang, C. Cai, G. Li, J. Hao and G. Yu, *Carbohydr. Polym.*, 2017, **176**, 29.
135. R. Gao, N. Wu, Y. Li and S. Chen, *J. Carbohydr. Chem.*, 2014, **33**, 471.
136. Y. Lv, X. Shan, X. Zhao, C. Cai, X. Zhao, Y. Lang, H. Zhu and G. Yu, *Mar. Drugs*, 2015, **13**, 3710.
137. A. Osorio-Madrazo, L. David, S. Trombotto, J.-M. Lucas, C. Peniche-Covas and A. Domard, *Biomacromolecules*, 2010, **11**, 1376.
138. Q. Chen, W. Xiao, L. Zhou, T. Wu and Y. Wu, *Polym. Degrad. Stab.*, 2012, **97**, 49.
139. C. Hao, L. Gao, Y. Zhang, W. Wang, G. Yu, H. Guan, L. Zhang and C. Li, *Mar. Drugs*, 2015, **13**, 1267.
140. L. Gao, C. Li, S. Wang, X. Zhao and H. Guan, *Chin. J. Mar. Drugs*, 2013, **32**, 21.
141. H. Guan, *J. Med. Res.*, 1999, **28**, 8.
142. Y. Xue, L. Ren, S. Li, L. Wang, X. He, X. Zhao, G. Yu, H. Guan and C. Li, *Carbohydr. Polym.*, 2016, **144**, 330.
143. C. Li, Y. Sun and H. Guan, *Chin. Bull. Life Sci.*, 2012, **24**, 1019.

144. M. Xin, L. Ren, Y. Sun, H. Li, H. Guan, X. He and C. Li, *Eur. J. Med. Chem.*, 2016, **114**, 33.
145. J. Cummings, G. Lee, T. Mortsdorf, A. Ritter and K. Zhong, *Alzheimer's Dementia*, 2017, **3**, 367.
146. R. Jiang, X. Du, X. Zhang, X. Wang, D. Hu, T. Meng, Y. Chen, M. Geng and J. Shen, *Acta Pharmacol. Sin.*, 2013, **34**, 1585.
147. H. Zhao, H. Liu, Y. Chen, X. Xin, J. Li, Y. Hou, Z. Zhang, X. Zhang, C. Xie, M. Geng and J. Ding, *Cancer Res.*, 2006, **66**, 8779.
148. J. Hu, M. Geng, K. Chen, Z. Ran, J. Xu and H. Guan, *Chin. J. Mar. Drugs*, 2000, **19**, 18.
149. M. Yu, Y. Wang, F. Ma, W. Yu, T. Jiang and Z. Lv, *Curr. Pharm. Biotechnol.*, 2017, **18**, 391.

CHAPTER 13

Glycan Production by Bacterial Fermentation

CHRISTOPH ALBERMANN*

University of Stuttgart, Institute of Microbiology, Allmandring 31, 70569 Stuttgart, Germany
*E-mail: christoph.albermann@web.de

13.1 Introduction

Oligosaccharides, in free or in conjugated form, play an important role in many biological processes. The increasing interest in studying the effects of specific glycan structures in biological functions, like cellular interaction, communication, or pathogenesis, has stimulated research in the synthesis of complex oligosaccharide structures. Over recent decades a wide range of chemical and enzymatic methods have been developed for the synthesis of almost all known types of mammalian glycan structures. However, although the synthesis of apparently any oligosaccharide can be achieved, the procedures are often very laborious and expensive and therefore inappropriate for synthesis on a kilogram or ton scale.

For some oligosaccharides from human milk (Figure 13.1) their biological function and their effects on the development of infants have been intensively investigated. An overview of structure and occurrence and about the physiological properties of human milk oligosaccharides are given in several excellent review articles.[1-4] Due to several positive aspects of human milk oligosaccharides for the health and wellbeing of infants, the production of these oligosaccharides would be highly desirable.

Chemical Biology No. 11
Synthetic Glycomes
Edited by Wanyi Guan, Lei Li and Peng George Wang

Published by the Royal Society of Chemistry, www.rsc.org

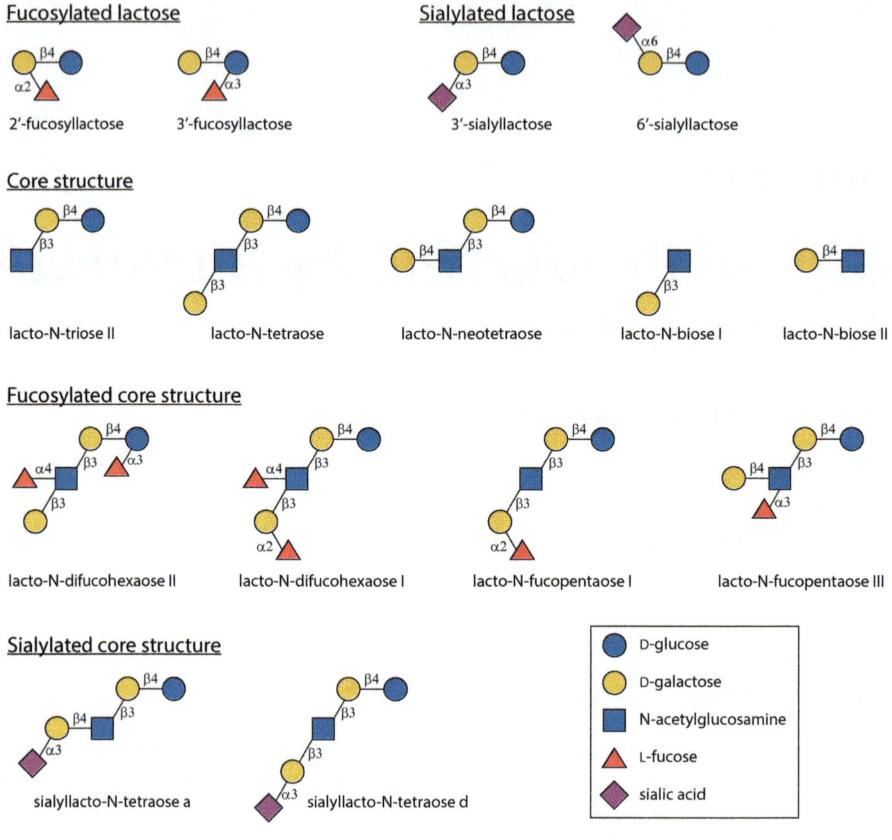

Figure 13.1 HMO core structures and fucosylated or sialylated lactoses.

For the large-scale formation of free oligosaccharides approaches *via* (chemo-)enzymatic synthesis or whole-cell biotransformation with recombinant bacterial cells have been described in the literature. This chapter focuses on only the *in vivo* synthesis of oligosaccharides. *In vivo* synthesis uses only one strain of growing cells in which all the required biosynthesis pathways are expressed. Necessary substrates that are supplemented to the culture are transported by suitable transporter proteins into the cells. This contrasts with other whole cell biotransformation approaches in which several permeabilized bacteria strains with different genotypes are applied. The basic process of the bacterial *in vivo* synthesis is outlined in Scheme 13.1.

13.1.1 Establishment of Metabolic Pathways for the Synthesis of Oligosaccharide in Bacteria

Beside chemical and enzyme-catalysed glycan synthesis the utilization of suitable recombinant bacterial host organisms constitutes another approach to gain access to complex oligosaccharide molecules. This approach is of

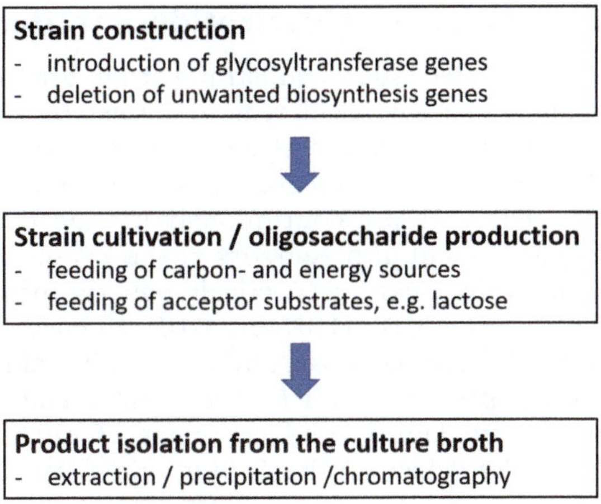

Scheme 13.1 General procedure of the glycan formation by *in vivo* synthesis.

particular interest for the large-scale synthesis of oligosaccharides, because it has the advantage over chemical synthesis that no protecting groups are needed and the advantage over the enzymatic synthesis with Leloir glycosyltransferases that necessary nucleotide-activated donor substrates are produced and regenerated by the metabolism of the bacterial strain.

But since no native bacterial strain is known to synthesize human milk oligosaccharides or other unconjugated glycans, the necessary biosynthesis pathways must be established in a suitable bacterial host. By the recombinant expression of biosynthesis enzymes of nucleotide-activated sugars together with suitable glycosyltransferases in non-pathogenic *Escherichia coli* strains, for example, the synthesis of several oligosaccharides has been demonstrated.[5–8]

Because of the large variety of tools for genetic engineering and the simple and fast cultivation, *E. coli* is currently the preferred host organism for the *in vivo* synthesis of oligosaccharides.

To allow the synthesis of oligosaccharides in bacterial cells the following requirements need to be fulfilled. (1) The expression of an active Leloir-glycosyltransferase corresponding to the available donor and acceptor substrates. Since most bioactive oligosaccharides are found in mammalian cells, the corresponding glycosyltransferase genes are the first choice for the heterologous biosynthesis. However, the expression of eukaryotic proteins in bacteria often failed to produce functional enzymes. Many pathogenic Gram-negative bacteria possess LPS *O*-antigen structures that are similar to mammalian glycans. Therefore, the utilization of the genes of glycosyltransferases from the *O*-antigen biosynthesis are another suitable source for the heterologous synthesis by *E. coli* strains. (2) The intracellular availability of an acceptor substrate in *E. coli*. The acceptor can either be an exogenous substrate, *e.g.* lactose, or an endogenous substrate, *e.g. N*-acetylglucosamine. (3) The synthesis of an activated donor substrate for a glycosyltransferase in the cytosol of the cell.

13.1.2 *In Vivo* Synthesis of Nucleotide-activated Sugars

The nucleotide-activated sugar substrates that are needed for the Leloir glycosyltransferase catalyzed reaction must be provided by the bacterial metabolism. In *E. coli*, the glycosyltransferase substrates, UDP-D-galactose (UDP-Gal), UDP-D-glucose (UDP-Glc) and UDP-*N*-acetyl-D-glucosamine (UDP-GlcNAc), are continuously produced by the growing cell, because they are essential precursors of the LPS-core structure and of the bacterial peptidoglycan, respectively. On the other hand, UDP-*N*-acetyl-D-galactosamine (UDP-GalNAc), GDP-L-fucose (GDP-Fuc), the nucleotide-activated form of L-fucose, and CMP-sialic acid (CMP-Neu5Ac) are not permanently present in all *E. coli* strains. The following paragraphs describe the availability and synthesis of nucleotide activated sugar substrates for the *in vivo* synthesis of mammalian oligosaccharides in bacterial cells.

GDP-Fuc—In *E. coli* strains GDP-Fuc is a natural intermediate of the colanic acid biosynthesis,[9,10] but the colanic acid biosynthesis pathway, and therefore the formation of GDP-Fuc, is only turned on under certain environmental conditions.

Therefore, an efficient *in vivo* synthesis of fucosylated oligosaccharides by *E. coli* requires an increased intracellular formation of this activated deoxyhexose.

In nature two enzymatic pathways exist for the formation of GDP-Fuc, one is the *de novo* pathway that is present in colanic acid producing *E. coli* strains. The *de novo* pathway starts from the central metabolite fructose-6-phosphate, which is converted by five enzymatic steps (ManA phosphomannose isomerase, ManB mannose-6-phosphate mutase, ManC mannose-1-phosphate guanylyltransferase, Gmd GDP-D-mannose 4,6-dehydratase, WcaG/Fcl GDP-Fuc synthase) to GDP-Fuc (see Figure 13.2). The second pathway to GDP-Fuc consists of the direct activation of L-fucose *via* the bifunctional enzyme L-fucose 1-kinase/GDP-fucose pyrophosphorylase (Fkp). This enzyme has been found in a variety of eukaryotic organisms, but so far only one bacterial strain, *Bacteroides fragilis* carries a *fkp* gene.[11,12] Cloning and overexpression of the *fkp* gene from *B. fragilis* in *E. coli* results in the conversion of intracellular L-fucose into GDP-Fuc. To allow the uptake and to prevent the degradation of exogenous L-fucose by *E. coli* the increased expression of the L-fucose permease gene (*fucP*) and the disruption of the L-fucose isomerase gene (*fucI*) and the L-fuculose kinase (*fucK*) is required.[13–15]

To generate an intracellular GDP-Fuc concentration by the *de novo* pathway that allows the formation of fucosylated oligosaccharides *via* a recombinant fucosyltransferase (FucT) two approaches have been successfully used. On the one hand the heterologous overexpression of the genes *manB*, *manC*, *gmd*, and *wcaG* using plasmids or chromosomally integrated expression cassette and on the other hand by inactivation of the *wcaJ* gene and overexpression of

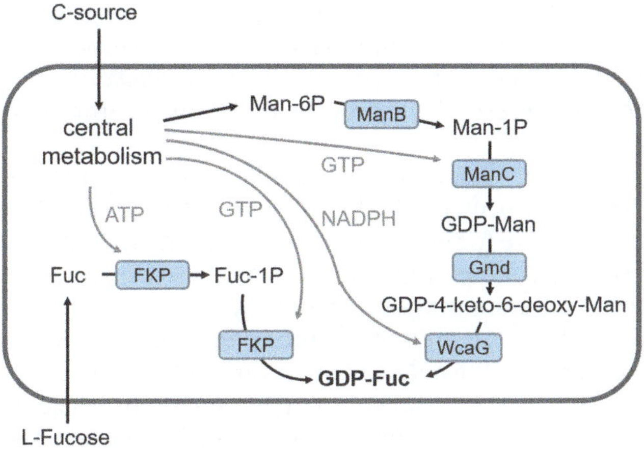

Figure 13.2 Bacterial pathways for the synthesis of the donor-substrate GDP-Fuc. *De novo* pathway as occurring in *E. coli* K-12. GDP-Fuc is synthesized from mannose-6-phosphate *via* four enzymatic steps using GTP and NADPH. Salvage synthesis *via* a bifunctional L-fucose kinase/L-fucose-1-phosphate guanylyltransferase identified in *B. fragilis* using ATP and GTP Abbreviations: Fkp, bifunctional L-fucose kinase/L-fucose-1-phosphate guanylyltransferase; Fuc, L-fucose; GDP, guanosine 5′-diphosphate; GDP-Fuc, guanosine 5′-diphospho-β-L-fucose; Gmd, GDP-mannose 4,6-dehydratase; GTP, guanosine 5′-triphosphate; Man, α-D-mannose; ManB, phosphomannomutase; ManC, mannose-1-phosphate guanylyltransferase; WcaG, GDP-Fuc synthase.

the positive regulator gene, *rcsA*, of the colanic acid biosynthesis. The disruption of the undecaprenyl-phosphate glucose phosphotransferase gene *wcaJ* prevents the formation of the fucose containing colonic acid.[14,16,17]

An increase of the intracellular synthesis of GDP-Fuc by the *de novo* pathway has, in addition, been achieved *via* the enhanced expression of the guanosine-inosine kinase (Gsk) and the increased supply of NADPH, respectively.[18,19]

Besides *E. coli* a recombinant GDP-Fuc *de novo* pathway has also been established in the yeast strain *Saccharomyces cerevisiae* and in the Gram-positive bacterium *Corynebacterium glutamicum*.[20,21]

UDP-Glc—Since UDP-Glc is a precursor molecule of the cell wall biosynthesis almost all bacteria have the ability to synthesize this donor substrate. So far, UDP-Glc has not directly been used in the *in vivo* synthesis of oligosaccharides, but as precursor in the intracellular supply of UDP-Gal and UDP-D-glucuronic acid.[22]

UDP-GlcNAc—like UDP-Glc, UDP-GlcNAc is a precursor of the cell wall biosynthesis in bacteria. Several experiments have utilized the intracellular UDP-GlcNAc supply for the *in vivo* synthesis of oligosaccharide.[23,24] Studies to increase the intracellular UDP-GlcNAc supply for the synthesis of oligosaccharides have so far not been described in literature.

UDP-GalNAc—For the *in vivo* synthesis this donor substrate could be synthesized in *E. coli* by the epimerization of UDP-GlcNAc using recombinant UDP-GlcNAc 4-epimerase (WbpP from *Pseudomonas aeruginosa* O6). UDP-GalNAc has successfully been utilized for the synthesis of ganglioside moiety.[25]

UDP-Gal—like UDP-Glc, UDP-Gal is a precursor of the LPS biosynthesis of many Gram-negative bacteria, including *E. coli* K-12 strains. The generation of this donor substrate for *in vitro* synthesis approaches has been intensively studied. Metabolic engineering studies to increase the intracellular concentration UDP-Gal have not yet been described.

CMP-sialic acid—Compared to the other nucleotide-activated sugars that are presented here, CMP-Neu5Ac is quite rare in bacteria. Only a few pathogenic strains like *Neisseria meningitidis* or *E. coli* K1 synthesize sialic acid containing glycans. To achieve the *in vivo* synthesis of CMP-Neu5Ac in *E. coli* K-12 strains, Neu5Ac can be synthesized by the conversion of UDP-GlcNAc using the enzymes NeuC and NeuB. Additionally, Neu5Ac can also be taken up from the medium *via* the transporter protein NanT. The activation of Neu5Ac is catalysed by the CMP-sialic acid synthase (NeuA) in the presence of CTP.[26]

The following sections of this chapter provide an overview of the *in vivo* synthesis of oligosaccharides with recombinant *E. coli* strains and will discuss the advantages and disadvantages of this approach.

13.2 *In Vivo* Synthesis of Oligosaccharides by *E. coli*

13.2.1 Fucosylated Oligosaccharides: 2′-Fucosyllactose (2′-FL) and 3-Fucosyllactose (3-FL)

2′-FL and 3-FL are the most abundant neutral oligosaccharides found in human milk.[27–33] It has been demonstrated that 2′-FL provides a health benefit to infants by its prebiotic and anti-adhesive activities.[34,35] As a result the production of fucosyllactose for nutraceutical and pharmaceutical purposes has gained much interest. In recent years, several groups have reported the synthesis of 2′- and 3-FL production with recombinant *E. coli* cells.[14,15,17,36–38]

The production of fucosyllactose in *E. coli* K-12 depends on the intracellular concurrent presence of lactose, GDP-Fuc, and the introduction of the corresponding FucT activity. A native α1–2-FucT or α1–3-FucT that accepts lactose as substrate has not been described in *E. coli* K-12 or *E. coli* B strains. These three components needed for the synthesis of fucosyllactose will be discussed.

Lactose is a main side product of the dairy industry and thus is an inexpensive substrate for biochemical applications.[5] The intestinal bacterium *E. coli* can take up lactose from the medium by the galactoside proton symporter, LacY.[39] The cytoplasmatic degradation of lactose is catalysed by the β-galactosidase, LacZ. In order to allow the intracellular synthesis of oligosaccharides with lactose as acceptor molecule in bacterial cells,

LacZ-activity has to be eliminated while the lactose uptake must be retained or even enhanced;[14,37,40,41] LacZ− LacY+ strains are therefore preferred as producer strains. Bacteria with residual β-galactosidase activity as in commercial λDE3 lysogen *E. coli* strains that carry the T7 RNA polymerase gene under control of the lacUV5 promoter, lead to a fairly reduced yield of fucosylated lactose.[13,38]

GDP-Fuc is a natural intermediate of colanic acid biosynthesis in *E. coli*,[9,10] but for the production of fucosylated oligosaccharides its synthesis has to be detached from the competing colanic acid biosynthesis. As mentioned in Section 13.1.2, GDP-Fuc is synthesized in nature by two different pathways, the *de novo* pathway and the salvage pathway (existing in mammalian cells and in some *Bacteroides* strains) (Figure 13.2). However, the salvage pathway requires the supplementation of L-fucose. This more rare and expensive sugar can be activated by the bifunctional Fkp enzyme L-fucose 1-kinase/GDP-fucose pyrophosphorylase, from *B. fragilis*.[11,12] Expression of the *fkp* gene from *B. fragilis* in *E. coli*, uptake of exogenous L-fucose *via* a permease (FucP) and mutations to prevent its degradation (Δ*fucIK*) result in an intracellular formation of GDP-Fuc.[13–15]

If the acceptor lactose and the donor GDP-Fuc are available in the cytoplasm the expression of a suitable Leloir glycosyltransferase is needed to yield a fucosylated lactose. By the recombinant expression of exogenous FucTs, like *futC* gene for an α1–2-FucT or *futA* for an α1–3-FucT from *Helicobacter pylori*, the *in vivo* synthesis of 2′-FL or 3-FL has been achieved in sufficient yields.[13,14,18,37,38] For this, the formation of GDP-Fuc was either reached *via* the *de novo* and/or the salvage pathway. However, the cultivation of strains that produce GDP-Fuc by the salvage pathway from exogenously supplied L-fucose resulted in a two-fold higher yield of 2′-FL compared to the use of the *de novo* pathway only.[14] The reason for this was found in the intracellular concentration of GDP-Fuc. *E. coli* cells showed a five times higher supply of GDP-Fuc, due to the expression of the *de novo* and the salvage pathway.[14] For the formation of GDP-Fuc by the *de novo* pathway, it is known that the GDP-D-mannose 4,6-dehydratase (Gmd) is strongly inhibited by the end product GDP-Fuc.[42] This shows that the supplies of the intracellular substrates can significantly influence the yield of the produced glycan. In comparison, the variation in the number of copies of the recombinant α1–2-FucT genes had little influence on the 2′-FL yield.[14]

Chin and co-workers[13] used both the *de novo* pathway and the salvage pathway for the *in vivo* synthesis of 2′-FL. In this case, the genes *fucI* and *fucK* for the L-fucose degradation were knocked out in the BL21Star(DE3) *lacZ*ΔM15-*lacYA* strain that carried both the *fkp* gene and the α1–2-FucT gene *wcfB* from *B. fragilis* together on the same expression plasmid. Used in batch fermentation with glycerol as the main carbon and energy source, this strain produced 2.08 g L^{-1} of 2′-FL from 1.11 g L^{-1} of L-fucose and 4.16 g L^{-1} of lactose consumed.[13] In fed-batch, this strain yielded 20.6 g L^{-1} of 2′-FL (27 g L^{-1} of L-fucose and 53.6 g L^{-1} of lactose consumed). Possibly due to the partially restored LacZ-activity in the *lacZ*ΔM15 mutant in conjunction with the (DE3) system, lactose was still consumed. An improved strain (*lacYA* genes in a clean Δ*lacZ* background) gave a maximal 2′-FL concentration of 23.1 g L^{-1} (consumption of 21.8 g L^{-1} of L-fucose and 44 g L^{-1} of lactose).[13] These 2′-FL titers are in the same range as a previous report on 2′-FL *via* the *de novo* route.[14]

Previously, the group of Seo based their approach on lactose feeding (the use of a lysogeny broth growth medium presumably was responsible for GDP-Fuc provision) and overexpression of the genes *manB*, *manC*, *gmd*, and *wcaG* in the *de novo* pathway. Maximal 2′-FL concentrations of 1.23 g L^{-1} (0.9 g 2′-FL per g of lactose) were reached in batch fermentations with a JM109 strain (lacking LacZ activity).[37]

The first attempts for the *in vivo* synthesis of fucosylated oligosaccharides relied on the plasmid-based expression of heterologous genes for the GDP-Fuc pathway and the FucT.[37,40,41] The use of high copy expression plasmids with strong promoters may be the best choice for the cloning and the short-term expression of recombinant genes, especially for the maximum overexpression of a given protein. In metabolic engineering applications, however, the maximum overexpression of one or a few proteins, may be unfavourable for long-term productivity.[43] The chromosomal integration of heterologous expression cassettes can be favourably compared to high-copy plasmids in terms of metabolic burden effects, structural instability, and segregational instability.[44] A stable chromosomal insertion also obviates the use of selection-markers, like antibiotics, that are commonly used for the maintenance of plasmids during cultivation. Especially since antibiotics are both costly and can hamper the product purification in food and pharmaceutical production processes.

Stable chromosomal integration of recombinant expression cassettes is a solution to avoid the use of plasmids as gene vectors and of antibiotics for maintenance of these plasmids.

The synthesis of 2′-FL in a plasmid free production strain (*E. coli* JM109 Δ*lacZ*, *lacY*+) has been established with an extra set of the *manB*, *manC*, *gmd* and *wcaG* genes of the *de novo* pathway which were placed as single or double gene cassettes into the chromosome; the expression of each gene was under the control of the IPTG-inducible P_{tac} promoter system.

The gene of the α1–2-FucT *futC* (*H. pylori*) was also integrated under a controlled *tac*-promoter. The growth of this strain using shake flask experiments with mineral medium containing glycerol as a carbon and energy source and a supply of lactose resulted in the formation of about 200 mg 2′-FL per g of cell dry weight (cdw). The insertion of a second chromosomal copy of *futC* led to a moderate increase of the yield (about 250 mg g^{-1} of cdw). This strain was then chosen for a fed-batch fermentation (final volume of 13.5 liters) with glycerol as carbon and energy source. Under this controlled fermentation condition the concentration of 2′-FL reached 20.28 g L^{-1} (total amount: 272 g). The 2′-FL formation per g of cell dry weight yielded 309 mg g^{-1}; space-time yield was 0.57 g L^{-1} h^{-1}. The synthesized 2′-FL was almost equally distributed between culture broth and the cell pellet,[14] this could suggest an efflux system in *E. coli* which transports 2′-fucosylactose from the cytoplasm into the culture medium. These values of oligosaccharide production might be even higher when a *fkp* gene for the salvage pathway is expressed (see above), but the required additional feed of L-fucose would be too costly for the given scale of fermentation.

Seo and co-workers used a strain of the *E. coli* B series (BL21) and engineered its *lac* genotype by first disrupting the chromosomal *lacZYA* operon and then by site-specific transposition of Tn7-*lacZ*ΔM15-*lacYA* into the chromosome.[36] For the expression of the GDP-Fuc *de novo* pathway genes and for the FucT gene *fucT2* a plasmid-based expression system had been used, respectively. To increase the solubility of the recombinant FucT an engineered *fucT2* gene leading to three N-terminally added aspartate residues was expressed.[36] In a cultivation with complex medium and lactose feeding in batch mode, the best BL21-based strain produced less 2′-FL (0.34 g L^{-1}) than a JM109-based strain (1.23 g L^{-1}). In a glycerol-based fed-batch fermentation and with lactose feeding, 2.6 g L^{-1} of 2′-FL was reached (0.063 g per g of lactose consumed).[36]

A more recent report used a modular approach to investigate the formation of 2′-FL and 3-FL with recombinant *E. coli* strains.[18] To optimize the production, the expression of the GDP-Fuc *de novo* pathway (plasmid-encoded *manA*, *manB*, *manC*, *gmd*, and *wcaG* genes) was combined with various genes encoding FucTs (from different species or subspecies of *Helicobacter*, *E. coli* and *Bacteroides*). The experiments were carried out both with an *E. coli* JM109(DE3)-derivative and with an *E. coli* BL21(DE3)-derivative. The highest 2′-FL production was gained by an engineered *futC* gene from *H. pylori* (0.37 g L^{-1} in batch with lactose feeding). For 3-FL, *futA* (also from an *H. pylori* origin) yielded 0.7 g L^{-1} which agrees with a former study;[17] *futC* and *futA* were chosen for further studies in a BL21(DE3)Δ*lacZ* strain. To increase the uptake of the acceptor substrate lactose into the cell, the expression of *lacY* was enhanced by the insertion of another gene copy (gene transcription was raised 11-fold). As a result, the 2′-fucosyllatose titer increased to 0.83 g L^{-1} (*futC*$^{+}$ strain) and the titer of 3-FL was 1.59 g L^{-1} (*futA*$^{+}$ strain).[17] An additional supply of GDP-Fuc to the culture both increased the formation of 2′-FL by about 30% and of 3-FL by about 23%. This again demonstrates the limiting supply of this nucleotide sugar by the recombinant *E. coli* cells. However, an exogenous supply of GDP-Fuc to the growth medium is, due to the high price of GDP-Fuc, not economical.

To further increase the supply of GDP-Fuc various genetic modifications were studied for their effects. Positive effects (alone or in combination with other genetic changes) were exerted by the deletion of the *lon* gene (the Lon protease degrades, among others, the positive regulator RcsA), overexpression of *rcsA*, and deletion of the *wcaJ* gene. Several modifications in the central metabolism especially those for improved NADPH production, were tested. Among combinations of these genes the one with *zwf-pntAB* resulted in the highest fucosyllactose titer. The most developed *E. coli* strain had the following genetic features: BL21Δ*lacZ*Δ*lon*Δ*wcaJ*–/+*manA,B,C,gmd*/*lacY*/*rcsA*/*futA*/*zwf-pntA*, with genes on four different plasmids. Finally, culture conditions such as medium composition and inducer were optimized. On a minimal medium with glucose and feeding with 12 g L^{-1} of lactose in batch mode, the highest titers were 9.12 g L^{-1} 2′-FL and 12.43 g L^{-1} 3-FL, respectively.[17]

From the metabolic engineering approaches that have been studied to increase the formation of fucosyllactoses it can be concluded that the supply of the donor-substrate GDP-Fuc is the most limiting step of the *in vivo* synthesis. The highest product titers are obtained by the use of the GDP-Fuc salvage pathway. However, the salvage pathway requires the supplementation of rare and expensive L-fucose to the culture.

13.2.2 Lacto-*N*-Tetraose (LNT), Lacto-*N*-neotetraose (LNnT) and Modified Derivatives Thereof

In contrast to the synthesis of fucosyllactose, in which only one glycosyltransferase and one activated donor substrate is needed, the *in vivo* synthesis of more complex glycans requires a coordinated sequence of several Leloir-glycosyltransferase catalysed reactions. In the case of the formation of human milk oligosaccharide core structures, LNT or LNnT (see Figure 13.1), the glycosyl transfer of *N*-acetylglucosaminyl and galactosyl moieties to the acceptor substrate, lactose, is required in the cytoplasm of recombinant bacteria. *E. coli* cells synthesize UDP-Gal and UDP-GlcNAc intracellularly during cell growth for the formation of the cell wall and the LPS from their central metabolic pathway. But *E. coli* K-12 cells do not express β1–3-*N*-acetylglucosaminyltransferase (GlcNAcT) and do not have β1–3- or β1–4-galactosyltransferase (GalT) activities. Therefore, genes like *lgtA* from *N. meningitidis* which encode this function (transfer of GlcNAc from UDP-GlcNAc onto lactose), were expressed from plasmids in β-galactosidase-negative *E. coli* strains. The cultivation of such strain in the presence of lactose, led to the formation and excretion of the trisaccharide lacto-*N*-triose II (LNT II, GlcNAcβ1–3Galβ1–4Glc) reaching up to 6 g L^{-1}.[40] Co-expression of *lgtB* from *N. meningitidis* (encoding a β1–4-GalT) in this strain then resulted in the formation of LNnT (Figure 13.1) when grown on glucose and fed with lactose. An earlier report had described that glycerol could not be used as a carbon source if a two plasmid system with *lgtA* and *lgtB* was used.[40]

In contrast to the precursor molecule LNT II the synthesized LNnT was detected only in the intracellular fraction of the culture, showing that this tetrasaccharide is not secreted by the *E. coli* cells. The intracellular accumulated LNnT, additionally served also as an acceptor for another GlcNAc-transfer reaction by LgtA to give a pentasaccharide which again is a substrate for another round of galactosyltransfer (LgtB) as lacto-*N*-neohexaose and even lacto-*N*-neooctaose were detected.[40] By the additional expression of a glucuronyltransferase (catalytic domain of the mouse glucuronyltransferase

GlcAT-P) and of a UDP-glucose dehydrogenase (*kfiD* from *E. coli* K5) in a LNnT producing *E. coli* strain, LNnT and its LacNAc extended version can be further decorated by glucuronic acid and this results in the formation of the nonsulfated HNK-1 moiety.[22]

LNT is one of the predominant oligosaccharides in human milk.[45] So far, it has been available by extraction from human milk or by chemical or enzymatic synthesis.[46,47] Baumgärtner and co-workers[23] described the first heterologous *in vivo* synthesis of LNT. This was achieved by the recombinant expression of the chromosomally inserted genes *lgtA* (synthetic gene and codon-optimized for *E. coli* expression) and *wbgO* (encoding the regio- and stereo-selective β1–3-GalT from *E. coli* O55:H7).[48] Chromosomal gene integration allows fermentations to be performed without the use of antibiotics or other selection markers that are undesired in the food industry.[23] Based upon an *E. coli* K-12 (from the W3110 series) which was *lacZA*-negative by the insertion of a copy of *lgtA*, a single copy of *lacY* was inserted into the fucose-degradation operon (*fucIK* locus). When grown on glucose or glycerol (each at 10 g L^{-1}) and fed with lactose (2 g L^{-1}) in shake flasks experiments, the strain formed the trisaccharide, LNT II. The titer was 1.9 g L^{-1} on glucose (10 g L^{-1}) and 2.46 g L^{-1} on glycerol (10 g L^{-1}) and LNT II was detected predominantly in the medium of the culture. The LNT II yield on lactose was 953 mg g^{-1} lactose (on glucose) and 1232 mg g^{-1} lactose (on glycerol). After introduction of the β1–3-GalT gene *wbgO*, two products were found in shake flask cultivations: LNT II was still dominant (1.89 mg and 2.09 mg on glucose and glycerol, respectively) and another oligosaccharide LNT was detected (219 mg on glucose and 162 mg on glycerol). LNT was found both intra- and extracellularly. From these results it can be deduced that the engineered *E. coli* host strains can supply both nucleotide-activated precursors, UDP-GlcNAc (for LNT II) and UDP-Gal (for LNT),[23] see Figure 13.3. However, the low yield of LNT in relation to the supplied amount of lactose (only *ca.* 5% of lactose was converted into LNT) pointed to either a limitation in WbgO activity, a deficient quantity of LNT II in the cytosol of the cell, or to an insufficient UDP-Gal supply. Since UDP-Gal is an essential intermediate of the galactose degradation pathway (Leloir pathway), it was studied whether supply of D-galactose can increase the availability of UDP-Gal and thereby boost the production of LNT. Indeed, the exogenous supply of D-galactose (2 g L^{-1}) to cells growing on glycerol (10 g L^{-1}) and supplemented by lactose (2 g L^{-1}) led to an increased formation of LNT while LNT II concentration was roughly unchanged. A combination of galactose and glucose had no significant improvement in LNT formation (lactose consumption and LNT II formation were even lower). The highest yields of LNT were reached in batch cultures with D-galactose (10 g L^{-1}) alone. While LNT II titers were around 1.5 g L^{-1}, LNT titers were 0.81 g L^{-1}.[49] In contrast to the *in vivo* synthesis of LnNT, more than 90% of the produced LNT was excreted from the cells if D-galactose was used as the sole carbon and energy source. The cytosolic concentrations of UDP-Gal were about

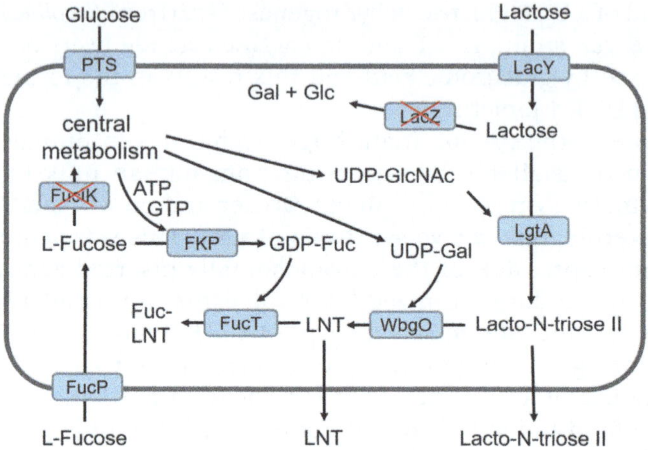

Figure 13.3 *In vivo* synthesis of fucosylated LNTs using Leloir glycosyltransferases and intracellularly generated nucleotide-activated sugars. Utilization of different FucTs leads to differently fucosylated LNT and elongated core structures (summarized as LNFX). Abbreviations: Fkp, bifunctional L-fucose kinase/L-fucose-1-phosphate guanylyltransferase; Fuc, L-fucose; FucP, L-fucose permease; FucT, fucosyltransferase; LacY, lactose permease; LgtA, β1–3-GlcNAcT; Fuc-LNT, fucosylated core structures with various lengths and kinds of fucosylation; PTS, bacterial phosphoenolpyruvate-dependent sugar phosphotransferase (glucose uptake system); WbgO, β1–3-GalT.

three times higher in galactose-grown cells then in cultures that were grown on glucose or glycerol only. This confirms the hypothesis that the availability of UDP-Gal is, in this case, limiting for the WbgO catalysed reaction. Using fed-batch conditions (with a final biomass of about 55 g L^{-1} cdw), cells were grown on D-galactose and were supplemented with lactose (total feed of 14.7 g L^{-1}). The supplied lactose was completely converted into LNT II and LNT resulting almost equal amounts (LNT: 12.7 g L^{-1}; LNT II: 13.7 g L^{-1}). A total amount of 173 g of LNT was produced; about 89% of LNT II and 65% of LNT were detected in the supernatant of the culture. Although the increase of the UDP-Gal supply enhanced the conversion of LNT II into LNT, the WbgO catalysed reaction was still not sufficient to convert LNT II completely. It remains to be examined whether a higher β1–3-GalT activity could lead to a faster conversion of LNT II into LNT before LNT II is secreted from the cell.

Beside the main products of this cultivation, LNT and LNT II, small amounts (*ca.* 1.9 g L^{-1}) of the disaccharide GlcNAcβ1–3Gal could be isolated from the culture broth. Most likely this glycan is a side product of the reaction of LgtA on galactose.[49] The use of D-galactose as acceptor substrate for the *in vivo* synthesis of glycans has already been demonstrated by the use of galactose kinase (GalK) negative *E. coli* strain.[16]

The LNnT and LNT moieties constitute the core structures of many mammalian glycans. In the milk of so-called secretor women the dominant human milk oligosaccharides, beside 2′-FL and LNT, were the fucosylated lacto-*N*-fucopentaose I (LNFP I) and lacto-*N*-difucohexaose I (LNDFH I) (see Figure 13.1).[45] To obtain such complex glycans by *in vivo* synthesis the engineering of LNnT or LNT producing *E. coli* cells by additional glycosyltransferases is required. A recombinant *E. coli* strain (*lacZ* negative, producing GDP-Fuc *via* a *de novo* pathway, expressing *futC* from *H. pylori* and *lgtA*, *lgtB* of *N. meningitidis*) resulted 3.0 g L^{-1} of lacto-*N*-neofucopentaose V (LNnFP V) when grown on glucose and fed with lactose in a fed batch bioreactor cultivation.[41] The lacto-*N*-neofucopentaose is also known as H antigen, a constituent of the ABO histo-blood group system.[38]

The formation of fucosylated LNT structures had so far only been demonstrated by the enzymatic synthesis using *H. pylori* α1–2- or α1–3-FucT.[50–52] The *in vivo* synthesis of fucosylated LNT derivatives in *E. coli* strains could be achieved by the cloning of a FucT gene and expression of the GDP-Fuc salvage pathway in LNT producing cells. Expression of chromosomally integrated *lgtA* and *wbgO* genes resulted in the formation of LNT II and LNT. In addition, *fkp* gene and genes for different FucTs (either *futC* for an α1–2-FucT or *fucT*14 for a codon-optimized α1–3/4-FucT,[53] both from *H. pylori*) were chromosomally expressed.[49] To avoid L-fucose degradation, the *fucIK* genes were deleted. In shake flasks with 10 g L^{-1} of glucose and 2 g L^{-1} of lactose, LNT II and LNT were formed.[54] L-fucose was added after lactose consumption to avoid the fucosylation of lactose by FucTs. The intracellular formation of LNFP I was demonstrated when *futC* from *H. pylori* was expressed and only if L-fucose was fed. In shake flask experiments, a mixture of products was obtained (2′-FL (265 mg L^{-1}), LNT II (1578 mg L^{-1}), LNT (63 mg L^{-1}), and LNFP I (271 mg L^{-1})), demonstrating the limitation in activity and specificity of the FutC catalysed fucosylation reaction.

The expression of an α1–3/4-FucT gene (with L-fucose addition as in the case of *futC*) resulted in the formation of novel fucosylated compounds (see Figure 13.3) in the cytosolic fraction of the shake flask culture. One product was LNDFH II, a difucosylated LNT molecule with fucosyl groups on the glucose (α1–3) and the *N*-acetylglucosamine (α1–4) moiety of the core structure. The second product was a tetrasaccharide (fucosylated LNT II derivative) with α1–3-bound fucosyl group on the glucose moiety. The most elongated oligosaccharide that could be detected (about 5 mg L^{-1}) was a difucosylated lacto-*N*-pentasaccharide with the fucosyl groups attached as in LNDFH II but with an extra GlcNAc moiety at the non-reducing end. Product concentrations were 547 mg L^{-1} of LNDFH II, 817 mg L^{-1} for the fucosylated LNT II, 60 mg L^{-1} for LNT, and 1046 mg L^{-1} for LNT II. The observation that fucosyl groups are α1–3- and α1–4-attached shows the significant α1–3-FucT activity of the *fucT*14 gene product from *H. pylori*.[54]

13.2.3 Sialyllactoses and Other Sialylated Oligosaccharides

Sialylated oligosaccharides contain a neuraminyl moiety that derives from the activated donor substrate CMP-Neu5Ac. For the heterologous formation of sialylated oligosaccharides, like sialyllactose, CMP-Neu5Ac must be provided as most bacteria, except for pathogenic bacteria, are unable to synthesize sialic acid (Neu5Ac).[55]

Samain and co-workers demonstrated the first production of 3′-sialyllactose (3′-SL) by a whole cell production within one recombinant *E. coli* K-12 strain.[40] The formation of CMP-Neu5Ac had been achieved by feeding Neu5Ac to *E. coli* cells which expressed the gene for a specific Neu5Ac transporter, NanT (see Figure 13.4). In order to avoid the intracellular degradation of Neu5Ac by the sialic acid aldolase (NanA), the gene for this enzyme was knocked out. Overexpression of the recombinant CMP-NeuAc synthase gene (*neuA*) from *N. meningitidis* resulted in the formation of CMP-Neu5Ac. Following the direct activation of L-fucose by Fkp, this formation of CMP-Neu5Ac is also called "salvage pathway" (Figure 13.4).[56]

The additional expression of a α2–3-sialyltransferase from *N. meningitidis* (NanT) in CMP-Neu5Ac providing cells allows for the sialylation of the acceptor substrate lactose. Lactose and sialic acid supplementation to a glycerol-grown fed-batch cultivation of *E. coli* JM107-derived cells (LacY+, LacZ−, NanT+, NanA−) resulted in a release of up to 1.1 g L^{-1} 3′-SL into the culture medium.[40]

To avoid the feeding of expensive starting material, sialic acid, the engineering of *E. coli* cells expressing the *de novo* pathway was described by the group of Samain. In *E. coli* K-12 cells the heterologous pathway starts from UDP-GlcNAc, an indispensable substrate of *E. coli*'s cell wall biosynthesis. A recombinant UDP-GlcNAc epimerase (gene *neuC*), that facilitates the formation of ManNAc, can be cloned from pathogenic *E. coli* K1 or *Neisseria* species. The sialic acid synthase (gene *neuB*) catalyses the reaction of ManNAc and phosphoenolpyruvate (PEP) to Neu5Ac (see Figure 13.4). To avoid the degradation of Neu5Ac by *E. coli*'s ManNAc kinase and Neu5Ac aldolase activities, the corresponding genes (*nanA, nanK*) need to be deleted. For the *in vivo* synthesis of CMP-Neu5Ac the genes *neuABC* from *Campylobacter jejuni* and the *nst* gene for an α2–3-sialyltransferase of *N. meningitidis* were plasmid-based cloned into these appropriately engineered *E. coli* K-12 strains. The strain was cultivated on glycerol and was fed continuously with the acceptor substrate lactose. The fermentation resulted in the extracellular accumulation of up to 15.5 g L^{-1} of 3′-SL. The total 3′-SL titer of the fed-batch culture reached a concentration of 25.5 g L^{-1}.[57] This demonstrates a ten times higher formation by the utilization of the *de novo* pathway over the CMP-Neu5Ac salvage pathway.[40]

Besides the formation of 3′-SL in CMP-Neu5Ac providing *E. coli* strains, the production of 6′-sialyllactose (6′-SL) has been demonstrated by the expression for an α2–6-sialyltransferase from *Photobacterium* sp. During fed-batch

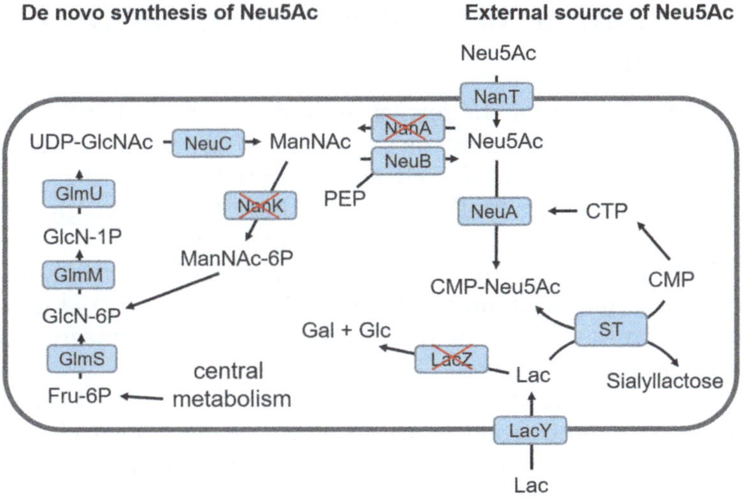

Figure 13.4 Whole-cell synthesis of sialyllactoses in *E. coli*. The required sialic acid (Neu5Ac) was either supplied (right part[40]) or generated *de novo* from glycerol *via* expression of the sialic acid synthase (NeuB) and the inactivation of the *N*-acetylmannosamine kinase (NanK) and the sialic acid aldolase (left part[57]). Utilization of an α2–3- or α2-6-ST resulted in the synthesis of 3′- or 6′-SL, respectively.[57,58] Abbreviations: CMP, cytidine 5′-monophosphate; CTP, cytidine 5′-triphosphate; Fru, fructose; Gal, D-galactose; Glc, D-glucose; GlcN, glucosamine; GlcNAc, *N*-acetyl-D-glucosamine; GlmM, phosphoglucosamine mutase; GlmS, glucosamine:fructose-6-phosphate aminotransferase; GlmU, glucosamine-1-phosphate acetyltransferase/*N*-acetylglucosamine-1-phosphate uridyltransferase; Gly, glycerol; Lac, D-lactose; LacY, lactose permease; ManNAc, *N*-acetyl-D-mannosamine; NagA, *N*-acetylglucosamine-6-phosphate deacetylase; NagB, glucosamine-6-phosphate deaminase; NanA, *N*-acetylneuraminate lyase; NanE, *N*-acetylmannosamine-6-phosphate epimerase; NanK, *N*-acetylmannosamine kinase; NanT, *N*-acetylneuraminate:H+ symporter; Neu5Ac, *N*-acetylneuraminic acid (sialic acid); NeuA, CMP-sialic acid synthase; NeuB, sialic acid synthase; NeuC, *N*-acetylglucosamine-6-phosphate epimerase; PEP, phosphoenolpyruvate; Pyr, pyruvate; ST, sialyltransferase; UDP, uridine 5′-diphosphate.

fermentation with a continuous supply of lactose, a product titer of up to 23 g L^{-1} in the culture broth and 11 g L^{-1} in the intracellular fractions was detected.[58] However, this *in vivo* synthesis also led to the formation of 6,6′-disialyllactose by the addition of a second Neu5Ac residue onto the terminal glucose of 6′-SL, and to 6′-Kdo-lactose (Kdo = 2-keto-3-deoxyoctonate).[58] This result shows that the applied sialyltransferase apparently had side activities with acceptor 6′-SL and donor CMP-Kdo. The donor substrate CMP-Kdo is an intermediate of core LPS biosynthesis and is permanently synthesized during growth of *E. coli*.

13.3 Limitations of the Bacterial *In Vivo* Synthesis of Glycans

The proof of principle for the formation of free oligosaccharides has been demonstrated in *E. coli* cells by the expression of appropriate glycosyltransferases and in the presence of nucleotide-activated donor and acceptor substrates. In most examples for the *in vivo* synthesis the desired glycans could be produced and purified in gram-scales quantities. The *in vivo* synthesis has not only been used for the formation of human milk oligosaccharides, but in particular the commercial interest for a large scale synthesis of human milk oligosaccharides has driven the research on metabolic engineering towards the *in vivo* synthesis of glycans.

One limitation of the *in vivo* synthesis lies in the capacity of transporting acceptor substrates and products in and out of the cells. The *in vivo* synthesis of LNT, for example, is achieved by two glycosylation steps starting from lactose.[23] The main product of the cultivation, however, is lacto-*N*-triose II which is for the most part detected extra-cellularly. The way that the trisaccharide is transported over the plasma membrane of *E. coli* is not yet described in the literature. But the fact that the acceptor substrate is withdrawn from the compartment in which the glycosylation reaction takes place, significantly reduces the yield of the desired product. In contrast to LNT, LNnT is not secreted from the cell. The increasing concentration of LNnT, however, facilitates further glycosylation reactions to lacto-*N*-neothexaose and lacto-*N*-neooctaose. This demonstrates that the accumulation or the release of the oligosaccharides from the cell plays a crucial role for the yield and the selectivity of the *in vivo* glycosylation reactions. Therefore, further investigations on the transport of glycans in and out of *E. coli* are necessary to optimize the *in vivo* synthesis.

Another challenge for the whole cell synthesis is the sufficient supply of nucleotide-activated sugars for the Leloir-glycosyltransferase catalysed reactions. Although the synthesis of all necessary donor substrates for the synthesis of human glycans can be managed in *E. coli*, an increase of the donor supply by means of metabolic engineering has so far only been demonstrated for the *de novo* synthesis of GDP-Fuc. However, the importance of the intracellular availability of donor substrates in sufficient amounts for glycosyltransferase reactions has been shown in many studies.[13–15,36]

Besides the total yield of the *in vivo* synthesis, the formation of unwanted side products is another challenge of the whole cell synthesis. The concurrent activity of several glycosyltransferases in *E. coli* is obviously the main reason for the formation of side products. Because of the low specificity of some glycosyltransferases different acceptor molecules can be glycosylated. To solve this problem a successive sequence of glycosylation reactions would be desired. Since GDP-Fuc and CMP-Neu5Ac are not essential for the growth of *E. coli* their salvage pathways can be used to trigger fucosylation or sialylation reactions. Thereby the formation of LNT, for example, can be separated in terms of time from the fucosylation or sialylation reactions. This can reduce the undesired formation of side products if a group of glycosyltransferases are expressed at the same time. Another approach to solve this would be the controlled expression of glycosyltransferases, for example, by the utilization of different inducible promoters. Especially with respect to the industrial production, the formation of oligosaccharide side products is most unwanted, because the isolation of oligosaccharide mixtures, in most cases, can only be accomplished by laborious chromatography methods.

E. coli is currently the best suitable host organism for the heterologous synthesis of natural products. A wide range of tools for the genetic engineering of the *E. coli* are available. The cultivation of *E. coli* can be performed in a simple mineral medium and is easily scalable. A drawback for the use of bacteria is the difficulty to express eukaryotic protein in a functional form. The successful expression of mammalian glycosyltransferases for the *in vivo* synthesis of glycans has only been demonstrated in a few studies.[22]

Another problem with the use of Gram-negative bacteria, like *E. coli*, in particular for application in pharma and food industries, is the presence of endotoxins (lipid-A) from the outer membrane of the cells. These toxins need to be removed during the purification of the oligosaccharides. Thus, to avoid the laborious and expensive cleaning to remove endotoxins, the use of safer production organisms, like *Saccharomyces cerevisiae, Bacillus subtilis, Corynebacterium glutamicum* or endotoxin-free *E. coli* strains,[59] is very much recommended. The heterologous expression of a GDP-Fuc pathway, for example, has already been demonstrated for *S. cerevisiae* and *C. glutamicum*.[20,21]

13.4 Conclusions and Outlook

The successful synthesis of a large variety of oligosaccharides, in particular human milk oligosaccharides, has been achieved in recent years by the use of whole cell biotransformation. Thus, 2′-FL,[14,27,38] 3-FL,[17] a mixture of LNT II and LNnT and longer saccharides,[17,40] LNT,[23,48] LNFP I or LNDFH II,[53] fucosylated LNnT,[38,41] and 3′-SL[7,40,56] have been shown.

Chemical and enzymatic synthesis has provided a larger variety of natural oligosaccharides and also of some unnatural derivatives in comparison to that accomplished by *in vivo* synthesis, but in most cases at low yields. Thus, in view of a scaled production of oligosaccharides, these approaches are less likely. The *in vivo* synthesis approach with recombinant bacteria cells has provided oligosaccharides at gram quantities in relatively simple systems (shake flask cultivation) in several cases and scale-up (fed-batch cultivation) has been successfully demonstrated. Besides academic research, several companies (*e.g.* BASF (Germany), Glycom (Denmark), Inbiose (Belgium), Jennewein-Biotechnology (Germany)) have adopted the whole cell fermentation approach for the industrial production of oligosaccharides, especially for the synthesis of 2′-FL.[60] Besides lactose, GlcNAc and galactose have been successfully used as acceptor molecules for glycan synthesis. The modification of the pathways by means of metabolic engineering is still at the beginning. Past attempts to synthesise glycans *in vivo* established the essential pathways and enzyme activities to show the proof of concept for a heterologous synthesis. The efficiency and yield of *in vivo* synthesis need to be improved to meet the economic requirements of the food and pharmaceutical industries. Besides the insufficient supply of precursor substrates, a significant limitation for the whole cell synthesis of even more complex glycan structures is the lack of suitable glycosyltransferases which can be expressed in a functional form in bacterial cells. Also, for large-scale production efficient purification methods are in great demand. As *E. coli* cells are the most commonly used host organism, downstream processing needs to remove any LPS to guarantee endotoxin-free production.

References

1. L. Bode, *Glycobiology*, 2012, **22**, 1147.
2. L. Bode, *Early Hum. Dev.*, 2015, **91**, 619.
3. A. Kobata, *Proc. Jpn. Acad., Ser. B*, 2010, **86**, 731.
4. R. Chen, *Appl. Microbiol. Biotechnol.*, 2018, **102**, 3017.
5. L. Bode, N. Contractor, D. Barile, N. Pohl, A. R. Prudden, G. J. Boons and Y. S. Jin, *Nutr. Rev.*, 2016, **74**, 635.
6. A. Ruffing and R. R. Chen, *Microb. Cell Fact.*, 2006, **5**, 25.
7. T. Endo, S. Koizumi, K. Tabata and A. Ozaki, *Appl. Microbiol. Biotechnol.*, 2000, **53**, 257.
8. T. Endo and S. Koizumi, *Curr. Opin. Struct. Biol.*, 2000, **10**, 536.
9. K. Andrianopoulos, L. Wang and P. R. Reeves, *J. Bacteriol.*, 1998, **180**, 998.
10. G. Stevenson, K. Andrianopoulos, M. Hobbs and P. R. Reeves, *J. Bacteriol.*, 1996, **178**, 4885.
11. M. J. Coyne, B. Reinap, M. M. Lee and L. E. Comstock, *Science*, 2005, **307**, 1778.
12. W. Yi, X. Liu, Y. Li, J. Li, C. Xia, G. Zhou, W. Zhang, W. Zhao, X. Chen and P. G. Wang, *Proc. Natl. Acad. Sci. U. S. A.*, 2009, **106**, 4207.
13. Y. W. Chin, N. Seo, J. H. Kim and J. H. Seo, *Biotechnol. Bioeng.*, 2016, **113**, 2443.
14. F. Baumgärtner, L. Seitz, G. A. Sprenger and C. Albermann, *Microb. Cell Fact.*, 2013, **12**, 40.
15. E. Hüfner, J. Parkot and S. Jennewein, *US Pat.* application, 300584, 2011.
16. C. Dumon, E. Samain and B. Priem, *Biotechnol. Prog.*, 2004, **20**, 412.
17. C. Dumon, C. Bosso, J. P. Utille, A. Heyraud and E. Samain, *ChemBioChem*, 2006, **7**, 359.
18. D. Huang, K. Yang, J. Liu, Y. Xu, Y. Wang, R. Wang, B. Liu and L. Feng, *Metab. Eng.*, 2017, **41**, 23.
19. W. H. Lee, S. Y. Shin, M. D. Kim, N. S. Han and J. H. Seo, *Appl. Microbiol. Biotechnol.*, 2012, **93**, 2327.
20. Y. W. Chin, J. B. Park, Y. C. Park, K. H. Kim and J. H. Seo, *Bioprocess Biosyst. Eng.*, 2013, **36**, 749.
21. P. Mattila, J. Räbinä, S. Hortling, J. Helin and R. Renkonen, *Glycobiology*, 2000, **10**, 1041.

22. E. Yavuz, S. Drouillard, E. Samain, I. Roberts and B. Priem, *Glycobiology*, 2008, **18**, 152.
23. F. Baumgärtner, J. Conrad, G. A. Sprenger and C. Albermann, *ChemBioChem*, 2014, **15**, 1896.
24. S. Cottaz and E. Samain, *Metab. Eng.*, 2005, **7**, 311.
25. T. Antoine, B. Priem, A. Heyraud, L. Greffe, M. Gilbert, W. W. Wakarchuk, J. S. Lam and E. Samain, *ChemBioChem*, 2003, **4**, 406.
26. M. Ringenberg, C. Lichtensteiger and E. Vimr, *Glycobiology*, 2001, **11**, 533.
27. M. Coulet, P. Phothirath, L. Allais and B. Schilter, *Regul. Toxicol. Pharmacol.*, 2014, **68**, 59.
28. E. Castanys-Muñoz, M. J. Martin and P. A. Prieto, *Nutr. Rev.*, 2013, **71**, 773.
29. A. L. Morrow, G. M. Ruiz-Palacios, M. Altaye, X. Jiang, L. Guerrero, J. K. Meinzer-Derr, T. Farkas, P. Chaturvedi, L. K. Pickering and D. S. Newburg, *J. Pediatr.*, 2004, **145**, 297.
30. P. Chaturvedi, C. D. Warren, M. Altaye, A. L. Morrow, G. Ruiz-Palacios, L. K. Pickering and D. S. Newburg, *Glycobiology*, 2001, **11**, 365.
31. R. M. Erney, W. T. Malone, M. B. Skelding, A. A. Marcon, K. M. Kleman-Leyer, M. L. O'Ryan, G. Palacios, M. D. Hilty, L. K. Pickering and P. A. Prieto, *J. Pediatr. Gastroenterol. Nutr.*, 2000, **30**, 181.
32. G. V. Coppa, P. Pierani, L. Zampini, I. Carloni, A. Carlucci and O. Gabrielli, *Acta Paediatr. Suppl.*, 1999, **88**, 89.
33. S. Thurl, J. Henker, M. Siegel, K. Tovar and G. Sawatzki, *Glycoconjugate J.*, 1997, **14**, 795.
34. E. Vazquez, A. Barranco, M. Ramirez, A. Gruart, J. M. Delgado-Garcia, E. Martinez-Lara, S. Blanco, M. J. Martin, E. Castanys, R. Buck, P. Prieto and R. Rueda, *J. Nutr. Biochem.*, 2015, **26**, 455.
35. G. M. Ruiz-Palacios, L. E. Cervantes, P. Ramos, B. Chavez-Munguia and D. S. Newburg, *J. Biol. Chem.*, 2003, **278**, 14112.
36. Y. W. Chin, J. Y. Kim, W. H. Lee and J. H. Seo, *J. Biotechnol.*, 2015, **210**, 107.
37. W. H. Lee, P. Pathanibul, J. Quarterman, J. H. Jo, N. S. Han, M. J. Miller, Y. S. Jin and J. H. Seo, *Microb. Cell Fact.*, 2012, **11**, 48.
38. S. Drouillard, H. Driguez and E. Samain, *Angew. Chem.*, 2006, **45**, 1810.
39. J. Abramson, I. Smirnova, V. Kasho, G. Verner, H. R. Kaback and S. Iwata, *Science*, 2003, **301**, 610.
40. B. Priem, M. Gilbert, W. W. Wakarchuk, A. Heyraud and E. Samain, *Glycobiology*, 2002, **12**, 235.
41. C. Dumon, B. Priem, S. L. Martin, A. Heyraud, C. Bosso and E. Samain, *Glycoconjugate J.*, 2001, **18**, 465.
42. L. Sturla, A. Bisso, D. Zanardi, U. Benatti, A. De Flora and M. Tonetti, *FEBS Lett.*, 1997, **412**, 126.
43. K. L. Jones, S. W. Kim and J. D. Keasling, *Metab. Eng.*, 2000, **2**, 328.
44. K. Friehs, *Adv. Biochem. Eng./Biotechnol.*, 2004, **86**, 47.
45. T. Urashima, S. Asakuma, F. Leo, K. Fukuda, M. Messer and O. T. Oftedal, *Adv. Nutr.*, 2012, **3**, 473S.

46. M. R. E Aly, E.-S. I. Ibrahim, E.-S. H. El Ashry and R. R. Schmidt, *Carbohydr. Res.*, 1999, **316**, 121.
47. T. Murata, T. Inukai, M. Suzuki, M. Yamagishi and T. Usui, *Glycoconjugate J.*, 1999, **16**, 189.
48. X. W. Liu, C. Xia, L. Li, W. Y. Guan, N. Pettit, H. C. Zhang, M. Chen and P. G. Wang, *Bioorg. Med. Chem.*, 2009, **17**, 4910.
49. F. Baumgärtner, G. A. Sprenger and C. Albermann, *Enzyme Microb. Technol.*, 2015, **75–76**, 37.
50. T. Miyazaki, T. Sato, K. Furukawa and K. Ajisaka, *Methods Enzymol.*, 2010, **480**, 511.
51. D. B. Stein, Y. N. Lin and C. H. Lin, *Adv. Synth. Catal.*, 2008, **350**, 2313.
52. S. L. Martin, M. R. Edbrooke, T. C. Hodgman, D. H. van den Eijnden and M. I. Bird, *J. Biol. Chem.*, 1997, **272**, 21349.
53. S. Rabbani, F. Corona and B. Ernst, *BioMetals*, 2009, **22**, 1011.
54. F. Baumgärtner, L. Jurzitza, J. Conrad, U. Beifuss, G. A. Sprenger and C. Albermann, *Bioorg. Med. Chem.*, 2015, **23**, 6799.
55. Y. Li and X. Chen, *Appl. Microbiol. Biotechnol.*, 2012, **94**, 887.
56. N. S. Han, T. J. Kim, Y. C. Park, J. Kim and J. H. Seo, *Biotechnol. Adv.*, 2012, **30**, 1268.
57. N. Fierfort and E. Samain, *J. Biotechnol.*, 2008, **134**, 261.
58. S. Drouillard, T. Mine, H. Kajiwara, T. Yamamoto and E. Samain, *Carbohydr. Res.*, 2010, **345**, 1394.
59. U. Mamat, K. Wilke, D. Bramhill, A. B. Schromm, B. Lindner, T. A. Kohl, J. L. Corchero, A. Villaverde, L. Schaffer, S. R. Head, C. Souvignier, T. C. Meredith and R. W. Woodard, *Microb. Cell Fact.*, 2015, **14**, 57.
60. http://www.nutrition.basf.com, http://www.inbiose.com, http://www.glycom.com, http://www.jennewein-biotech.de.

CHAPTER 14

Solid-phase Glycan Synthesis

FABIAN PFRENGLE*[a,b]

[a]Department of Biomolecular Systems, Max Planck Institute of Colloids and Interfaces, Am Mühlenberg 1, 14476 Potsdam, Germany; [b]Institute of Chemistry and Biochemistry, Freie Universität Berlin, Arnimallee 22, 14195 Berlin, Germany
*E-mail: Fabian.Pfrengle@mpikg.mpg.de

14.1 Introduction

The ability to access principally any glycan of choice in a quick and reliable manner would dramatically facilitate the study of glycans regarding both their functional roles in biology and their properties as materials.[1] Conventional solution-phase chemistry can deliver these glycans, but is usually cumbersome and time-consuming.[2] Automated solid-phase synthesis may overcome these drawbacks and has already revolutionized the scientific fields of proteomics and genomics. Peptides and oligonucleotides are routinely accessed through automated solid-phase synthesis and can be ordered from commercial vendors who use automated synthesizers for their custom syntheses.[3,4] Developments in the automated synthesis of glycans, which operates on the same principle of successive addition of building blocks to a solid phase, have been lacking behind due to the structural complexity of glycans. A much higher diversity of reagents and reaction parameters is required for their synthesis. Many different linkages connect the individual building blocks that include, at least in the case of bacteria and plants, a

Chemical Biology No. 11
Synthetic Glycomes
Edited by Wanyi Guan, Lei Li and Peng George Wang

Published by the Royal Society of Chemistry, www.rsc.org

vast number of different monosaccharides. There are two significant challenges in moving from solution-phase to solid-phase glycosylation chemistry: (1) Side products such as those resulting from incomplete conversion or imperfect stereoselectivity during glycosylation reactions cannot be removed by chromatography after the individual reactions. (2) Reactions at the solid-phase usually require a significantly larger excess of coupling agent than the respective solution-phase reactions.

The first solid-phase synthesis of a glycan was reported in 1971,[5] but it was only in the last decade that impressive advances towards solving these issues have been made. In 2001, the Seeberger laboratory reported the automation of the solid-phase oligosaccharide synthesis process (automated glycan assembly).[6] Since then, this technology has been constantly advanced, resulting in the introduction of new improved instruments[7,8] and numerous successful syntheses of complex glycans.[9] In this chapter solid-phase syntheses of all major classes of glycans are described. The chapter focusses on solid-phase syntheses of glycans in which the glycosyl acceptor is bound to the solid-phase through a linker at the anomeric position, as this has been the method of choice for modern automated glycan assembly as well as for all recently described solid-phase syntheses of oligosaccharides. Other synthetic strategies have been extensively covered in previous reviews.[10–12] After general considerations, solid-phase syntheses of different classes of glycans are described.

14.2 Automated Glycan Assembly

The automation of solid-phase glycan synthesis has tremendously advanced this method, because the manual conduction of countless individual reactions becomes unnecessary and many more solid-phase syntheses can be performed by the same operator. Since the introduction of automated solid-phase glycan synthesis (automated glycan assembly) in 2001,[6] a vast number of structurally diverse glycans have been prepared with the help of automated oligosaccharide synthesizers.[9] Oligosaccharide synthesizers are able to carry out temperature-controlled reactions and washes in a fully automated fashion, similar to peptide and oligonucleotide synthesizers. Automated synthesizers apply standardized protocols, generating comparability between the experiments performed by different laboratories. With the recent commercialization of these synthesizers, an increasing number of laboratories will have access to this technology and the universe of accessible glycans will further increase quickly.[8]

The main preventive cause for a more wide-spread use of automated oligosaccharide synthesizers is, at least currently, their cost. A potential alternative setup has recently been reported by Demchenko and coworkers who used a stock HPLC for automated glycan synthesis.[13] In this approach the resin has been packed into an HPLC-column and activated glycosyl donors have been pumped through the column to glycosylate the resin-bound acceptor. In this way, a β1–6-glucan hexasaccharide has been successfully assembled.

14.3 Linkage to the Solid Support

Solid-phase glycan synthesis relies on the iterative addition of carbohydrate building blocks to a growing glycan chain. Either the glycosyl donor or the glycosyl acceptor may in principle be attached *via* a cleavable linker to the solid support. Modern solid-phase glycan synthesis makes use of the acceptor-bound approach which is more flexible in terms of the glycosyl donors that can be used. Furthermore, the first carbohydrate building block is attached to the linker *via* the anomeric position rather than any of the other hydroxyl groups. This provides the opportunity to use a standard building block in the glycosylation of the linker hydroxyl group rather than attaching an individually designed building block to the linker at any of the other positions through *e.g.* esterification.

14.3.1 Solid Support

Many kinds of solid supports have been explored for solid-phase glycan synthesis. Polystyrene resins, in particular Merrifield resin, were found most reliable due to their compatibility with many solvents, mechanical stability, good loading capability, and low cost. Controlled pore glass and other resins including TentaGel, JandaGel, ArgoGel, and Wang resin gave very good results in specific cases, but suffered from low general applicability.

14.3.2 Linker to the Solid Support

The ideal linker connecting glycan with solid support has to fulfill the following requirements: (i) It must be chemically inert to all conceivable glycosylation and deprotection conditions. (ii) It has to be selectively cleaved at the end of the synthesis. Despite these strict prerequisites, a large number of different linkers have been developed (Figure 14.1). Alkene-linkers were the first to be used in solid-phase oligosaccharide synthesis. In their pioneering study in 1971, Frechet and Schuerch employed an allylic moiety (**1**) that was cleaved by ozonolysis after glycan synthesis was complete.[5] Later on, the Seeberger laboratory developed an octenediol-linker (**2**) that can be cleaved by olefin metathesis using Grubbs 1 catalyst and ethylene gas.[14] This linker was the basis for the successful automated synthesis of numerous complex oligosaccharides. To avoid the need for using ethylene gas and low turnover numbers of the catalyst due to covalent inactivation, others have subsequently introduced self-cleavable alkene-linkers (**3**, **4**).[15,16] In these cases, an intramolecular olefin metathesis reaction produces a cyclopentene moiety on the resin, preventing covalent attachment of the catalyst to the solid phase. The disadvantage of alkene-linkers is their incompatibility with thioglycoside donors as they are not inert to the electrophilic reagents used for their activation. Another very commonly used predetermined breaking point in the linker is an ester which is stable to

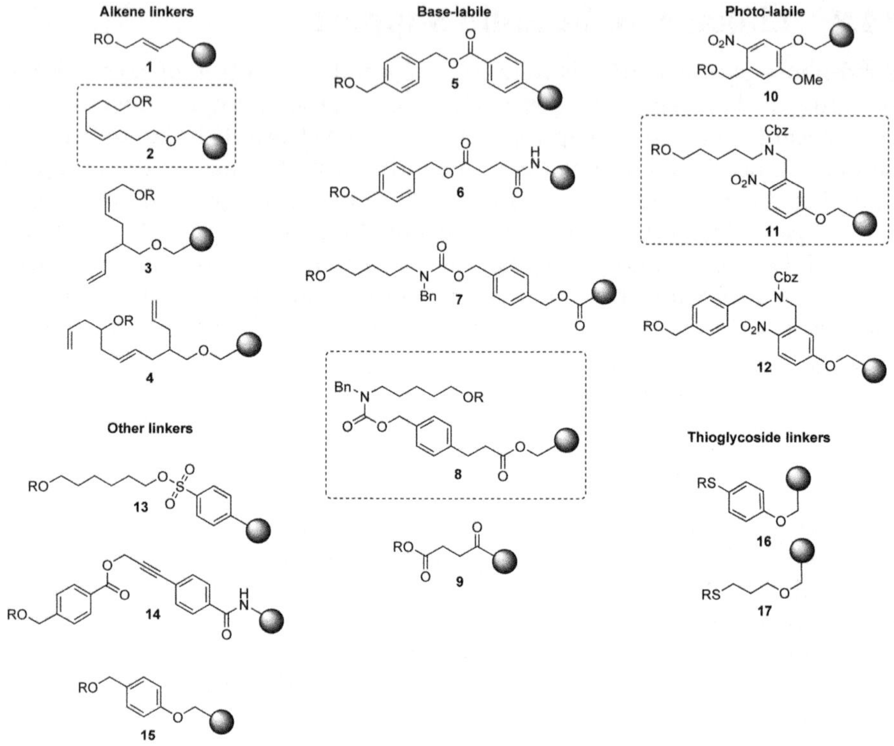

Figure 14.1 Selection of cleavable linkers for solid-phase glycan synthesis. Boxed linkers are used in current automated glycan assembly protocols.

the acidic conditions employed during the glycosylation reactions, but can easily be cleaved after assembly of the glycan by methanolysis. The ester moiety may be spaced from the anomeric carbon of the first attached sugar residue by a 1,4-di(hydroxymethyl)benzene unit (**5, 6**) to provide free reducing sugars after a final hydrogenolysis reaction.[17,18] Alternatively, an additional alkyl spacer linked to the 1,4-di(hydroxymethyl)benzene through a benzyl protected carbamate moiety (**7**, **8**) has been introduced.[7,19] During methanolysis both the ester and carbamate functionalities are cleaved and a Cbz-protected amino group is formed. The Cbz group can be subsequently removed by hydrogenolysis, resulting in an aminopentyl linker that allows direct conjugation of the assembled glycans to surfaces for the production of microarrays and to proteins *e.g.* for vaccine development. A direct ester connection between linker and anomeric carbon of the reducing sugar (**9**) is not advisable since after cleavage from the resin a free reducing sugar in the otherwise still protected assembly product is formed, complicating analysis and purification.[20] An even higher degree of orthogonality is provided by photocleavable linkers based on the *p*-nitrophenol moiety that are inert to electrophilic reagents as well as acids and bases.[21] The first photocleavable linker (**10**) for the solid-phase synthesis of a disaccharide was already reported in 1973.[22] In this case, the glycan was directly linked to a hydroxymethyl substituent of the nitrophenyl group. As the obtained cleavage yield was rather low, this linker has not found further application in solid-phase glycan synthesis. Instead, new linkers have recently been developed that connect glycan and the photo-labile unit either through a pentyl (**11**)[23] or a hydroxymethylbenzene (**12**)[24] spacer. These linkers can be efficiently cleaved in a continuous flow photo-reactor and provide aminopentyl-functionalized and free reducing glycans, respectively.

The three types of linkers described in the previous paragraph are the most commonly used, with the boxed linkers from Figure 14.1 being the standard linkers for automated glycan assembly. Other examples for linkers used to connect the growing glycan chain with the solid support through the anomeric position of the reducing end sugar include sulfonate- (**13**)[25] and alkyne-based (**14**)[26] linkers, and a hydroxymethylbenzene linker (**15**)[27] with no additional labile group. The anomeric linkage to this linker can be cleaved using strong Lewis acid. Thioglycoside linkers (**16**) and (**17**) provide the opportunity to directly transfer the assembled glycan from the solid support to acceptor substrates such as amino acids, but have not been explored extensively yet.[28,29] This approach might be useful for the construction of libraries containing glycosylated small molecules.

14.4 Glycosyl Donors

14.4.1 Leaving Group

Choosing the right glycosyl donor is essential for a successful glycosylation reaction, both in solution- and in solid-phase chemistry. The most commonly used types of glycosyl donors in solid-phase glycan synthesis are glycosyl imidates, thioglycosides, and glycosyl phosphates. But also carboxybenzyl glycosides and glycosyl thioimidates have been applied in specific cases. Very good results have been obtained with all of the leaving groups and none of them seem generally superior over the others. Thioglycosides have come in handy due to the opportunity of introducing the thioether moiety early in glycosyl donor synthesis and using it as an anomeric protecting group. They were thus applied in many recent automated glycan assemblies. However, in some cases glycosyl phosphates provided better results.

14.4.2 Protecting Groups

Maybe even more important than the leaving group is the choice of protecting groups for the glycosyl donor. The protecting groups have a very strong influence on the donor's reactivity as well as on the stereochemical outcome of the glycosylation reactions. In solid-phase glycan synthesis, two general types of protecting groups can be defined: permanent and temporary protecting groups. Permanent protecting groups remain untouched during the solid-phase synthesis and are only removed in solution after the assembled glycan is cleaved from the solid support. Temporary protecting groups on the other hand are selectively removed during glycan assembly on the solid phase to enable the next glycosylation reaction. Benzyl ethers and simple esters such as acetates and benzoates are very suitable as permanent protecting groups as they remain intact during the removal of more specialized temporary protecting groups. These include carbonates and functionalized ethers and esters such as Nap ethers, Allyl ethers, and levulinoyl esters. Both types of protecting groups can be further categorized into participating and non-participating protecting groups when installed at the C-2 position of the glycosyl donor. Participating protecting groups typically favor the formation of 1,2-*trans*-linkages due to the neighboring group participation effect. Non-participating protecting groups enable the synthesis of 1,2-*cis*-bonds if the whole ensemble of protecting groups is carefully selected.[30] Temporary protecting groups for solid-phase glycan synthesis are particularly useful if they provide the opportunity for reaction progress monitoring by UV-spectroscopy such as the 9-fluorenylmethoxycarbonyl (Fmoc) group.

*14.4.2.1 Formation of 1,2-*cis*-Linkages*

Due to the missing opportunity to remove side products after the individual glycosylation reactions, high stereoselectivity is demanded for every step of the solid-phase synthesis. This requirement is particularly problematic for syntheses of glycans containing multiple 1,2-*cis*-glycosidic bonds as they are not always formed with high stereoselectivity. Some reactions such as α-fucosylations were typically found to give good 1,2-*cis*-selectivity while others such as α-xylosylations gave poor selectivity. In the case of the important α-glucose and α-galactose linkages the protecting groups at remote positions of the glycosyl donors can be carefully tuned in order to promote the selective formation of 1,2-*cis*-linkages.[31] Judicious placement of acetyl groups in the C3- and/or C6-position of glucose building blocks enabled for example the automated glycan assembly of an α-glucan hexasaccharide containing exclusively 1,2-*cis*-configured glycosidic bonds (Scheme 14.1).[32] Three differently protected thioglycoside building blocks were added to a Merrifield resin equipped with the photocleavable linker, and a glucan oligosaccharide with three α1–6- and two α1–3- glycosidic bonds was obtained in good overall yield. In total a set of five glucan oligosaccharides containing α1–3-, α1–4-, and α1–6-linkages were prepared in the fully deprotected form.

Studies working towards a general method for the construction of 1,2-*cis*-linkages on the solid phase have been undertaken by the Boons laboratory. Boons and coworkers have developed a chiral auxiliary that, when installed at the C2-position of the glycosyl donor, forms a cyclic anomeric sulfonium ion upon activation, blocking the top phase of the donor and forcing the nucleophile to attack from the bottom.[33] The use of this chiral

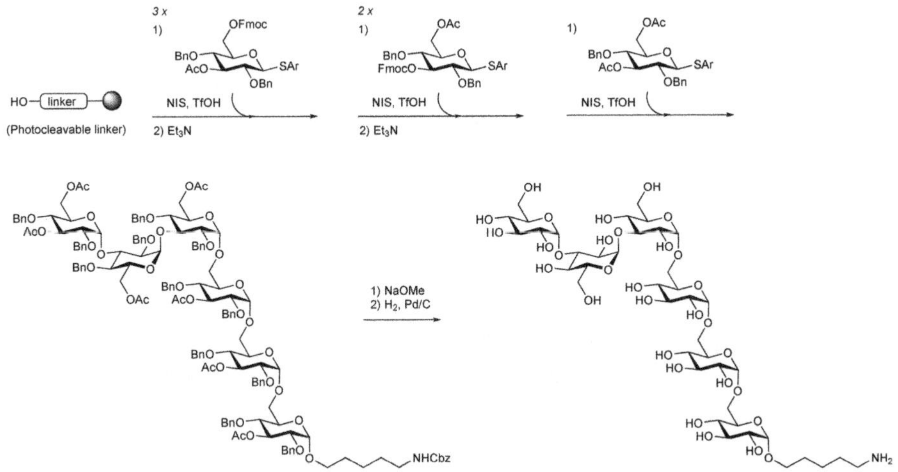

Scheme 14.1 Automated glycan assembly of an α-glucan oligosaccharide containing five 1,2-*cis*-glycosidic bonds enabled by building blocks equipped with remote participating protecting groups. (Ar = 2-Methyl-5-*tert*-butylphenyl).

auxiliary strongly promoted the formation of 1,2-*cis*-linkages on the solid phase and an α-glucan pentasaccharide was successfully produced by the successive addition of three different glycosyl imidates to a Merrifield resin equipped with a base-labile linker (Scheme 14.2).[34] The synthesis relied on the use of allyloxycarbonyl (Alloc) and Fmoc as temporary protecting groups in the C3- and C6-position respectively. Since the steric bulk of the C2-auxiliary prevented glycosylation in the C3-position, the auxiliary had to be replaced by an acetyl group using Lewis acid and Ac_2O, before Alloc deprotection and further glycosylation could be performed to introduce branching.

A particularly challenging linkage to construct in carbohydrate chemistry is the β-mannosidic bond as its formation does not benefit from the anomeric effect, and the neighboring participation mechanism of C2 ester protecting groups promotes the formation of α-linkages. A possibility to favor β-mannosylation is to lock the C4- and C6-hydroxyl groups of mannose in form of a cyclic acetal, as introduced by the Crich laboratory.[35] Seeberger and coworkers successfully applied this methodology in the automated glycan assembly of a trisaccharide containing two β-mannosidic linkages.[36] A benzylidene protected mannose building block carrying a carboxybenzyl leaving group was added twice to a mannose residue linked to Merrifield resin *via* an alkene linker, and a trimannoside containing two β-mannosidic linkages was obtained with remarkably high stereoselectivity (Scheme 14.3). However, the cyclic acetal must be regarded as a permanent protecting group since the selective opening of the acetal to further elongate either the C4- or C6-position of the mannose proved very difficult on the solid phase.

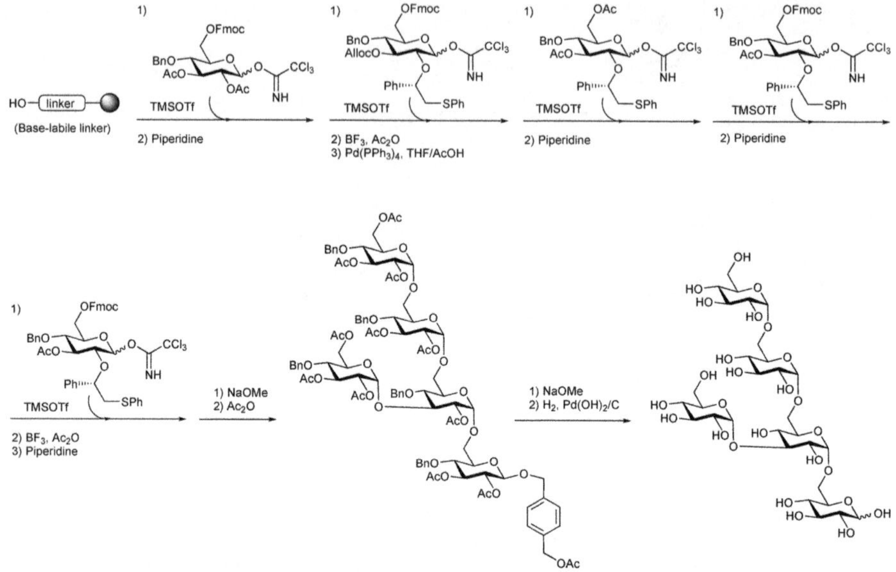

Scheme 14.2 Solid-phase synthesis of a branched α-glucan oligosaccharide containing four 1,2-*cis*-glycosidic bonds enabled by building blocks equipped with a chiral auxiliary in the 2-position.

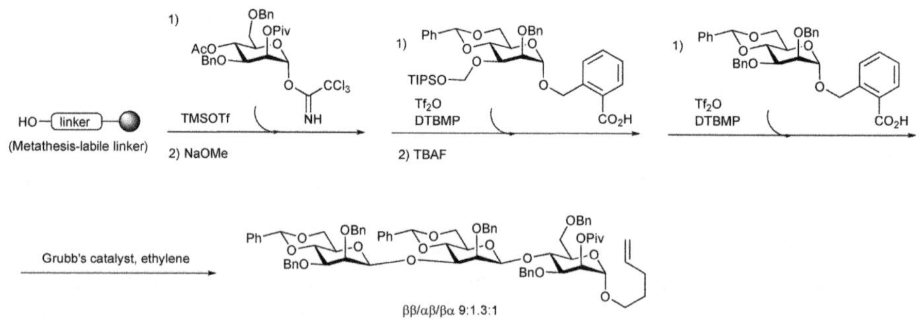

Scheme 14.3 Automated glycan assembly of a trisaccharide containing two β-mannosidic linkages using C4/C6-locked benzylidene protected mannose building blocks.

14.5 Monitoring Glycosylation Efficiency

Monitoring the efficiency of individual reactions on a solid phase is notoriously difficult but highly desirable, as it strongly facilitates the optimization of individual reactions and allows the operator to abort the synthesis of large molecules as soon as an unsatisfactory reaction is recognized. Both direct and indirect methods to monitor glycosylation reactions on the solid phase have been developed. Direct on bead analysis is in principle possible by NMR and IR spectroscopy.[37,38] The combined use of a fluorinated linker and fluorinated protecting groups for example provided the opportunity to calculate the amount of glycan linked to the solid phase after every reaction by integration of the signals corresponding to the linker and protecting groups in gel-phase ^{19}F-NMR spectroscopy.[37] However, the necessity to introduce labels such as fluoride or ^{13}C dramatically limits the pool of usable protecting groups and/or increases the costs. An indirect assessment of the efficiency of glycosylation reactions is much easier. The amount of protecting group cleaved off of the resin after a glycosylation reaction directly corresponds to the conversion in the previous glycosylation reaction. When Fmoc groups are removed using NEt_3, dibenzofulvene is formed that can be detected by simple UV-spectroscopy.[39] This is part of the reason why Fmoc has become the standard temporary protecting group for automated glycan assembly. Other protecting groups such as the nitrophtalimidobutyric acid ester (NPB) that produce UV/Vis-active compounds upon cleavage have been developed as well, but have not yet found wide application in solid-phase oligosaccharide syntheses.[40]

14.6 Catch and Release Technologies for Product Purification

The main side products formed during solid-phase glycan syntheses are stereoisomers arising from anomeric mixtures, and deletion sequences arising from incomplete glycosylation reactions. The latter can be removed without using chromatographic separation methods if the target glycan is linked *via* the last introduced building block to a functionalized scavenger resin. Many different linkers and chemistries have been developed for the immobilization on scavenger resins to enable a catch and release purification.[41–43] This technology is particularly helpful if a chromatographically inseparable mixture of products is obtained after the solid-phase synthesis. Seeberger and coworkers for example have made use of an aminoalkyl linker attached through an ester group to the 6-position of the last mannoside donor used in the automated glycan assembly of a 30-mer polymannoside.[44] In this way the 30-mer sequence was selectively tagged and all deletion sequences could be washed away. Subsequent base-induced cleavage of the safety catch linker with concomitant loss of all ester protecting groups afforded the desired 30-mer as the only product. Similar to peptide chemistry, where many different methods have also been developed, catch and release protocols have not found wide application in current automated solid-phase synthesis protocols. New approaches that require minimal numbers of chemical transformations and avoid the need for specialized building blocks may establish catch and release protocols as a standard technology in the solid-phase synthesis of glycans in the future.

14.7 Solid-phase Synthesis of Different Glycan Classes

Glycans from many different classes have been synthesized by solid-phase chemistry. Researchers have mostly focused on mammalian glycans, but more recently also syntheses of bacterial, fungal, and plant glycans have increasingly been pursued.

14.7.1 Mammalian Glycans

14.7.1.1 N-*Glycans*

During the last two decades different groups have developed solid-phase syntheses of *N*-glycans. The Schmidt laboratory reported the first synthesis of the common core pentasaccharide of *N*-glycans in 2002 (Scheme 14.4).[17] The synthesis was based on a base-labile linker attached to Merrifield resin and trichloroacetimidate chemistry. Formation of the β-mannosidic linkage on resin was avoided by using a disaccharide building block consisting of a glucosamine and the central core mannose residue. Fmoc was used as the standard temporary protecting group for elongation of the glycan and phenoxyacetyl (PA) as a temporary protecting group for introducing the mannose branch in the 3-position of the central core mannose residue.

An alternative route to *N*-glycans was developed by Tanaka and coworkers who used *N*-phenyltrifluoroimidate chemistry on a *JandaJel* resin equipped with a base-labile linker (Scheme 14.5).[18] Also this synthesis relied on Fmoc as the temporary protecting group for chain elongation, but branching was accomplished using a 4-azido-3-chlorobenzyl (AzClBn) group which can be removed using PBu_3 followed by 2,3-dichloro-5,6-dicyano-1,4-benzoquinone

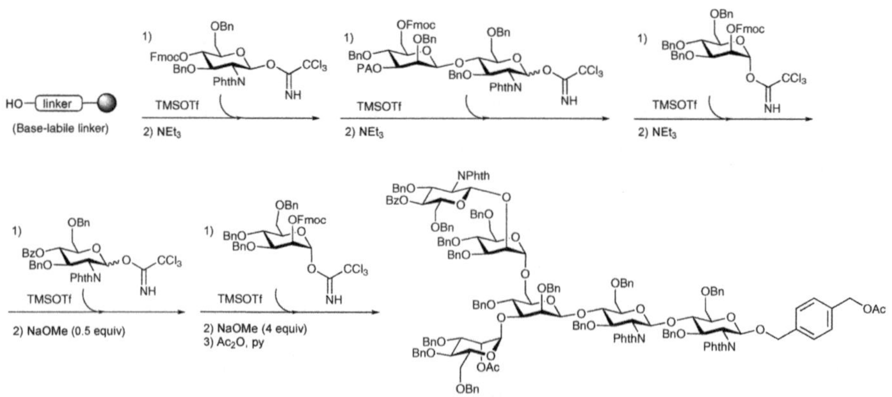

Scheme 14.4 Solid-phase synthesis of an extended *N*-glycan core pentasaccharide using a disaccharide building block having the β-mannosidic linkage preinstalled.

(DDQ). Two disaccharide building blocks were employed in this synthesis: one building block having the β-mannosidic linkage preinstalled and one building block consisting of a galactose and an α2–6-linked sialic acid monosaccharide. The latter one enabled the introduction of a terminal sialic acid residue, which is usually difficult to achieve on the solid phase using a sialic acid monosaccharide donor.

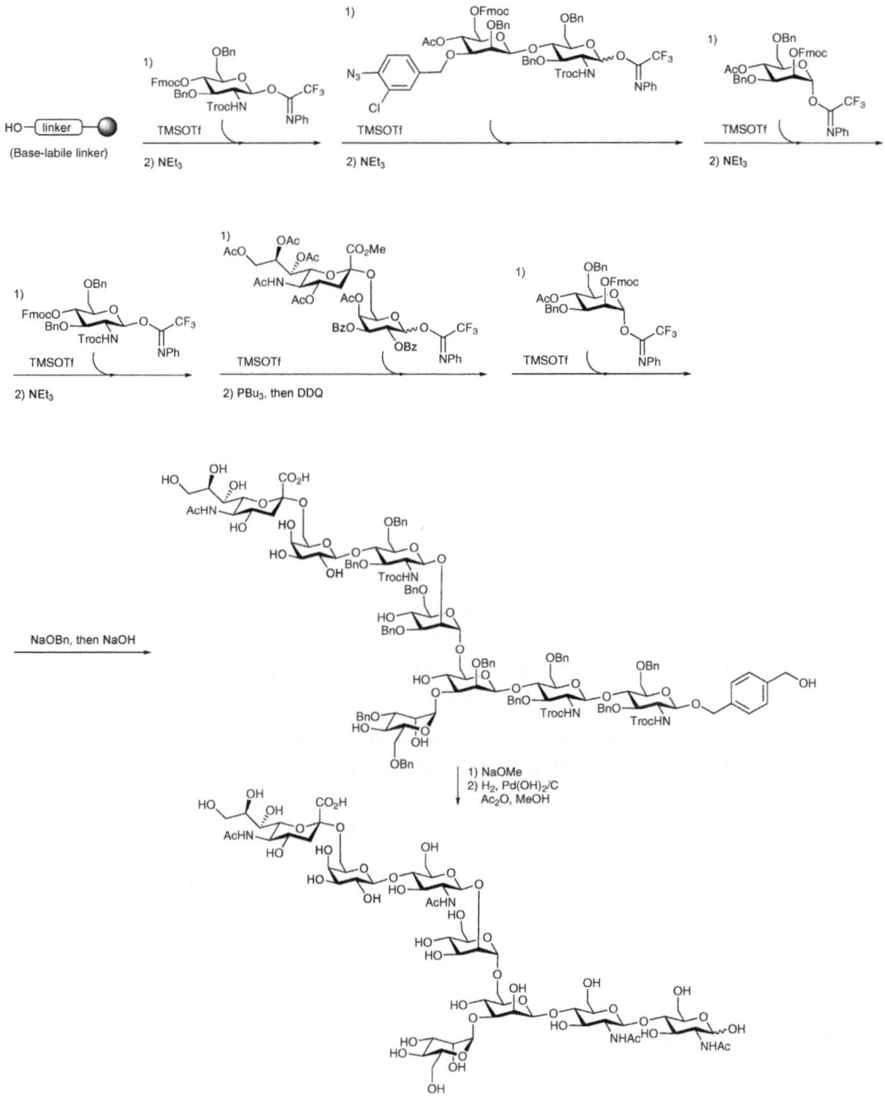

Scheme 14.5 Solid-phase synthesis of a sialylated *N*-glycan heptasaccharide using two disaccharide building blocks having the difficult β-mannosidic and α2–6-sialic acid linkages preinstalled.

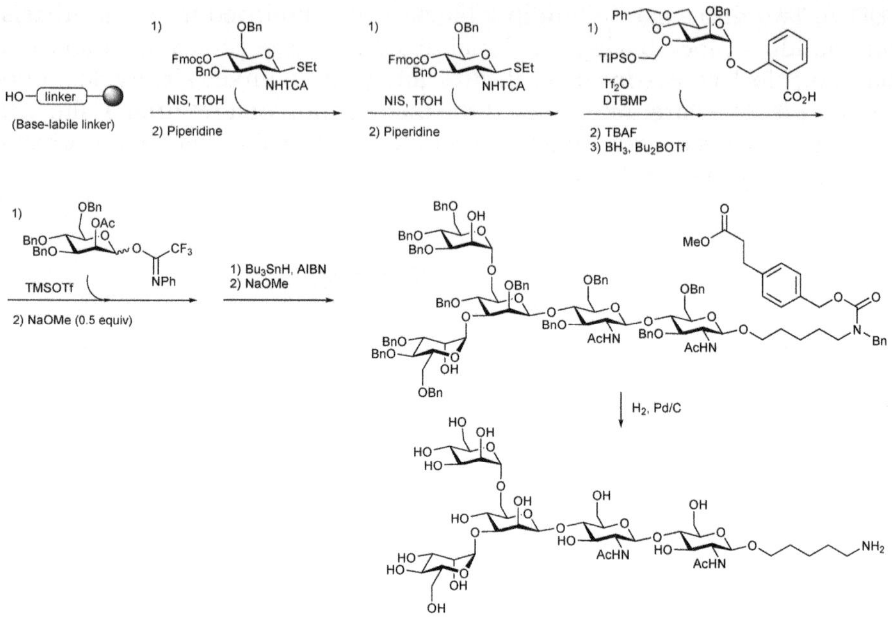

Scheme 14.6 Automated glycan assembly of the *N*-glycan core pentasaccharide using exclusively monosaccharide building blocks.

The only synthesis of the *N*-glycan core pentasaccharide that is based solely on monosaccharide building blocks was reported by the Seeberger laboratory.[7] In their automated glycan assembly on a Merrifield resin equipped with the base-labile linker, Seeberger and coworkers employed a benzylidene locked donor which provided good stereoselectivity in the difficult β-mannosylation reaction (Scheme 14.6). As a temporary protecting group in the 3-position of the mannose building block was used a [(triisopropylsilyl) oxy]methyl (TOM) ether, which can be removed using tetrabutylammonium fluoride. However, this strategy provided rather low overall yield due to difficulties in reductive benzylidene opening on the solid phase required for further glycosylations.

14.7.1.2 *Blood Group Determinants*

Oligosaccharides constituting blood group determinants including AB0-, Lewis-, and H-type antigens are differentially expressed on erythrocytes of different individuals. A number of Lewis- and H-type oligosaccharides have been prepared by automated glycan assembly. A Le^x-Le^y chimera nonasaccharide containing three 1,2-*cis* glycosidic linkages for example was already efficiently prepared in 2004 by iterative addition of phosphate donors to a Merrifield resin equipped with the octenediol linker.[45] In this synthesis, pivaloyl (Piv) and trichloroacetyl protecting groups were used as permanent protecting groups in the 2-position when neighboring group participation was required and Fmoc and levulinoyl (Lev) proved very reliable as temporary protecting groups (Scheme 14.7). These temporary protecting groups are still routinely used in current automated glycan assembly protocols.

Very recently, the automated glycan assembly of other blood group determinants such as the H-type I antigen was reported.[46] In this case, the photocleavable linker was used instead of the octenediol linker since it is easily cleaved and delivers conjugation-ready glycans for biological studies (Scheme 14.8). The glycosylation protocols were improved compared to the previous syntheses, so that thioglycosides could be used with the same efficiency as the glycosyl phosphates. Thioglycosides are typically solids and easier to

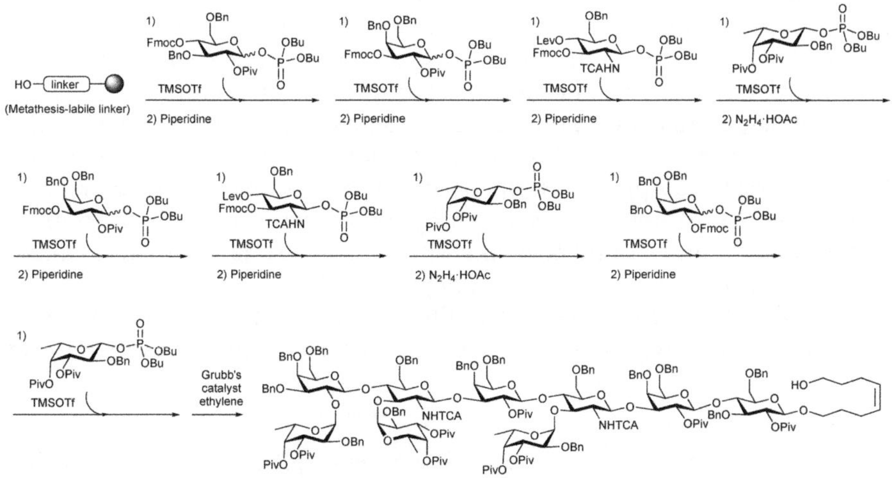

Scheme 14.7 Automated glycan assembly of a Lewis blood group oligosaccharide (Le^x-Le^y chimera).

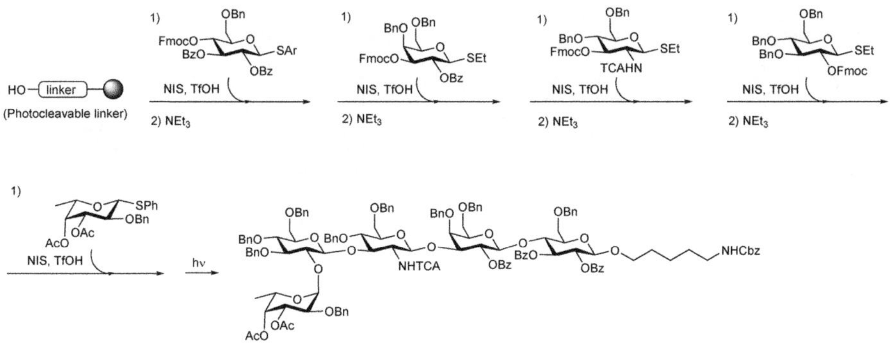

Scheme 14.8 Automated glycan assembly of an H-type blood group oligosaccharide (H-type I). (Ar = 2-Methyl-5-*tert*-butylphenyl).

store long term than phosphates. The Piv protecting groups that were often installed in the 2-position of the building blocks were replaced by benzoyl esters due to their more facile removal under basic conditions in the end of the synthesis.

14.7.1.3 Glycosaminoglycans

Glycosaminoglycans (GAGs) are important components of the extracellular matrix that are constructed from disaccharide repeating units rich in sulfated amino sugars and uronic acids. They are highly negatively charged and very hydrophilic. Due to these properties they can serve in the body as lubricants and shock absorbers. The solid-phase synthesis of glycosaminoglycans is particularly challenging, due to the low reactivity of uronic acid donors and the chemical lability of sulfate groups. Successful syntheses of chondroitin sulfate[23] and dermatan sulfate[47] oligosaccharides were accomplished *via* an on resin sulfation strategy, but their deprotection proved difficult. A chondroitin sulfate hexasaccharide was for example prepared by addition of galactosamine and glucuronic acid donors in the form of glycosyl phosphates (Scheme 14.9).[23] Iterative addition of these building blocks to a Merrifield resin equipped with the photocleavable linker provided precursor containing Lev groups at the primary hydroxyl groups of the galactosamine residues. Lev removal and reaction with sulfurtrioxide afforded a sulfated hexasaccharide that was not further deprotected.

The only reported synthesis of a fully deprotected glycosaminoglycan oligosaccharide is a keratan sulfate tetrasaccharide which was prepared from

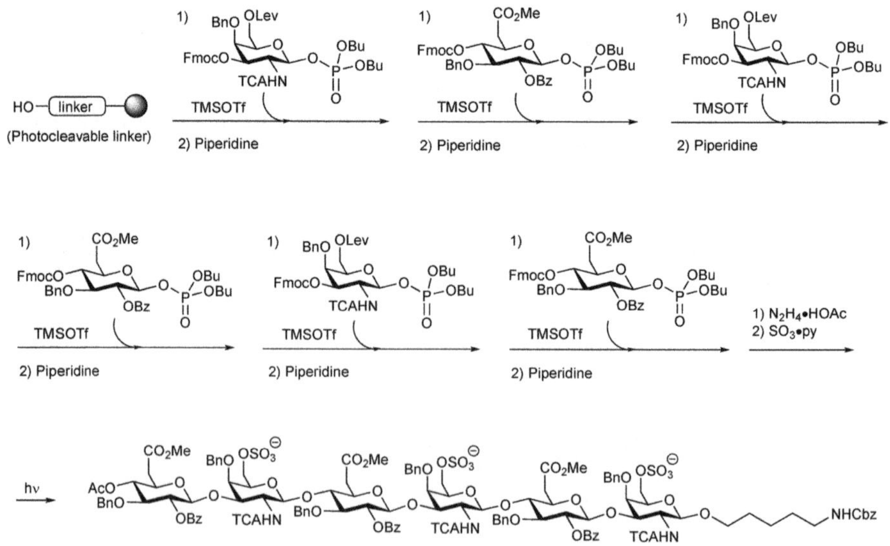

Scheme 14.9 Automated glycan assembly of a protected chondroitin sulfate hexasaccharide.

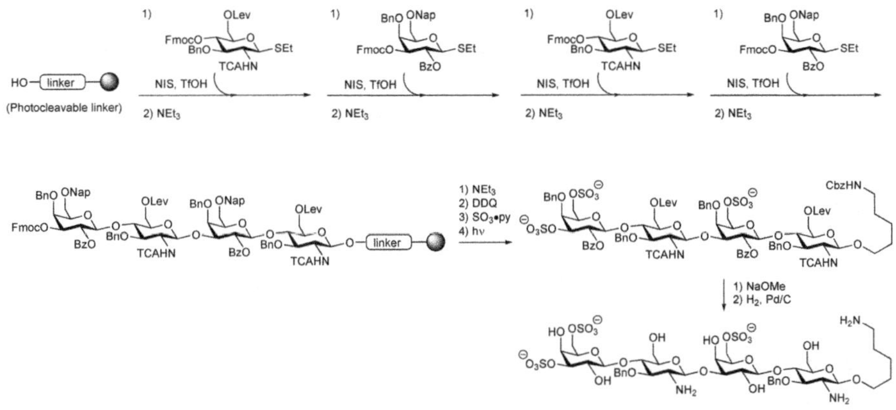

Scheme 14.10 Automated glycan assembly of a fully deprotected keratan sulfate tetrasaccharide.

galactose and glucosamine building blocks (Scheme 14.10).[48] In this case three orthogonal temporary protecting groups were required to enable selective sulfation of the galactose moieties. While the primary hydroxyl groups of the glucosamine units were protected with a Lev ester, the primary hydroxyl in the galactose units were protected as 2-methylnaphtyl (Nap) ethers. This enabled their selective removal using DDQ and sulfation on the solid phase on resin before cleavage of the photolabile linker and further deprotection.

14.7.1.4 *Cancer-associated Antigens*

Cancer-associated antigens are interesting synthetic targets due to their potential to act as the antigenic component of carbohydrate-based vaccines. Cancer-associated antigens Gb1 and Globo-H have been prepared by automated glycan assembly.[49] Six different building blocks derived from four monosaccharides were iteratively added to a Merrifield resin equipped with the octenediol linker to assemble the Globo-H hexasaccharide (Scheme 14.11). They were equipped either with a phosphate or an *N*-phenyltrifluoroacetimidate leaving group and provided, depending on the protecting group used in the 2-position, the desired 1,2-*cis* and 1,2-*trans* glycosidic linkages with high stereoselectivity.

14.7.1.5 *Synthesis of Sialylated Glycans*

Sialic acid is a common monosaccharide found at the tips of mammalian glycans. However, the chemical synthesis of sialic acid containing glycans is difficult due to issues with reactivity and stereoselectivity. Still, some success was achieved in the formation of α2–6-linkages using sialic acid monosaccharide donors in automated glycan assembly (Scheme 14.12).[50] The choice of protecting groups turned out to be crucial for good conversion in the

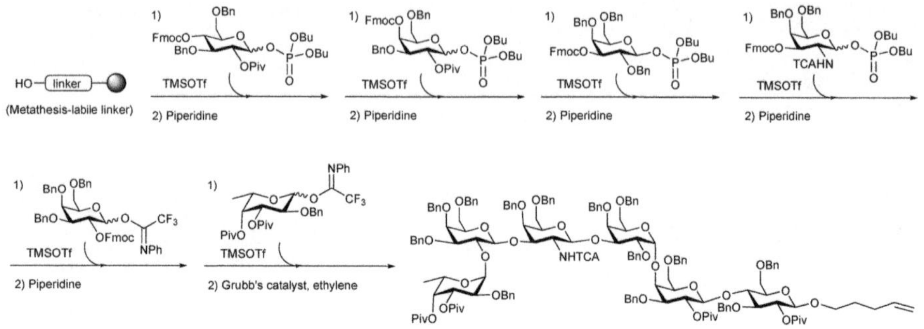

Scheme 14.11 Automated glycan assembly of a cancer-associated antigen (Globo-H).

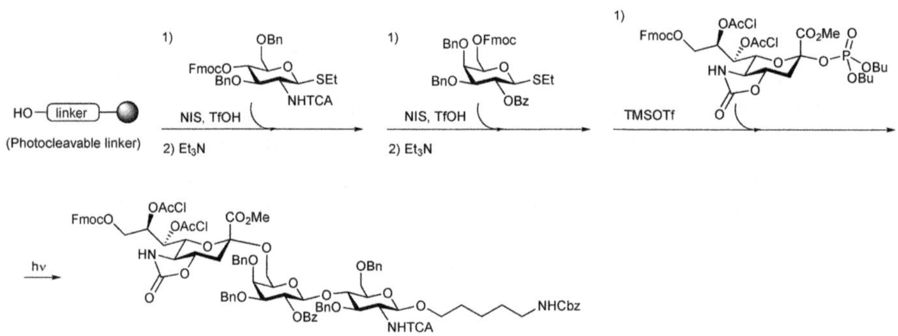

Scheme 14.12 Automated glycan assembly of α2–6-sialyl-*N*-acetyllactosamine using a sialic acid monosaccharide building block.

sialylation reaction. A cyclic 4*O*,5*N*-oxazolidinone protecting group was used to promote the formation of α-linkages and Fmoc and chloroacetyl groups were used as permanent protecting groups. Addition of this sialic acid donor to a Merrifield resin equipped with the photocleavable linker in combination with a glucosamine and a galactose building block provided for example α2–6-sialyl-*N*-acetyllactosamine in good overall yield. The synthesis of α2–3-linkages, however, was not possible.

One possibility to circumvent the need of introducing the sialic acid on the solid support is to pre-form a disaccharide building block containing either an α2–3- or α2–6-glycosidic bond.[51] These are the two most common ways that sialic acid is linked to an underlying oligosaccharide. Another, more preferable, option is to first assemble a non-sialylated oligosaccharide on the solid phase and then add enzymatically the sialic acid in solution. In this way, Seeberger and coworkers have recently prepared a collection of five different α2–3-sialylated oligosaccharides including the ganglioside oligosaccharide GM1b (Scheme 14.13).[52]

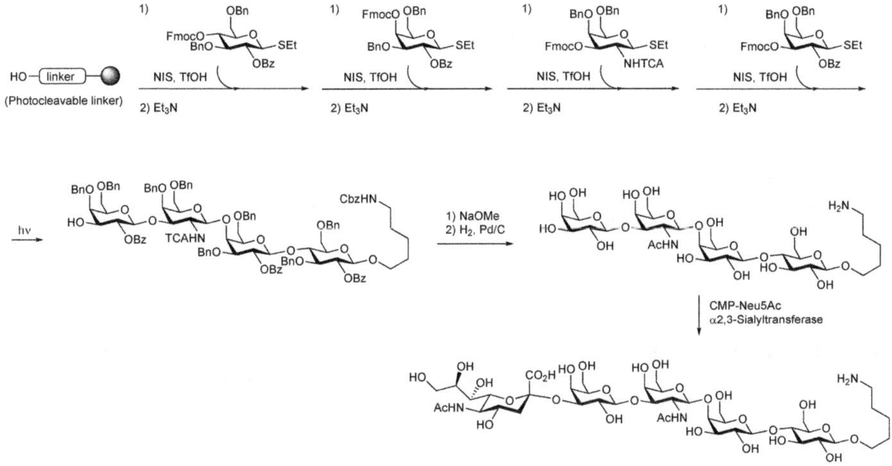

Scheme 14.13 Automated glycan assembly of α2–6-sialylated oligosaccharides using a combined approach of automated glycan assembly and enzymatic sialylation.

14.7.2 Synthesis of Bacterial Glycans

Bacterial glycans are important biological tools as they can serve as substrates for studying enzymes involved in bacterial cell wall synthesis, and are promising compounds for vaccine development. The large number of different monosaccharides and linkages found in bacterial glycans has prevented researchers from pursuing the solid-phase synthesis on a grand scale, since it is often not possible to draw on established building blocks. However, with time more and more building blocks based on rare monosaccharides will be developed and the solid-phase synthesis of bacterial glycans will become more attractive.

The first bacterial glycans prepared by solid-phase synthesis were already reported in 1987.[53] β1–5-linked oligogalactofuranosides from the mycobacterial cell wall have been assembled by iterative addition of galactofuranosyl chlorides to a solid-phase bound serine residue. After basic hydrolysis oligogalactofuranosides up to a heptasaccharide were obtained in good overall yields. It was only recently that further solid-phase syntheses of bacterial glycans were performed. The Seeberger laboratory reported the automated glycan assembly of different tetrasaccharide fragments of the capsular polysaccharide of *Streptococcus pneumonia* serotype 8.[54] These tetrasaccharides were constructed from one glucuronic acid based glycosyl phosphate and three glucose and galactose thioglycoside building blocks (Scheme 14.14). One of the prepared oligosaccharides, when conjugated to a carrier protein, induced an antibacterial immune response in rabbits and thus represents a promising candidate for the development of a semisynthetic glycoconjugate vaccine.

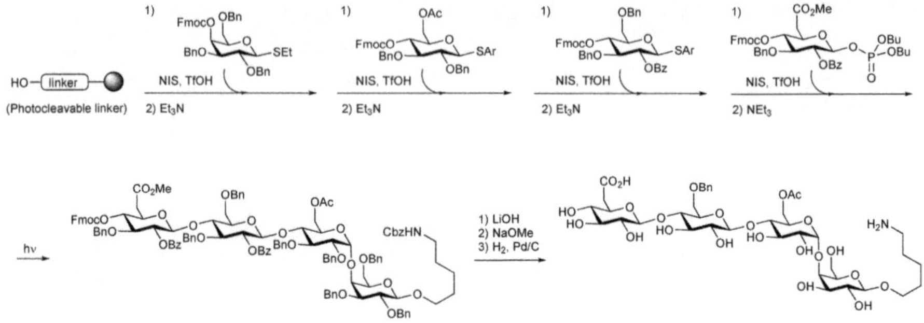

Scheme 14.14 Automated glycan assembly of a tetrasaccharide fragment of the capsular polysaccharide of *Streptococcus pneumonia* serotype 8.

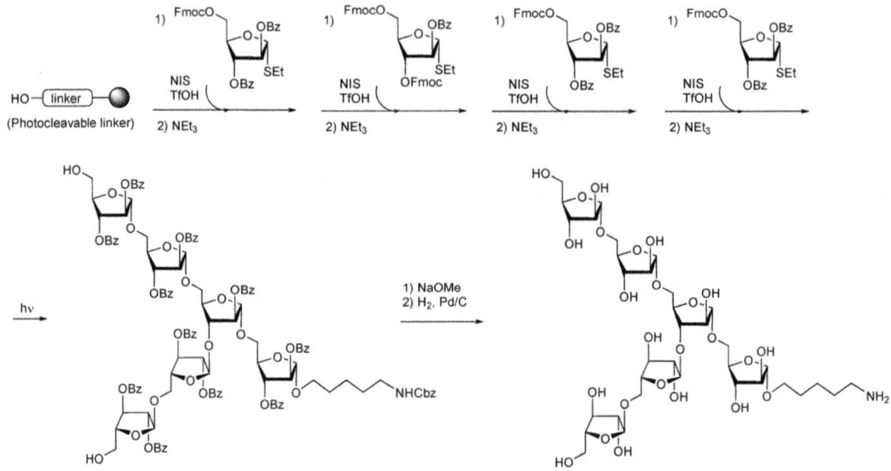

Scheme 14.15 Automated glycan assembly of a branched mycobacterial arabinan hexasaccharide.

Another bacterial glycan synthesized by automated glycan assembly was based on the cell wall polysaccharide of Group A streptococcus (GAS). Addition of *N*-phenyltrifluoroacetimidate activated disaccharide building blocks to a Merrifield resin equipped with a base-labile linker provided oligorhamnans with alternating α1–3- and α1–2-glycosidic bonds up to 16 monosaccharides in lengths.[55] Arabinans from the mycobacterial cell wall were prepared by automated glycan assembly as well (Scheme 14.15).[56] The required arabinofuranose donors proved to be very efficient building blocks for automated glycan assembly, probably due to their notoriously high reactivity. The same arabinose building blocks have also been applied in the synthesis of arabinomannan oligosaccharides.[8] All described bacterial oligosaccharides have been prepared in the fully deprotected form.

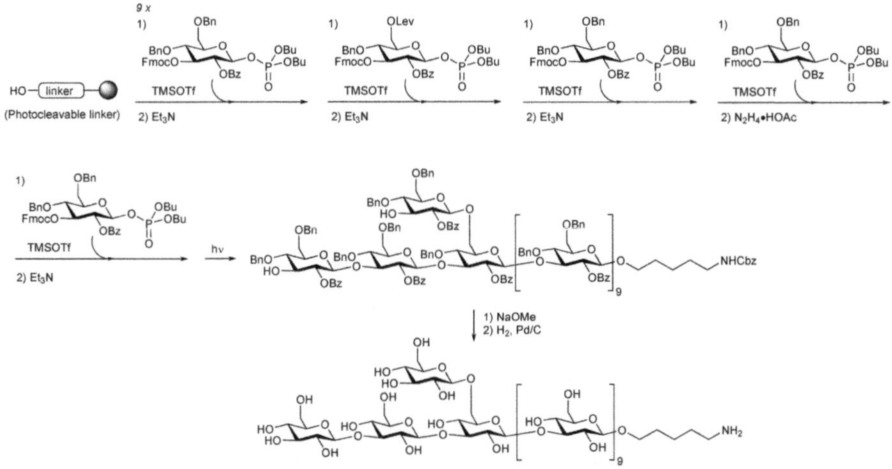

Scheme 14.16 Automated glycan assembly of a 13-mer β1–6-branched β1–3-glucan oligosaccharide.

14.7.3 Synthesis of Fungal Glycans

Fungal glycans are of interest to glycobiology research due to their ability to stimulate receptors on human immune cells. β1–3-Glucan for example binds to the C-type lectin dectin-1 and this interaction induces phagocytosis and the production of proinflammatory cytokines.[57] Seeberger and coworkers have produced linear and β1–6-branched β1–3-glucan oligosaccharides up to the size of a 13-mer by automated glycan assembly (Scheme 14.16).[58] The syntheses were based on the iterative addition of Fmoc-protected glucosyl phosphate building blocks to a Merrifield resin equipped with the photocleavable linker.

14.7.4 Synthesis of Plant Glycans

While extensive research has been devoted to the synthesis of mammalian and bacterial glycans, plant carbohydrates have received less attention. This may be for two reasons: (1) Plant carbohydrates are mostly of high structural complexity, rendering their synthesis very challenging; (2) The importance of synthetic plant carbohydrates for biological research has not yet been equally recognized as in the case of mammalian and bacterial glycans, although synthetic plant glycans are powerful molecular tools for investigating the structure and function of plant cell walls. Therefore, it does not come as a surprise that no solid-phase syntheses of plant glycans have been reported until recently when Pfrengle and coworkers performed a number of syntheses of plant cell wall glycans.[59] They reported the automated glycan assembly of a library containing 19 different xylan oligosaccharides.[60] Xylan is the second most abundant polysaccharide in plant

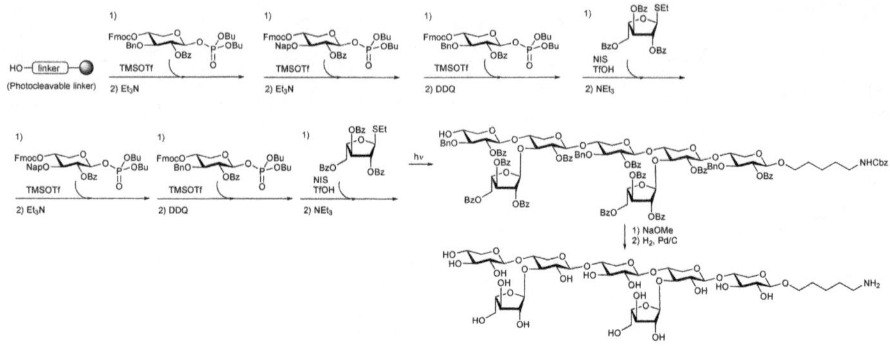

Scheme 14.17 Automated glycan assembly of a plant arabinoxylan oligosaccharide.

biomass and has thus great potential as a renewable resource for the production of fuels and new materials.[61] Xylan polysaccharides consist of β1–4-linked xylopyranoses which are partially substituted with arabinofuranose, glucuronic acid, and/or acetyl groups. Fmoc and Nap served for example as the temporary protecting groups in the automated glycan assembly of an arabinoxylan heptasaccharide containing two arabinofuranose substituents (Scheme 14.17).[62] While Fmoc was installed at the 4-position of the two xylose building blocks used for linear elongation of the backbone, Nap was additionally placed in the 3-position of one of the xylose building blocks to enable branching with arabinose.

Another group of oligosaccharides prepared by automated glycan assembly are fragments of the plant cell wall polysaccharide xyloglucan which plays an important role in plant cell wall architecture.[63] Xyloglucans have a cellulosic β1–4-glucan backbone which is substituted with α-linked xylopyranoses at most of the primary hydroxyl groups of the glucose units. The xyloses in turn may be further substituted with β1–2-linked galactose and other monosaccharides. A particular challenge in the chemical synthesis of xyloglucan oligosaccharides represents the α-xylosidic bond which is usually formed with low stereoselectivity.[64] To avoid constructing this bond on the solid phase, Pfrengle and coworkers used a disaccharide building block having the α-xylosidic linkage preinstalled in the automated glycan assembly of xyloglucan oligosaccharides (Scheme 14.18).[65] Formation of the β1–2-linkage between xylose and galactose required a temporary non-participating protecting group in the 2-position of the xylose residue. A *p*-methoxybenzyl (PMB) group served this purpose and was selectively cleaved on the solid phase using DDQ. However, care must be taken when using PMB in solid-phase carbohydrate chemistry as it may be labile to the strongly acidic conditions used for some glycosylation reactions. Other plant cell wall oligosaccharides that have been prepared by automated glycan assembly are mixed-linkage glucan[66] and arabinogalactan[67,68] oligosaccharides which have been prepared in a similar manner as the xylan and xyloglucan oligosaccharides described above.

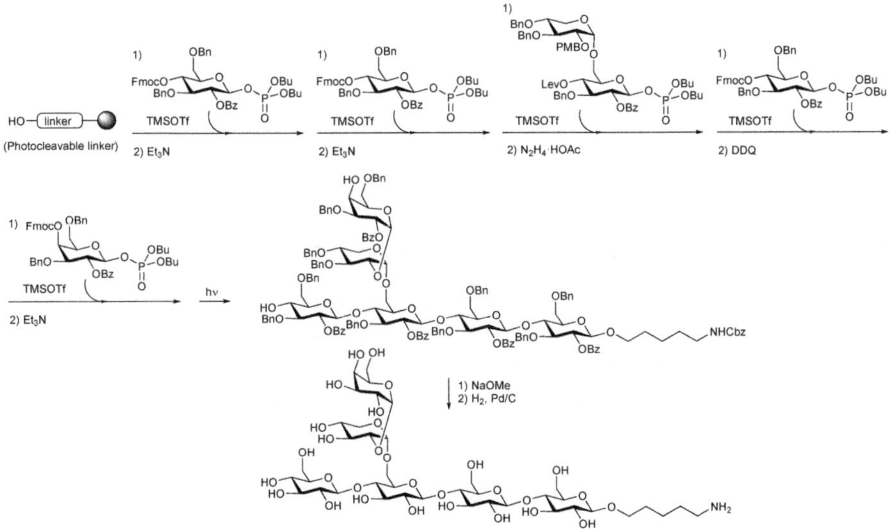

Scheme 14.18 Automated glycan assembly of a galactosylated xyloglucan hexasaccharide using a disaccharide building block having the difficult α-xylosidic linkage preinstalled.

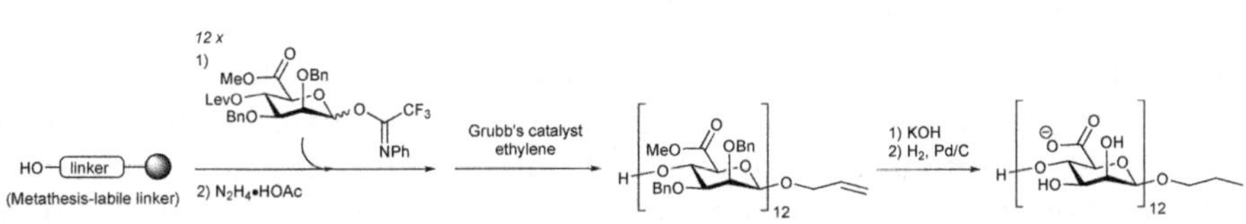

Scheme 14.19 Automated glycan assembly of a mannuronic acid alginate dodecasaccharide using a highly β-selective mannuronic acid building block.

14.7.5 Synthesis of Algal Oligosaccharides

Polymannuronic acid alginate is a constituent of many algal cell walls. It is composed of a linear arrangement of β1–4-linked mannuronic acid monomers. Oligosaccharide fragments of this polysaccharide were reported to have immunomodulatory properties.[69] The Codee lab described the automated glycan assembly of mannuronic acid alginates up to a 12-mer (Scheme 14.19).[70] Key to the success of this challenging synthesis was the fact that glycosylations with mannuronic acid donors are highly 1,2-*cis* selective while the corresponding mannosyl donors are 1,2-*trans* selective. A butenediol linker was chosen for attachment of the growing glycan chain to a Merrifield resin, preventing the use of thioglycoside building blocks. Instead, *N*-phenyltrifluoroacetimidates, which can be prepared from the respective thioglycosides on a multigram scale, served as the glycosyl donors and Lev was used as the temporary protecting group for chain elongation.

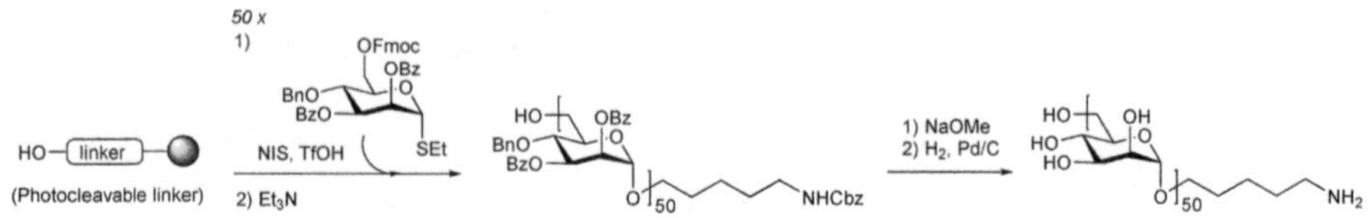

Scheme 14.20 Automated glycan assembly of a 50-mer polymannoside.

14.8 Synthesis of Large Oligomers

The size of molecules that can be prepared can serve as a benchmark of how powerful a technique is. A 50-mer oligomannoside is the largest glycan prepared by automated glycan assembly to date.[71] A mannose protected with two benzyl and one benzoyl group was found to be the ideal building block to be added to a Merrifield resin equipped with the photocleavable linker to provide an α1–6-linked polymannoside in a 102 step synthesis (Scheme 14.20). As separation of the product from potential deletion sequences becomes increasingly difficult with glycan lengths, acetylation steps were introduced to cap any free hydroxyl groups remaining after the individual glycosylation reactions.

14.9 Conclusions

The technology of solid-phase glycan synthesis has developed in the last almost 50 years to a point where large and complex glycans can be routinely assembled with high efficiency, as demonstrated by the many impressive syntheses described in this chapter. The key achievement to accelerate these developments was the automation of solid-phase glycan synthesis (automated glycan assembly), permitting researchers to perform many more syntheses in the same time. Significant advances in the last couple of years have been the synthesis of glycans containing multiple 1,2-*cis*-glycosidic bonds and of long homooligomers. Furthermore, automated glycan assembly has proven useful for the synthesis of collections of structurally related oligosaccharides that were for example applied in plant cell wall biology.[72] However, there are a number of remaining challenges to be addressed before principally any glycan of choice can be efficiently prepared. The set of building blocks available for solid-phase synthesis must be expanded to include rare monosaccharides that are required for the synthesis of many bacterial and plant glycans. Further research into the construction of difficult glycosidic linkages such as in β-mannosides and α-xylosides is needed and glycosylation efficiency must be improved for reactions where many equivalents of expensive building blocks are required such as in the case of uronic acid donors. These developments will further increase the immense impact that automated glycan assembly has already made on research in glycobiology, in particular by providing essential molecular tools for the population of glycan arrays or antigenic components for glycoconjugate vaccines.

Acknowledgements

Financial support from the Max Planck society and the German Research Foundation (DFG, Emmy Noether program PF850/1-1) is most gratefully acknowledged.

References

1. A. Varki and P. Gagneux, in *Essentials of Glycobiology*, ed. A. Varki, R. D. Cummings, J. D. Esko, P. Stanley, G. W. Hart, M. Aebi, A. G. Darvill, T. Kinoshita, N. H. Packer, J. H. Prestegard, R. L. Schnaar and P. H. Seeberger, Cold Spring Harbor Laboratory Press, Cold Spring Harbor, NY, USA, 3rd edn, 2017.
2. D. B. Werz, *Methods Mol. Biol.*, 2012, **808**, 13.
3. F. Albericio, *Curr. Opin. Chem. Biol.*, 2004, **8**, 211.
4. C. B. Reese, *Org. Biomol. Chem.*, 2005, **3**, 3851.
5. C. Schuerch and J. M. Frechet, *J. Am. Chem. Soc.*, 1971, **93**, 492.
6. O. J. Plante, E. R. Palmacci and P. H. Seeberger, *Science*, 2001, **291**, 1523.
7. L. Krock, D. Esposito, B. Castagner, C.-C. Wang, P. Bindschadler and P. H. Seeberger, *Chem. Sci.*, 2012, **3**, 1617.
8. H. S. Hahm, M. K. Schlegel, M. Hurevich, S. Eller, F. Schuhmacher, J. Hofmann, K. Pagel and P. H. Seeberger, *Proc. Natl. Acad. Sci. U. S. A.*, 2017, **114**, E3385.
9. P. H. Seeberger, *Acc. Chem. Res.*, 2015, **48**, 1450.
10. K. Tanaka and K. Fukase, in *Solid-Phase Organic Synthesis: Concepts, Strategies, and Applications*, ed. P. H. Toy and Y. Lam, John Wiley & Sons, Inc., Hoboken, 2012, pp. 489–530.
11. P. H. Seeberger and W.-C. Haase, *Chem. Rev.*, 2000, **100**, 4349.
12. C. S. Bennett, *Org. Biomol. Chem.*, 2014, **12**, 1686.
13. S. G. Pistorio, S. S. Nigudkar, K. J. Stine and A. V. Demchenko, *J. Org. Chem.*, 2016, **81**, 8796.
14. R. B. Andrade, O. J. Plante, L. G. Melean and P. H. Seeberger, *Org. Lett.*, 1999, **1**, 1811.
15. A. R. de Jong, A. G. Volbeda, B. Hagen, H. van den Elst, H. S. Overkleeft, G. A. van der Marel and J. D. C. Codée, *Eur. J. Org. Chem.*, 2013, **2013**, 6644.
16. L. Knerr and R. R. Schmidt, *Eur. J. Org. Chem.*, 2000, **2000**, 2803.
17. X. Wu, M. Grathwohl and R. R. Schmidt, *Angew. Chem., Int. Ed.*, 2002, **41**, 4489.
18. K. Tanaka, Y. Fujii, H. Tokimoto, Y. Mori, S.-i. Tanaka, G.-m. Bao, E. R. O. Siwu, A. Nakayabu and K. Fukase, *Chem. - Asian J.*, 2009, **4**, 574.
19. P. Czechura, N. Guedes, S. Kopitzki, N. Vazquez, M. Martin-Lomas and N.-C. Reichardt, *Chem. Commun.*, 2011, **47**, 2390.
20. G. Excoffier, D. Gagnaire and J.-P. Utille, *Carbohydr. Res.*, 1975, **39**, 368.
21. K. C. Nicolaou, N. Winssinger, J. Pastor and F. DeRoose, *J. Am. Chem. Soc.*, 1997, **119**, 449.
22. U. Zehavi and A. Patchornik, *J. Am. Chem. Soc.*, 1973, **95**, 5673.

23. S. Eller, M. Collot, J. Yin, H. S. Hahm and P. H. Seeberger, *Angew. Chem., Int. Ed.*, 2013, **52**, 5858.
24. M. Wilsdorf, D. Schmidt, M. P. Bartetzko, P. Dallabernardina, F. Schuhmacher, P. H. Seeberger and F. Pfrengle, *Chem. Commun.*, 2016, **52**, 10187.
25. C. T. Tanifum, J. Zhang and C.-W. T. Chang, *Tetrahedron Lett.*, 2010, **51**, 4323.
26. M. Izumi, K. Fukase and S. Kusumoto, *Synlett*, 2002, **2002**, 1409.
27. H. Shimizu, Y. Ito, O. Kanie and T. Ogawa, *Bioorg. Med. Chem. Lett.*, 1996, **6**, 2841.
28. J. Rademann and R. R. Schmidt, *J. Org. Chem.*, 1997, **62**, 3650.
29. S. H. Nguyen, A. H. Trotta, J. Cao, T. J. Straub and C. S. Bennett, *Org. Biomol. Chem.*, 2012, **10**, 2373.
30. S. S. Nigudkar and A. V. Demchenko, *Chem. Sci.*, 2015, **6**, 2687.
31. A. V. Demchenko, E. Rousson and G. J. Boons, *Tetrahedron Lett.*, 1999, **40**, 6523.
32. H. S. Hahm, M. Hurevich and P. H. Seeberger, *Nat. Commun.*, 2016, **7**, 12482.
33. J.-H. Kim, H. Yang, J. Park and G.-J. Boons, *J. Am. Chem. Soc.*, 2005, **127**, 12090.
34. T. J. Boltje, J. H. Kim, J. Park and G. J. Boons, *Nat. Chem.*, 2010, **2**, 552.
35. D. Crich and S. Sun, *J. Am. Chem. Soc.*, 1998, **120**, 435.
36. J. D. C. Codee, L. Krock, B. Castagner and P. H. Seeberger, *Chem. - Eur. J.*, 2008, **14**, 3987.
37. M. Mogemark, M. Elofsson and J. Kihlberg, *Org. Lett.*, 2001, **3**, 1463.
38. T. Kanemitsu, C.-H. Wong and O. Kanie, *J. Am. Chem. Soc.*, 2002, **124**, 3591.
39. X. Wu, M. Grathwohl and R. R. Schmidt, *Org. Lett.*, 2001, **3**, 747.
40. K.-S. Ko, G. Park, Y. Yu and N. L. Pohl, *Org. Lett.*, 2008, **10**, 5381.
41. J. Wu and Z. W. Guo, *J. Org. Chem.*, 2006, **71**, 7067.
42. E. R. Palmacci, M. C. Hewitt and P. H. Seeberger, *Angew. Chem., Int. Ed.*, 2001, **40**, 4433.
43. H. Ando, S. Manabe, Y. Nakahara and Y. Ito, *Angew. Chem., Int. Ed.*, 2001, **40**, 4725.
44. O. Calin, S. Eller and P. H. Seeberger, *Angew. Chem., Int. Ed.*, 2013, **52**, 5862.
45. K. Routenberg Love and P. H. Seeberger, *Angew. Chem., Int. Ed.*, 2004, **43**, 602.
46. H. S. Hahm, C.-F. Liang, C.-H. Lai, R. J. Fair, F. Schuhmacher and P. H. Seeberger, *J. Org. Chem.*, 2016, **81**, 5866.
47. J. Kandasamy, F. Schuhmacher, H. S. Hahm, J. C. Klein and P. H. Seeberger, *Chem. Commun.*, 2014, **50**, 1875.
48. H. S. Hahm, F. Broecker, F. Kawasaki, M. Mietzsch, R. Heilbronn, M. Fukuda and P. H. Seeberger, *Chem*, 2017, **2**, 114.
49. D. B. Werz, B. Castagner and P. H. Seeberger, *J. Am. Chem. Soc.*, 2007, **129**, 2770.
50. C.-H. Lai, H. S. Hahm, C.-F. Liang and P. H. Seeberger, *Beilstein J. Org. Chem.*, 2015, **11**, 617.
51. D. Esposito, M. Hurevich, B. Castagner, C.-C. Wang and P. H. Seeberger, *Beilstein J. Org. Chem.*, 2012, **8**, 1601.

52. R. J. Fair, H. S. Hahm and P. H. Seeberger, *Chem. Commun.*, 2015, **51**, 6183.
53. G. H. Veeneman, S. Notermans, R. M. J. Liskamp, G. A. van der Marel and J. H. van Boom, *Tetrahedron Lett.*, 1987, **28**, 6695.
54. B. Schumann, H. S. Hahm, S. G. Parameswarappa, K. Reppe, A. Wahlbrink, S. Govindan, P. Kaplonek, L.-a. Pirofski, M. Witzenrath, C. Anish, C. L. Pereira and P. H. Seeberger, *Sci. Transl. Med.*, 2017, **9**, eaaf5347.
55. A. Geert Volbeda, J. van Mechelen, N. Meeuwenoord, H. S. Overkleeft, G. A. van der Marel and J. D. C. Codée, *J. Org. Chem.*, 2017, **82**, 12992.
56. J. Kandasamy, M. Hurevich and P. H. Seeberger, *Chem. Commun.*, 2013, **49**, 4453.
57. G. D. Brown, *Nat. Rev. Immunol.*, 2006, **6**, 33.
58. M. W. Weishaupt, S. Matthies and P. H. Seeberger, *Chem. - Eur. J.*, 2013, **19**, 12497.
59. F. Pfrengle, *Curr. Opin. Chem. Biol.*, 2017, **40**, 145.
60. D. Senf, C. Ruprecht, G. H. M. de Kruijff, S. O. Simonetti, F. Schuhmacher, P. H. Seeberger and F. Pfrengle, *Chem. - Eur. J.*, 2017, **23**, 3197.
61. A. Ebringerova and T. Heinze, *Macromol. Rapid Commun.*, 2000, **21**, 542.
62. D. Schmidt, F. Schuhmacher, A. Geissner, P. H. Seeberger and F. Pfrengle, *Chem. - Eur. J.*, 2015, **21**, 5709.
63. Y. B. Park and D. J. Cosgrove, *Plant Cell Physiol.*, 2015, **56**, 180.
64. P. Dallabernardina, F. Schuhmacher, P. H. Seeberger and F. Pfrengle, *Org. Biomol. Chem.*, 2016, **14**, 309.
65. P. Dallabernardina, C. Ruprecht, P. J. Smith, M. G. Hahn, B. R. Urbanowicz and F. Pfrengle, *Org. Biomol. Chem.*, 2017, **15**, 9996.
66. P. Dallabernardina, F. Schuhmacher, P. H. Seeberger and F. Pfrengle, *Chem. - Eur. J.*, 2017, **23**, 3191.
67. M. P. Bartetzko, F. Schuhmacher, P. H. Seeberger and F. Pfrengle, *J. Org. Chem.*, 2017, **82**, 1842.
68. M. P. Bartetzko, F. Schuhmacher, H. S. Hahm, P. H. Seeberger and F. Pfrengle, *Org. Lett.*, 2015, **17**, 4344.
69. M. Iwamoto, M. Kurachi, T. Nakashima, D. Kim, K. Yamaguchi, T. Oda, Y. Iwamoto and T. Muramatsu, *FEBS Lett.*, 2005, **579**, 4423.
70. M. T. C. Walvoort, H. van den Elst, O. J. Plante, L. Kröck, P. H. Seeberger, H. S. Overkleeft, G. A. van der Marel and J. D. C. Codée, *Angew. Chem., Int. Ed.*, 2012, **51**, 4393.
71. K. Naresh, F. Schumacher, H. S. Hahm and P. H. Seeberger, *Chem. Commun.*, 2017, **53**, 9085–9088.
72. C. Ruprecht, M. P. Bartetzko, D. Senf, P. Dallabernadina, I. Boos, M. C. F. Andersen, T. Kotake, J. P. Knox, M. G. Hahn, M. H. Clausen and F. Pfrengle, *Plant Physiol.*, 2017, **175**, 1094.

CHAPTER 15

Reverse Synthesis of Natural Glycans

YUYANG ZHU, DAVID F. SMITH AND XUEZHENG SONG*

Department of Biochemistry, Emory Comprehensive Glycomics Core, Emory University School of Medicine, Atlanta, GA, USA 30322
*E-mail: xsong2@emory.edu

15.1 Introduction

Glycosylation is one of the most important post-translational modifications occurring widely in various biological systems. Glycosylation includes not only the glycosidic bond formation to produce oligosaccharides and polysaccharides (glycans) from monosaccharides, but also the bioconjugation of glycans with other biomolecules such as proteins and lipids. Accompanied by DNA, RNA and proteins, naturally occurring glycans are one of the indispensable biopolymers affecting many aspects of biology such as cell signaling, pathogen recognition, protein folding, activation of cell–cell interactions, *etc.* (Figure 15.1).[1]

Although study of the biological roles of glycans has received considerable interest in recent years,[1–6] the study of glycans or glycoscience, has lagged far behind the study of other biomolecules, such as proteins/peptides and nucleic acids. This is due to the structural complexity of naturally occurring glycans. While nucleic acids and proteins are linear polymers, glycans are often highly branched molecules. The structural features of a glycan include not only sequences of monosaccharides, but also linkage positions and anomeric stereochemistry of glycosidic bonds. While the biosynthesis of

Chemical Biology No. 11
Synthetic Glycomes
Edited by Wanyi Guan, Lei Li and Peng George Wang

Published by the Royal Society of Chemistry, www.rsc.org

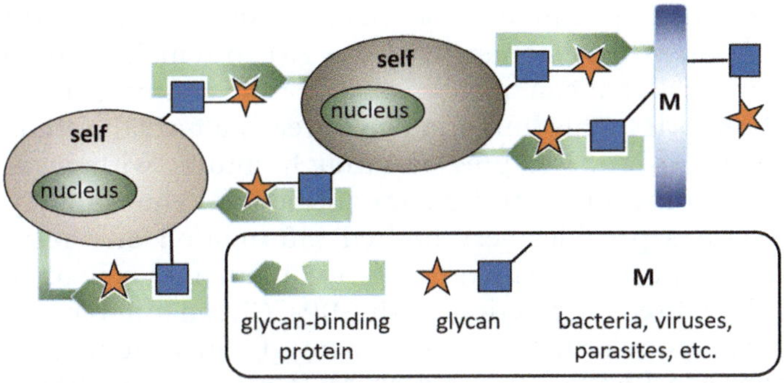

Figure 15.1 General classification of the biological roles of glycans. A simplified and broad classification is presented, especially emphasizing the roles of organism-intrinsic and organism-extrinsic glycan-binding proteins in recognizing glycans. There is some overlap between the categories, *e.g.*, some structural properties involve specific recognition of glycans. Binding shown on the left of the central "self" cell represents intrinsic recognition, and extrinsic recognition is represented by binding shown to the right of that cell. Molecular mimicry of host glycans by the external entity adds further complexity to potential roles. Adapted from ref. 1 with permission from Oxford University Press, Copyright The Authors 2016.

proteins and nucleic acids are template driven and determined by genetic codes, the biosynthesis of glycans requires multiple glycosyltransferases with different substrate and linkage specificities. As a result, genomics and proteomics, the systematic study of nucleic acids and proteins respectively, have advanced greatly in the last decades. Whole genome sequencing can be achieved through a number of next generation sequencing (NGS) platforms at affordable costs. Nucleic acids can be routinely synthesized from a virtual sequence using automatic DNA synthesizer or polymerase chain reaction (PCR). Similarly, protein/peptide can be sequenced using modern mass spectrometry (MS) in a high throughput fashion. Automatic peptide synthesis can be carried out as common practice in many laboratories or through many commercial service providers, and large proteins can be expressed using recombinant technologies. Using these high throughput techniques for genomics and proteomics, virtual sequence information of proteins and nucleic acids can be quickly obtained using only a small amount of biological material. From a virtual sequence, actual protein/peptide or nucleic acids can be synthesized for functional study. This free-flow between virtual sequence information and actual biologic material is essential for modern genomics and proteomics, which unfortunately does not exist for glycomics yet. Despite many techniques for characterizing glycans including nuclear magnetic resonance (NMR), high-performance liquid chromatography (HPLC)[7–9] and mass spectrometry,[10] high throughput glycan structural analysis is not yet available. Besides the inherent structural complexity of glycans, the lack of defined glycans as structural standards greatly impeded the development of high throughput analytics for glycomics.[4,5]

Despite recent improvements of chemical and enzymatic synthesis of glycans,[11–17] the synthesis of any complex glycan remains a time consuming, laborious task that can only be carried out in a limited number of specialized laboratories. This limitation also greatly affects the study of glycan function, which is largely expressed through binding with glycan binding proteins (GBPs) including anti-glycan antibodies. Thus, simply profiling glycan structures provides very limited information on their biological functions. One approach to identifying glycan function is to obtain glycans and study their direct interactions with GBPs. Glycan microarray[18–20] assays have become a popular tool to screen many glycan structures to elucidate GBP binding specificities. In a glycan microarray, many different glycan structures are immobilized onto a microarray slide through either covalent or non-covalent attachment. The slide is assayed with relevant GBPs or antibodies, which are fluorescently tagged either directly or through a secondary binding. For example, a biotinylated plant lectin can be applied onto the glass slide with a number of glycans immobilized at different locations. After washing off the unbound lectin, fluorescently tagged streptavidin is applied on the slide to light up the spots where specific lectin-glycan binding occurs. The fluorescent image is processed to identify the bound glycans and the comparison of bound and unbound structures provides information on the binding specificity of the lectin. Besides efficient immobilization of glycans on the glass slide, the success of a glycan microarray is largely determined by the structural diversity and biological relevance of the glycans. It is well recognized that one of the major obstacles for glycoscience is the lack of access to large numbers of biomedically relevant complex glycans in sufficient quantities to drive both the structural analysis methods and functional studies. Currently no methods are available to amplify the glycome from any source making lack of sufficient material for study a major problem for glycoscience. Therefore, synthetic targets should be carefully chosen based on their potential biological functions. Unfortunately, without first obtaining certain natural glycan structures and studying their functions, it is difficult to choose these targets for synthesis. Currently the synthetic approach is also relatively low throughput and cannot meet the fast growth of interest in glycoscience.

An alternative approach to synthesis is the release and purification of natural glycans. In analytic glycomics, natural glycans are routinely released from biological materials and derivatized for HPLC and MS characterization. This process usually involves special enzymes or harsh chemical treatment, both of which can only be applied to small amounts of biological samples. Since glycans are minor components of cells, this traditional process can only provide a small amount of glycans which are generally adequate for analytical purposes, and since current glycan microarray preparation requires only a small amount of material, we and others have used this approach to prepare natural glycan microarrays for the study of their specific bindings with GBPs and viruses. However, the amounts of glycan material generated using a traditional glycomics approach significantly limits more detailed structural and functional study beyond MS profiling and glycan microarray preparation. We reasoned that without a useful amplification technology, the only remaining approach would be to use large quantities of natural products as starting material. This approach, however, required the development of simple, inexpensive chemical approaches that could process kilogram levels of biological material and generate quantities of natural glycans in quantities comparable to chemoenzymatic synthesis. Since these glycans would be produced from more complex glycoconjugates, we term this approach "reverse synthesis". In this chapter, we will review the "reverse synthesis" of natural glycans, including *N*- and *O*-glycans from glycoproteins and glycans from glycosphingolipids, from the traditional glycomics approach to the current large scale chemical approach.

15.2 *N*-Glycans Release

15.2.1 Enzymatic Approaches

N-Glycans are attached to certain asparagine residues of the protein backbone (Figure 15.2). *N*-glycans are the most well-studied class of glycans and several enzymes are found to cleave *N*-glycans off glycoproteins. Peptide-*N*-Glycosidase F (PNGase F) is the most popular reagent to remove *N*-glycans from glycoproteins and glycopeptides.[21,22] PNGase F is an amidohydrolase which can hydrolyze most *N*-linked complex, hybrid, or high mannose oligosaccharides to generate deaminated peptide or protein and 1-amino glycan. The 1-amino glycan can be further hydrolyzed to free reducing glycans or directly captured by amine-reactive reagents for HPLC or MS analysis.[21,22] However, *N*-glycans containing α1–3-fucose at the chitobiose core, which are commonly found in plants and insects, cannot be released by PNGase F.[23] PNGase A isolated from almond meal was found to have broader substrate specificity including core α1–3-fucosylated glycopeptides and glycoproteins.[24] Recent reports on PNGases demonstrate that PNGase H$^+$,[25] F-II[26] and Yl[27] can combine the advantages of PNGase F and A. Very recently, PNGase Ar, a commercially available recombinant form of rice vacuolar PNGase, shows the capability to release GalαFucα1,3 reducing

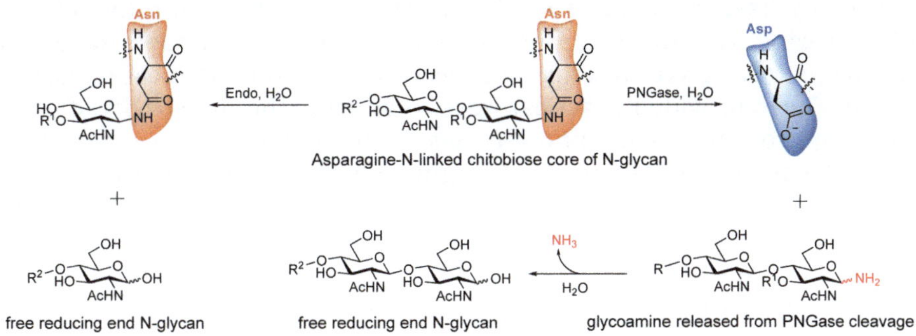

Figure 15.2 PNGase and Endo hydrolysis pathways on asparagine-linked *N*-glycan, R^1 = H, fucose, R^2 = H, other oligosaccharide chains.

terminal core from *Caenorhabditis elegans*.[28] The release of *N*-glycans using PNGase F and A from glycopeptides and glycoproteins usually requires denaturing of proteins for higher deglycosylation efficiency. Endoglycosidases cleaving the β1–4-linked di-*N*-acetylchitobiose core is another option for *N*-glycan release, and such enzymes include Endo H isolated from *Streptomyces plicatus* and endoglycosidases F from *Elizabethkingia meningoseptica* (Endo F1, F2 and F3). In opposition to PNGase F, endoglycosidases F from *Elizabethkingia meningoseptica* (Endo F1, F2 and F3) are more robust towards protein/peptide conformation and are more efficient for releasing glycans from native proteins and peptides. Although Endo F and Endo H cleave *N*-glycans at the same position, they have different substrate specificities related to fine structures of the *N*-glycans. Endo F1 can hydrolyze high mannose and hybrid structures without core α1–3-fucose presence. Endo F2 mainly cleaves biantennary *N*-glycans and it can tolerate core α1–3-fucose. Endo F3 can react with both bi- and triantennary glycans attached to the peptide and has efficient performance to cleave core α1–3-fucose of biantennary structures. Endo H usually cleaves high mannose glycans and can also cut core α1–3-fucose structures.[29]

Another enzymatic approach is the full digestion of proteins using proteases such as pronase. These enzymes cleave all peptide bonds with broad specificities leaving glycan-peptide linkages intact. The resulting glycoamino acid can be isolated from the pronase digest and labeled with fluorescent tags such as Fmoc.[30] Although this approach is significantly cheaper than using PNGases or endoglycosidases and can be potentially applied to higher scale, full digestion of glycoproteins to glycoamino acids is often hard to control and seldom 100% efficient resulting in heterogeneous glycopeptides with short peptide chains.

15.2.2 Chemical Approaches

Although enzymatic approaches to release *N*-glycans from glycoproteins and glycopeptides have contributed greatly to the development of the glycosciences, these enzymes are expensive and often require pre-treatment such as reduction/alkylation of glycoproteins and trypsin digestion, and many have restricted specificities that prevent quantitative release of *N*-glycans. Chemical approaches have long been pursued to release *N*-glycans from glycoproteins and recent reports on chemical release of glycans have provided practical ways to address the drawbacks of enzymatic approaches. Hydrazinolysis has been shown to release *N*-glycans from glycoproteins. The method originally uses anhydrous hydrazine as reagents.[31] Due to the toxicity of anhydrous hydrazine, hydrazine hydrate has also been employed as a satisfactory substitute for the release of *N*-glycans.[32] Nevertheless the release of *N*-glycans by hydrazinolysis requires high temperature and long time, which may affect the structural integrity of *N*-glycans through peeling reactions. *N*-glycans have also been released under basic conditions along with *O*-glycans.[33] Under a set of optimized mild alkaline conditions, *N*-glycans without core α1–3-fucose can be released by selective hydrolysis of *N*-glycopeptide with no detectable deacetylation or peeling reaction. The core α1–3-fucosylated *N*-glycans can be truncated and derivatized with 1-phenyl-3-methyl-5-pyrazolone (PMP) *in situ*.[34] More recently, we have reported two different chemical approaches for large scale release of *N*-glycans. In the first approach, we developed a novel "chemoenzymatic" method named "threshing and trimming (TaT)" to release *N*-glycans (Figure 15.3).[35] Firstly, glycoproteins, tissues, and organs were treated with pronase (threshing step) to create a pool of *N*-glycoamino acids and glycopeptides with short peptide moieties. Secondly, *N*-bromosuccinimide (NBS) was used to "trim" these short glycopeptides to generate aglycon products as free reducing glycans, nitriles, or aldehyde under different reaction conditions. Released free reducing end glycans and aldehydes can be further derivatized with bifunctional fluorescent 2-amino-*N*-(2-aminoethyl)benzamide (AEAB) for HPLC purification, MALDI-TOF-MS analysis and microarray preparations. *N*-glycan nitriles, which are released by NBS under neutral pH at room temperature, can be labeled with 2-aminobenzamide (2-AB) by palladium/carbon (Pd/C) catalyzed *N*-alkylation for further biological function study. TaT provides an approach for *N*-glycan release without using specialty enzymes, hazardous chemical reagents, and harsh reaction conditions so that it can be scaled up for larger amounts of biological materials and the structural integrity of released glycans can be retained.

Although TaT is a robust method to release *N*-glycans from glycopeptides, the first step Pronase digestion requires long time and *O*-glycans were not

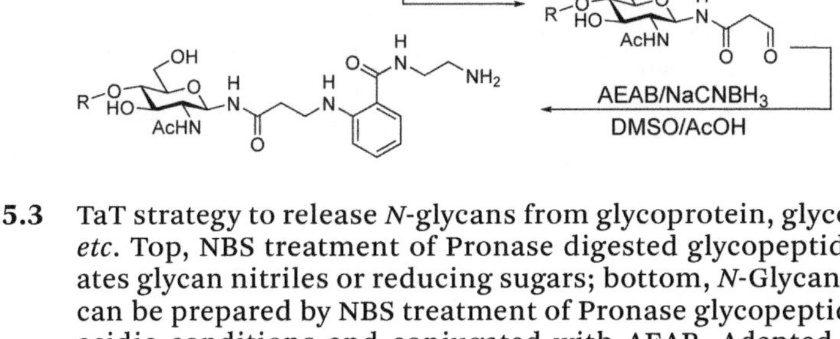

Figure 15.3 TaT strategy to release *N*-glycans from glycoprotein, glycopeptides *etc.* Top, NBS treatment of Pronase digested glycopeptides generates glycan nitriles or reducing sugars; bottom, *N*-Glycan aldehyde can be prepared by NBS treatment of Pronase glycopeptides under acidic conditions and conjugated with AEAB. Adapted from ref. 35 with permission from American Chemical Society, Copyright 2014.

efficiently released after the whole treatment procedure. It is possible that *O*-glycans inhibit pronase digestion at proximity to amide linkages and prevent the "trimming" by NBS. Inspired by NBS treatment, we examined other halogen-related oxidants and made a surprising discovery that sodium hypochlorite (NaClO) (household bleach) is an efficient oxidative reagent to liberate nearly all species of glycans from crude biological materials (*N*-, and *O*-glycans, GSLs).[36] This method is named oxidative release of natural glycans (ORNG). After homogenizing natural materials such as animal/plant tissues with water using a Warring blender, household bleach is added and stirred for less than 30 minutes at room temperature. After acid precipitation, free reducing *N*-glycans were released into the solution, which can be purified from other impurities using a number of chromatography techniques including size exclusion, anion exchange and hydrophobic/hydrophilic interaction. Purified glycans can be fluorescently tagged by reductive amination and separated into individual components by multidimensional HPLC. The process is fast and simple to scale up and can be applied to kilograms of materials. In our recent study, the ORNG method was applied to a kilogram of commercial soy protein isolates to prepare several high mannose and paucimannose *N*-glycans with individual yields up to 125 milligram.[37] By analyzing the side products of ORNG reaction with soyprotein sample, we observed in addition to *N*-glycans, glycans with a single GlcNAc at the reducing end, glyconitriles and glycans with erythrose at the reducing end. Based on these observations we proposed that the reaction pathway involves a glycan-isocyanate intermediate. This glycan-isocyanate species can be further hydrolyzed to free reducing glycans as the major product. Meanwhile, after isocyanate converting to glycosyamine, glycosyamine can be chlorinated and degraded through a six-membered ring pericyclic rearrangement to form glycans with erythrose reducing end or glyconitrile (Figure 15.4).

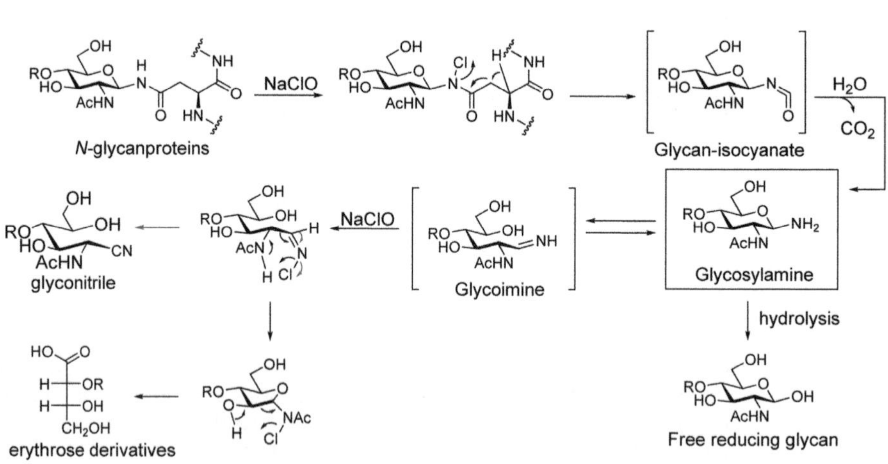

Figure 15.4 Proposed mechanisms of ORNG method to release *N*-glycan. Adapted from ref. 37 with permission from Elsevier, Copyright 2018.

15.3 *O*-Glycans Release

15.3.1 Enzymatic Approaches

O-Glycans attached to serine or threonine residues of proteins, especially mucin-type *O*-GalNAc-containing glycans, are another major post-translational modification of proteins. The study of *O*-glycans is considered more challenging for several reasons. First, while *N*-glycosylation follows the consensus sequence Asn-X-Ser/Thr (X including all amino acid except proline), there is no such sequence known for *O*-glycosylation. Second, there is no enzyme with broad specificities cleaving *O*-glycans off the protein backbone. The only known *O*-glycanase, endo-α-*N*-acetylgalactosaminidase (endo-GalNAc-ase), can only cleave mucin type core-1 *O*-glycan (Galβ1-3GalNAc).[38] This significantly limits the current study of *O*-glycosylations.

15.3.2 Chemical Approaches

Despite the fact that there are no well-established enzymatic methods for *O*-glycans as those for *N*-glycans, chemical approaches to release *O*-glycans have been well studied in both solutions or on a membrane immobilized with *O*-glycoproteins.[39,40] Several chemical approaches have been explored for releasing *O*-glycans over the past several decades. The most commonly used method is reductive β-elimination. Glycoproteins were mixed with sodium or potassium hydroxide and sodium borohydride. After sodium hydroxide hydrolyzes *O*-glycans from the glycoprotein, sodium borohydride reduces glycans to form alditols to minimize glycan peeling reaction. Without *in situ* reduction, peeling reaction for *O*-GalNAc glycans under basic conditions is much more abundant than that for *N*-glycans. This is due to the 3-O-substitution of GalNAc-Ser/Thr behaving as a good leaving group for β-elimination. Besides a strong base, other mild bases such as dimethylamine can be utilized to release *O*-glycans.[41,42] To increase the efficiency of reductive β-elimination, glycoproteins can be first digested to single amino acid level using extensive protease.[43] Since the final product of reductive β-elimination is alditol, this approach prevents further derivatization of released *O*-glycans, either for UV or fluorescence detection during HPLC separation or for microarray-based functional study. To avoid formation of non-reactive alditols and to further derivatize released glycans for structural and functional studies, non-reductive β-elimination methods have been developed. Aqueous ethylamine[44] and lithium hydroxide[45] were used to maintain the released *O*-glycan without further reducing. These two reagents, however, caused low releasing efficiency with a high level of peeling products. Combined with ammonium carbonate, ammonium hydroxide was used for *O*-glycan release through β-elimination and the inhibition of peeling reaction is ascribed to the formation of glucosylamine carbamates.[33] Moreover, employing ammonium carbamate can further reduce or eliminate peeling reactions during *O*-glycan release and prevent the deletion of acid-labile sialic acid residues.[46] Nevertheless, peeling reaction is unlikely to be fully eliminated during ammonium-based *O*-glycan release.[47] β-Elimination using hydrazine has been explored extensively to cleave *O*-glycans and retain the reducing end. To prevent peeling products of *O*-glycans as other non-reductive methods, several procedures are explored such as using anhydrous reagents, exchanging buffer with 0.1% trifluoroacetic acid[48] or low-molarity (100 to 5 mM) ethylenediaminetetraacetic acid (EDTA).[49] While these approaches may work well at microscale to generate reducing *O*-glycans for HPLC and MS analysis, they often involve toxic reagents, harsh conditions and lengthy procedures, preventing their application in reverse synthesis of *O*-glycans at large scales.

Since *O*-glycosidic linkage is much more stable towards oxidation by bleach than amide bonds, bleach essentially degrades peptide chains to expose *O*-glycans making the ORNG method a useful process for releasing *O*-glycans, as well.[36] Therefore, larger amounts of NaClO are needed to degrade peptide chains more completely to release *O*-glycans efficiently compared to the amount of bleach used for *N*-glycan release. MALDI-TOF-MS analysis indicated that due to the stability of *O*-glycosidic linkages to oxidation, the released products possess a glycolic acid (serine-linked) or lactic acid (threonine-linked) aglycon (*O*-glycan-acids), instead of free reducing glycans as for *N*-glycans. Although sodium hypochlorite is considered as a strong oxidant, it is a fairly mild reagent towards glycan structures as glycans are in high oxidation state. Therefore, many common labile *O*-glycan functional groups, such as *O*-acetylation, sulfation, are fairly stable towards bleach treatment. For example, while 9-*O*-acetylated sialic acid (Neu5,9Ac2) in bovine submaxillary mucin (BSM) does not survive any base-catalyzed β-elimination methods, it is well preserved after ORNG treatment and was stable to partial methylation of carboxylic acids. Sulfated glycans in PSM were identified as well using the ORNG process. The released *O*-glycan-acids can be easily derivatized by activation with 1-ethyl-3-(3-dimethylaminopropyl)carbodiimide and *N*-hydroxysuccinimide (EDC-NHS) and a fluorescent linker with an amino group, such as mono-9-fluorenyl-methoxycarbonyl (mono-Fmoc) ethylenediamine for further

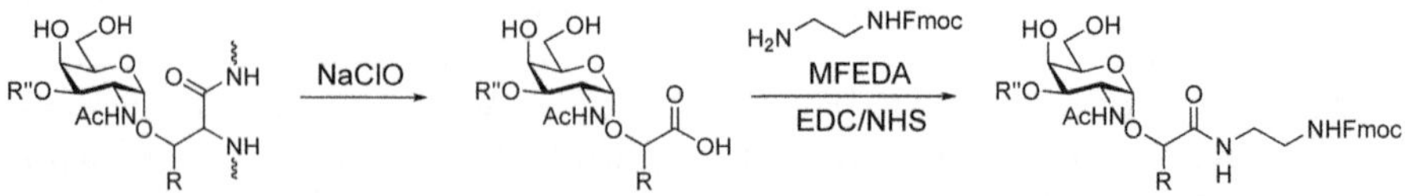

Figure 15.5 NaClO treatment of glycoproteins to release *O*-glycans and labeling with Fmoc tag. Adapted from ref. 36 with permission from Springer Nature, Copyright 2016.

MALDI-TOF-MS analysis and HPLC separation to build up *O*-glycan libraries for microarray preparation (Figure 15.5).

15.4 Glycans Release from Glycosphingolipids

15.4.1 Enzymatic Approaches

Glycosphingolipids (GSLs) are a subclass of glycolipids widely distributed in the cell membranes. GSLs are comprised of a hydrophilic glycan chain (head) and hydrophobic lipid sphingosine or ceramide (tail). Exo- and endoglycosidases have been discovered that cleave glycans from GSLs.[50] While exoglycosidases sequentially shorten glycans from their non-reducing end, endoglycosidases cleave the internal glycosidic bond within the chain. To release entire glycans from GSLs, endoglycoceramidase II enzyme (EGCase II) is commonly applied.[51] However, the action point of EGCase II on the GSLs is limited to Galβ1–4Glcβ1-1′Cer core structure. EGALC enzyme purified from *Rhodococus equi* can cleave GSLs with Galβ1–6Galβ1-1′Cer core structure only.[52] Later, a newly separated endoglycoceramidase named EGCase I from *Rhodococcus triatomea* was found to have broad specificity with GSLs.[53]

15.4.2 Chemical Approaches

Taking advantage of the C=C double bond in the sphingosine moiety, chemical release of free reducing glycans from GSLs usually utilize ozonolysis[54] or osmium tetraoxide to oxidize[55] the C=C double bond followed by base-catalyzed β-elimination. To prevent the possibly detrimental base treatment, we have developed two approaches to release, tag and separate GSLs for microarray preparation through covalent immobilization. One approach is to preserve a larger component of the lipid moiety and insert a fluorescent tag by conjugating the fluorescent tag with ozonized GSLs directly.[19] The conjugated glycans can either be immobilized covalently on *N*-hydroxysuccinimide (NHS) ester or epoxy slides, or non-covalently on nitrocellulose slides. The other approach is to gently heat ozonized GSLs under neutral pH, which also releases free reducing glycans efficiently.[56] These methods, however, are most effective on relatively pure preparations of glycosphingolipids. Using our ORNG approach, we discovered that bleach can be applied to both gangliosides purified by organic solvent extraction as well as homogenized total brain tissue in aqueous conditions.[36] MALDI-TOF-MS analysis suggested that ORNG resulted in the formation of cyanomethyl glycosides rather than free reducing glycans. Similarly, NBS can also cleave gangliosides at 65 °C to generate glycan nitriles, however, this reaction does not work directly on homogenized brain tissue. The liberated cyanomethyl glycosides can be further derivatized by nickel chloride ($NiCl_2$) and sodium borohydride to form aminoethyl glycosides, or by 2-amino benzamide (AB) to form fluorescence tag conjugated products (Figure 15.6).

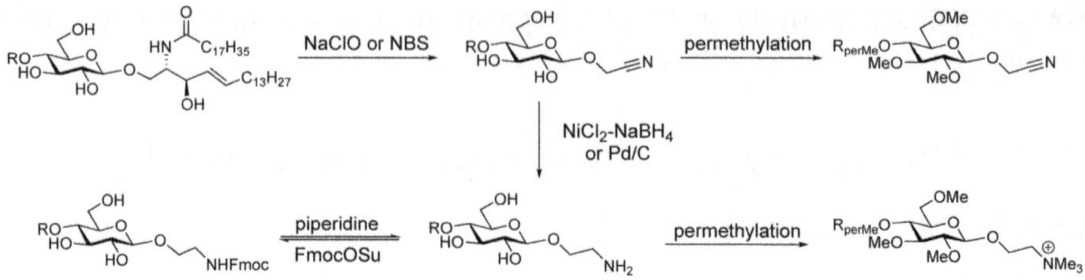

Figure 15.6 Release and tagging of glycans from GSLs by NaClO or NBS. Adapted from ref. 36 with permission from Springer Nature, Copyright 2016.

15.5 Conclusions

Here we have reviewed the "reverse synthesis" of complex glycans from more complex glycoconjugates. Both enzymatic and chemical approaches have been discussed. Both methods have their advantages and should be carefully selected based on the goal of the proposed study. When a large amount of natural glycans are desired, chemical approaches, especially the new ORNG approach provides the greatest potential. The ORNG approach uses the most affordable chemical reagents (household bleach), mild reaction condition, and simple operation to quickly release glycans from several major classes of glycoconjugates. Depending on the source of the glycoconjugates, the released glycans possess different aglycons, providing potential to isolate and separate all different classes of glycans. One important advantage of the ORNG approach is the scalability to multi-kilograms of crude biological material, which essentially provides an upfront amplification of glycan structures. As a result, it provides great potential to obtain multi-milligrams to grams of biomedically relevant, complex glycans at a scale similar to common laboratory synthetic approaches.

Conflict of Interest

D.F.S and X.S are co-founders of NatGlycan LLC, which commercializes the ORNG process.

Acknowledgements

This work was supported by NIH Common Fund Glycoscience (U01GM116254) and STTR grant R41GM122139 (XS and DFS). X. Song and D. F. Smith acknowledge the support of the Emory Comprehensive Glycomics Core (ECGC), which is subsidized by the Emory University School of Medicine and is one of the Emory Integrated Core Facilities.

References

1. A. Varki, *Glycobiology*, 2017, **27**, 3.
2. J. Stevens, O. Blixt, J. C. Paulson and I. A. Wilson, *Nat. Rev. Microbiol.*, 2006, **4**, 857.
3. D. F. Smith, X. Z. Song and R. D. Cummings, in *Methods in Enzymology, Vol 480: Glycobiology*, ed. M. Fukuda, Elsevier Academic Press Inc, San Diego, 2010, vol. 480, p. 417.
4. S. Gaunitz, G. Nagy, N. L. B. Pohl and M. V. Novotny, *Anal. Chem.*, 2017, **89**, 389.
5. G. Lu, C. L. Crihfield, S. Gattu, L. M. Veltri and L. A. Holland, *Chem. Rev.*, 2018, **118**, 7867.
6. L. R. Ruhaak, G. Xu, Q. Li, E. Goonatilleke and C. B. Lebrilla, *Chem. Rev.*, 2018, **118**, 7886.
7. L. Royle, M. P. Campbell, C. M. Radcliffe, D. M. White, D. J. Harvey, J. L. Abrahams, Y.-G. Kim, G. W. Henry, N. A. Shadick, M. E. Weinblatt, D. M. Lee, P. M. Rudd and R. A. Dwek, *Anal. Biochem.*, 2008, **376**, 1.
8. C. E. Doneanu, W. Chen and J. C. Gebler, *Anal. Chem.*, 2009, **81**, 3485.
9. L. R. Ruhaak, G. Zauner, C. Huhn, C. Bruggink, A. M. Deelder and M. Wuhrer, *Anal. Bioanal. Chem.*, 2010, **397**, 3457.
10. J. Zaia, *Chem. Biol.*, 2008, **15**, 881.
11. O. Gornik, T. Keser and G. Lauc, in *Sample Preparation Techniques for Soil, Plant, and Animal Samples*, ed. M. Micic, Springer, New York, 2016, , DOI: 10.1007/978-1-4939-3185-9_27, p. 377.
12. L. L. Kiessling and R. A. Splain, *Annu. Rev. Biochem.*, 2010, **79**, 619.
13. C.-H. Hsu, S.-C. Hung, C.-Y. Wu and C.-H. Wong, *Angew. Chem., Int. Ed.*, 2011, **50**, 11872.
14. T. J. Boltje, T. Buskas and G.-J. Boons, *Nat. Chem.*, 2009, **1**, 611.
15. L.-X. Wang and B. G. Davis, *Chem. Sci.*, 2013, **4**, 3381.
16. P. M. Danby and S. G. Withers, *ACS Chem. Biol.*, 2016, **11**, 1784.
17. L. Krasnova and C.-H. Wong, *Annu. Rev. Biochem.*, 2016, **85**, 599.
18. X. Song, B. Xia, S. R. Stowell, Y. Lasanajak, D. F. Smith and R. D. Cummings, *Chem. Biol.*, 2009, **16**, 36.
19. X. Song, Y. Lasanajak, B. Xia, J. Heimburg-Molinaro, J. M. Rhea, H. Ju, C. Zhao, R. J. Molinaro, R. D. Cummings and D. F. Smith, *Nat. Methods*, 2011, **8**, 85.
20. J. Y. Hyun, J. Pai and I. Shin, *Acc. Chem. Res.*, 2017, **50**, 1069.
21. T. H. Plummer Jr, J. H. Elder, S. Alexander, A. W. Phelan and A. L. Tarentino, *J. Biol. Chem.*, 1984, **259**, 10700.
22. A. L. Tarentino, C. M. Gomez and T. H. Plummer Jr, *Biochemistry*, 1985, **24**, 4665.
23. V. Tretter, F. Altmann and L. Maerz, *Eur. J. Biochem.*, 1991, **199**, 647.
24. N. Takahashi, *Biochem. Biophys. Res. Commun.*, 1977, **76**, 1194.
25. T. Wang, Z. P. Cai, X. Q. Gu, H. Y. Ma, Y. M. Du, K. Huang, J. Voglmeir and L. Liu, *Biosci. Rep.*, 2014, **34**, 673.

26. G. Sun, X. Yu, C. Bao, L. Wang, M. Li, J. Gan, D. Qu, J. Ma and L. Chen, *J. Biol. Chem.*, 2015, **290**, 7452.
27. K. J. Lee, J. Y. Gil, S.-Y. Kim, O. Kwon, K. Ko, D.-I. Kim, D. K. Kim, H. H. Kim and D.-B. Oh, *J. Biochem.*, 2015, **157**, 35.
28. S. Yan, J. Vanbeselaere, C. Jin, M. Blaukopf, F. Woels, I. B. H. Wilson and K. Paschinger, *Anal. Chem.*, 2018, **90**, 928.
29. A. J. Fairbanks, *Chem. Soc. Rev.*, 2017, **46**, 5128.
30. X. Song, Y. Lasanajak, C. Rivera-Marrero, A. Luyai, M. Willard, D. F. Smith and R. D. Cummings, *Anal. Biochem.*, 2009, **395**, 151.
31. Z. Yosizawa, T. Sato and K. Schmid, *Biochim. Biophys. Acta*, 1966, **121**, 417.
32. S. Nakakita, W. Sumiyoshi, N. Miyanishi and J. Hirabayashi, *Biochem. Biophys. Res. Commun.*, 2007, **362**, 639.
33. Y. Huang, Y. Mechref and M. V. Novotny, *Anal. Chem.*, 2001, **73**, 6063.
34. J. Yuan, C. Wang, Y. Sun, L. Huang and Z. Wang, *Anal. Biochem.*, 2014, **462**, 1.
35. X. Song, H. Ju, C. Zhao and Y. Lasanajak, *Bioconjugate Chem.*, 2014, **25**, 1881.
36. X. Song, H. Ju, Y. Lasanajak, M. R. Kudelka, D. F. Smith and R. D. Cummings, *Nat. Methods*, 2016, **13**, 528.
37. Y. Zhu, M. Yan, Y. Lasanajak, D. F. Smith and X. Song, *Carbohydr. Res.*, 2018, **464**, 19.
38. H. Iwase and K. Hotta, *Glycoprotein Analysis in Biomedicine*, Humana Press, Totowa, New Jersey, 1993.
39. P. H. Jensen, D. Kolarich and N. H. Packer, *FEBS J.*, 2010, **277**, 81.
40. S. Mulagapati, V. Koppolu and T. S. Raju, *Biochemistry*, 2017, **56**, 1218.
41. G. Zauner, C. A. M. Koeleman, A. M. Deelder and M. Wuhrer, *Biochim. Biophys. Acta, Gen. Subj.*, 2012, **1820**, 1420.
42. S. Maniatis, H. Zhou and V. Reinhold, *Anal. Chem.*, 2010, **82**, 2421.
43. J. A. Goetz, M. V. Novotny and Y. Mechref, *Anal. Chem.*, 2009, **81**, 9546.
44. W. Chai, T. Feizi, C.-T. Yuen and A. M. Lawson, *Glycobiology*, 1997, **7**, 861.
45. K. Yamada, S. Hyodo, M. Kinoshita, T. Hayakawa and K. Kakehi, *Anal. Chem.*, 2010, **82**, 7436.
46. Y. Miura, K. Kato, Y. Takegawa, M. Kurogochi, J.-I. Furukawa, Y. Shinohara, N. Nagahori, M. Amano, H. Hinou and S.-I. Nishimura, *Anal. Chem.*, 2010, **82**, 10021.
47. G. Yu, Y. Zhang, Z. Zhang, L. Song, P. Wang and W. Chai, *Anal. Chem.*, 2010, **82**, 9534.
48. A. H. Merry, D. C. A. Neville, L. Royle, B. Matthews, D. J. Harvey, R. A. Dwek and P. M. Rudd, *Anal. Biochem.*, 2002, **304**, 91.
49. R. P. Kozak, L. Royle, R. A. Gardner, D. L. Fernandes and M. Wuhrer, *Anal. Biochem.*, 2012, **423**, 119.
50. Y.-T. Li and S.-C. Li, *Anal. Biochem.*, 1999, **273**, 1.
51. Y.-T. Li, C.-W. Chou, S.-C. Li, U. Kobayashi, Y.-h. Ishibashi and M. Ito, *Glycoconjugate J.*, 2009, **26**, 929.

52. Y. Ishibashi, T. Nakasone, M. Kiyohara, Y. Horibata, K. Sakaguchi, A. Hijikata, S. Ichinose, A. Omori, Y. Yasui, A. Imamura, H. Ishida, M. Kiso, N. Okino and M. Ito, *J. Biol. Chem.*, 2007, **282**, 11386.
53. S. Albrecht, S. Vainauskas, H. Stockmann, C. McManus, C. H. Taron and P. M. Rudd, *Anal. Chem.*, 2016, **88**, 4795.
54. H. Wiegandt and G. Baschang, *Z. Naturforsch., B: J. Chem. Sci.*, 1965, **20**, 164.
55. S. Hakomori, *J. Lipid Res.*, 1966, **7**, 789.
56. X. Song, D. F. Smith and R. D. Cummings, *Anal. Biochem.*, 2012, **429**, 82.

CHAPTER 16

Novel Technologies for Quantitative O-Glycomics and Amplification/Preparation of Cellular O-Glycans†

ZHONGHUA LI[a], TATIANA A. CHERNOVA[a] AND TONGZHONG JU*[a,b]

[a]Department of Biochemistry, Emory University School of Medicine, Atlanta, GA 30322, USA; [b]Office of Biotechnology Products, Center for Drug Evaluation and Research, U.S. Food and Drug Administration, Silver Spring, MD 20993, USA
*E-mail: Tongzhong.ju@fda.hhs.gov

16.1 Introduction

Mucin-type *O*-glycosylation is characterized by *O*-GalNAc linked on Ser/Thr residues in mucins and glycoproteins.[1,2] Recent evidence also shows that tyrosine (Tyr) residues in proteins can also be *O*-glycosylated.[3,4] In animal cells, it has been estimated that more than 83% of proteins traversing through secretory pathway carry one or more *O*-glycans.[1] *O*-glycans on glycoproteins, either secreted or membrane bounded play many critical roles in

†The views expressed in this article are those of the author and do not necessarily reflect the official policy or position of the U.S. Food and Drug Administration and the Department of Health and Human Services, nor does mention of trade names, commercial products, or organizations imply endorsement by the U.S. Government.

Chemical Biology No. 11
Synthetic Glycomes
Edited by Wanyi Guan, Lei Li and Peng George Wang

Published by the Royal Society of Chemistry, www.rsc.org

biological systems, such as signal transduction, cell adhesion, development, and immunity.[2] Importantly, the *O*-glycome changes in a tissue/cell-specific manner, and it is often altered in cells at disease states.[5,6] Consequently, the alteration of *O*-glycome often impact cellular properties and functions.[2,6] The roles of glycans in biology, which is termed functional glycomics, is a new era of life science. Apparently, it is necessary to comprehend *O*-glycome in order to reveal the functional *O*-glycomics. Understanding of the *O*-glycome, or analysis of *O*-glycan, however, often requires their release from glycoproteins prior to Ultra Performance Liquid Chromatography (UPLC) and/or mass spectrometry (MS).[7,8] In contrast to *N*-glycan release by reliable endo-glycosidases, no such comparable method exists for *O*-glycans.[9] Instead, release of *O*-glycans from glycoproteins often relies on β-elimination, which has many limitations, such as non-specificity and low efficiency, and often requirement for large amounts of material. Furthermore, quantitative analysis of *O*-glycans or quantitative *O*-glycomics is an unmet demand.[10] The preparation of complex glycans for biological studies is even more challenging, since only limited numbers and amounts of *O*-glycans can be chemically or chemo-enzymatically synthesized[11] or isolated from natural sources.[12]

It is worth noting that there are a defined number of human genes and enzymes involved in the biosynthesis of the *O*-glycome, yet the diversity and size of the human *O*-glycome are unknown.[8] The structural complexity, diversity, as well as heterogeneity of *O*-glycans make it impossible to sequence them in a similar fashion to DNA/RNA or proteins. Furthermore, the structures of glycans change throughout animal development, from fertilization to aging of the organism, and specifically with cell transformation, as in the case of tumor cells.[6] Changes in glycans can benefit the survival and progression of tumor cells. But in most cases, we know little regarding how these changes are linked to biological readouts. The study of the relationship between structure and function needs to have feasible tools. Moreover, fully understanding functional glycomics, and ultimately applying glycomics for improving human health, requires not only an understanding of their structure information, but also how to synthesize or prepare them.

16.2 *O*-Glycan Biosynthetic Pathways and Their Known Biological Functions

The *O*-glycosylation is initiated by the addition of GalNAc to Ser/Thr in polypeptides or proteins by a family of polypeptide α-GalNAc Transferases (ppGalNAcTs) mainly localized in the early Golgi apparatus to form GalNAc-α1-Ser/Thr, the Tn antigen.[13] The Tn antigen of glycoproteins, or mucins does not normally appear on the cell surface; it is usually further elongated and branched by a series of glycosyltransferases. Following its formation in the Golgi, the Tn antigen can serve as an acceptor for at least three other Golgi glycosyltransferases leading to three pathways[2,3] (Figure 16.1): (i) Core 1 pathway: Core 1 β1–3-galactosyltransferase (Core 1 β3GalT or T-synthase)

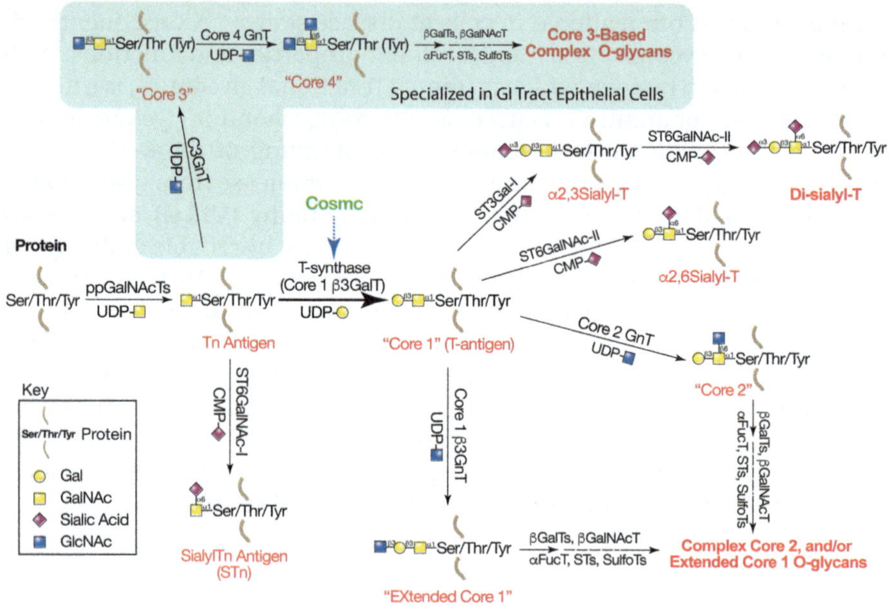

Figure 16.1 The mucin-type *O*-glycan biosynthetic pathways. Mucin-type *O*-glycosylation (*O*-glycosylation) of proteins initiates mainly in the Golgi when polypeptide α-GalNAc-transferases (ppGalNAcTs) transfer GalNAc from UDP-GalNAc to Ser/Thr/Tyr in proteins traversing secretory pathway to form Tn antigen (GalNAcα-Ser/Thr/Tyr). The Tn antigen is normally further modified by Core 1 β3GalT (C1GALT1, T-synthase) which transfers Gal from UDP-Gal to GalNAcα-Ser/Thr in glycoproteins to form Core 1 structure, also known T antigen, Galβ1,3GalNAcα-Ser/Thr in all cell types. Cosmc which is encoded by Cosmc (*C1GALT1C1*) on Xq24 is the specific molecular chaperone required for the biogenesis of active T-synthase *in vivo*. In all cell types, the T antigen is usually sialylated by sialyltransferases (STs) to form mono-sialylated Core 1, or di-sialyl Core 1 structures. In addition, the T antigen can also be further modified by Core 2 β6-*N*-acetylglucosaminyltransferase (C2GnT) and/or by Core 1 β3-*N*-acetylglucosaminyltransferase (C1GnT) as well as other various glycosyltransferases and sulfotransferases to synthesize Core 1-based complex *O*-glycans. In gastrointestinal tract epithelial cells, the Tn antigen can also be modified by β3GnT6 or Core 3 synthase (C3GnT) by transferring GlcNAc from UDP-GlcNAc to form Core 3, GlcNAcβ1,3GalNAcα-Ser/Thr which can be further modified by various glycosyltransferases and sulfotransferases to synthesize Core 3-based complex *O*-glycans. The STn antigen (Neu5Acα2,6GalNAcα-Ser/Thr) results from the actions of ST6GalNAc-I, which transfers Neu5Ac from CMP-Neu5Ac to the Tn antigen. Mutation in Cosmc results in an inactive T-synthase leading to Tn and/or STn antigens on the cell surface.

to synthesize the Core 1 structure, also known as T- or TF-antigen, (ii) Core 3 pathway: Core 3 β1–3-*N*-acetylglucosaminyltransferase (Core 3 β3GnT, C3GnT, *β3GnT6*) to synthesize the Core 3 structure, and (iii) SialylTn pathway: ST6GalNAc-I to synthesize the sialyl-Tn antigen. Among them, the Core 1 pathway is the most common one that occurs in every animal cell through an action of the Core 1 β3GalT (*C1GalT1*) to form the Core 1 disaccharide, Galβ1–3GalNAcα-Ser/Thr.[14,15] Since Core 1 disaccharide is also known as T- or TF (Thomsen-Friedenreich) antigen, Core 1 β3GalT is also named T-synthase. T-synthase is the key enzyme in mucin type *O*-glycan biosynthesis expressed by all cells in humans and mice.[15,16] Importantly, biosynthesis of active T-synthase requires an ER-localized molecular chaperone Cosmc (*Co*re 1 β3Galactosyltranerase *S*pecific *M*olecular *C*haperone) which assists its correct folding. While human T-synthase is encoded by a gene *C1GALT1* on 7p21.3, Cosmc is encoded by an X-linked gene *Cosmc* (*C1GalT1C1*) on *Xq24*.[17,18] Core 1 structure is usually modified into three types of *O*-glycans: Sialyl-Core 1 (or Sialyl-T), Core 2 *O*-glycans, and extended Core 1 *O*-glycans (Figure 16.1). Sialylation of Core 1 structure by ST3Gal-I yields α2–3Sialyl-T antigen, SAα2–3Galβ1–3GalNAcα-Ser/Thr,[19] which can then be further sialylated by ST6GalNAc-II to yield DiSialyl-T antigen (DiSialyl-Core 1), SAα2–3Galβ1–3(SAα2–6)GalNAcα-Ser/Thr;[20] or sialylation of Core 1 by ST6GalNAc-II forms the 2–6-sialylated Core 1 (2–6Sialyl-T), Galβ1–3(SAα2–6)GalNAcα-Ser/Thr.[21] Elongation of Core 1 structure leads to form extended Core 1 *O*-glycan, GlcNAcβ1–3Galβ1–3GalNAcα-Ser/Thr by Core 1 β3GnT;[22] and branching of Core 1 by the Core 2 β1–6-GlcNAcT (Core 2 GnT) generates Core 2 structure, Galβ1–3(GlcNAcβ1–6)GalNAcα1-Ser/Thr[23] (Figure 16.1). Core 2 and Extended Core 1 can be further modified by other βGalTs, αFucTs, GnTs, STs, and even sulfotransferases to form Core 1-based complex *O*-glycans, such as Sialyl-Lewis^X (SLe^X) on Core 2, Sulfo-Sialyl-Lewis^X on extended Core 1, and polyLactosamine structures, *etc.* The second common pathway from the Tn antigen is to synthesize Core 3 structure, GlcNAcβ1–3GalNAcα1-Ser/Thr by C3GnT mostly found in the gastrointestinal tract (Figure 16.1). Some Core 3 can be branched by a β6GlcNAc through an action of Core 4 GnT to form the Core 4 structure, GlcNAcβ1–3(GlcNAcβ1–6)GalNAcα1-Ser/Thr. Core 3 and 4 structures are usually further modified by other βGalTs, αFucTs, GnTs, STs, and even sulfotransferases to form Core 3-based complex *O*-glycans[24] (Figure 16.1). The third pathway from the Tn antigen is through ST6GalNAc-I which sialylates the Tn antigen to form Sialyl-Tn (STn), SAα2–6GalNAcα-Ser/Thr. STn is not detectable in most normal tissues other than the submaxillary gland of cows and sheep.[25] If formed, STn is a terminal structure. In addition to Core 1–4 structures in *O*-glycans abovementioned, there are other Core structures in *O*-glycans, Core 5 (GalNAcα1–3GalNAcα1-Ser/Thr), Core 6 (GlcNAcβ1–6GalNAcα1-Ser/Thr), Core 7 (GalNAcα1–6GalNAcα1-Ser/Thr), Core 8 (Galα1–3GalNAcα1-Ser/Thr) (Table 16.1). For these rare Core 5–8 structures, Core 6 can be synthesized by an isoenzyme of Core 2 GnT, a candidate Core 6 GnT, while the glycosyltransferases responsible for synthesizing Core 5, 7 and 8 are not yet known.

Table 16.1 Core structures of mucin-type *O*-glycans.

Name of cores	Structures
Core 1	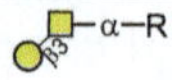
Extended Core 1	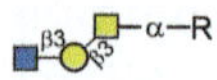
Core 2	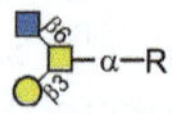
Core 3	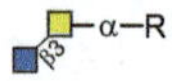
Core 4	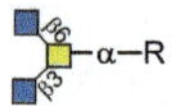
Core 5	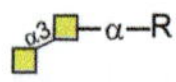
Core 6	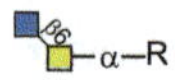
Core 7	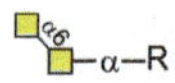
Core 8	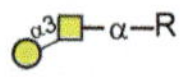

Table 16.2 Representative common complex *O*-glycans.

Types	Structures
Core 1 based	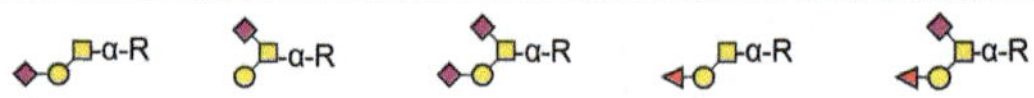
[a]Core 2 & [a]extended Core 1 based	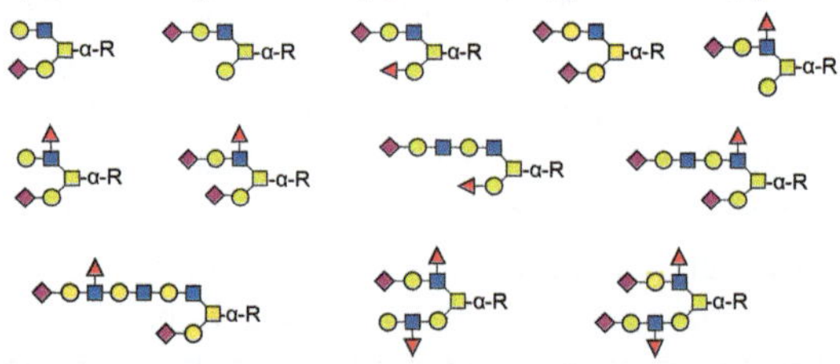
[a]Core 3 based	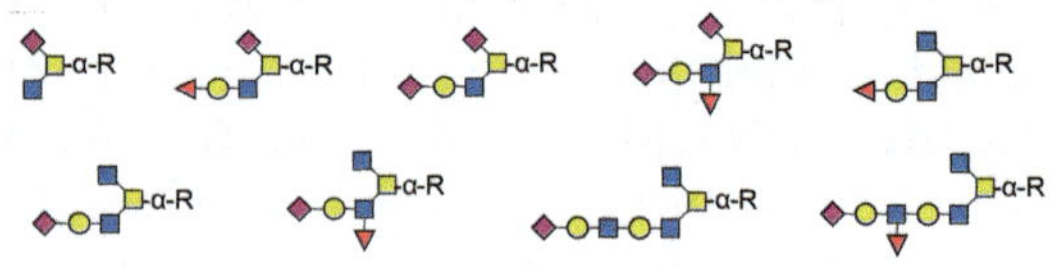

[a]Many Core 2-, extended Core 1- and 3-based structures could be sulfated: 6-*O*-Sulfation of GlcNAc or 3-*O*-Sulfation of Gal.

Generally, all animal cells contain Core 1-based complex *O*-glycans, such as sialyl Core 1, SLe^X [SAα2–3Galβ1–4(Fucα1–3)GlcNAcβ], SLe^a [SAα2–3Galβ1–3(Fucα1–4)GlcNAcβ], Sulfo-SLe^X, on Core 2, and even extended Core 1 *O*-glycan structures[2,5] (Table 16.2). In the GI tract, Core 3-based and Core 1-related structures are shared on mucins at different Ser/Thr residues, due to expression of both T-synthase and Core 3 GnT in the same cells[26] (Figure 16.1 and Table 16.2). There are also unique *O*-glycan structures from specialized cells/tissues, such as Wehi-3 cells, HUVEC, and HL-60 cells (Table 16.3).[27]

Unique structures	Core 1-based	Core 2-based
Di-sialyl Motif 	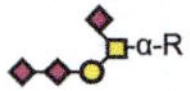	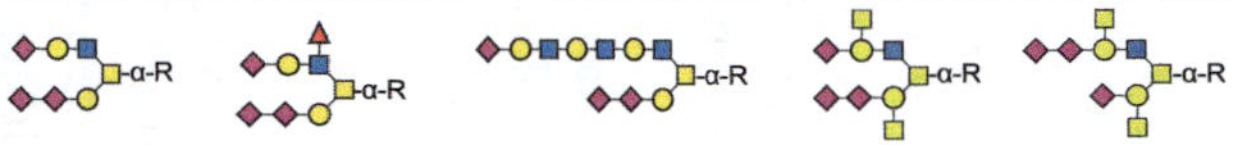
Sd[a]/Cad Antigen[b] 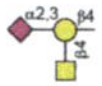	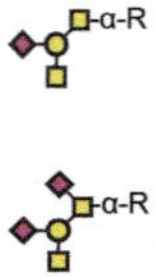	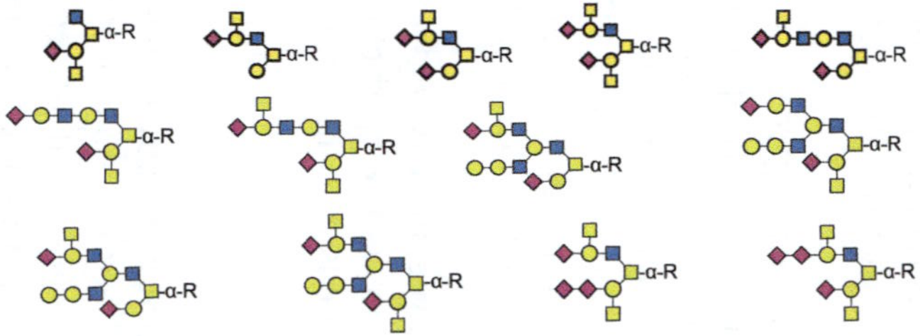
I Antigen 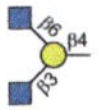		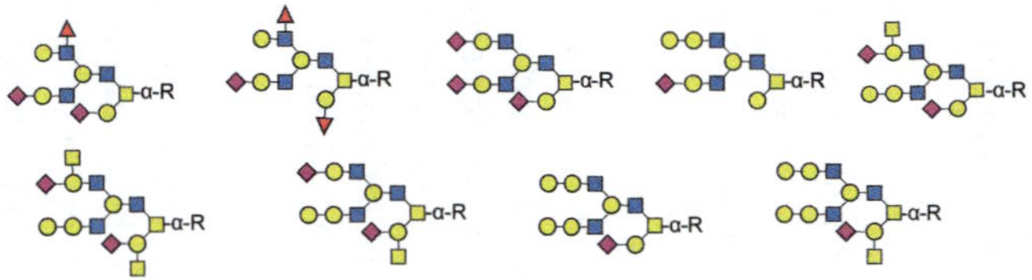
α-Gal[a] 		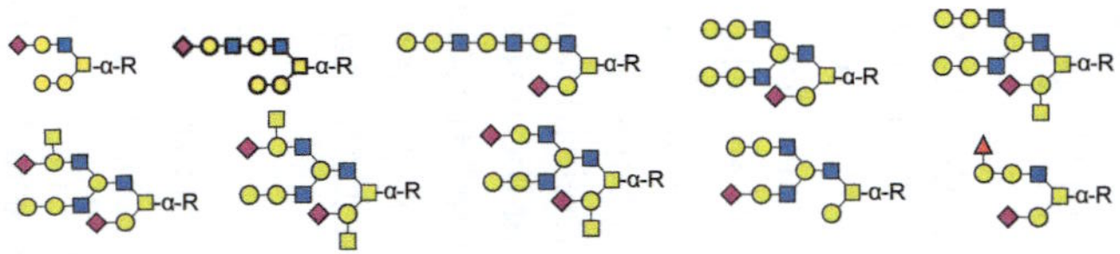
Vim-2 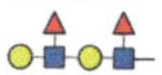		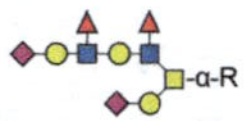

[a]From Wehi-3 cells.
[b]Many Core 3-based structures often carry Sd[a]/Cad antigens.

Why do animal cells evolve this sophisticated machinery to decorate proteins with these complex carbohydrates? The answer is because *O*-glycans on glycoproteins and mucins play a broad range of important biological functions:

1. Directly participating in cell–cell interactions:[28–30] for example, the SLe^X on Core 2 *O*-glycans of P-Selectin Glycoprotein Ligand-1 (PSGL-1) is one of the required components for interactions with P-selectin which mediate the rolling of leukocytes on endothelial cells; Sulfo-SLe^X on extended Core 1 *O*-glycans (MCA-79 epitope) is required for the interactions of L-selectin ligands with L-selectin mediating the lymphocytes' homing.
2. Stabilizing *O*-glycoproteins: for example, complex *O*-glycans on mucin-like protein podoplanin which plays critical roles in lymphogenesis,[31] and on the major platelet glycoproteins GPIbα, GP-VI and integrins which are important for biogenesis and biological functions of platelets,[32] as well as on others including LDL receptor[33] and FGF23[34,35] stabilize the protein backbones. Loss of complex *O*-glycans or *O*-glycosylation on these proteins leads to a proteolytic degradation and hence impairs their biological functions.
3. Protecting epithelia of the GI tract: for example, heavily *O*-glycosylated mucins including MUC2 in GI tracts[36,37] protects the epithelia against pathogens and other harmful substances. Loss of complex *O*-glycans results inflammation and even colitis.

It is worth noting that tyrosine residues in proteins can also carry *O*-GalNAc, or mucin type *O*-glycans, as recently reported.[3,4] For example, Halim *et al.* reported that amyloid precursor protein (APP) carries a sialyl-Core 1 structure on one tyrosine residue (Tyr10) in addition to multiple Thr-residues.[3] One recent study showed the GalNAcα1-*O*-Tyr, similar to GalNAcα1-*O*-Ser/Thr, can also be recognized by α-GalNAc-binding Lectins, such as HPA, VVL, and SBA, and by several anti-Tn antibodies, as well as by human macrophage galactose-type lectin-1 (MGL-1) and mouse MGL-2(mMGL-2).[38] Several questions regarding the pathways and biological functions for Tyr-*O*-glycans, however, require further investigation. Which ppGalNAcT(s) is responsible for initiation of the Tyr *O*-glycosylation? Are the Tyr-*O*-glycans elongated or branched in the same fashion as or in a similar fashion to Ser/Thr-*O*-glycans? Do Tyr-*O*-glycans have similar functions to Ser/Thr-*O*-glycans?

Nevertheless, compelling evidence has shown that the structures of *O*-glycans are linked to their biological functions. Yet, unlike *N*-glycans which can be released from glycoproteins by endo-glycosidases, such as PNGase F and feasibly analyzed, there are obstacles in deducing *O*-glycan structures. Recent development of CORA technology overcomes the obstacle by bypassing the *O*-glycan releasing step through supplying cells with a surrogate, Benzyl-α-*N*-acetylgalactosamine (Bn-α-GalNAc) which is a common

O-glycan precursor.[27] CORA has demonstrated its high sensitivity, reproducibility, feasibility and amplification nature, providing tools for us to assess and amplify cellular *O*-glycans. In this chapter, we describe how to advance current technology CORA to a quantitative CORA (qCORA) for quantitative *O*-glycomics and preparative CORA (pCORA) for preparation of cellular *O*-glycans.

16.3 Quantitative *O*-Glycomics

Analysis of glycans synthesized by cells in different biological systems and the search for *O*-glycan biomarkers associated with diseases require sensitive and adaptable technologies that can capture the changes in quality (structures) and quantity of *O*-glycans. Analytical techniques often encounter challenges in structural elucidation and quantification of *O*-glycans due to (1) their diversity and complexity, (2) their low abundance, (3) their insensitive detectable chemical feature, and (4) the absence of efficient and unbiased tools for releasing them from their carriers.

16.3.1 Technologies for *O*-Glycome Analysis

Modern methods of glycomic analyses largely evolve several separation and mass spectrometric techniques, including capillary electrophoresis, liquid chromatography (different modes), matrix assisted laser desorption ionization mass spectrometry (MALDI-MS), and electrospray ionization mass spectrometry (ESI-MS).[39] Typically, prior to separation and analysis, glycans need to be released from glycoproteins, and free glycans are then fluorescently labeled for electrophoretic or chromatographic analysis. Free glycans can also be permethylated, extracted, and analyzed, or often merely profiled by MS and MS^n without separation. The MS data can usually derive glycan composition and predict structures based on the presumed biosynthetic pathways,[39] while the MS^n enables structures to be assigned more accurately. For those glycans with complex structures, other data, such as exo-glycosidase digestion in addition to MS and MS^n is often needed.

16.3.1.1 Techniques for O-*Glycan Release*

For analysis of its structure, glycan release is often a critical step to eliminate the heterogeneity contributed from peptides/proteins. The techniques are devised by enzymatic and/or chemical procedures. Unlike *N*-glycans which can be cleaved from glycoproteins by specific endo-glycosidases, such as peptide *N*-glycosidase (PNGase) F and PNGase A,[40,41] *O*-glycans cannot be generally released from glycoproteins by enzymatic methods because there is no such endoglycosidase available. An exception is endo-α-*N*-acetylgalactosaminidase (*O*-glycanase), which has a high substrate specificity to remove the Core 1 disaccharides, and less efficiently, the Core 3 structures from glycoproteins.[42,43] Thus, *O*-glycans have mostly been subjected to chemical digestion termed β-elimination. The reagents in traditional or reductive β-elimination include NaOH and $NaBH_4$. Hydroxide anions not only catalyze the cleavage of *O*-glycosidic bond between Ser/Thr and GalNAc to release *O*-glycans but also continue the similar reaction on glycans and thus often cause the glycan degradation, referred to as the "peeling" reactions. $NaBH_4$ will immediately reduce *O*-glycans released from the reaction to their alditol form to minimize the "peeling" reaction. Reductive β-elimination technique combined with matrix assisted laser desorption ionization-time of flight (MALDI-TOF) analysis was used to characterize the structure of the neutrophil *O*-glycome,[44] *O*-glycan profile of parent and glycosylation mutant CHO cell lines,[44,45] and *O*-glycomic alteration of mouse ES cells upon differentiation into embryoid bodies or extraembryonic endodermal cells.[46] Recently an one-pot *O*-glycome analytical method namely β-elimination in the presence of pyrazolone analogs (BEP) was developed.[47] Due to a substitution of $NaBH_4$ with 1-phenyl-3-methyl-5-pyrazolone (PMP), the derived PMP-*O*-glycans were easily isolated from the reaction mixture, yet a similar or a lower level of peeling reaction still occurred.[48,49] The β-elimination method has a major limitation that *in situ* reduction or conjugation results in a loss of the original hemiacetal group at the reducing end of the carbohydrate, which prevents further derivatization suitable for enrichment, chromatographic, and mass spectrometric analyses. To overcome the disadvantage, novel β-elimination approaches, so-called "non-reductive" β-elimination have been developed to yield reducing sugars, which can be further conjugated with fluorescence dyes (2-AA, 2-AB, 2-PA, DAP)[50] for relative quantitative MS.[9] For example, ammonium-based alkali-catalyzed non-reductive β-elimination was used to profile the *O*-glycans of human ES cells.[51] Yet, non-reductive β-elimination also has some drawbacks: low efficiency/yield, biased, and non-specific (*N*-glycans could be released).

Hydrazinolysis is another chemical de-*O*-glycosylation method that produces *O*-glycans in a non-reduced form.[52–55] Non-reductive *O*-glycans were fluorescently tagged with 2-aminopyrydine (2-PA) and then analyzed on high performance liquid chromatography (HPLC), MALDI-MS in combination with exoglycosidase digestion. This method was used to report the *O*-glycomic profile of human iPS cells and human dermal fibroblasts.[56] Notably, hydrazinolysis not only requires a high quality reagent, non-aqueous hydrazine which is hazardous, but also to cleave *N*-glycans from glycoproteins.

Recently, Song *et al.* developed an oxidative release of natural glycans (ORNG) method, which uses household bleach (NaClO) to efficiently release and prepare all types of free reducing *N*-glycans and *O*-glycans from different natural sources.[57] The released glycans can be directly derivatized with fluorescent tags, then separated and prepared for functional glycomics. However, the oxidation condition needs to be controlled or finely optimized in order to be adapted to an analytical-suitable method since oxidation can cause glycan degradations and undesired reactions, which is not applicable for analyzing a small amount of material.

16.3.1.2 The Derivatization of Glycans with Fluorescence Tagging and Quantification of Tagged Glycans with H(U)PLC

Derivatization of glycans is often pursued in order to enhance the detection and facilitate the separation, but quantitative glycomics is generally limited due to lack of glycan standards. Nevertheless, researchers have developed three fundamental strategies for quantitative glycomics: (1) metabolic radioactive sugar labeling with ^{3}H- or ^{14}C-monosaccharides in the cell culture[58] which was utilized extensively in an early era of glycobiology but is rarely used nowadays; (2) fluorescence tagging the reducing end of glycans released from glycoconjugates and analyzed by HPLC/UPLC or electrophoresis; (3) quantitative mass spectrometry. Each strategy includes several different approaches with respective merits and limitations.

Since every glycan has only one single reducing end which is reactive to an amine group of many different fluorescent compounds, every molecule of glycan carries one fluorescence tag after labeling. This allows their quantity to be measured by comparing it to a control (quantitative standard) such as a monosaccharide with the same tag. There are two classes of fluorescence dyes:

(1) Fluorescent dyes with a charge for electrophoretic separation: this group include aminopyrene-3,6,8-trisulfonic acid (APTS),[59,60] 8-aminonaphthalene-1,3,6-trisulfonic acid (ANTS) and 2-aminoacridone (AMAC).[61–63] The disadvantage is that the standards for acquiring structural information are not always available.

(2) Fluorescence dyes for hydrophilic interaction liquid chromatography (HILIC) or hydrophobic interaction (or reverse-phase, RP) chromatography: 2-PA, 2-amino-benzamide (2-AB), and 2-aminobenzoic acid (2-AA) are commonly used[64–67] to tag glycans released from glycoconjugates. The HILIC-HPLC or -UPLC will separate glycans by relying on the differences in their structures, while RP-interaction counts on the hydrophobic group (the dye), and separation of glycans with different structures on RP-columns will be still based on the overall hydrophobicity of tagged glycans to which the glycan moiety has a negative contribution. Again, it still needs relevant standards to acquire structural information of glycans.

16.3.1.3 Mass Spectrometry (MS)

The high sensitivity and fragmentation capability of MS and unique fragmentation patterns of structurally-different saccharides make MS the most popular modern approach for analyzing saccharides. In the past three decades, significant advances in mass spectrometry techniques such as MALDI-MS and ESI-MS have provided a bright opportunity for glycomic studies, owing to their superiority in rapid glycan sequencing and quantification.[68,69]

Although MS analysis by itself is not quantitative, derivatization of free glycans allows quantitation analysis by MS. Derivatization can be achieved by introducing a group with a unique property (such as fluorescence) to a glycan reducing end or masking all hydrophilic groups with the same hydrophobic groups (such as methyl-group).

(1) Quantitative-MS of permethylated glycans: Permethylation converts highly polar groups (–OH,–NH, and –COOH) of the glycans to non-polar (–OCH_3, –NCH_3, and –$COOCH_3$) groups, respectively.[10,70,71] This increases MS detection sensitivity, allows simultaneous detection of neutral and acidic glycans in MALDI-MS positive mode with high ionization efficiency, and simplifies the interpretation of tandem mass spectra and structural elucidation. Currently, permethylation is one of the most popular approaches for glycan derivation and has been successfully employed to quantitatively assess glycomic changes in different cancers.[72]

(2) Quantitative glycomics through stable isotopic labeling of released glycans: Relative quantitation using stable isotope labeling has become a convenient and versatile strategy for the quantitative analysis of various glycan samples. With this strategy, the abundance of each glycan in one sample can be promptly determined in terms of relative quantity compared to the identical glycan in another sample. This approach employs stable isotope labeling to introduce signature mass tags to glycans that can be used by a mass spectrometer to quantify each analyte and to determine the sample from which it originates. Usually, two samples are labeled as "light" and "heavy", respectively. Equal amounts of samples are mixed and subjected to LC/MS analysis. The same glycans will appear as doublet peaks with a unique mass difference as the "light" tag *versus* the "heavy" tag. The relative quantity of each glycan in both samples can be calculated based on the intensity ratio between the "light" and "heavy" isotopic MS signal. Several approaches based on the reagents have been employed for the quantification of glycans: iodomethane-*d*3 stable isotopic labeling *via* permethylation;[73,74] stable isotope labeling by reductive amination with [$^{13}C_6$]Aniline,[75] or deuteriomethyl (*d*4) forms of 2-anthranilic acid,[76] or tetraplex tagging reagent,[77] or hexadeuterium 2-PA;[78] ^{18}O-isotopic labeling of *N*-glycans released by endo-β-*N*-acetylglucosaminidase in heavy water;[79] and stable isotopic labeling of Asn-Glycans with [$^{13}C_6$]4-phenethylbenzohydrazide.[80] These are frequently applied to *N*-glycan analysis.

(3) Quantitative glycomics through metabolic stable isotopic labeling:

The metabolic stable isotopic labeling of glycans refers to the incorporation of stable isotopic labels during the process of cellular metabolism and glycan synthesis.

(a) The isotopic detection of amino sugars with glutamine:[81] In the hexosamine biosynthetic pathway, the side chain amide of Gln is the source of nitrogen in the production of glucosamine moiety in UDP-GlcNAc. When cells were cultured in the presence of amide-^{15}N-Gln and amide-^{14}N-Gln media respectively, ^{15}N and ^{14}N were incorporated into glycans. Samples could be mixed together at the beginning of sample preparation and analysis. However, the incorporation of isotopic reagents is limited by the availability of a biological sample, and the different glycans contain different amounts of GlcNAc, therefore comparison of the glycans in two samples is not straightforward. Alternatively, incorporating an unnatural synthetic monosaccharide into a glycoprotein is an attractive metabolic labeling method for glycans. However, this is not easily achievable because of the high complexity of glycan biosynthesis, especially with unknown or uncharacterized pathways.

(b) CORA–*C*ellular *O-gly*come *R*eporter/*A*mplification: a novel method for profiling and amplifying mucin-type *O*-glycans from living cells, was recently developed.[27] CORA uses peracetylated benzyl-α-*N*-acetylgalactosamine [Bn-α-GalNAc$(Ac)_3$] as *O*-glycan precursor and endogenous different *O*-glycosyltransferases in the secretory pathway to produce and amplify a large variety of free *O*-glycan derivatives from living cells. Bn-α-GalNAc which is used as an acceptor for T-synthase, and C3GnT activity assay,[14,82,83] is also used as an "inhibitor" of cellular *O*-glycans at high concentrations (2~5 mM).[84] Bn-α-GalNAc acts as a surrogate of GalNAc-α1-*O*-Ser/-Thr (Tn antigen), the precursor for *O*-glycan synthesis.[84] In order to be efficiently taken up by cells, Bn-α-GalNAc needs to be peracetylated to Bn-α-GalNAc$(Ac)_3$. Bn-α-GalNAc$(Ac)_3$ crosses the cell membrane[85,86] and gets into cytosol, is deacetylated by cytosolic esterases and converted back to Bn-α-GalNAc. Bn-α-GalNAc is then transported into the Golgi apparatus where it is elongated and branched by different glycosyltransferases utilizing sugar donors to form a variety of Bn-*O*-glycans in the same fashion as glycosylation of glycoproteins. The matured Bn-*O*-glycans are finally secreted into the media for isolation and MS analysis.[27] CORA bypasses *O*-glycan release and amplifies the *O*-glycome from cultured cells, resulting in 100–1000x increased sensitivity over traditional methods. Using CORA technology, the *O*-glycomes of 16 cell lines, including adherent cells, suspension cells, cancer cells, primary cells, human cells, and mouse cells were clarified clearly by MS. Compared to *O*-glycome profiles from β-elimination-MS, CORA not only identified most of the structures seen in β-eliminated samples, but also revealed many new structures, for example, I-antigen (Wehi-3 cells and HUVECs), extended Core 1 (HL-60 cells), VIM-2 (HL-60 cells and HUVECs) and disialic acid motif (Wehi-3 cells).[27] This technology can feasibly be advanced to quantitative CORA (qCORA) for quantitative *O*-glycomics.

16.3.2 qCORA – Quantitative CORA

A combination of the advantages of stable isotope labeling with CORA, termed quantitative CORA (qCORA) will provide sensitive, high-throughput, comparative/quantitative analysis of cellular *O*-glycome. qCORA takes a strategy by "metabolic stable isotopic labeling of *O*-glycomes of cultured cells (SILOC)" with $[^{13}C_6]$ or $[^2H_7]$Bn-α-GalNAc$(Ac)_3$ and their regular compounds (Figure 16.2). In this approach, any two types of cell or the same cell line cultured at different conditions can be metabolically labeled with regular compound $[^{12}C_6]$ or $[^1H_7]$ Bn-α-GalNAc$(Ac)_3$ (light) and $[^{13}C_6]$ or $[^2H_7]$Bn-α-GalNAc$(Ac)_3$ (heavy) respectively; the mixture of the "light" and "heavy" media will be then processed and finally analyzed in a Mass Spectrometer. The $[^{12}C_6]$ and $[^{13}C_6]$Bn-*O*-glycans with a difference of 6-Da in Mass, or $[^1H_7]$ and $[^2H_7]$Bn-*O*-glycans with a difference of 7-Da in Mass can be easily compared (Figure 16.3). Hence, the qualitative differences (compositions/structures) and quantitative differences (amount/abundance) of *O*-glycans between those cells will be readily resolved. An approach named ICORA (Isotope-Cellular *O*-glycome Reporter Amplification) was recently developed and applied to analyze specific differences in *O*-glycan biosynthesis and alterations between colorectal cancer lines using $[^2H_7]$Bn-α-GalNAc(Ac_3) and statistically interrogated for changes in *O*-glycan abundance.[87]

qCORA by SILOC has unprecedented merits for *O*-glycomics in addition to the advantages of CORA. (1) It directly quantitatively compare *O*-glycomes and promptly monitors the changes between cells: since "light" $[^{12}C_6]$ or $[^1H_7]$ Bn-α-GalNAc$(Ac)_3$ and "heavy" $[^{13}C_6]$ or $[^2H_7]$Bn-α-GalNAc$(Ac)_3$ are unbiasedly and equally decorated by glycosyltransferases, and the products, "light" and "heavy" Bn-*O*-glycans are processed and analyzed in the same conditions, relative abundance of the glycans from different cells can be confidently determined. (2) It can identify novel *O*-glycans based on the defined mass shift: the mass confirmation of "heavy" *O*-glycan species to the "light" ones allows confidence in assigning each *O*-glycans, especially for the unique structures. Without confirmation, some novel *O*-glycans might be missed. In addition, qCORA can assess qualitative (structural) changes in *O*-glycans better than CORA.

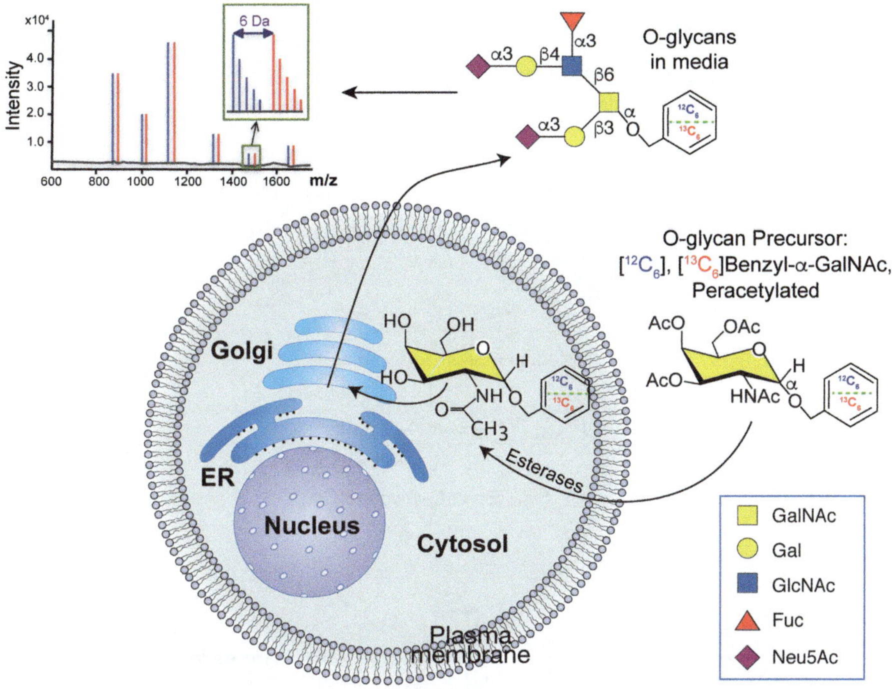

Figure 16.2 Overview of SILOC – "metabolic *S*table *I*sotopic *L*abeling of *O*-glycomes of cultured *C*ells": Cells are incubated either with peracetylated "light" (regular) [$^{12}C_6$] Benzyl-α-GalNAc(Ac)$_3$ or with its stable isotope labeled counterpart, [$^{13}C_6$] Benzyl-α-GalNAc(Ac)$_3$. *O*-glycan precursors cross the plasma membrane and enter the cell. Cytosolic esterases generate [$^{12}C_6$]- or [$^{13}C_6$] Bn-α-GalNAc, which is transported into the Golgi and modified by endogenous *O*-glycosyltransferases during anterograde trafficking. Elaborated [$^{12}C_6$]- or [$^{13}C_6$]Bn-*O*-glycans are secreted into the media, purified, analyzed and compared by Mass Spectrometer (MS).

16.4 pCORA–Amplification and Preparation of Cellular *O*-Glycans Using Cells as Factories

The vast diversity and complexity of the structures make it challenging to amplify and synthesize the *O*-glycans. Currently, there are three main approaches to prepare *O*-glycans: chemical synthesis,[88,89] chemo-enzymatic synthesis,[90,91] and isolation of *O*-glycans from natural sources,[92] which is also generally true for preparation of other types of oligosaccharides. The chemical method can synthesize relatively large quantities of pure glycans (3~6 monosaccharides), whereas, for highly complex *O*-glycans, it is still challenging due to the low reproducibility of the procedure, low yield, and high cost. Therefore, complicated chemical methods in glycan synthesis are more frequently replaced by the chemo-enzymatic method, which combines

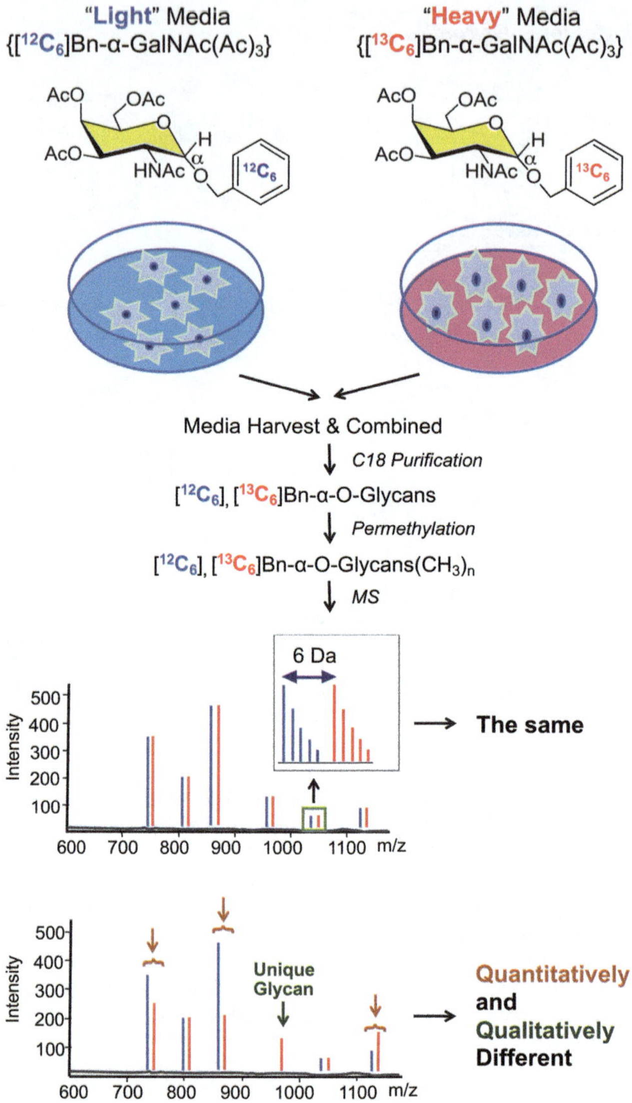

Figure 16.3 The Procedure of SILOC. Two different cell lines or the same cell line cultured at two different conditions are labeled in the "light" [$^{12}C_6$]Bn-α-GalNAc(Ac)$_3$ and "heavy" [$^{13}C_6$]Bn-α-GalNAc(Ac)$_3$ containing media respectively for an appropriate time. The media from two cell lines or two conditions is mixed in a 1 : 1 ratio, the Bn-*O*-glycans are isolated by C18 Cartridges, permethylated and analyzed by MALDI/TOF or mass spectrometry. The *O*-glycans with the same structures appear as doublet peaks with a 6-Da difference in mass. The relative ratio of these doublet peaks represents the relative abundance of this *O*-glycan in these two cell lines or cells cultured at two conditions. Single peak is the unique *O*-glycan lacking in one of the cell lines or one of the culture conditions. The stable isotopic material [$^{2}H_7$]Bn-α-GalNAc(Ac)$_3$ and its pair, regular Bn-α-GalNAc(Ac)$_3$, ([$^{1}H_7$]Bn-α-GalNAc(Ac)$_3$) can also be used to substitute [$^{13}C_6$]Bn-α-GalNAc(Ac)$_3$ and [$^{12}C_6$]Bn-α-GalNAc(Ac)$_3$ respectively. In this case, the mass of the *O*-glycans with the same compositions and structures will show as doublet peaks with a 7-Da mass difference.

the chemical method with glycosyltransferase catalyzed enzymatic reactions. This approach can synthesize a number of complex *O*-glycans and *N*-glycan structures.[93,94] It is still a big challenge to make complex branched *O*-glycans in a reasonable quantity by this method. Of note, chemical synthesis and chemo-enzymatic synthesis are both target-driven, yet, lots of natural *O*-glycans are still unknown in terms of structure. Therefore, isolation of *O*-glycans from natural sources is an important approach. This approach, however, still involves releasing the *O*-glycans from glycoproteins/mucins, which is the same hurdle as *O*-glycan analysis.

CORA is a novel approach for amplifying *O*-glycome in live cells, using the native sugar donors and endogenous *O*-glycosyltransferases, to produce Bn-*O*-glycans.[27] Given that CORA can amplify and profile cellular *O*-glycome efficiently, it could be applied for preparation of *O*-glycans. However, the inherent property of benzyl (Bn) group of the Bn-α-GalNAc precursor limits the application of CORA for preparation of *O*-glycans, since its UV absorption signal is not sensitive enough to monitor the purification step. In addition, the aglycone, Benzyl-group is not easily derivatized for installation of functional groups for monitoring and purification or printing on a glass slide to make glycan-microarrays. Hence, it is necessary to advance CORA to a preparative-CORA (pCORA) for preparation of *O*-glycans by replacing the Benzyl with groups with the desired properties.

In order to monitor *O*-glycans prepared from CORA, the benzyl group of Bn-α-GalNAc(Ac)$_3$ needs to be replaced by other groups. From previous studies, we have learned that T-synthase, the key glycosyltransferase which is responsible for synthesizing Core 1 structures, can utilize many different artificial substrates (R-α-GalNAc) as its acceptors, ranging from the small R, such as methyl, to benzyl (Bn), *p*-nitrophenyl (pNP), and even 4-methylumbelliferone (4-MU) which are bulky groups.[14,82,95] That information offers some direction on selecting the candidates of substituent groups to replace the Benzyl group (Figure 16.4).

An attractive group of candidates is the one having a fluorescence property such as 2-PA,[65] 2-AA,[67] 2-AB,[96] 2,6-diaminopyridine (DAP),[97] and 2-amino-*N*-(2-aminoethyl)-benzamide (AEAB),[98] which are commonly used to label the carbohydrates for quantification and purification on HPLC. In addition, these compounds all contain an active NH_2 group which can be covalently immobilized onto *N*-hydroxysuccinimide (NHS)-activated glass slides for making Glycan-microarrays. So, if these fluorescence-containing compounds can serve as *O*-glycan precursors, their derivatized *O*-glycans can be directly separated on multidimensional HPLC by monitoring their fluorescence, the purified *O*-glycans can then be directly immobilized to make microarrays. An alternative strategy is to use the Bn derivatives in Bn-α-GalNAc. By replacing the "H" group of the benzene, these compounds possess a small functional group, such as azide (N_3), alkynyl, and ketone, which are widely used as the biorthogonal chemical reporters to incorporate into the target biomolecules through the cell's own biosynthetic machinery.[99] Particularly, azide-containing monosaccharides have been applied to incorporation into glycoconjugates by glycan biosynthetic pathways. For instance, the azido

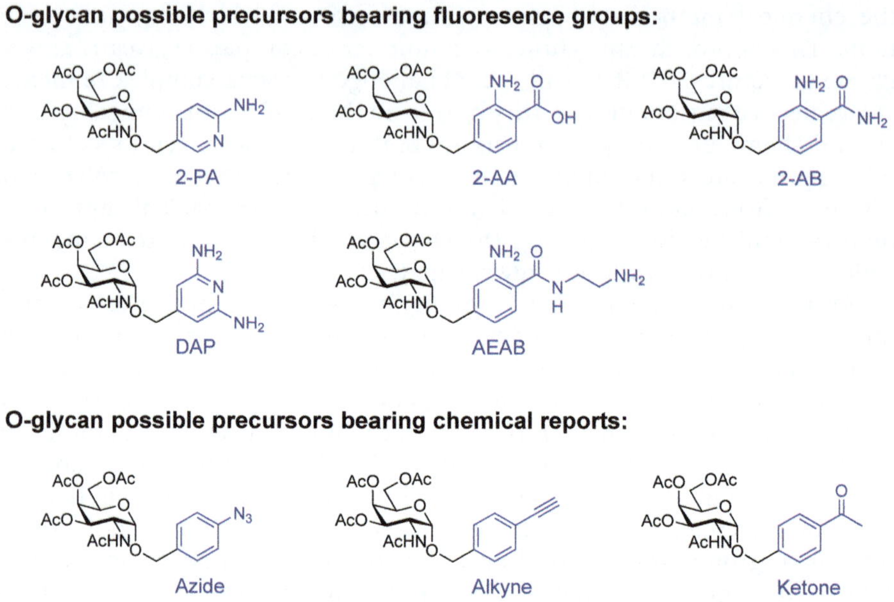

Figure 16.4 Structures of possible *O*-glycan precursors. (1) The fluorescent group of compounds includes 2-PA, 2-AA, 2-AB, DAP and AEAB; (2) biorthogonal chemical report group of compounds includes azide, alkyne, and ketone (X-).

analog of GalNAc, termed GalNAz,[100] was metabolically introduced into the mucin *O*-glycoproteins; azido analogs of sialic acid (SiaNAz) and ManNAc (ManNAz)[101] were metabolized and converted to azido sialosides in glycoproteins appearing on cell surfaces among others.[99] Since these chemical reporters are very small and stable in cells, these Benzyl derivatives, X-Bn-α-GalNAc (where X- is azide (N_3), alkynyl, and ketone) will possibly displace Bn-α-GalNAc and efficiently produce X-Bn-*O*-glycans. Then the X-Bn-*O*-glycans can be covalently derivatized in a highly selective fashion with a bifunctional fluorescent tag, such as AEAB. Then the fluorescent tagged Bn-*O*-glycans can be feasibly purified from the culturing media and immobilized on glass slides to make cellular *O*-glycan-microarrays; or they can even be individually purified on multidimensional HPLC and then immobilized on glass slides to make defined *O*-glycan-microarrays. Since cells can be cultured at a large scale with a repeated feeding/harvesting process, *O*-glycans can be robustly produced. In this case, live cells function as factories to manufacture *O*-glycans. Knowing the special capability each cell line has, panels of defined *O*-glycans are feasibly prepared through this preparative-CORA (pCORA). The whole process is illustrated in Figure 16.5.

Of note, CORA technology, including qCORA and pCORA, has its merits, and also limitation. qCORA is a simple, sensitive, and straightforward method for quantitative and qualitative *O*-glycomics; pCORA amplifies

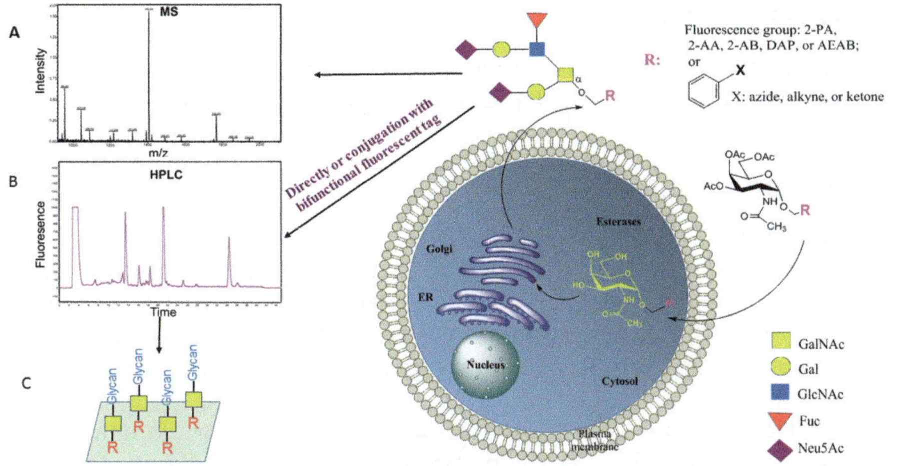

Figure 16.5 Overview of preparative-CORA. Peracetylated α-GalNAc derivative (R-α-GalNAc$(Ac)_3$) which contains a bifunctional fluorescence group or Bn derivative (X-Bn) can enter cells, be deacetylated, transported into the Golgi and modified to form R-α-*O*-glycans, which are secreted into the media, purified, and analyzed by MS (A). If the functional group (R) is fluorescent functional group, it will be directly separated and purified by HPLC; if the functional group (R) is chemical report contained X-Bn, it will be conjugated with a bifunctional fluorescent tag, and the fluorescently tagged *O*-glycans can be analyzed and separated by HPLC (B). The fluorescently labeled *O*-glycans are purified by multidimensional HPLC, if necessary, to obtain the Tagged *O*-glycan Library that can be printed on glass slides to make *O*-glycan Microarrays (C). Adapted from ref. 27 with permission from Springer Nature, Copyright 2015.

and produces cellular *O*-glycans that can be easily purified, derivatized for functional *O*-glycomics studies. In principle, CORA is a metabolic labeling method, hence live cells and tissue/cell cultures are required.

16.5 Perspective

Functional *O*-glycomics concerning the biological functions of *O*-glycans not only requires a novel technology to qualitatively and quantitatively analyze *O*-glycans, but also needs a tool to prepare a significant amount of *O*-glycans of interest to study and confirm their functions. qCORA offers a feasible technology for quantitative *O*-glycomics, while pCORA provides a tool to amplify and prepare *O*-glycans by adapting cells to *O*-glycan factories. qCORA and pCORA can meet the demands and therefore will facilitate the cellular *O*-glycomics and functional *O*-glycomics studies.

Conflict of Interest

The authors declare no conflict of interest.

Acknowledgements

We thank Dr Ashutosh Rao for their helpful suggestions on the manuscript. This work was supported by NIH Common Funds 3U01CA20781 to T.J.

References

1. C. Steentoft, S. Y. Vakhrushev, H. J. Joshi, Y. Kong, M. B. Vester-Christensen, K. T. Schjoldager, K. Lavrsen, S. Dabelsteen, N. B. Pedersen, L. Marcos-Silva, R. Gupta, E. P. Bennett, U. Mandel, S. Brunak, H. H. Wandall, S. B. Levery and H. Clausen, *EMBO J.*, 2013, **32**, 1478.
2. T. Ju, V. I. Otto and R. D. Cummings, *Angew. Chem., Int. Ed. Engl.*, 2011, **50**, 1770.
3. A. Halim, G. Brinkmalm, U. Ruetschi, A. Westman-Brinkmalm, E. Portelius, H. Zetterberg, K. Blennow, G. Larson and J. Nilsson, *Proc. Natl. Acad. Sci. U. S. A.*, 2011, **108**, 11848.
4. S. Y. Vakhrushev, C. Steentoft, M. B. Vester-Christensen, E. P. Bennett, H. Clausen and S. B. Levery, *Mol. Cell. Proteomics*, 2013, **12**, 932.
5. T. Ju, Y. Wang, R. P. Aryal, S. D. Lehoux, X. Ding, M. R. Kudelka, C. Cutler, J. Zeng, J. Wang, X. Sun, J. Heimburg-Molinaro, D. F. Smith and R. D. Cummings, *Proteomics: Clin. Appl.*, 2013, **7**, 618.
6. S. R. Stowell, T. Ju and R. D. Cummings, *Annu. Rev. Pathol.*, 2015, **10**, 473.
7. J. Furukawa, N. Fujitani and Y. Shinohara, *Biomolecules*, 2013, **3**, 198.
8. R. D. Cummings and J. M. Pierce, *Chem. Biol.*, 2014, **21**, 1.
9. G. Zauner, R. P. Kozak, R. A. Gardner, D. L. Fernandes, A. M. Deelder and M. Wuhrer, *Biol. Chem.*, 2012, **393**, 687.
10. Y. Mechref, Y. Hu, J. L. Desantos-Garcia, A. Hussein and H. Tang, *Mol. Cell. Proteomics*, 2013, **12**, 874.
11. L. X. Wang and J. V. Lomino, *ACS Chem. Biol.*, 2012, **7**, 110.
12. X. Song, J. Heimburg-Molinaro, R. D. Cummings and D. F. Smith, *Curr. Opin. Chem. Biol.*, 2014, **18**, 70.
13. E. P. Bennett, U. Mandel, H. Clausen, T. A. Gerken, T. A. Fritz and L. A. Tabak, *Glycobiology*, 2012, **22**, 736.
14. T. Ju, R. D. Cummings and W. M. Canfield, *J. Biol. Chem.*, 2002, **277**, 169.
15. T. Ju, K. Brewer, A. D'Souza, R. D. Cummings and W. M. Canfield, *J. Biol. Chem.*, 2002, **277**, 178.
16. L. Xia, T. Ju, A. Westmuckett, G. An, L. Ivanciu, J. M. McDaniel, F. Lupu, R. D. Cummings and R. P. McEver, *J. Cell Biol.*, 2004, **164**, 451.
17. Y. Wang, T. Ju, X. Ding, B. Xia, W. Wang, L. Xia, M. He and R. D. Cummings, *Proc. Natl. Acad. Sci. U. S. A.*, 2010, **107**, 9228.
18. T. Ju and R. D. Cummings, *Proc. Natl. Acad. Sci. U. S. A.*, 2002, **99**, 16613.
19. W. Gillespie, S. Kelm and J. C. Paulson, *J. Biol. Chem.*, 1992, **267**, 21004.
20. Y. C. Lee, M. Kaufmann, S. Kitazume-Kawaguchi, M. Kono, S. Takashima, N. Kurosawa, H. Liu, H. Pircher and S. Tsuji, *J. Biol. Chem.*, 1999, **274**, 11958.
21. N. Kurosawa, N. Kojima, M. Inoue, T. Hamamoto and S. Tsuji, *J. Biol. Chem.*, 1994, **269**, 19048.

22. J. Mitoma, B. Petryniak, N. Hiraoka, J. C. Yeh, J. B. Lowe and M. Fukuda, *J. Biol. Chem.*, 2003, **278**, 9953.
23. M. F. Bierhuizen and M. Fukuda, *Proc. Natl. Acad. Sci. U. S. A.*, 1992, **89**, 9326.
24. T. Iwai, N. Inaba, A. Naundorf, Y. Zhang, M. Gotoh, H. Iwasaki, T. Kudo, A. Togayachi, Y. Ishizuka, H. Nakanishi and H. Narimatsu, *J. Biol. Chem.*, 2002, **277**, 12802.
25. N. Kurosawa, S. Takashima, M. Kono, Y. Ikehara, M. Inoue, Y. Tachida, H. Narimatsu and S. Tsuji, *J. Biochem.*, 2000, **127**, 845.
26. J. M. Yang, J. C. Byrd, B. B. Siddiki, Y. S. Chung, M. Okuno, M. Sowa, Y. S. Kim, K. L. Matta and I. Brockhausen, *Glycobiology*, 1994, **4**, 873.
27. M. R. Kudelka, A. Antonopoulos, Y. Wang, D. M. Duong, X. Song, N. T. Seyfried, A. Dell, S. M. Haslam, R. D. Cummings and T. Ju, *Nat. Methods*, 2016, **13**, 81.
28. J. B. Lowe, *Curr. Opin. Cell Biol.*, 2003, **15**, 531.
29. M. Martinez, M. Joffraud, S. Giraud, B. Baisse, M. P. Bernimoulin, M. Schapira and O. Spertini, *J. Biol. Chem.*, 2005, **280**, 5378.
30. D. A. Carlow, K. Gossens, S. Naus, K. M. Veerman, W. Seo and H. J. Ziltener, *Immunol. Rev.*, 2009, **230**, 75.
31. J. Fu, H. Gerhardt, J. M. McDaniel, B. Xia, X. Liu, L. Ivanciu, A. Ny, K. Hermans, R. Silasi-Mansat, S. McGee, E. Nye, T. Ju, M. I. Ramirez, P. Carmeliet, R. D. Cummings, F. Lupu and L. Xia, *J. Clin. Invest.*, 2008, **118**, 3725.
32. Y. Wang, S. M. Jobe, X. Ding, H. Choo, D. R. Archer, R. Mi, T. Ju and R. D. Cummings, *Proc. Natl. Acad. Sci. U. S. A.*, 2012, **109**, 16143.
33. M. Kuwano, T. Seguchi and M. Ono, *J. Cell Sci.*, 1991, **98**(Pt 2), 131.
34. I. Chefetz and E. Sprecher, *Biochim. Biophys. Acta*, 2009, **1792**, 847.
35. K. T. Schjoldager and H. Clausen, *Biochim. Biophys. Acta*, 2012, **1820**, 2079.
36. M. Van der Sluis, B. A. De Koning, A. C. De Bruijn, A. Velcich, J. P. Meijerink, J. B. Van Goudoever, H. A. Buller, J. Dekker, I. Van Seuningen, I. B. Renes and A. W. Einerhand, *Gastroenterology*, 2006, **131**, 117.
37. H. Kawashima, *Biol. Pharm. Bull.*, 2012, **35**, 1637.
38. R. Gibadullin, D. W. Farnsworth, J. J. Barchi Jr and J. C. Gildersleeve, *ACS Chem. Biol.*, 2017, **12**, 2172.
39. J. Hsu, S. J. Chang and A. H. Franz, *J. Am. Soc. Mass Spectrom.*, 2006, **17**, 194.
40. T. H. Plummer, J. H. Elder, S. Alexander, A. W. Phelan and A. L. Tarentino, *J. Biol. Chem.*, 1984, **259**, 700.
41. T. Suzuki, C. Huang and H. Fujihira, *Gene*, 2016, **577**, 1.
42. R. A. Dwek, C. J. Edge, D. J. Harvey, M. R. Wormald and R. B. Parekh, *Annu. Rev. Biochem.*, 1993, **62**, 65.
43. D. Koutsioulis, D. Landry and E. P. Guthrie, *Glycobiology*, 2008, **18**, 799.
44. P. Babu, S. J. North, J. Jang-Lee, S. Chalabi, K. Mackerness, S. R. Stowell, R. D. Cummings, S. Rankin, A. Dell and S. M. Haslam, *Glycoconjugate J.*, 2009, **26**, 975.
45. S. J. North, H. H. Huang, S. Sundaram, J. Jang-Lee, A. T. Etienne, A. Trollope, S. Chalabi, A. Dell, P. Stanley and S. M. Haslam, *J. Biol. Chem.*, 2010, **285**, 5759.

46. A. V. Nairn, K. Aoki, M. dela Rosa, M. Porterfield, J. M. Lim, M. Kulik, J. M. Pierce, L. Wells, S. Dalton, M. Tiemeyer and K. W. Moremen, *J. Biol. Chem.*, 2012, **287**, 37835.
47. G. Zauner, C. A. Koeleman, A. M. Deelder and M. Wuhrer, *Biochim. Biophys. Acta*, 2012, **1820**, 1420.
48. C. Wang, P. Zhang, W. Jin, L. Li, S. Qiang, Y. Zhang, L. Huang and Z. Wang, *J. Proteomics*, 2017, **150**, 18.
49. C. Wang, W. Fan, P. Zhang, Z. Wang and L. Huang, *Proteomics*, 2011, **11**, 4229.
50. J. Furukawa, N. Fujitani, K. Araki, Y. Takegawa, K. Kodama and Y. Shinohara, *Anal. Chem.*, 2011, **83**, 9060.
51. S. Natunen, T. Satomaa, V. Pitkanen, H. Salo, M. Mikkola, J. Natunen, T. Otonkoski and L. Valmu, *Glycobiology*, 2011, **21**, 1125.
52. T. Patel, J. Bruce, A. Merry, C. Bigge, M. Wormald, A. Jaques and R. Parekh, *Biochemistry*, 1993, **32**, 679.
53. A. H. Merry, D. C. A. Neville, L. Royle, B. Matthews, D. J. Harvey, R. A. Dwek and P. M. Rudd, *Anal. Biochem.*, 2002, **304**, 91.
54. H. Tateno, M. Toyota, S. Saito, Y. Onuma, Y. Ito, K. Hiemori, M. Fukumura, A. Matsushima, M. Nakanishi, K. Ohnuma, H. Akutsu, A. Umezawa, K. Horimoto, J. Hirabayashi and M. Asashima, *J. Biol. Chem.*, 2011, **286**, 20345.
55. S. Takasaki, T. Mizuochi and A. Kobata, *Methods Enzymol.*, 1982, **83**, 263.
56. K. Hasehira, H. Tateno, Y. Onuma, Y. Ito, M. Asashima and J. Hirabayashi, *Mol. Cell. Proteomics*, 2012, **11**, 1913.
57. X. Song, H. Ju, Y. Lasanajak, M. R. Kudelka, D. F. Smith and R. D. Cummings, *Nat. Methods*, 2016, **13**, 528.
58. S. Diaz and A. Varki, *Curr. Protoc. Mol. Biol.*, 2009, **87**, 17.4.1.
59. F. T. Chen and R. A. Evangelista, *Electrophoresis*, 1998, **19**, 2639–2644.
60. A. Guttman, F. T. Chen, R. A. Evangelista and N. Cooke, *Anal. Biochem.*, 1996, **233**, 234.
61. P. Jackson, *Methods Enzymol.*, 1994, **230**, 250.
62. P. Jackson, *Anal. Biochem.*, 1991, **196**, 238.
63. P. Jackson, *Mol. Biotechnol.*, 1996, **5**, 101.
64. N. Tomiya, M. Kurono, H. Ishihara, S. Tejima, S. Endo, Y. Arata and N. Takahashi, *Anal. Biochem.*, 1987, **163**, 489.
65. K. Deguchi, T. Keira, K. Yamada, H. Ito, Y. Takegawa, H. Nakagawa and S. Nishimura, *J. Chromatogr. A*, 2008, **1189**, 169.
66. K. R. Anumula and S. T. Dhume, *Glycobiology*, 1998, **8**, 685.
67. L. R. Ruhaak, C. Huhn, W. J. Waterreus, A. R. de Boer, C. Neususs, C. H. Hokke, A. M. Deelder and M. Wuhrer, *Anal. Chem.*, 2008, **80**, 6119.
68. J. Zaia, *OMICS*, 2010, **14**, 401.
69. M. J. Kailemia, L. R. Ruhaak, C. B. Lebrilla and I. J. Amster, *Anal. Chem.*, 2014, **86**, 196.
70. R. Madiyalakan, R. A. DiCioccio and K. L. Matta, *Carbohydr. Res.*, 1984, **129**, 298.

71. I. Ciucanu and C. E. Costello, *J. Am. Chem. Soc.*, 2003, **125**, 16213.
72. Y. Mechref, Y. L. Hu, J. L. Desantos-Garcia, A. Hussein and H. X. Tang, *Mol. Cell. Proteomics*, 2013, **12**, 874.
73. P. Kang, Y. Mechref, Z. Kyselova, J. A. Goetz and M. V. Novotny, *Anal. Chem.*, 2007, **79**, 6064.
74. Y. Hu, J. L. Desantos-Garcia and Y. Mechref, *Rapid Commun. Mass Spectrom.*, 2013, **27**, 865.
75. B. Xia, C. L. Feasley, G. P. Sachdev, D. F. Smith and R. D. Cummings, *Anal. Biochem.*, 2009, **387**, 162.
76. A. M. Hitchcock, C. E. Costello and J. Zaia, *Biochemistry*, 2006, **45**, 2350.
77. M. J. Bowman and J. Zaia, *Anal. Chem.*, 2007, **79**, 5777.
78. J. Yuan, N. Hashii, N. Kawasaki, S. Itoh, T. Kawanishi and T. Hayakawa, *J. Chromatogr. A*, 2005, **1067**, 145.
79. R. Chaerkady, P. J. Thuluvath, M. S. Kim, A. Nalli, P. Vivekanandan, J. Simmers, M. Torbenson and A. Pandey, *Clin. Proteomics*, 2008, **4**, 137.
80. S. H. Walker, J. Budhathoki-Uprety, B. M. Novak and D. C. Muddiman, *Anal. Chem.*, 2011, **83**, 6738.
81. R. Orlando, J. M. Lim, J. A. Atwood 3rd, P. M. Angel, M. Fang, K. Aoki, G. Alvarez-Manilla, K. W. Moremen, W. S. York, M. Tiemeyer, M. Pierce, S. Dalton and L. Wells, *J. Proteome Res.*, 2009, **8**, 3816.
82. Y. Gao, R. P. Aryal, T. Ju, R. D. Cummings, G. Gahlay, D. L. Jarvis, K. L. Matta, J. Z. Vlahakis, W. A. Szarek and I. Brockhausen, *Biochim. Biophys. Acta*, 2013, **1830**, 4274.
83. I. Brockhausen, G. Moller, A. Pollex-Kruger, V. Rutz, H. Paulsen and K. L. Matta, *Biochem. Cell Biol.*, 1992, **70**, 99.
84. J. Huang, J. C. Byrd, W. H. Yoon and Y. S. Kim, *Oncol. Res.*, 1992, **4**, 507.
85. A. K. Sarkar, T. A. Fritz, W. H. Taylor and J. D. Esko, *Proc. Natl. Acad. Sci. U. S. A.*, 1995, **92**, 3323.
86. M. M. Fuster, J. R. Brown, L. Wang and J. D. Esko, *Cancer Res.*, 2003, **63**, 2775.
87. M. R. Kudelka, A. V. Nairn, M. Y. Sardar, X. Sun, E. L. Chaikof, T. Ju, K. W. Moremen and R. D. Cummings, *Glycobiology*, 2018, **28**, 214.
88. X. Zhu and R. R. Schmidt, *Angew. Chem., Int. Ed. Engl.*, 2009, **48**, 1900.
89. P. H. Seeberger, *Chem. Commun.*, 2003, 1115.
90. L. X. Wang and M. N. Amin, *Chem. Biol.*, 2014, **21**, 51.
91. T. Murata and T. Usui, *Biosci., Biotechnol., Biochem.*, 2006, **70**, 1049.
92. X. Song, J. Heimburg-Molinaro, R. D. Cummings and D. F. Smith, *Curr. Opin. Chem. Biol.*, 2014, **18**, 70.
93. Z. Wang, Z. S. Chinoy, S. G. Ambre, W. Peng, R. McBride, R. P. de Vries, J. Glushka, J. C. Paulson and G. J. Boons, *Science*, 2013, **341**, 379.
94. T. J. Boltje, T. Buskas and G. J. Boons, *Nat. Chem.*, 2009, **1**, 611.
95. T. Ju, B. Xia, R. P. Aryal, W. Wang, Y. Wang, X. Ding, R. Mi, M. He and R. D. Cummings, *Glycobiology*, 2011, **21**, 352.
96. L. Royle, M. P. Campbell, C. M. Radcliffe, D. M. White, D. J. Harvey, J. L. Abrahams, Y. G. Kim, G. W. Henry, N. A. Shadick, M. E. Weinblatt, D. M. Lee, P. M. Rudd and R. A. Dwek, *Anal. Biochem.*, 2008, **376**, 1.

97. B. Xia, Z. S. Kawar, T. Ju, R. A. Alvarez, G. P. Sachdev and R. D. Cummings, *Nat. Methods*, 2005, **2**, 845.
98. X. Z. Song, B. Y. Xia, S. R. Stowell, Y. Lasanajak, D. F. Smith and R. D. Cummings, *Chem. Biol.*, 2009, **16**, 36.
99. J. A. Prescher and C. R. Bertozzi, *Nat. Chem. Biol.*, 2005, **1**, 13.
100. H. C. Hang, C. Yu, D. L. Kato and C. R. Bertozzi, *Proc. Natl. Acad. Sci. U. S. A.*, 2003, **100**, 14846.
101. E. Saxon and C. R. Bertozzi, *Science*, 2000, **287**, 2007.

CHAPTER 17

Current Stage of Commercially Available Glycans to Support Realization of Biologic Drugs

JENIFER L. HENDEL* AND DARYL L. FERNANDES

Ludger Limited, Culham Science Centre, Oxford, OX14 3EB, UK
*E-mail: jenifer.hendel@ludger.com

17.1 Introduction

Glycans are ubiquitous in nature. They are found throughout the cells and tissues of plants and animals and are implicated in virtually all physiological processes in the human body.[1] The abundance and importance of glycans coupled with advances in glycochemistry is allowing them to be used in a highly diverse range of industries including biotechnology, synthetic chemistry, health, energy, and materials science. Applications of innovative glycoscience range from glycans being used for biomass conversion into new plastics,[2] to the development of novel medicines.[3] This chapter covers the biotechnology industry where glycans can play critical roles in the medical and commercial success or failure of glycoprotein therapeutics. Our focus is the availability and use of commercial analytical glycan standards to support the realization of biopharmaceuticals. Here we define 'realization' as all the processes required for the design, development, manufacturing and regulatory compliance of the drug.

Chemical Biology No. 11
Synthetic Glycomes
Edited by Wanyi Guan, Lei Li and Peng George Wang

Published by the Royal Society of Chemistry, www.rsc.org

Glycans are the most visible post-translational modification of biologic drugs as well as being their greatest source of within-batch molecular heterogeneity and batch-to-batch variability. Drug glycosylation patterns can also greatly influence clinical performance of drug products, particularly *in vivo* safety and efficacy profiles, as well as drug manufacturability and the cost per therapeutic dose.[4]

Given this, biopharma companies are facing increasing pressures from regulatory authorities to properly design, measure and control the glycosylation patterns of their therapeutics. However, this is challenging and problems in producing biopharmaceuticals with well-characterized, consistent glycosylation have caused considerable delays in, and even denial of, regulatory approval. There have been many examples of such problems over the years including those for both novel therapeutics and biosimilars. Most of these challenges result from the inherent complexity and variability of glycosylation. They include product inconsistencies throughout the drug lifecycle (particularly during scale-up) and out-of-specification (OOS) failures during biomanufacturing.[5,6] The importance of addressing these challenges is underlined by the numbers of glycosylated therapeutics being developed – in the past two years 27 out of 28 of the biologic drugs approved by the FDA have been glycoproteins.[7]

The financial cost of glycosylation problems to the industry has been considerable, with billions of dollars wasted. The human costs to patients and their families – particularly through subsequent drug shortages and inaccessibility due to high drug prices – are incalculable. Fortunately, technological advances have helped to greatly reduce these problems over the past decade. Glycochemists have played important roles in this process, particularly through their work developing better glycoanalysis methods.

This chapter focuses on one aspect of the contribution that glycochemists continue to make to improve the biopharma industry – advancements in the provision of well-characterized, commercial-grade glycan standards to support reliable structural analysis of drug *N*- and *O*-glycosylation patterns.

We will overview the classes and uses of currently available glycan standards for both qualitative and quantitative glycoanalysis work. In addition, we will give case studies of problems with glycosylation that have occurred in the biopharma industry and how they could have been avoided through proper use of glycan standards. We will also touch on ways to enhance the reliability and costs for high throughput glycomics during Quality-by-Design (QbD) based biopharmaceutical realization. This is an increasingly important area – particularly for developers of biosimilars who must closely match glycosylation patterns of the originator's drugs and get their products approved and into the market in a timely way or face commercial failure. We will then summarize our thoughts on how glycochemists could further enhance the repertoire of commercially-available glycan standards in order to aid development and manufacturing of safer, more effective and more affordable biologic drugs.

17.1.1 Changes in the Therapeutic Glycoprotein Industry

The biopharma industry is currently experiencing a time of considerable growth and change, it is shifting from a position at the frontiers of science to becoming a mature technical business.[8] Many challenges arise through this transition including an increased demand by patients and drug payers for improved access to more affordable biopharmaceuticals, opening out of the markets to competition and increasing complexity of supply chain and operations as markets expand. Methods for drug realization are also undergoing considerable changes with greater systemization and adoption of risk-based approaches to quality issues largely driven by the need for companies to increase efficiencies, lower costs, reduce time to market and improve quality compliance as scrutiny from regulators intensifies.

The approved and marketed glycoprotein drugs span a broadening range of structure–activity relationship (SAR) classes. These encompass human-engineered versions of naturally occurring bioactive glycoproteins such as hormones (*e.g.* Follicle-Stimulating Hormone, FSH, which is used to treat infertility), cytokines (such as erythropoietin that increases red blood cells for patients with kidney dysfunction) and blood clotting factors (*e.g.* Factor VIII for treating haemophilia). Currently, the largest class of glycoprotein drugs are the monoclonal antibodies (mAbs). These are designed as enhanced variants of natural immunoglobulins and fall into two categories depending on the principal modes of action (MoA): neutralizing mAbs that clear out foreign agents *via* the immune system and others with 'effector functions'. These are created to seek and bind to target (*e.g.* cancer) cells then initiate complex immune cascades that lead to target killing and elimination.

The glycoprotein therapeutics are developed to interact with their molecular targets in precise ways. This gives them tremendous medical value because the therapeutic actions can, in principle, be localized to dysfunctional molecular pathways and biological systems. This increases their clinical efficacy and lowers the risk of side reactions compared to broader-acting drugs. However, there is a high financial cost for this precision – biologic drugs are too expensive for most people. For example, patients in the US report that one year's treatment with the anti-breast cancer mAb Herceptin typically costs in excess of $80k, while at $100k – $300k per year Genzyme's drug Myozyme (for treatment of Pompe's Disease) made the Forbes list of the top ten most expensive drugs in the world.[9]

The high prices these drugs command reflect: (a) the increased costs for development and manufacturing of such large, structurally complex molecules and (b) the monopoly positions that drug innovators can hold. However, this situation is changing because the biopharma market is opening out to competition in the form of biosimilars.[10]

The advent of biosimilars is stimulating rapid growth of the biotherapeutics market. As official copies of previously licensed biopharmaceuticals these products constitute both a regulatory and commercial construct designed to widen market competition, drive prices down and increase patient access.

However, the great structural complexity of these biologic drugs coupled with commercial secrecy for the manufacturing processes generally precludes competitors from making exact copies of the innovator's molecule. Glycosylation patterns are particularly difficult to copy faithfully. Given this, regulators require biosimilar companies to produce drugs that have exact copies of the protein portion of the original drug (this is relatively straightforward to achieve) and to reproduce as closely as possible the glycosylation patterns and other post-translational modifications of the innovator's product (which is considerably more challenging). The aim is for the original drug and approved biosimilars to represent variants with similar physicochemical characteristics and clinical safety and efficacy profiles as judged by comprehensive comparability studies.[11] Such comparability of structures and functions would allow medics and healthcare payers to switch between the original drug and a biosimilar without compromising the patient.

The commercial freedom to produce biosimilars has arisen because the patents protecting the first generation of highly lucrative glycoprotein therapeutics have been expiring steadily. This 'patent cliff' will stimulate biosimilars development for many years although for some biopharmaceuticals clearing the patents is likely to be more problematic than it has been for small molecule therapeutics.[12]

The increased availability of biosimilars is expected to have greatest impact in emerging markets, where access to the originator products remains beyond the reach of many patients. The expectation is that an increased access to the copy drugs will go hand-in-hand with reduction in originator revenues.[8] In response, just as in other industries such as airlines and cars, the originators are themselves developing their own 'low cost' versions of their premium (drug) products. In addition, originators are developing 'biobetters' – versions of their first generation drugs with tangible improvements to clinical performance but with the same mode of action.[13] Given the contribution of glycosylation to safety and efficacy, both faithful copying and skillful optimization of glycan patterns are central to the development of biosimilars and biobetters.

17.1.2 The Significance of Glycosylation for Therapeutic Function

Given that glycans play critical roles in the clinical performance of most glycoprotein therapeutics, the skill with which companies design, optimize, measure and control their drug's glycosylation will impact its passage along the regulatory pathways as well as its commercial success. The glycosylation patterns will typically impact bioactivity, stability, biological half-life and immunogenicity of these drugs.[14,15] Glycosylation also correlates with 'microheterogeneity' and the formation of glycoforms. These are isoforms of the drug that have structurally identical protein moieties but different patterns of glycans across the various glycosylation sites. In effect, each vial of a glycoprotein drug will be a mixture of diverse glycoforms with structure–function properties and therefore acting like different drugs.

This complexity causes significant challenges to biopharma companies who must demonstrate consistent high quality for their product through intensive biochemical characterization of the drug substance. This includes a thorough examination of both glycan and glycoform distributions and the potential impacts of variations in each of these on drug function *in vivo*. For example, fucosylation of monoclonal antibodies relying on Fc effector functions can significantly reduce levels of antibody-dependent cell mediated cytotoxicity (ADCC).[16] Another example is that there is a positive correlation between increased sialylation and the anti-inflammatory activity of a glycoprotein.[17]

A systematic approach to dealing with drug glycan patterns should involve early identification and prioritization of the Glycosylation Critical Quality Attributes (GCQAs). These are the glycosylation features that most influence the drug's clinical performance and the ones that must be designed, optimized, measured and controlled throughout the product's lifecycle. Their discovery arises from detailed knowledge of: (a) the glycosylation machinery of cell lines used for drug production, (b) the production process, (c) the drug's mode of action, and (d) structure–activity relationships of the drug product in patients. The structural complexity of glycans means there would typically be dozens of possible Glycosylation Quality Attributes (GQAs) that could be considered and prioritized to produce the GCQA list. Prioritization can be done by assessing 'criticality' – *i.e.* the impact of a change in each GQA on patient safety and therapeutic efficacy. GQA changes include the presence, absence or modification in the level of the attribute either, across the whole set of glycoforms in the drug, or in each of the constituent glycoforms. Developing the initial list of prioritized GCQAs is a major task for a biopharma company and requires a combination of glycan structural analyses, functional tests, clinical data and detailed knowledge of the drug's MoA. Furthermore, the prioritized GCQA list should be updated as more relevant knowledge is gained throughout the product lifecycle. This discipline greatly reduces the risk of regulatory, commercial and medical problems (such as adverse reactions) relating to glycosylation.

Knowledge gained in developing the prioritized GCQA list can also improve the chances of commercial success. For both Herceptin and Myozyme, as with other glycoprotein therapeutics, the cost per therapeutic dose and therefore profitability and price to patients could have been greatly improved during early stage development through selection of cell lines that produced drugs enriched in the safe, high activity glycoforms.[18]

17.1.3 Regulation of Drug Glycosylation

The following account of Myozme illustrates how a lack of rigorous monitoring and control of a drug's glycosylation patterns can lead to failure to win regulatory approval, loss of millions of dollars for the company and disrupt access to a life-saving drug by patients. Myozyme (alglucosidase alfa) was a drug developed and marketed by Genzyme as an enzyme replacement treatment for Pompe's disease, a rare and gradually debilitating disorder caused by a deficiency in the lysosomal enzyme alpha-glucosidase (GAA). This drug was initially produced in a 160 L production facility and, in order to meet a growing demand for the product, Genzyme built a dedicated 2000 L facility for its large-scale production.[6] The application to sell the product from the 2000 L plant was rejected by the United States Food and Drug Administration (FDA) regulators on the grounds that this version of the drug was no longer the same as the one produced in the 160 L tank. The FDA cited that the differences in glycosylation, in this case variation in mannose-6-phosphate (M6P), indicated that the scaled up product was not biologically equivalent to the original material which could impact drug efficacy. Their reasoning was that M6P is essential for targeting and transportation of newly synthesized hydrolases to the liposome and lower levels of M6P in the scaled-up drug were likely to cause reduction in clinical efficacy. This was despite Genzyme's argument that they had already conducted a clinical trial on the scaled-up drug and were able to demonstrate its safety and effectiveness.

However, the FDA prevailed and Genzyme had to abandon their commercialization route for Myozyme, and instead expand their Lumizyme program and submit an additional biologics regulatory submission to the FDA with new clinical trial data. This episode was very costly for Genzyme and caused significant delays in drug availability. The latter would have had a serious impact on Pompe's patients who, if not treated with enzyme replacement therapy, suffer increased mortality and morbidity.[19]

This was one of the first times that a drug batch was rejected on the basis of altered glycosylation. The lessons learned from Genzyme's mistake were twofold; (1) the regulatory authorities were starting to use the established technical guidelines on glycosylation to judge biopharmaceutical products more carefully; and (2) glycosylation could change enough during the bio-manufacturing processes and scale-up that they could fall out of specification. Drug manufacturers responded to this demand for more rigorous analysis of glycosylation by starting to characterize product glycosylation in relation to the drugs biological and clinical activity, and also by monitoring relevant glycan characteristics throughout production.

Regulatory agencies worldwide require state-of-the-art glycan analyses to be used when characterizing a therapeutic glycoprotein.[20] These methods must be both robust and reproducible to ensure accuracy and consistency. For *N*-glycans, both the International Conference on Harmonization (ICH) guidelines and European Medicines Agency (EMA) guidelines suggest that the degree of mannosylation, galactosylation, fucosylation and sialylation (total amount and types) should be reported and that distribution of the main glycan structures should be determined.[21,22] They also state that if available, additional information on the following should be provided: *N*-glycan site occupancy, lactosamine repeats, *N*- and *O*-acetylation, *O*-site occupancy and *O*-glycan profile, monosaccharide composition, and presence of immunogenic epitopes (galactose-α1–3-galactose (α-gal) and *N*-glycolylneuraminic

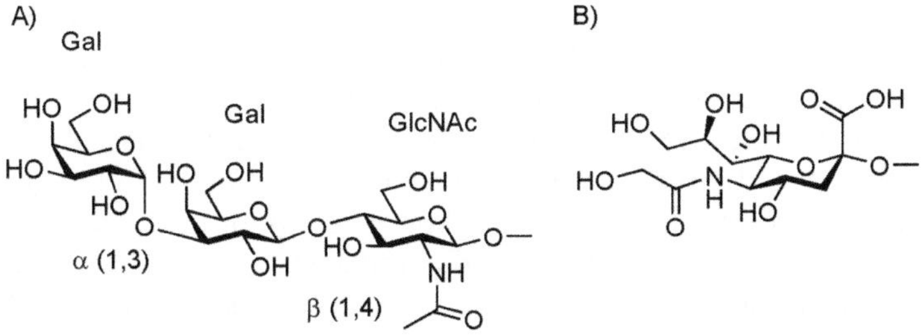

Figure 17.1 Immunogenic epitopes (A) galactose-α1–3-galactose (α-gal) and (B) Neu5Gc.

acid (Neu5Gc), Figure 17.1). Reports must contain all necessary analytical information, including detailed technical description of glycan preparation, analysis methods, and parameters that are used to process the data.

17.1.4 Control of Glycosylation During Biomanufacturing

A significant challenge facing glycoprotein manufacturers is to obtain consistent, clinically desired glycosylation. Glycosylation is a post-translational modification and the biosynthesis of glycoproteins results in the formation of a multitude of compounds that have an identical protein backbone, but differ in glycan structure and/or in occupancy of the glycosylation sites.[23] The heterogeneous nature of glycoprotein production means that every therapeutic drug is actually a collection of molecular structures. In manufacturing, the majority of therapeutic proteins are produced in bioreactor-based mammalian cell culture (*e.g.* Chinese hamster ovary (CHO) or murine myeloma transfectomas) and a select few are produced in other expression systems (*e.g. Escherichia coli*).[24] During bioprocessing the glycosylation of the expressed protein can be influenced by the selected expression system as well as bioreactor conditions (nutrient levels, pH, oxygen content or for example cell density).[25,26] Additionally, cell cultures can produce glycoproteins with glycans containing α-gal and Neu5Gc which can be immunogenic even in small quantities.[27,28]

It is clear that out of the many characteristics of a therapeutic protein, glycosylation stands out as one of the most difficult parameters to control owing to its complexity and sensitivity to the manufacturing process. It is for these reasons that the biopharmaceutical industry is moving towards better optimization of culture conditions, as well as stringent quality control measures in efforts to make drugs with better defined glycosylation profiles. One example of this shift in paradigm is the implantation of quality by design (QbD). QbD is a risk-based scientific approach to the entire drug design and development process.[4,18,29] It requires the manufacturer to have an in-depth understanding of how the drug works and to actively apply this knowledge at every step in production. In this way quality is built into the process, rather than testing it after the product manufacturing is complete. This approach is particularly well-suited to glycan analysis and involves three components; (1) Define the desired clinical or *in vivo* product performance. (2) Identify the GCQAs and which process parameters affect them. (3) Design, analyze, and control the manufacturing by measuring the products' critical quality and performance attributes throughout the entire process to ensure that the defined target product is realized.[30] A key component to the success of QbD is having the required glycan standards in place to ensure that the identified GCQAs are being appropriately and accurately measured during the entire process.

17.1.5 Characterization of Glycosylation

The characterization of glycoproteins is not a trivial task. As mentioned previously, one therapeutic drug is composed of a heterogeneous pool of glycoproteins and for characterization this translates into the analysis of a mixture of glycans (at least 15) on a very small scale (picomole quantities of each glycan species). This is further complicated by the stereo- and regioisomeric structural diversity of glycans; there are both isomeric (compounds with the same molecular formula but different chemical structure) and isobaric (compounds with the same mass but of different elemental composition) glycans to be identified. The ability to accurately identify isomers in protein glycoforms can have a huge implication on patient safety and this was exemplified by the Cetuximab case. In 2004, ImClone and Bristol–Meyers Squibb were investigating Cetuximab, a monoclonal antibody, which is specific for the epidermal growth factor receptor and was in clinical trials for the treatment of metastatic colorectal cancer.[28] In a subset of the patients an IgE immune response induced severe allergic symptoms such as anaphylaxis. The characterization of Cetuximab glycosylation later revealed that, out of 21 distinct oligosaccharide structures, approximately 30% had one or more α-1–3 linked galactosyl residues. It was concluded that, in individuals who have IgE to α-gal ($\geq$0.35 IU mL^{-1}), anaphylaxis reactions were likely to occur. If a thorough characterization of the protein glycosylation, complete with glycan standards containing the common immunogenic epitopes, had flagged the α-gal structures prior to patient involvement in the trial, then the desensitization protocol required for Cetuximab to be administered safely could have been implemented from the beginning.

In general, there are two purposes for characterization (1) to obtain a thorough assessment of glycan distribution with as much structural detail as possible (for example for initial characterization) and (2) abbreviated testing for comparability with a reference material, for example, in batch testing for routine monitoring and quality control.[31] In every aspect of glycan characterization glycan standards are used for assessing system suitability, determining analytical process efficiency, quantifying samples and aiding in structural elucidation during glycoanalysis.

Glycan analysis can be performed on the intact glycoprotein, glycopeptides generated from a proteolytic digest of the glycoprotein or intact glycans released from the protein. The most common method for glycosylation analysis during bioprocess development is the analysis of liberated intact glycans. For the scope of this chapter, a general overview of the approaches that are commonly used with released glycans are provided, for information on the other methods for glycan analysis there are various reviews covering the topic.[32–35]

There are two major types of protein glycosylation: *N*-, asparagine (Asn)-linked and *O*-, serine/threonine (Ser/Thr)-linked. The goal of glycan profiling is to best determine the relative intensity (%) of individual glycans present in the overall glycan profile of a glycoprotein. Several complementary analytical approaches are available.[36,37] Generally, the first step in glycan analysis is to cleave the carbohydrate moiety from the protein backbone using an enzymatic or chemical reaction. The gold standard method for release of *N*-linked glycans is peptide-*N*-glycosidase F (PNGase F), an endoglycosidase that cleaves the glycosidic bond between the Asn residue of the protein and the GlcNAc of the glycan. While PNGaseF is suitable for most therapeutic glycoproteins, it cannot cleave *N*-glycans with core α1–3 fucosylation. If core α1–3 fucosylation is present, which is commonly the case in plant and insect cells, peptide-*N*-glycosidase A(PNGase A) can be used to cleave the glycosidic bond between the Asn residue of the protein and the GlcNAc of both *N*-linked glycans with and without α1–3-linked core fucose residues.[38] Due to the lack of intrinsic chromophores, it is common to derivatize the glycans with a fluorescent label prior to analysis. The traditional method involves the reaction of the *N*-glycan which contains a single reducing end site (aldol) with an excess of fluorescent label *via* reductive amination, such that each *N*-glycan will be attached to one fluorophore.[39] There are also instant labelling methods which involve the reaction of the glycosylamine form of the *N*-glycan with an *N*-hydroxysuccinimidyl(NHS)-carbamate activated fluorescent tag.[40] Alternatively, the glycans can be chemically modified by permethylation to facilitate detection.[41] The chemically modified glycans can then be analyzed using different techniques such as (ultra) high performance liquid chromatography ((U)HPLC), capillary electrophoresis (CE) or by mass spectrometry (MS) approaches like electrospray ionization mass spectrometry (ESI-MS) or time-of-flight mass spectrometry (MALDI-TOF MS).[31,37]

The analysis of *O*-glycans is more challenging. *O*-Glycans tend to have higher structural variability in comparison to *N*-glycans and in addition, there is no PNGase F-like enzyme equivalent for their deglycosylation. *O*-Glycanase (endo-β-*N*-acetylgalactosaminidase), the closest enzyme equivalent to PNGaseF, has very strict substrate specificity and will only cleave the glycosidic bond between disaccharide galactosyl-β1–3-*N*-acetylgalactosamine and serine/threonine of the peptide backbone. Therefore, the analysis of *O*-glycosylation usually involves the release of *O*-glycans using chemical methods including hydrazinolysis or alkaline (reductive and non-reductive) β-elimination, followed by their chemical modification and analysis using the above-mentioned techniques.[42]

All of these methods provide reproducible results, but orthogonal methods are required to confirm glycan identity and structure.[43] Frequently (U)HPLC and CE methods are combined with MS to more thoroughly identify the constituents of the eluted peaks. Additionally, exhaustive exoglycosidase digestions can be performed to sequentially release selected monosaccharides in a controlled and predictable manner which is then followed by analysis on (U)HPLC to further confirm structural identity.

In addition to the above-mentioned approaches, there are also methods that quantify the total mass of specific monosaccharides within the glycan pool. For neutral monosaccharide analysis, glycoproteins are treated with acid to release the monosaccharides, they are then fluorescently labelled and separated by chromatography to quantify the absolute amounts of fucose (Fuc), mannose (Man), galactose (Gal), *N*-acetylglucosamine (GlcNAc), and *N*-acetylgalactosamine (GalNAc). Sialic acid levels can also be quantified by high performance anion exchange chromatography with pulsed amperometric detection (HPAEC-PAD) or weak acid release, 1,2-diamino-4,5-methylenedioxybenzene (DMB) labelling and (U)HPLC analysis.[44]

There has been significant progress made in glycan analysis owing to multiple advancements; new fluorescent reagents have made glycan analysis faster and easier, improvements in separation technology have resulted in more effective columns and the development of more sensitive detection instrumentation including advanced mass spectrometers is providing higher-resolution data.[34] Additionally, the availability of commercial kits which contain the necessary reagents for release and labelling of monosaccharides, sialic acids, and *N*- and *O*-glycans have made it easier for laboratories to adopt technologies for glycan analysis. However, even with the advancement in glycan characterization tools, multiple inter-laboratory studies have shown that there is still a lack in consistency of the data produced during glycan analysis.[45,46] These problems highlight the existing need for commercially available well-characterized glycan reference standards.

17.2 Commercially Available Glycans

Glycan standards, especially pure samples, are vital tools for the detection, identification, and quantification of drug glycan patterns. According to multiple sources,[33,47–50] the current opinion in both the biopharmaceutical industry and the glycomics field is that there are a limited number of commercially available standards. The following sections will look to address this view, exploring the market by seeing what standards are available, how they are made in industry, how they are used, and will conclude with a commentary on current and future perspectives.

17.2.1 Importance of Appropriate Commercial Sources for Glycans in the Biopharmaceutical Industry

Commercial-grade qualitative and quantitative glycan standards are a key part of reliable drug glycoanalysis schemes required for compliance with regulations. These are continuing to be tightened, in part due to lessons learned from Myozyme, Cetuximab and many other cases where serious problems were caused by inadequate characterization of drug glycans. In particular, materials and outsourced services used by biopharma companies must be qualified as part of the regulated Pharmaceutical Quality System that is implemented in Good Manufacturing Practices (GMP).[51] Consequently, the production and quality systems used by manufacturers of glycan standards must be improved continually to match the ever-more stringent demands of their clients and the industry regulators.

17.2.2 Comment on Commercial Glycan Providers

Worldwide, there are only a handful of companies that offer glycan standards. The following is a non-exhaustive list of these providers: Ludger Ltd., Prozyme, Dextra Laboratories Ltd., Toronto Research Chemicals Inc., Carbosynth Ltd., Cambridge Isotope Laboratories Inc., Sigma-Aldrich Company Ltd., Chemily Glycoscience, Elicityl and Asparia Glycomics. This is a niche industry and most suppliers are small to medium size enterprises. However, there is growing appreciation for the importance of glycans in human health and biologics.[52] This is evidenced by the uptake of glycan characterization platforms by mainstream analytical instrument companies (*e.g.* Waters Ltd., Agilent, Sciex) and the growing support from funding agencies for the development of glycan standards and glycoproducts.

17.2.3 Types of Commercially Available Glycan Standards

A survey of commercial glycan providers revealed that the currently available products can be separated into the following categories; individual *N*-glycans, individual *O*-glycans, glycan libraries, monosaccharides, glycoproteins and glycopeptides. These categories represent the types of compounds that would often be encountered during glycan analysis of biopharmaceutical glycoproteins during in-depth characterization or during quality control.

17.2.3.1 Individual O*-Glycans*

The commercially available *O*-linked glycans are linear or biantennary and contain an initial GalNAc residue. *O*-Glycans are grouped into eight core structures, however, the limited selection of commercial *O*-glycan standards

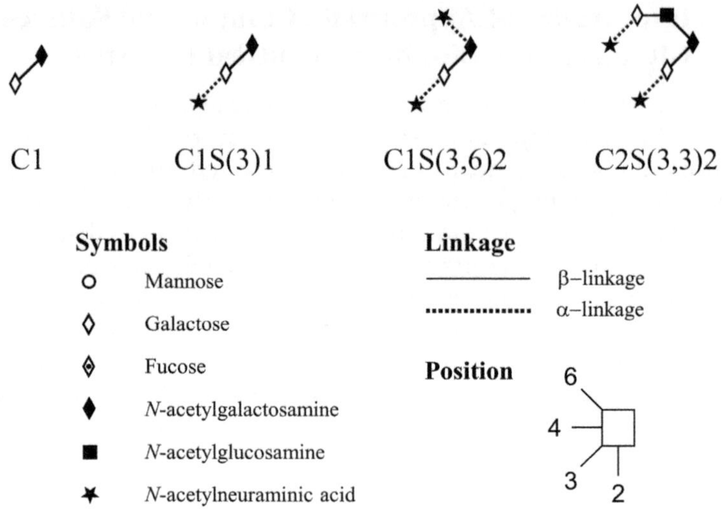

Figure 17.2 Common *O*-glycan standards that are commercially available. Glycans are named and drawn[53] according to Oxford Notation.[54]

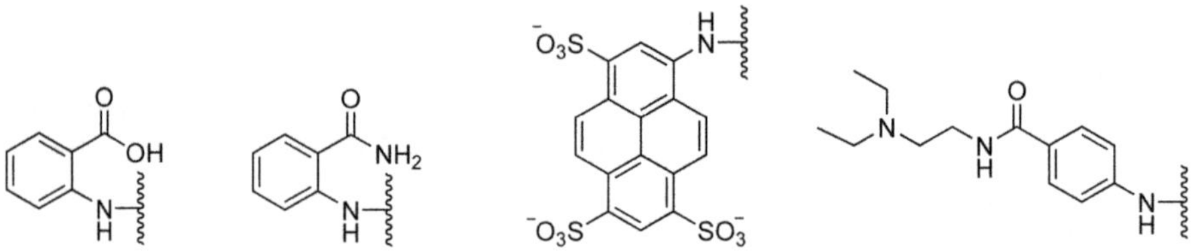

2-AA 2-AB APTS Procainamide

Figure 17.3 Frequently used fluorescent labels for glycoanalysis.

generally belong to core 1 and 2 (Figure 17.2). These glycans are provided with a free reducing terminus (aldol), or are derivatized as a fluorescently labelled glycoside. The fluorescent labels that are commonly used in the biopharmaceutical industry are 2-AA, 2-AB, APTS, and procainamide (Figure 17.3). A selection of isotopically labelled *O*-glycan standards are available as well.

17.2.3.2 Individual N*-Glycans*

The *N*-glycans that are commercially available are found in most glycoproteins (Figure 17.4). They can be grouped into the following categories: oligomannoses, and complex glycans that are fucosylated, sialylated, and di-, tri- and tetra-antennary. The glycans are available as molecules with a free reducing terminus (aldol), or chemically derivatized. The common chemical

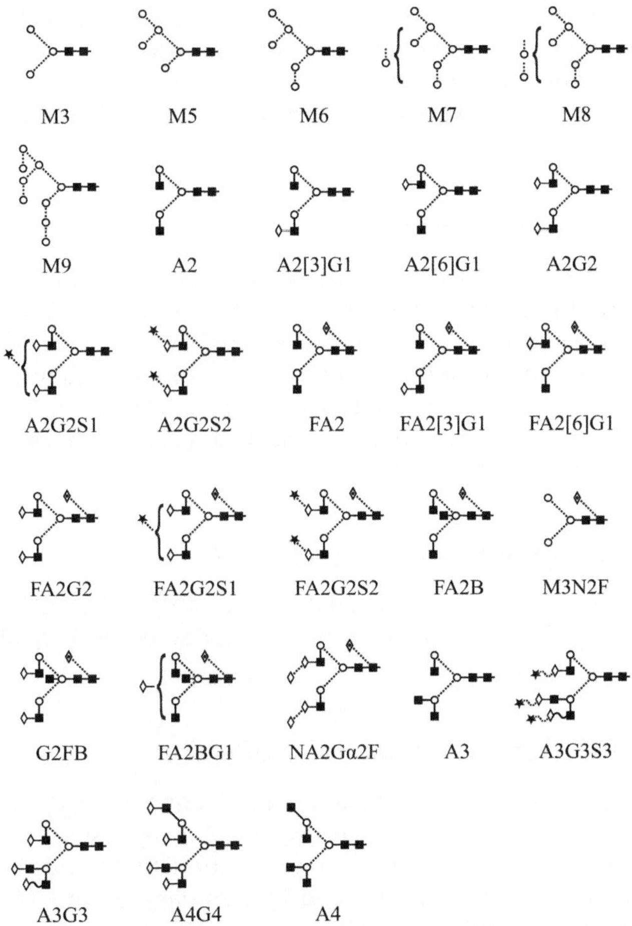

Figure 17.4 Common *N*-glycan standards that are commercially available. Glycans are named and drawn[53] according to Oxford Notation.[54]

derivatives of *N*-glycans are fluorescently labelled glycosides (2-AA, 2-AB, APTS, procainamide, *etc.*), permethylated, or isotopically labelled glycans. They are available as both qualitative and quantitative standards.

17.2.3.3 Glycan Libraries

The commercially available glycan libraries are made up of a mixture of species from a specific type of glycoprotein or as panels with common characteristics. The following are examples of glycan libraries produced from a specific glycoprotein: (i) *N*-glycans released from human IgG antibody glycoprotein. This library contains fucosylated and bi-antennary glycans with variable sialylation. (ii) *O*-Glycans released from fetuin glycoprotein (purified from fetal calf serum). This library contains mainly core 1 and core 2

O-glycans with variable sialylation. On the other hand, panels containing glycans with common characteristics are also available and can include, for example, a mixture of biantennary *N*-glycans containing α2–3-linked sialic acid or a mixture of tetraantennary oligosaccharides containing α2–6-linked sialic acid.

17.2.3.4 Monosaccharides

Monosaccharides are available as either a quantitative standard or a qualitative standard. Examples of these types of standards are: sialic acid quantified standards (Neu5Ac, Neu5Gc), sialic acid reference panels which contain a mixture of sialic acids found in humans and animals, quantitative xylose standards, and a quantitative monosaccharide panel which contains glucosamine (GlcN), galactosamine (GalN), galactose (Gal), mannose (Man), glucose, (Glc) and fucose (Fuc) monosaccharides.

17.2.3.5 Glycoproteins and Glycopeptides

A selection of glycoproteins are available including mAbs, IgGs and bovine fetuin. Glycopeptide standards are also available as both qualitative and quantitative standards.

17.2.3.6 Analysis of Commercial Availability

There are approximately 30 *N*-glycan standards and 10 *O*-glycan standards widely available as purified individual glycans. The highest sellers among the list of *N*-glycans are the ones related to common GCQAs and these include; oligomannoses (M5, M6 and M9); and biantennary glycans (A2 and A2G2 and their core fucosylated and sialylated variants). Among the available standards there is a lack of availability of hybrid *N*-glycans, antennary fucosylated *N*-glycans, sialylated *N*-glycans (acetylated variants, linkage variants (α2–3- *vs.* α2–6-)) and core fucosylated oligomannosides.

A price analysis on a selection of 2-AB labelled *N*-glycans and unlabelled *N*-glycans was carried out to provide a general benchmark. The average price for a 2-AB labelled *N*-glycan standard is $130/100 pmol and this average is valid across the range of available standards and vendors. The price for unlabelled *N*-glycans is more variable and is dependent on the *N*-glycan. For example, the average price for an unlabelled FA2G2S1 *N*-glycan standard is $580/20 μg with a price range among providers of approximately $400–760/20 μg.

17.2.4 Production of Commercial Glycan Standards

The preparation of homogeneous glycans is a major bottleneck in the production of commercial glycan standards. This barrier is due to the immense structural diversity of glycans and a lack of scalable production methods.

The laboratory techniques that are commonly used to prepare commercial glycan standards include isolation from natural sources, synthetic methods and enzymatic methods.

17.2.4.1 Isolation from Natural Sources

The traditional and most commonly used industrial method for the preparation of glycan standards is by isolation from natural sources. The procedure often begins with the enzymatic or chemical deglycosylation of naturally abundant glycoproteins followed by a series of purification steps.[55,56] The sources used for this process are both plant-based (*e.g.* beans)[57–59] and animal-based (*e.g.* bovine serum, hen egg yolk).[60,61] Each source provides a unique selection of glycans, for example, edible beans are an abundant source of high mannose *N*-glycans.[57,59] The release of the glycans from the glycoprotein or peptide generally involves PNGase F, PNGase A, acid, or hydrazinolysis. In certain cases a proteolytic digest (*e.g.* trypsin) is also required prior to glycan release. Various separation and clean-up steps are required to produce standards of suitable purity. The most common approach is to use a set of separation and clean-up methods consecutively and/or iteratively; for example ion-exchange resins (anion/cation) are used for peptide/protein removal and glycan separation based on charge, then hydrophilic interaction chromatography (HILIC), gel filtration/size exclusion chromatography and graphitized carbon resins are used for glycan separation. Each step in this process can take multiple hours to multiple days and generally provides microgram to milligram quantities of glycan. The best practice is to run analysis at each step of the process in addition to the full structural elucidation of the final product, this ensures high purity and results in well-characterized glycan standards. The main drawback to using natural sources for glycan production is that while the starting materials are cost effective, the use of processing enzymes is expensive, and the task of clean-up and purification is time-consuming and laborious. However, because of the abundance of biologically relevant glycoforms present in natural sources, current research is still being devoted to improving the extraction and purification processes.[49,62–64]

17.2.4.2 Total Synthesis of Glycans

A purely synthetic approach to glycan production is the best way for obtaining a homogeneous glycan standard. One of the most challenging aspects of glycan synthesis is that the strategy must afford both regiospecific (modifying one specific hydroxyl group in the presence of many others) and stereospecific formation of glycosidic bonds. The other chapters in this book and the literature describe examples of synthetic strategies for obtaining glycans.[50,65] Unfortunately, because traditional synthetic routes are often time consuming and technically challenging, they are not currently the most viable option for *N*-glycan production in industry. Even though *O*-glycans seem better suited for total synthesis, in comparison to *N*-glycans, because of their smaller size (di- to hexasaccharides), many reported large scale methods involve some aspect of enzyme catalysis.[66,67] Over the last 20 years, significant advancements have been made by synthetic carbohydrate chemists and this has greatly improved the number and diversity of glycans that can be made chemically.[50,68] The commercial availability of functionalized starting materials and building blocks (glycosyl acceptors and donors) has also made synthesis more accessible for preparation of oligosaccharides. In general, for future techniques to be widely implemented in industry they must be both scalable and economically efficient. Additionally, the progress towards automation of glycan synthesis may offer additional benefits in due time, however, automated tools are still in their early stages and require more development before being adopted by industry.[69,70]

17.2.4.3 Enzymatic and Chemo-Enzymatic Production of Glycans

The chemo-enzymatic or enzymatic approach to glycan production involves the use of a purified oligosaccharide precursor (produced either synthetically or isolated from a natural source) as a starting material for the production of a target glycan using glycosyltransferases or glycosidases for modification. This method offers flexibility by taking advantage of structurally diverse starting materials and the specificity of enzymes.[50,62,68,71–75] These techniques have been used to produce highly complex glycans including asymmetrically branched *N*-glycans that are difficult to obtain using synthetic methods.[76] However, they are not yet suitable for commercial scale preparation. The drawback of the approach is that the cost, availability, purity and specificity of the commercially available glycoenzymes limit the scalability of the process.

17.2.4.4 Quality Control

Regardless of the path taken to obtain the glycan standard, the commercial product must be subjected to robust quality control (QC) and have an accompanying certificate of analysis (C of A). The QC of each lot of material involves thorough testing to ensure accurate identity, and high chemical purity. QC testing often consists of multiple chemical analyses and includes (U)HPLC and NMR for chemical purity determination, and MS and NMR for identity confirmation. If the glycan standard is to be provided as a quantified material, methods such as quantitative NMR,[77] and reference to either isotopically labelled materials or other quantified standards can be used to determine amount.

17.2.5 Use of Commercial Glycan Standards

The generation of reliable data is the objective of any analytical laboratory. A key component in a well-designed analytical strategy is the use of standards. When used appropriately, standards can help with characterization,

		Process Control													System suitability						Reference					Quantification*			
		Release & labelling				Release				Labelling				Exo	(U)HPLC				MS	CE									
Standards		SA	MC	N	O	SA	MC	N	O	SA	MC	N	O		SA	MC	N	O			SA	MC	N	O	GU	SA	MC	N	O
Glycoprotein/ Glycopeptide	mAbs	•	•	•		•	•	•																					
	Fetuin	•	•	•	•	•	•	•	•																				
	Glycopeptides	•	•	•		•	•	•													•	•	•			•	•		
Monosaccharides (including sialic acids)	Neu5Ac									•											•					•			
	Neu5Gc									•											•					•			
	Neu5,9Ac$_2$									•											•								
	Sialic acid panel									•					•						•								
	Monosaccharide panel										•					•						•					•		
	Xylose										•											•					•		
Native glycans	*N*-glycans											•		•			•		•				•					•	
	O-glycans												•	•				•	•					•					•
	GHP											•	•				•	•					•	•	•				
Fluorescently labelled glycans	*N*-glycans													•			•		•	•			•	•				•	
	O-glycans													•				•	•	•			•	•					•
	GHP																•	•					•	•	•				
	α-gal glycan													•															
Permethylated glycans	*N*-glycans																		•										
	O-glycans																		•										
Isotopically labelled glycans	*N*-glycans																		•									•	
	O-glycans																		•										•

SA- sialic acid analysis, MC- monosaccharide analysis, N- *N*-glycan analysis, O- *O*-glycan analysis, Exo- exoglycosidase sequencing, GU- glucose units, GHP- glucose homopolymer
* requires the use of a quantitative glycan standard

Figure 17.5 The applications of glycan standards for both qualitative and quantitative glycoanalysis work.

they allow quantitation, and they help provide vital information on the performance of an analytical process. For example, in a QbD framework, glycan standards that match the GCQAs which have been identified for optimized drug function can be utilized to ensure that the analytical methods chosen for characterization provide the necessary data required to monitor the drug realization process. The data acquired when using standards ultimately allow the analyst and any reviewer to judge whether a specific stage in the process or the sample itself is in or out of specification. The reliability of data coming from the glycan analysis of biopharmaceuticals is especially important due to the implications for safety and efficacy of the therapeutic.

Glycan standards are used for different applications during the analysis of sialic acids, monosaccharides, *N*-glycans and *O*-glycans. In general, the use of glycan standards can be broken down into four broad categories: system suitability standards, process standards, reference standards, and quantitative standards. Figure 17.5 gives a visual summary of how commercially available standards can be utilized during glycan analysis.

17.2.5.1 System Suitability Standards

System suitability standards enable an analyst to test the holistic functionality of an analytical system (*e.g.* chromatographic, mass spectroscopic and/or CE) and evaluate whether it is adequate for its intended use.

System suitability is widely recognized as a critical component in chemical analysis and is frequently referred to in governmental regulations and guidance policies (*e.g.* ICH(Q2R1)[78]). While USP and ICH require system suitability criteria to be set, it is up to the method developer or analyst to determine which tests will provide the most assurance that the method is running as expected. Some possible system suitability test criteria include parameters on resolution, retention, precision or response. A common way to establish these parameters is during method validation. For example, the outcome from the evaluation of robustness could be a series of system suitability parameters (*e.g.* resolution test). Most importantly, the system-suitability criteria should be set so that they can be met easily if the method is working right but will fail if there is a problem. An advantage of using commercially available glycan standards is that the data provided in the C of A can help direct the analyst to analytical parameters that could be used for system suitability criteria.

The glycan standards that are commonly implemented to test system suitability depend on both the assay and the analytical platform. A few examples of how commercial glycan standards can be used for system suitability testing are provided hereafter.

For monosaccharide and sialic acid analysis there are panels (mixtures) of multiple monosaccharides or sialic acids that can be employed to test system suitability. When using these products, the acceptance criteria can be that the analytical profiles for three replicates of the mixture overlay with minimal drift in retention time, and also that there is minimal shift in the retention time of each peak in the sample over the course of the analysis.

During *N*-glycan and *O*-glycan analysis there are many options for system suitability testing with glycan standards. In general, any of the *O*- and *N*-glycan standards or glucose homopolymer (GHP), either in their native form or labelled, can be used to assess system suitability. For example, the use of fluorescently labelled GHP is common for (U)HPLC. Here, the criteria are that the profile from the GHP should be similar to the profile shown in the C of A. In addition, the peaks should be symmetrical and well resolved, and at least two fluorescence profiles of GHP should overlay. For CE a mixture of APTS-labelled *N*-glycans are often used as system suitability standards. The choice of standards usually includes glycans which are known to be difficult to resolve using CE (*e.g.* FA2B, A2, FA2G1, A2G2).[79]

17.2.5.2 Process Standards

Process standards or process controls are used to verify that part of or an entire process has worked correctly. There are four main categories for processes standards in glycoanalysis: release, labelling, release followed by labelling and exoglycosidase sequencing.

Glycoproteins and glycopeptides are used as process controls for glycan release, and consecutive release followed by labelling. Glycoprotein standards that are widely available are IgG, mAbs and fetuin. IgG and mAbs have low levels of *N*-glycosylation (2% by weight) and low sialylation. Fetuin is a glycoprotein containing both *N*-linked and *O*-linked glycans (approximately

50% by weight) and is highly sialylated. These differences are key in understanding when to use them as suitable controls during endoglycosidase release (*N*-glycans), hydrazinolysis or beta elimination (generally *O*-glycans) and acid release (during monosaccharide/sialic acid analysis). Fetuin, IgG and mAbs can also be used as process controls for proteolytic digestion during glycopeptide mapping.

Native glycans are generally suitable process controls for labelling methods using reductive amination with fluorescent tags or when performing derivatizations like permethylation or ethyl esterification. Monosaccharides, such as hexoses, hexosamines and sialic acids, are suitable fluorescent labelling controls for their respective analyses.

Native or fluorescently labelled glycans are common process controls to use when using exoglycosidase sequencing for structure elucidation. Your choice of standard will depend on the analytical detection method (HPAE-PAD/fluorescence/MS) and which exoglycosidase are being used. These process controls allow you to assess if the exoglycosidase is cleaving the right monosaccharide with the precise glyosidic linkage. For example, standards containing the α-gal epitope are essential process controls when using an α-galactosidase. This gives confidence in the function of the enzyme and in the resulting characterization; being able to unequivocally and confidently identify an α- from a β-galactose in a glycan structure has major implications for the corresponding glycoprotein therapeutic (recall the Cetuximab story).

17.2.5.3 Reference Standards

The main use of a reference standard is to allow for characterization by comparison. A straightforward example of characterization by comparison to a reference standard or mixture of reference standards is sialic acid and monosaccharide identification. Defined reference panels of well-resolved monosaccharides and sialic acids are analyzed along with the 'unknown' samples. This allows for identification of each species based on comparison of retention times. A sialic acid reference panel (SRP) containing a mixture of sialic acids that are found in humans and animals is used as a reference standard (Neu5Gc, Neu5Ac, Neu5,7Ac$_2$, Neu5Gc9Ac, Neu5,9Ac$_2$, and Neu5,7,(8),9Ac$_3$) during sialic acid analysis. A mixture of common monosaccharides (GlcN, GalN, Gal, Man, Glc, Fuc, Xyl) is used as a reference panel for monosaccharide analysis.

The use of glycans as reference standards can take two forms. In one form, characterization can be accomplished by the direct comparison of the chromatographic or electrophoretic retention time of an unknown to that of an isolated glycan standard or glycan library whose structures have been fully characterized. This method requires that the standards are run on the same system at the same time for comparison of retention times.

If glycan libraries or isolated standards are not available for direct comparison, a GHP ladder can be used as an external standard during the analysis of glycans by (U)HPLC or CE after fluorescent labelling. This standard gives a characteristic ladder profile from monomeric glucose to approximately 20-mers of glucose oligosaccharides, dependent upon the chromatographic conditions employed. The elution position of each peak in this ladder is expressed as a glucose unit (GU) and is used to assign GU values to peaks in the released glycan pool by comparison. These GU values are reproducible and predictive as each monosaccharide in a glycan contributes a set increment to the GU value.[80] This allows for primary assignment of structure by comparison of GU values for unknown glycans with glycan standards whose GU values are in databases or reported in the literature. It is important to remember that with both of these methods the purposed identity of a glycan must be confirmed with further orthogonal structural analysis (*e.g.* MS or exoglycosidase sequencing).

17.2.5.4 Quantitative Standards

Quantitative glycan standards are used to determine the absolute amount of an analyte in a sample. They can also be used to quantify the efficiency of a process.

Quantitative standards can be used internally or externally. The prerequisite for an internal standard is that it does not react with the test material, does not contain impurities with the same retention time as that of the analytes, and is stable and resolved from the component(s) to be quantified (analytes). Examples of this are isotopically labelled glycans used for MS and quantified chitotriose for CE or (U)HPLC. In these cases, the standard of known concentration is spiked into the sample, run on an analytical platform for analysis and the quantity of glycan analyte is inferred by comparison of peak height or peak area to that of the standard (known amount). The internal standard is generally used to quantify analytes within the one sample to which it was added. External quantitative standards are run as additional samples in the sample set and can be used to quantitate many samples rather than just one. Quantitation is carried out by comparison of peak area or peak height or by use of a calibration curve. For example, sialic acids (Neu5Ac and Neu5Gc) and monosaccharide panels are available as quantitative standards and are dispensed to an exact amount. Preparation of serial dilutions of the quantitated standards provide calibration curves. The peak areas or peak heights in the test samples are then compared to the calibration curves to provide quantitative data on the amount of each sialic acid or monosaccharide.

Any quantitative standard that can be taken through a process can be used to access its efficiency. An example of this is the use of quantitative sialylglycopeptide as a process control. This glycopeptide contains a biantennary *N*-linked glycan terminating in two sialic acids (A2G2S2). This standard can be used to check the efficiency of glycan release, labelling and recovery and is particularly useful when doing quantitative studies to ensure reliable data is being produced. This standard can be used with either acidic release for monosaccharide or sialic acid quantitation or with PNGase F, and can be followed by labelling with the appropriate fluorescent labels if required. Quantitative native glycan standards can be also used for determining the labelling reaction process efficiency.

17.3 Commentary

The current status of commercial glycan standards can be summarized by saying that availability is characterized by abundance – glycans that are ‘easy’ to isolate from natural sources are the most widely available. For *N*-glycans this means that, out of a group whose diversity is estimated to exceed 20 000 structures, there are approximately 30 commercially available pure, well-characterized glycan standards. Within this selection the majority of *N*-glycans can be categorized as either biantennary symmetrical glycans or oligomannose glycans. The market lacks asymmetrically branched multiantennary *N*-glycan standards. Compared with *N*-glycans, the availability of *O*-glycan standards is more limited. Of the *O*-glycans that are commercially available, the majority are relatively short oligosaccharides belonging to the core 1 and core 2 families. The commercial glycans that are available are provided in their native form and are also available as chemically derivatized glycans to match the commonly used glycoanalysis methods (*e.g.* 2-AB or 2-AA labelled M7). The range of quantitative standards available is growing, now in addition to quantified monosaccharides, there are both glycopeptides and glycans. The quantitative glycans include chitotriose, M8 and a large selection of isotopically labelled native glycans. Sialylglycopeptide is an example of a commercially offered quantitative glycopeptide standard.

One of the biggest challenges facing commercial glycan providers is production. While significant advancement in the chemical and enzymatic synthesis of glycans has taken place, new methodologies are not yet ready or suitable for translation from the academic laboratory to a scalable industrial process. The result of this has been a reliance on natural sources for glycan production. While isolating glycans from natural sources allows for nature to do the difficult (bio) synthesis, the challenge lies in the separation and purification of a crude mixture of glycans in both a scalable and efficient manner. The demand from consumers is for glycan standards with high purity, which are well-characterized and quality controlled. This means that iterative purification steps are needed to produce glycan standards with acceptable purity and with each purification there is loss in overall yield. After extensive purification, in-depth characterization and rigorous quality control are also required. The time and effort to produce each glycan standard is reflected in their cost with an average price for native glycan standards of approximately \$580/20 μg.

Regardless of the expense, access to standards during drug development is essential. Commercial glycan standards are key tools for; assessing system suitability, determining analytical process efficiency, quantifying samples, and aide in structural elucidation during glycoanalysis.

The consensus in both the biopharmaceutical and glycomics fields is that there is a lack of commercially available glycans and that this has hindered glycoscience research. In order to fully understand what this statement means, it should be deconstructed. It could simply mean that the glycan standards that are needed are not available (30 out of 20 000 is not a huge amount). On the other hand, the issue could also include a combination of the following; that the standards needed are available but are not of the required quality; or that the standards needed are available but are too expensive; or that the standards that are needed are available but are not marketed well; or finally, that there is lack of knowledge as to which standards are actually needed. The overall issue may be composed of each one of these problems. The only way to move forward on this issue is by collaborating; industry to industry and industry to academia. Glycan providers need to work with biopharmaceutical companies to establish which glycan standards will best enable drug realization with tighter, well-characterized glycoforms. Biopharma should be very interested in collaboration to support this goal as they are now facing extreme pressure not only from regulatory authorities, but also from competitors who are developing biosimilars. In addition, academia and industry must work together to support the translation of academic research findings into commercial processes for glycan production that are scalable and reliable. If commercial glycan providers can identify and communicate their 'real-world problems' to academics, then the ability of academia to deliver outside-the-box solutions will be leveraged. This synergy will hopefully lead to an efficient streamlined approach for producing high-purity commercial grade glycans.

17.4 Future Perspectives

The future outlook for the advancement of commercial glycans to support biopharmaceutical realization is promising. The importance of glycoscience to biotech, pharma and biosimilar industries, as well as the healthcare sector is becoming more recognized. The increased appreciation for the industrial potential of glycoscience was catalyzed in part by the publication of two reports, one from the United States and one from Europe, outlining 'roadmaps' for taking advantage of the opportunities presented by glycoscience in the future.[47,48] There has since been an increase in grant based support to bolster growth in the sector for both collaborations between industry and industry, and also industry and academia. CarboMet (Metrology of Carbohydrates for Enabling European BioIndustries)[81] and IBCarb (Glycoscience Tools for Biotechnology and Bioenergy)[82] are examples of these grant funded programmes and networks involving industrial and academic collaboration whose focus is in some way geared towards commercial exploitation and advancement of glycan standards. For example, IBCarb lists 'fast and efficient synthesis of carbohydrates including scale up' as one of their specific objectives. Here the network seeks to identify target glycan structures and to develop the glycoenzymes, which are currently only used in academic laboratories and on small scale, into robust biocatalysts for scale-up.

It is also important to acknowledge that the field is advancing even with its current challenges. An example of this advancement is glycoscience playing a major role in the improvement of erythropoietin (EPO) biopharmaceuticals. EPO is a glycoprotein that is used for the treatment of anaemia. Glycans contribute to 40% of its molecular weight and its glycosylation greatly impacts its therapeutic activity. Glycan engineering was used with an aim to increase its sialic acid content and led to the production of a hyperglycosylated variant. This novel EPO therapeutic was found to have a prolonged serum half-life after injection (increased from 7 hours to 22 hours) greatly reducing administration frequency.[83,84] There is no doubt that both the availability of commercial glycan standards (specifically sialic acid standards) along with the improvement in glycoanalysis techniques greatly contributed to this achievement.[85]

17.5 Conclusions

While there has been significant advancement in the area of biopharmaceutical glycoanalysis, the range of commercially available glycans which are used as standards represents a small proportion of the glycome. There is an urgent demand from both the biopharmaceutical industry and the glycomics field for more commercially available glycan standards that are structurally defined and homogeneous. The justification for this demand is that the progress of glycoscience related to therapeutic glycoproteins and human health directly depends on the availability of these standards. The problem with this demand is that it is somewhat ‘chicken-or-the-egg’: pure glycan standards are needed to probe structure–activity relationships in drug development and disease mechanisms but in order to prioritize which glycans to synthesize or isolate there is the requirement to know which structures are biologically relevant. One hindrance in meeting the need for a wider selection of commercially available glycan standards is that the field is still relatively young and small, with a low number of scientists involved, and with limited scientific popularity of the fundamental glycosciences. Owing to this in part, the methods for efficient isolation of glycans from natural sources and also the chemical and enzymatic synthetic strategies to make these complicated molecules on a large-scale are still developing. In order to increase the number and range of commercially available glycans, the goal is for industry and academia to work together to devise and develop simplified and automated synthetic or enzymatic methods that are high-yielding and cost-efficient. The new technologies will need to be scalable to ease translation into a commercial manufacturing process. Expanding the repertoire of commercial glycans for biopharmaceutical drug realization will not only impact the area of glycoprotein therapeutics but also will have an immense capacity to contribute to the emerging field of medical glycomics and precision medicine.[86,87]

At Ludger Ltd. our medical glycomics platforms are based on the technical frameworks built from our learnings with biopharmaceutical glycosylation analysis.

Glycochemists, as expert makers and users of glycan standards, are in a position to make a large impact. We have a rare set of knowledge and chemistry skills which are vital to an emerging and underpopulated glycomics field which has many multi-disciplinary applications. This skill set provides us with an opportunity to not only work as a scientist within the field, but also as teachers and advisors in other disciplines where glycomics is not well understood. Since glycosylation is implicated in nearly all biological processes, it is easy to comprehend how increasing the availability of commercial glycan standards has the potential to impact wider society leading to an improvement in human health and well-being.

Acknowledgements

The authors would like to thank their colleagues at Ludger Ltd., specifically Daniel Spencer, Simon Peel and Rad Kozak for the valuable discussions about glycan standards, and Matthew Doherty for his help with graphic design.

References

1. A. Varki, *Glycobiology*, 2017, **27**, 3.
2. Y. Zhu, C. Romain and C. K. Williams, *Nature*, 2016, **540**, 354.
3. D. H. Dube and C. R. Bertozzi, *Nat. Rev. Drug Discovery*, 2005, **4**, 477.
4. P. Zhang, S. Woen, T. Wang, B. Liau, S. Zhao, C. Chen, Y. Yang, Z. Song, M. R. Wormald, C. Yu and P. M. Rudd, *Drug Discovery Today*, 2016, **21**, 740.
5. P. W. Tebbey, A. Varga, M. Naill, J. Clewell and J. Venema, *MAbs*, 2015, **7**, 805.
6. G. Mack, *Nat. Biotechnol.*, 2008, **26**, 592.
7. https://www.fda.gov/Drugs/DevelopmentApprovalProcess/DrugInnovation/default.htm, accessed 21 Jan 2018.
8. R. Otto, A. Santagostin and U. Schrader, https://www.mckinsey.com/industries/pharmaceuticals-and-medical-products/our-insights/rapid-growth-in-biopharma, accessed 19 Jan 2018.
9. https://www.forbes.com/2010/02/19/expensive-drugs-cost-business-healthcare-rare-diseases_slide/#4c53cc44f95b, accessed 12 Feb 2018.
10. C. H. Song and J. W. Han, *SpringerPlus*, 2016, **5**, 1.
11. M. Weise, M.-C. Bielsky, K. De Smet, F. Ehmann, N. Ekman, G. Narayanan, H.-K. Heim, E. Heinonen, K. Ho, R. Thorpe, C. Vleminckx, M. Wadhwa and C. K. Schneider, *Nat. Biotechnol.*, 2011, **29**, 690.
12. J. Preston, https://medcitynews.com/2017/01/drug-patents-expiring-2017-smoke-mirrors/, accessed 10 Feb 2018.

13. E. Moorkens, N. Meuwissen, I. Huys, P. Declerck, A. G. Vulto and S. Simoens, *Front. Pharmacol.*, 2017, **8**, 314.
14. R. J. Solá and K. Griebenow, *BioDrugs*, 2011, **24**, 9.
15. T. S. Raju, *Curr. Opin. Immunol.*, 2008, **20**, 471.
16. R. L. Shields, J. Lai, R. Keck, L. Y. O'Connell, K. Hong, Y. Gloria Meng, S. H. A. Weikert and L. G. Presta, *J. Biol. Chem.*, 2002, **277**, 26733.
17. R. M. Anthony and J. V. Ravetch, *J. Clin. Immunol.*, 2010, **30**, 9.
18. D. L. Fernandes, *Biopharmaceutical Production Technology*, Wiley-VCH Verlag GmbH & Co. KGaA, 2012, pp. 545–583.
19. D. Güngör, J. M. de Vries, W. C. J. Hop, A. J. J. Reuser, P. A. van Doorn, A. T. van der Ploeg and M. L. C. Hagemans, *Clin. Ther.*, 2011, **33**, S11.
20. S. A. Berkowitz, J. R. Engen, J. R. Mazzeo and G. B. Jones, *Nat. Rev. Drug Discovery*, 2012, **11**, 527.
21. EMA, Guideline on development, production, characterisation and specification for monoclonal antibodies and related products, http://www.ema.europa.eu/docs/en_GB/document_library/Scientific_guideline/2016/08/WC500211640.pdf, accessed 20 Dec 2018.
22. ICH, *Q6B: Specifications: Test Procedures and Acceptance Criteria for Biotechnological/Biological Products*, International Conference on Harmonisation of Technical Requirements for Registration of Pharmaceuticals for Human Use, 1999.
23. P. Hossler, S. F. Khattak and Z. J. Li, *Glycobiology*, 2009, **19**, 936.
24. B. Kelley, *mAbs*, 2009, **1**, 443.
25. M. Butler, *Appl. Microbiol. Biotechnol.*, 2005, **68**, 283.
26. F. Torkashvand and B. Vaziri, *Iran. Biomed. J.*, 2017, **21**, 131.
27. D. Ghaderi, R. E. Taylor, V. Padler-Karavani, S. Diaz and A. Varki, *Nat. Biotechnol.*, 2010, **28**, 863.
28. J. W. Steinke, T. A. E. Platts-mills and S. P. Commins, *J. Allergy Clin. Immunol.*, 2016, **135**, 589.
29. I. J. Del Val, C. Kontoravdi and J. M. Nagy, *Biotechnol. Prog.*, 2010, **26**, 1505.
30. A. S. Rathore, *Trends Biotechnol.*, 2009, **27**, 546.
31. E. K. Read, J. T. Park and K. A. Brorson, *Biotechnol. Appl. Biochem.*, 2011, **58**, 213.
32. A. Planinc, J. Bones, B. Dejaegher, P. Van Antwerpen and C. Delporte, *Anal. Chim. Acta*, 2016, **921**, 13.
33. J. B. Briggs, *Analytical Characterization of Biotherapeutics*, John Wiley & Sons, Inc., 2017, pp. 257–332.
34. R. O'Flaherty, I. Trbojević-Akmačić, G. Greville, P. M. Rudd and G. Lauc, *Expert Rev. Proteomics*, 2018, **15**, 13.
35. Z. Roth, G. Yehezkel and I. Khalaila, *Int. J. Carbohydr. Chem.*, 2012, **2012**, 640923.
36. K. Mariño, J. Bones, J. J. Kattla and P. M. Rudd, *Nat. Chem. Biol.*, 2010, **6**, 713.
37. C. Huhn, M. H. J. Selman, L. R. Ruhaak, A. M. Deelder and M. Wuhrer, *Proteomics*, 2009, **9**, 882.

38. J. Q. Fan and Y. C. Lee, *J. Biol. Chem.*, 1997, **272**, 27058.
39. L. R. Ruhaak, G. Zauner, C. Huhn, C. Bruggink, A. M. Deelder and M. Wuhrer, *Anal. Bioanal. Chem.*, 2010, **397**, 3457.
40. M. A. Lauber, Y.-Q. Yu, D. W. Brousmiche, Z. Hua, S. M. Koza, P. Magnelli, E. Guthrie, C. H. Taron and K. J. Fountain, *Anal. Chem.*, 2015, **87**, 5401.
41. I. Ciucanu and F. Kerek, *Carbohydr. Res.*, 1984, **131**, 209.
42. G. Zauner, R. P. Kozak, R. A. Gardner, D. L. Fernandes, A. M. Deelder and M. Wuhrer, *Biol. Chem.*, 2012, **393**, 687.
43. J. M. Prien, H. Stöckmann, S. Albrecht, S. M. Martin, M. Varatta, M. Furtado, S. Hosselet, M. Wang, T. Formolo, P. M. Rudd and J. E. Schiel, *State-of-the-Art and Emerging Technologies for Therapeutic Monoclonal Antibody Characterization Volume 2. Biopharmaceutical Characterization: The NISTmAb Case Study*, American Chemical Society, 2015, ch. 4, pp. 185–235.
44. R. R. Townsend, *Chromatography in Biotechnology*, American Chemical Society, 1993, pp. 7–86.
45. M. L. A. De Leoz, D. L. Duewer and S. E. Stein, *Glycobiology*, 2016, **26**, 1461.
46. S. Thobhani, C. T. Yuen, M. J. A. Bailey and C. Jones, *Glycobiology*, 2009, **19**, 201.
47. *Future of European Glycoscience: A Roadmap For Glycoscience in Europe*, 2014.
48. National Research Council, *Transforming Glycoscience: A Roadmap for the Future*, The National Academies Press, Washington, DC, 2012, DOI: 10.17226/13446.
49. X. Song, H. Ju, Y. Lasanajak, M. R. Kudelka, D. F. Smith and R. D. Cummings, *Nat. Methods*, 2016, **13**, 528.
50. L. Krasnova and C.-H. Wong, *Annu. Rev. Biochem.*, 2016, **85**, 599.
51. http://www.who.int/biologicals/vaccines/good_manufacturing_practice/en/, accessed 10 Feb 2018.
52. C. Sheridan, *Nat. Biotechnol.*, 2007, **25**, 145.
53. A. Ceroni, K. Maass, H. Geyer, R. Geyer, A. Dell and S. M. Haslam, *J. Proteome Res.*, 2008, **7**, 1650.
54. D. J. Harvey, A. H. Merry, L. Royle, M. P. Campbell, R. A. Dwek and P. M. Rudd, *Proteomics*, 2009, **9**, 3796.
55. Z. Yosizawa, T. Sato and K. Schmid, *Biochim. Biophys. Acta, Gen. Subj.*, 1966, **121**, 417.
56. S. Nakakita, W. Sumiyoshi, N. Miyanishi and J. Hirabayashi, *Biochem. Biophys. Res. Commun.*, 2007, **362**, 639.
57. M. Kimura, T. Hara and Y. Kimura, *Biosci., Biotechnol., Biochem.*, 2011, **75**, 155.
58. I. B. Wilson, R. Zeleny, D. Kolarich, E. Staudacher, C. J. Stroop, J. P. Kamerling and F. Altmann, *Glycobiology*, 2001, **11**, 261.
59. Y. Lu, J. Ye and F. Wold, *Anal. Biochem.*, 1993, **209**, 79.

60. A. Seko, M. Koketsu, M. Nishizono, Y. Enoki, H. R. Ibrahim, L. R. Juneja, M. Kim and T. Yamamoto, *Biochim. Biophys. Acta, Gen. Subj.*, 1997, **1335**, 23.
61. D. Witzigmann, P. Detampel, F. Porta and J. Huwyler, *RSC Adv.*, 2016, **6**, 97636.
62. B. S. Hamilton, J. D. Wilson, M. A. Shumakovich, A. C. Fisher, J. C. Brooks, A. Pontes, R. Naran, C. Heiss, C. Gao, R. Kardish, J. Heimburg-Molinaro, P. Azadi, R. D. Cummings, J. H. Merritt and M. P. Delisa, *Sci. Rep.*, 2017, **7**, 1.
63. B. Sun, W. Bao, X. Tian, M. Li, H. Liu, J. Dong and W. Huang, *Carbohydr. Res.*, 2014, **396**, 62.
64. L. Liu, A. R. Prudden, G. P. Bosman and G. J. Boons, *Carbohydr. Res.*, 2017, **452**, 122.
65. X. Zhu and R. R. Schmidt, *Angew. Chem., Int. Ed. Engl.*, 2009, **48**, 1900.
66. L. A. Marcaurelle and C. R. Bertozzi, *Glycobiology*, 2002, **12**, 69R.
67. A. K. Misra, M. Fukuda and O. Hindsgaul, *Bioorg. Med. Chem. Lett.*, 2001, **11**, 2667.
68. X. Cao, J. Zhou, Y. Guo, Y. Liu, L. Li and P. G. Wang, *Chemical Biology of Glycoproteins*, The Royal Society of Chemistry, 2017, pp. 188–208.
69. P. H. Seeberger, *Acc. Chem. Res.*, 2015, **48**, 1450.
70. S. G. Pistorio, S. S. Nigudkar, K. J. Stine and A. V. Demchenko, *J. Org. Chem.*, 2016, **81**, 8796.
71. L. Li, Y. Liu, C. Ma, J. Qu, A. D. Calderon, B. Wu, N. Wei, X. Wang, Y. Guo, Z. Xiao, J. Song, G. Sugiarto, Y. Li, H. Yu, X. Chen and P. G. Wang, *Chem. Sci.*, 2015, **6**, 5652.
72. O. Blixt and N. Razi, *Methods Enzymol.*, 2006, **415**, 137.
73. Z. Wang, Z. S. Chinoy, S. G. Ambre, W. Peng, R. McBride, R. P. de Vries, J. Glushka, J. C. Paulson and G.-J. Boons, *Science*, 2013, **341**, 379.
74. M. M. Palcic, *Curr. Opin. Chem. Biol.*, 2011, **15**, 226.
75. K. M. Koeller and C.-H. Wong, *Chem. Rev.*, 2000, **100**, 4465.
76. Y. Maki, R. Okamoto, M. Izumi, T. Murase and Y. Kajihara, *J. Am. Chem. Soc.*, 2016, **138**, 3461.
77. F. Malz and H. Jancke, *J. Pharm. Biomed. Anal.*, 2005, **38**, 813.
78. *ICH Topic Q2 (R1) Validation of Analytical Procedures : Text and Methodology*, 2005, http://www.ich.org/fileadmin/Public_Web_Site/ICH_Products/Guidelines/Quality/Q2_R1/Step4/Q2_R1__Guideline.pdf.
79. Á. Szekrényes, S. S. Park, M. Santos, C. Lew, A. Jones, T. Haxo, M. Kimzey, S. Pourkaveh, Z. Szabó, Z. Sosic, P. Feng, C. Váradi, F. de l'Escaille, J.-B. Falmagne, P. Sejwal, T. Niedringhaus, D. Michels, G. Freckleton, M. Hamm, A. Manuilov, M. Schwartz, J. K. Luo, J. van Dyck, P. K. Leung, M. Olajos, Y. Gu, K. Gao, W. Wang, J. Wegstein, S. Tep and A. Guttman, *MAbs*, 2016, **8**, 56.
80. G. R. Guile, P. M. Rudd, D. R. Wing, S. B. Prime and R. A. Dwek, *Anal. Biochem.*, 1996, **240**, 210.
81. https://carbomet.eu/, accessed 15 Jan 2018.

82. http://ibcarb.com/, accessed 12 Dec 2017.
83. A. M. Sinclair and S. Elliott, *J. Pharm. Sci.*, 2005, **94**, 1626.
84. M. K. Azharuddin, G. McKay and M. Fisher, *Practical Diabetes*, 2012, **29**, 163.
85. R. I. Thomson, R. A. Gardner, K. Strohfeldt, D. L. Fernandes, G. P. Stafford, D. I. R. Spencer and H. M. I. Osborn, *Anal. Chem.*, 2017, **89**, 6455.
86. G. Lauc, M. Pezer, I. Rudan and H. Campbell, *Biochim. Biophys. Acta, Gen. Subj.*, 2016, **1860**, 1574.
87. B. Adamczyk, T. Tharmalingam and P. M. Rudd, *Biochim. Biophys. Acta, Gen. Subj.*, 2012, **1820**, 1347.

CHAPTER 18

Glycan Microarrays with Semi-synthetic Neoglycoconjugate Probes in Understanding Glycobiology

ANGELINA S. PALMA[a] AND WENGANG CHAI*[b]

[a]UCIBIO-REQUIMTE, Department of Chemistry, Faculty of Science and Technology, NOVA University of Lisbon, Portugal; [b]Glycosciences Laboratory, Faculty of Medicine, Imperial College London, Hammersmith Campus, London W12 0NN, UK
*E-mail: w.chai@imperial.ac.uk

18.1 Introduction

There has been a major resurgence of interest in the biological and biomedical roles of the diverse glycan sequences in the form of glycoconjugates on glycolipids, glycoproteins and proteoglycans, or on polysaccharides and secreted oligosaccharides. It has now been well recognized that glycan–protein interactions are closely linked to cell physiology and cell–cell communication, and play a critical role in many biological and biomedical processes. Such interactions take place within cells, at the cell surface, in the extracellular matrix or in secretions. The

Chemical Biology No. 11
Synthetic Glycomes
Edited by Wanyi Guan, Lei Li and Peng George Wang

Published by the Royal Society of Chemistry, www.rsc.org

understanding of glycan–protein interactions at molecular level has been of prime interest to glycobiologists and central to development of glycomic approaches for glycan-based therapies and diagnostics. The knowledge gained from this research at the interface of chemistry and biology is attracting researchers in diverse fields, including immunology, infection and oncology.

The detection and characterization of glycan–protein interactions have been very challenging. With development of bioinformatics tools, structural biology methods and accessible databases, predictions can be made about the probable, or possible, glycan-binding activities of proteins on the basis of protein-sequence homologies and structures.[1,2] Nevertheless, direct binding experiments are required for validation and characterization of these predictions and to discover novel mechanisms of ligand binding.[3] Although polysaccharide interaction with proteins may require longer chains, *e.g.* glucan interaction with Dectin-1,[3,4] heparin binding to basic fibroblast growth factor[5] and chondroitin sulfate A interaction with malaria infected red blood cells,[6] most glycan recognition epitopes/determinants are below hexa- or heptasaccharides.[7]

Traditional methods for revealing glycan binding specificities have included assays of inhibition of binding with monosaccharides and simple oligosaccharides. Free glycans in their monomeric form can be examined only as inhibitors of such interactions. However, for many glycan-recognition systems, the large amount of glycan material that is required for inhibition assays is usually prohibitive. Indirect approaches based on selective chemical or enzymatic modification of glycoconjugates on target cells have also been frequently employed. For precise assignments of specificities, it is necessary to perform quantitative inhibition experiments with a series of structurally defined glycans. Although these approaches have been used with success, it is not always feasible as glycans derived from glycoconjugates are frequently heterogeneous and difficult to obtain in sufficient purity and quantity. Once removed from the protein backbone or lipid carrier, the amount of glycans required for inhibition are substantially greater than those available due to the loss of the cooperative effect of multivalency, as explained below.

At molecular level the glycan–protein interaction is of low affinity[8] and is of relatively low selectivity among glycan molecules with similar structures.[9] The dissociation constants (KD) are generally in the μM to mM range apart from some of the glycosaminoglycan (GAG) interactions with GAG-binding proteins. It has been gradually realized that these features are different from many observed bio-activities at cellular levels, which may exhibit a highly selective and tight binding for a specific glycan sequence. The multivalent presentation of both glycan ligands and glycan-binding proteins in the cellular context is considered an underlying reason for this difference.[10] Multivalent interactions with high avidity take place between multiple glycan chains with a similar motif displayed on protein backbones or after clustering of glycolipids on the surface of one cell, and multimeric or multidomain forms of glycan-binding proteins displayed at the surface of another cell.[11,12]

Thus, multivalent interactions between the glycan and the recognition protein are often required for sensitive detection in the binding experiments. The implications of naturally-occurring multivalency with glycoconjugates led to different designs of multivalent glycan molecules for sensitive detection of glycan–protein interactions. This knowledge led to the development of the idea for clustered and/or multivalent presentation of glycan probes in the 1980s. There has been a considerable effort in using different methods to achieve multivalent presentation of these glycan epitopes/determinants and these include the synthesis of glycodendrimers,[13] glyconanoparticles[14] and neoglycoconjugates.[15,16]

There has been also a need for a microscale technology that would allow the detection and characterization of the functional epitopes on the glycan sequences at a glycomic scale, and to discover novel ligands isolated from natural sources. Over the last decade, glycan microarrays have become essential tools for high sensitivity detection and specificity assignment of glycan–protein interactions and have revolutionized the unravelling of these interactions in health and disease processes, infectious and non-infectious. Glycan microarrays have the advantages of multivalent or clustered presentation of glycan probes, high-sensitivity detection and high-throughput analysis with multiple probes (hundreds at present into thousands in the near future). A variety of immobilization strategies have been described for generation of the microarrays. This chapter is focused on the neoglycoconjugate-based microarray systems, including the neoglycolipid (NGL)-based microarrays developed in the Glycosciences Laboratory at Imperial College, the first microarray system for sequence-defined glycans,[17] and the neoglycoprotein (NGP)-based microarrays.[18,19] We discuss construction and application of glycan microarrays using NGL and NGP as the probes.

18.2 Methods for Glycan Microarray Construction

Glycans (oligosaccharides) are of hydrophilic nature and can be readily dissolved in aqueous solutions in which biological assays are often carried out. Therefore solid-phase binding assay methods using glycans in their native form are not suitable. Conjugation to a lipid or protein molecule allows efficient immobilization on solid surfaces (*e.g.* silica-gel TLC and plastic microtiter well surface) by adsorption effect for either direct binding assays or as a receptor for inhibition assays. The solid-phase TLC-plate binding assay using natural glycolipids[20] or neoglycolipids[21] is considered as the early array format for investigation of glycan–protein interactions. The more recently developed miniaturized microarray technology allows thousands of test samples to be spotted and immobilized on a single microscopic slide with a modified surface. Microarrays developed to date have covered a wide range of naturally-occurring glycan molecules and allowed a diversity of protein–glycan interactions to be evaluated. Following on from the initial report on a proof-of-concept microarray system with structurally defined glycans[17] in 2002, many methods have been developed for microarray construction using different principles for printing and immobilizing glycan molecules. All these methods can generally fall into two categories: covalent or non-covalent immobilization (Figure 18.1A and B). In both approaches, manipulation of the slide surfaces and/or the glycan molecules has to be carried out.

18.2.1 Non-covalent Immobilization

Non-covalent immobilization is largely employed for polysaccharides and glycoconjugates. Hydrophobic slide surfaces have been a popular choice, *e.g.* nitrocellulose membrane-coated glass slides[28,29] and polystyrene slides,[30] as glycolipids, glycoproteins and some polysaccharides (particularly the neutral ones) are, to different degrees, hydrophobic, compared with oligosaccharides and can be directly immobilized by hydrophobic adsorption. However, the immobilization efficiency for different types of polysaccharides or the same polysaccharides with different sizes can be different.[28] Non-covalent immobilization *via* other principles has also been described, *e.g.* fluorine–fluorine interaction using fluorous chemistry for both the slide surface and sugar molecules,[31] biotin–streptavidin interaction using biotinylated sugar for streptavidin surface[32] and charge-based interaction using poly-lysine-modified surface for sulphated glycosaminoglycans.[33]

An application using the non-covalent immobilization principle is the development of glycan microarrays of NGL probes.[16,22] In this approach, glycans from different sources, natural or synthetic, are conjugated to an amino-lipid of the phosphatidyl ethanolamine type by reductive-amination or oxime-ligation to generate NGL probes with amphipathic properties for arraying on a nitrocellulose surface through non-covalent hydrophobic adsorption[34] (Figure 18.1A). Another example is preparation of microarrays of plant oligosaccharides derivatized to the protein BSA for adsorption to a nitrocellulose surface[19] (Figure 18.1A). In this approach, the oligosaccharides are coupled to BSA by reductive-amination, producing a ring-opened monosaccharide residue at the reducing end of the oligosaccharide linking to the protein.

18.2.2 Covalent Immobilization

Covalent immobilization is typically used for synthetic glycans which carry, at the reducing termini, short alkyl or aryl chain linkers terminated with an amino group. Reducing terminal derivatization of natural

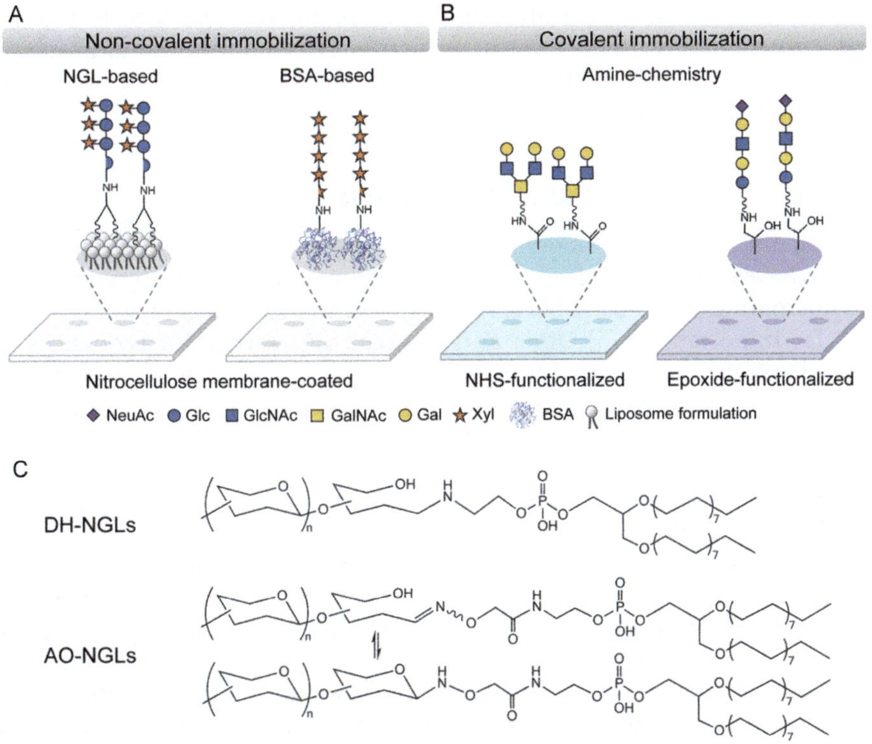

Figure 18.1 Graphic representation of examples of immobilization strategies used to generate glycan microarrays. (A) Non-covalent immobilization onto nitrocellulose-coated glass slides of reducing oligosaccharides derivatized by reductive amination to an aminolipid, to prepare neoglycolipids (NGLs),[22] or to BSA, to prepare neoglycoproteins.[19] (B) Covalent immobilization of synthetic oligosaccharides derivatized at the reducing end to an amino-terminating linker onto *N*-hydroxysuccinimide (NHS)-functionalized glass slides[23] or onto epoxide-functionalized glass slides.[24] (C) NGL probes prepared from reducing oligosaccharides by reductive amination (DHPE, DH-NGLs)[22] and by oxime ligation (AOPE, AO-NGLs);[25] the derivatization of the oligosaccharide by oxime ligation produces an equilibrium between the open- and closed-ring form of the reducing monosaccharide. Examples of oligosaccharide structures in the different libraries are shown using the symbol nomenclature according to Varki *et al.*, 2015.[26] Reproduced from ref. 27 with permission from The Royal Society of Chemistry.

reducing glycans is also used before covalent immobilization. These amino-terminating glycans are printed on chemically-modified glass slide surfaces. The chemical functionality introduced at the reducing termini of the glycan chains does not generally alter the biological properties of the molecules as other parts of the glycan sequences remain largely unaltered. Although some published methods require chemical treatment in a specialized carbohydrate laboratory of the slide surface to make it active for covalent linking to the reducing terminal functional groups of the glycans, variously surface-functionalized slides are now commercially available for selection.

18.2.2.1 Amino-functionalized Glycans

Considerable amounts of amino-terminating mono- and oligosaccharide probes have been synthesized chemically or chemo-enzymatically and used in the microarrays described above. However, glycan probes from natural sources are also important as: (i) larger and complex oligosaccharide sequences are difficult to be synthesized even with the modern chemical or enzymatic approaches;[35] (ii) oligosaccharides from the actual natural source can be considered to represent the glycome as there may be some sequences yet to be discovered. In the past few years, there have been some efforts in the development of methods for isolation and purification of glycans from natural sources and their immobilization on microarrays. HPLC has been proved as the most successful means for separation of oligosaccharides and glycopeptides. For high sensitivity detection reducing terminal derivatization with a fluorescent label is often required for oligosaccharides. Derivatization of reducing sugars with aromatic amines *via* reductive-amination has been very popular and extensively used for isolation and purification of oligosaccharides by HPLC. However, the reducing end aldehyde functionality is also required for linking to functional group on the slide surface for covalent immobilization. Therefore, adaptation of methods suitable for both HPLC and microarray construction has been the key to success.

Fluorescent diaminopyridine has been used as a linker between the reducing sugar and the functionalized slide surface.[36] The reducing oligosaccharide can be linked to one of the amino groups by reductive-amination to facilitate detection in HPLC fractionation. The isolated oligosaccharide–diaminopyridine conjugate can then be immobilized on the functionalized slide surface using the remaining amino group.

18.2.2.2 Functionalized Glass-slide Surface

The use of active esters of carboxylic acids, *e.g.* *N*-hydroxysuccinimide (NHS) esters, for linking to the amino groups of amino acids in solid-phase peptide synthesis has been a routine procedure and this has been well adopted for construction of glycan microarrays. There have been several reports using glass slides with NHS-activated surface for linking with the amino groups of amino-modified sugars[23] and glycopeptides.[37] This strategy has been used by the U.S. Consortium for Functional Glycomics (CFG) for various types of oligosaccharides synthesized with non-reducing terminal amino groups[23] (Figure 18.1B). Other types of surface-modified glass slides, *e.g.* aldehyde modified glass surface, have also been used for amino-modified sugars.[38] Epoxide-functionalized glass slides are very reactive towards amino groups as the epoxide surface can link not only to the amino groups of glyco- and neo-glycoproteins[18] and amino-modified oligosaccharides[39] (Figure 18.1B), but also efficiently to the secondary amines resulted from reductive-amination commonly used for derivatization of reducing sugars.[40] Other surface-modified glass slides include thiol-functionalized glass for maleimide-linked sugars[41] and photo-activable glass for poly-lysine conjugated poly- and oligosaccharides.[42]

18.3 Neoglycoconjugates

Neoglycoconjugates, including NGLs and NGPs, were developed to address the multivalency issues of glycans and both types of neoglycoconjugates use natural glycans conjugated with a natural lipid or protein as the reagents to form artificial glycolipid or glycoprotein, the neoglycolipid (NGL),[16] or the neoglycoprotein (NGP),[15] respectively. One NGP molecule has multiple copies of the same glycan chains whereas a NGL has one glycan chain on one lipid and can be presented to the protein in the clustered state. These neoglycoconjugate presentations confer the multivalency effect for interaction with recognition proteins.

18.3.1 Preparation of Multilayer Structured Composites

18.3.1.1 Development of the Neoglycolipid Technology

Tagging each glycan chain with a hydrophobic lipid tail confers an amphipathic property. Therefore, unlike free glycans, the NGLs formed can be immobilized such that their glycan chains are arranged as a clustered display on solid matrices (such as on the surface of TLC plates, plastic microtitre wells, or nitrocellulose-coated glass slides). The clustering and resulting multivalency increase the avidity of interactions with glycan-binding proteins and might also give rise to cooperativity[43] in the binding process.

The NGLs can also be incorporated into liposomes, singly or as mixtures, and examined in a multivalent state as inhibitors of glycan–protein interactions. This requires orders of magnitude less material than inhibition studies that use free glycans.[44] Cooperativity between an oligosaccharide ligand and a sulphotyrosine ligand for binding to P-selectin (a carbohydrate-binding cell-adhesion molecule) was shown using this approach.[44] Above all, as can be done for natural glycolipids, the NGL technology provides a way to resolve glycan mixtures using high-performance TLC (HPTLC) before binding experiments are carried out.[22]

A major component of the NGL technology is the ability to determine glycan sequences directly from the silica-gel surface of TLC plates using mass spectrometry.[45,46] This can be carried out in conjunction with binding experiments in which glycan-binding proteins are overlaid following the resolution of NGLs by HPTLC. In the mass spectra of NGLs, intense molecular ions and informative sequence ions are obtained at low picomole levels. From these ions, molecular mass, monosaccharide composition and information on monosaccharide sequence, branching pattern, sulphate and phosphate substitutions can be deduced.[22]

The full potential of NGL technology was shown in the early work with two of the glycan-binding proteins of the innate immune system, conglutinin and mannose-binding protein. NGLs derived from *N*-glycan mixtures that were released from glycoproteins and resolved by TLC were amenable not only to ligand-binding experiments, but also to sequence assignment by mass spectrometry *in situ*.[47] This and subsequent work[48] established that commonly occurring *N*-glycans, such as high mannose chains, are ligands for both conglutinin and mannose-binding protein, but that the availability of an *N*-glycan for binding is markedly influenced by the polypeptide on which it is presented.

Using the NGL technology, many other oligosaccharide sequences have been assigned roles as recognition structures. Typical examples include the discovery of unsuspected, sulphated glycan ligands on epithelial glycoproteins for the selectins[49] and assignment of the carbohydrate sequence of the prostate cancer-associated antigen F77.[50] In the section below, we briefly describe the preparation of NGLs and their use for construction of an advanced glycan microarray system.

18.3.1.2 Preparation of Neoglycolipid Probes

Generating NGL probes involves conjugation of reducing glycans to an amino-phospholipid (Figure 18.1C). The aldehyde form of the hemiacetal group at the reducing end of glycans is joined to the amino group of the lipid by reductive-amination. A natural amino-phospholipid, dipalmitoyl phosphatidylethanolamine (DPPE)[16] was initially selected for this purpose. As the acyl linkage of the fatty acid chains were found in some cases to be not completely stable, a synthetic amino-phospholipid with ether-linked dihexadecyl hydrocarbon chains (DHPE) was used.[22] To aid HPLC detection, a fluorescent anthracene-modified DHPE (ADHP) was developed.[51] As reductive-amination results in ring opening of the core monosaccharides and this often limits applicability to short-chain glycans, a modified lipid reagent with amino-oxy functionality (AOPE) was also developed. Reducing glycans can then be conjugated with AOPE *via* oxime ligation.[25]

NGL technology can be applied to the complete range of glycoconjugates. For *N*-glycans, their release using *N*-glycanases or hydrazinolysis give reducing sugars which can be converted directly into NGLs.[47] Release of *O*-glycans is conventionally carried out by the classic reductive alkaline hydrolysis[52] and the reduced *O*-glycans are obtained which cannot be conjugated directly with the aminolipid. In this case, a mild periodate oxidation procedure has been developed and optimized, which splits the reduced core sugar and generates reactive aldehyde at the reduced end of each chain, which is required for conjugation without degrading diols in the saccharide rings.[45,53] Various methods for non-reductive chemical release of *O*-glycans have been developed[54] and although the methods need perfecting[55] the released reducing sugars can be conjugated directly to the lipid in the same way as the *N*-glycans. For glycolipids, the glycan chains can be released by enzyme[56] or ozonolysis[57] to produce reducing sugars for conjugation. Alternatively, glycolipids can be treated in the same way as NGLs for structural and functional analyses. For polysaccharides, either mammalian sulphated glycosaminoglycans or plant and bacterial polysaccharides, various enzymatic[58] and chemical methods[4] can be employed for partial depolymerization and reducing glycan fragments can be obtained for direct conjugation.

For preparation of the microarray, the purified and quantified NGL probes are robotically dispensed onto nitrocellulose-coated glass slides at low fmol levels in a liposome formulation in the presence of carrier lipids.[59] This mode of presentation allows an element of mobility, simulating to some extent the cell surface display of glycans. A particularity of this microarray method is that naturally isolated or synthetic glycolipids are printed together with the NGLs. The applicability of the microarray system to glycolipids and to oligosaccharides derived from them, makes it possible to clinch the role of carbohydrate and ceramide moieties in recognition.[60]

18.3.2 Neoglycoproteins

18.3.2.1 Development of Neoglycoprotein Strategy

NGP is made by covalent linking of mono- or oligosaccharides to a natural protein or glycoprotein. However, a glycoprotein with its glycan chains chemically or enzymatically elongated is not considered as neoglycoprotein. The applications of NGPs for investigating the contribution of the glycan chain to the functions mediated by glycoproteins are only now beginning to be explored. The ways in which neoglycoproteins have been used to study lectins and glycan-specific antibodies could be extended to the study of other glycan-binding proteins such as glycosyltransferases. NGPs may also serve as soluble inhibitors of processes that are apparently glycoconjugate-mediated, such as cell–cell adhesion. Still another possible approach to investigate the role of a glycan in the behavior of cells would be to attach new sugars to the cell surface following methods developed to prepare NGPs. In the section below, we briefly describe the preparation of NGPs and the desired characteristics of modification reactions. We also discuss the properties of NGPs and elaborate on their applications and modification reactions.

Similar to NGLs, NGPs can also be used in many conventional assays employed in the past for investigation of glycan–protein binding (*e.g.* ELISA and Western blot) and also as multivalent inhibitors or immobilized receptors (as the coat) in ELISA-competition assays.

18.3.2.2 Preparation of Neoglycoprotein Probes

Many methods have been developed for conjugating glycans to natural proteins to produce NGPs. The most commonly used proteins are bovine and human serum albumin (BSA and HSA, respectively) and the widely used conjugation method is reductive-amination. As for NGLs, the reducing terminal aldehyde (in the form of hemiacetal) is coupled to the amino group of the lysine residue of a natural protein (*e.g.*, BSA) *via* reductive-amination (Figure 18.2A)[19,61] using sodium or tetrabutylammonium cyanoborohydride as the reducing agent in a borate buffer at pH 8. As the NGLs, reductive-amination leads to ring opening of the reducing-end monosaccharide residue. Although this is not a problem for longer glycan chains, for those shorter than tetra- or trisaccharides this ring opening can affect the recognition epitopes.

To overcome the ring opening problem various methods have been developed. An intermediate short chain linker was introduced between the reducing end of the glycan and the amino group of the protein. At one end, the linker can have a functionality to be able to conjugate with the aldehyde of the glycan without reduction maintaining the ring-closed form of the core monosaccharide. At the other end of the linker, various functionalities can be introduced to facilitate conjugation with the amino group of the protein. For example, linkers containing a free carboxyl can be coupled to free amines on proteins *via* formation of an activated ester, such as an *N*-hydroxysuccinimide ester (Figure 18.2B)[18] or *p*-nitrophenyl esters.[62] Linkers containing a free thiol can be conjugated to maleimide-derivatized proteins or dehydroalanine-containing proteins.[63] Other conjugation methods include copper(I)-catalyzed azide-alkyne cycloaddition[64] and conjugation of two amines with a squaric decyl ester.[65] Typically, the resulting product contains a distribution of conjugates, with varying attachment sites and conjugation ratios. For example, coupling to lysine residues might produce a product with an average conjugation ratio of 20 glycans per molecule of protein, with a significant amount of conjugates containing 18, 19, 21, or 22 glycan chains per molecule of protein, and the exact lysine residues that are modified can vary from one protein to another. Nonetheless, they each have only one type of glycan chain attached.

There are two types of microarray platforms using NGP probes. Similarly to the NGL microarray system described above, non-covalent immobilization onto nitrocellulose-coated slide *via* physical adsorption has also been used for NGP to generate microarrays of plant oligosaccharides.[19] The oligosaccharides are coupled to BSA by reductive-amination, producing a ring-opened core monosaccharide residue linking to the protein.[19] An alternative method for construction of NGP microarray has been covalent immobilization onto epoxide-modified glass slides *via* interaction between the amino group of the lysine residue of BSA (or HSA) and the epoxide of the slide surface[18] (Figure 18.2C).

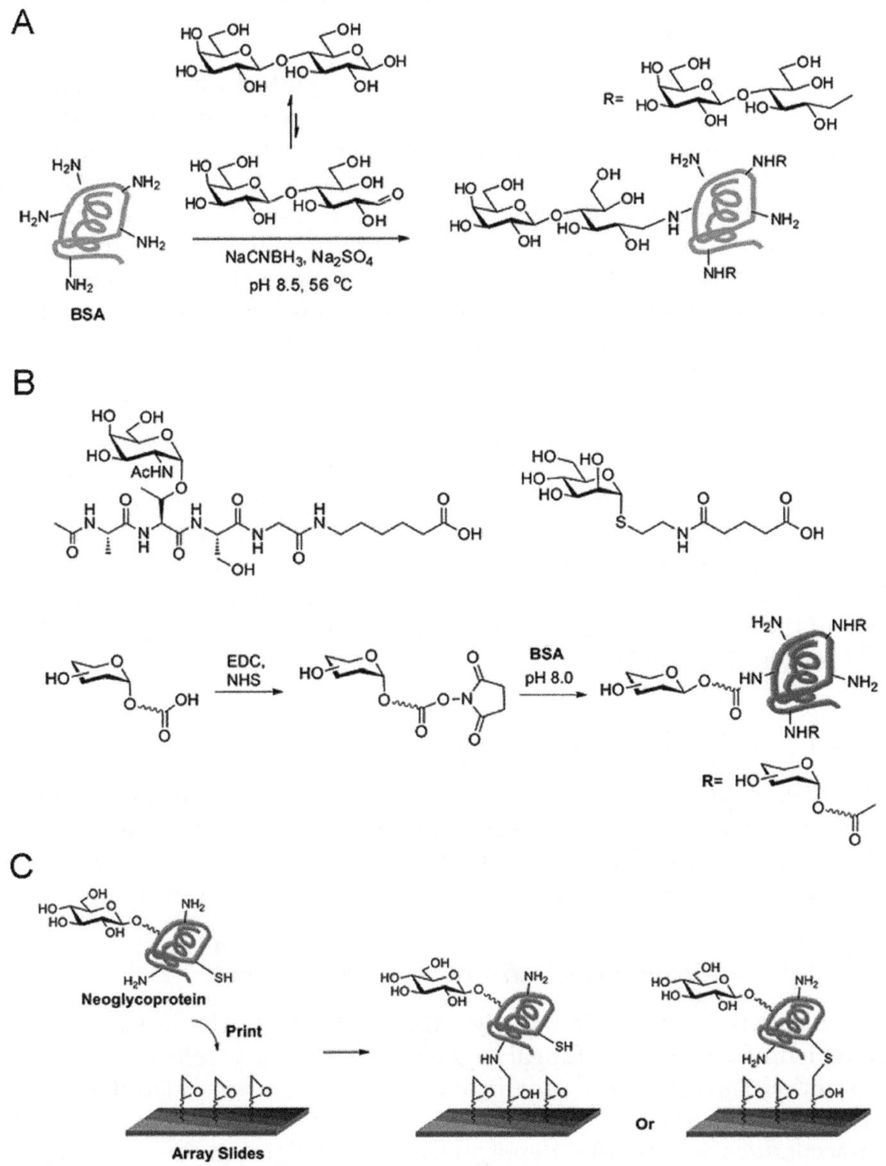

Figure 18.2 Different strategies for preparation of NGP probes and printing on epoxide functionalized slides. (A) Coupling *via* reductive amination. (B) Coupling *via* EDC activation: examples of sugar acids and reaction sequence. (C) Immobilization of NGP onto epoxide-coated slides. Reproduced from ref. 61 with permission from Springer Nature, Copyright 2012.

18.4 Applications of Neoglycoconjugate-based Microarrays

The glycan microarrays constructed have been probed with glycan-binding proteins (isolated from natural sources or recombinantly produced), including mammalian lectins and microbial lectins (viral hemagglutinin or capside proteins, bacterial and parasite adhesins), antibodies and carbohydrate-binding modules (CBMs). The readout of the binding assay is typically by fluorescence using for imaging a fluorescence microarray scanner. Different strategies can be applied for detection of protein binding, depending on the availability of specific antibodies to the protein or to a specific tag (for recombinant proteins), fluorescence or other labelling methods such as biotinylation.[59,66] Typical examples of fluorescence labels compatible with microarray scanners (*e.g.* λ_{Ex} and λ_{Em} between 450–650 nm) are Cyanine 3, Cyanine 5, FITC and AlexaFluor647. A fluorescence reagent commonly used for detection is the AlexaFluor647-labelled streptavidin that is used as a final readout for biotinylated proteins or detection antibodies. The microarrays can also be probed with biological fluids such as serum for detection of immune response. There has been success in the binding analysis using whole viruses[67,68] and virus-like particles (VLPs).[60] One of the best examined cases is the influenza virus binding to sialylated oligosaccharide probes.[23]

18.4.1 NGL-based Microarrays

The NGL-based microarray system has been applied to investigations of many biological and biomedical recognition systems (reviewed in ref. 69 and 70). Here we highlight some examples where the NGL-based microarrays were distinctive from other platforms for glycan-ligand discovery.

18.4.1.1 Endogenous Glycan-recognition Systems

Dectin-1 is a C-type lectin-like signalling receptor of the innate immune system directed against fungal β-glucans, with a key role in anti-fungal immunity. The oligosaccharide ligands for Dectin-1 were identified as linear β1–3-linked glucose oligomers using NGL-microarrays from glucan oligosaccharides derived after depolymerization of β-glucans that the receptor targets.[3] Following on from this study, and in order to assign the specificity of Dectin-1 interaction with glucans, a comprehensive series of gluco-oligosaccharides with different chain lengths, linkages and sequences were generated from partially depolymerized glucan polysaccharides. These gluco-oligosaccharides were converted into NGL probes and a 'glucome' microarray was generated (detailed below under the topic of microbial CBMs and Figure 18.5).[4] The microarray analyses revealed unique features for Dectin-1: restricted specificity (calcium-independent) toward fungal-type β1–3-linked gluco-oligosaccharides and a minimum chain length requirement of 11–12 glucose units, which is unusually long for a lectin (likely to be a helical epitope). No binding was detected to any of the other type of linkages, with either α- or β-configuration[4] nor to any of the mammalian-type probes included in the microarrays. A typical microarray result for Dectin-1 is shown in Figure 18.3A.

The NGL microarray also served to elucidate an endogenous ligand for malectin, a highly conserved protein in mammals. Structural studies of malectin had revealed a similarity in 3-dimensional structure to bacterial CBMs and unpredicted binding to glucose oligomers. Location of malectin in the endoplasmic reticulum (ER) led to the design of a NGL-microarray that included gluco-oligosaccharides and also glucosylated high-mannose *N*-glycans of the type that occur in the ER. These *N*-glycans are intermediates in the early steps of protein *N*-glycosylation and are targets for lectins involved in glycoprotein quality control. Microarray analyses corroborated the glucan-binding property of malectin, and additionally revealed a unique selectivity of malectin binding to a di-glucosyl high-mannose *N*-glycan (Figure 18.3B). Malectin was the first protein known to selectively bind this ER-glycan intermediate, which pointed to its function in the early steps *N*-glycosylation pathway.[71]

18.4.1.2 Virus–Host Interactions

Various glycan microarrays have been used for determination of receptor-binding specificities of known and emerging influenza viruses.[73,74] As an example, the NGL-based microarray was used in detection of binding by the pandemic influenza A (H1N1) 2009 virus (Figure 18.4). The results indicated binding to α2–3-sialyl in addition to α2–6-sialyl sequences, a distinctive receptor-binding feature compared to that of seasonal human H1N1 influenza virus.[68] This difference was exaggerated in hemagglutinin D222G mutants of the 2009 pandemic virus isolated from fatal cases,[75] thus providing a link between the ability of the virus to bind to α2–3-sialyl sequences and severity of disease.

NGL-based microarrays have been applied, in conjunction with structural studies, to assign glycan ligands for the adhesive proteins of polyoma- and reoviruses.[76–79] Highly restricted binding of human JC virus VP1 was observed to the α2–6-sialyl tetrasaccharide sequence, LSTc,[76] contrasting with the binding of B-Lymphotropic Polyomavirus (LPyV) to short α2–3-sialyl *N*-acetyllactosamine or lactose.[79] These findings led to structural analysis of the protein-ligand complexes using X-ray crystallography that revealed different modes of sialyl glycan recognition. In a recent study on human BK polyomavirus, a structure-guided mutation of amino acid at position 68 from lysine to serine, as in the simian SV40 virus, enabled the switch of the specificity of BKPyV from b-series gangliosides (*e.g.* GD3 and GD1b) to GM1.[78] NGL microarray analyses revealed that the mutant is specific for the ‘human’

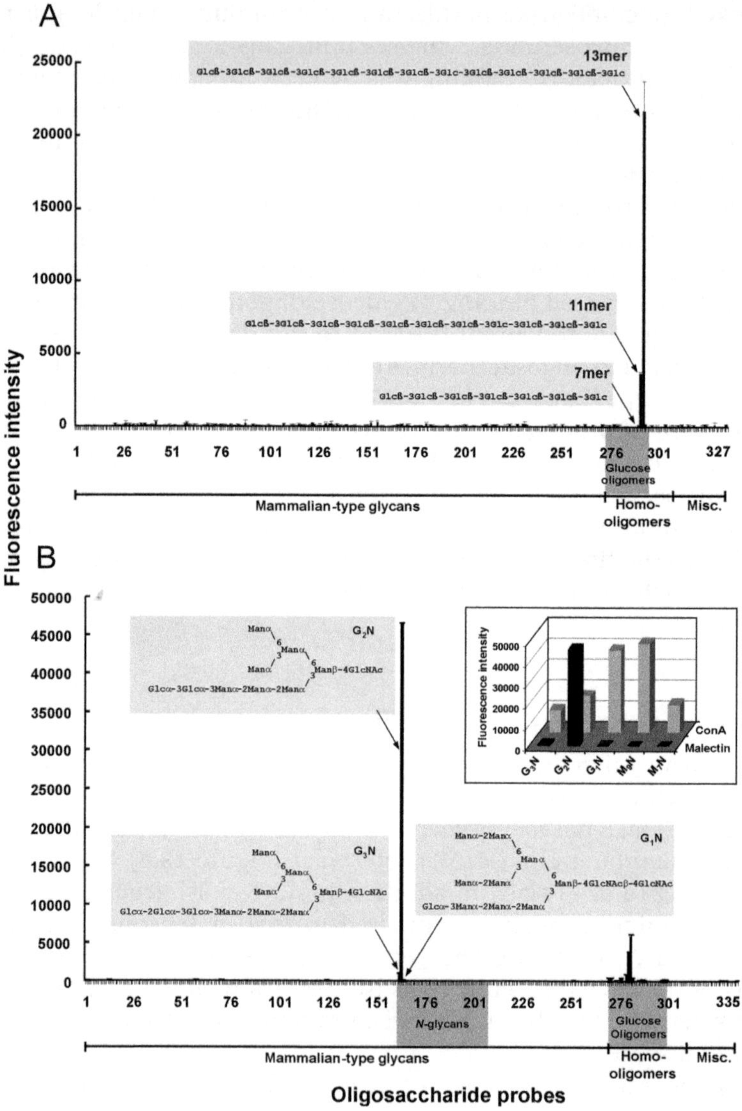

Figure 18.3 Microarray analyses of Dectin-1 and Malectin. Oligosaccharide probes (lipid-linked) were printed in duplicate on nitrocellulose-coated glass slides at 2 and 7 fmol/spot. Binding signals at 7 fmol/spot are shown. (A) Analysis of the binding of murine Dectin-1 tested as an IgG-Fc chimera. Binding was detected exclusively to β1–3-linked glucose sequences (11 and 13mers) derived from a glucan polysaccharide, in accordance with earlier data.[3] (B) Analysis of the binding of hexa-His-tagged *Xenopus laevis* malectin. High selectivity of malectin binding to the Glc2-*N*-glycan was revealed. Inset: expanded region of the microarray highlighting the selective binding of malectin to the Glc2-*N*-glycan probe in contrast with the broad binding profile of Concanavalin A. Abbreviations for the probes in the inset: G3N, Glc3Man7(D1)GlcNAc*;

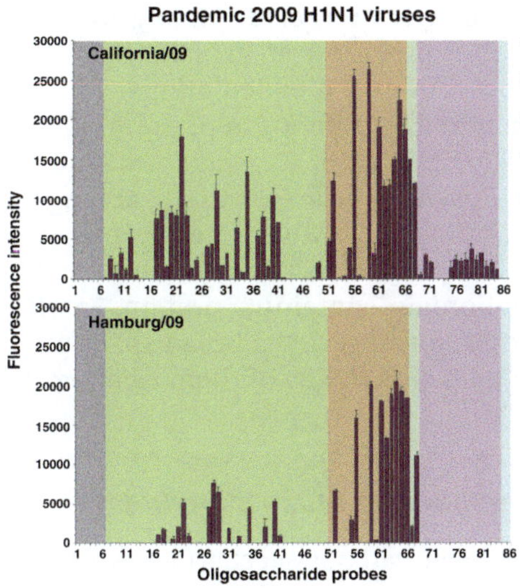

Figure 18.4 Analysis of seasonal and pandemic H1N1 influenza viruses. Reproduced from ref. 69 with permission from Elsevier, Copyright © 2014 The Authors.

sialic acid Neu5Ac, in contrast with SV40 VP1, which has a preference for Neu5Gc form of GM1,[60] characteristic of simian species and many other non-human mammals. The specificity of the mutant thus sheds light on the influence of sialic acid on species tropism.

18.4.1.3 Parasite Adhesins

NGL microarray has played an important role in studies of adhesive molecules, known as micronemal proteins (MICs), of *Toxoplasma gondii* and other apicomplexan parasites.[80] The results shed light on the way that the apicomplexan parasites use their MICs to decipher and select glycans of the host cells that they invade, with implications for host/tissue tropisms and pathobiology of infections. The information obtained so far is important for the rational design of therapeutic molecules. Studies began with the structural identification of a novel fold, microneme adhesive repeat (MAR), on the MIC1 of *T. gondii* (*Tg*MIC1). Ligands for *Tg*MIC1 MAR domains

G2N, Glc2Man7(D1)GlcNAc; G1N, Glc1Man9GlcNAc2; M9N, Man9GlcNAc2; and M7N, Man7(D1)GlcNAc2. These five probes were arrayed at positions 162–165 and at position 168, respectively.*Glc3Man7(D1) GlcNAc, Glc2Man7(D1)GlcNAc and Man7(D1)GlcNAc2 refer to Man7-related *N*-glycans that contain the outer most mannose residue in the Manα1–2Manα1–2Manα1–3Man arm, namely the D1 mannose residue. Reproduced from ref. 72 with permission from WALTER DE GRUYTER GMBH & CO. KG, Copyright 2009.

remained unknown until microarray analyses revealed binding to a wide range of sialyl glycan sequences,[80–83] thus providing an explanation for the ability of this parasite to infect almost all warm blooded animals, including humans. Among the proteins analyzed, a differing repertoire of sialyl glycan ligands have been identified for MIC13 of *T. gondii* and MIC1 of *Neospora caninum*, also for MIC3 of *Eimeria tenella* (EtMIC3). Among the interesting findings, is the lack of binding by EtMIC3 to *N*-glycolyl sialyl (Neu5Gc) sequences, suggesting an adaptation to the chicken host that lacks Neu5Gc.[82] A study of the 'apple'-domain-containing protein, TgMIC4, which is closely associated with *Tg*MIC1, has revealed a broad spectrum of galactose-terminating oligosaccharide ligands (many of them asialo-forms of the TgMIC1 ligands).[83]

18.4.1.4 Microbial Carbohydrate-binding Modules (CBMs)

The biodegradation of plant and fungal-cell wall polysaccharides is a crucial process for life sustainability. Cellulolytic microorganisms are highly efficient in this process by assembling modular architectures of carbohydrate-active enzymes with appended non-catalytic carbohydrate-binding modules (CBMs). The structural proximity promotes the CBM function in enhancing the enzyme activity. CBMs sequences range from 30 to 200 amino acids and are classified and divided into families based on sequence similarity annotated in the CAZY database (http://www.cazy.org). Given its wide variety of ligand specificities, CBMs are an excellent model to study protein-carbohydrate recognition mechanisms.

For screening of glucan-binding properties of CBMs, a sequence-defined 'glucome' NGL microarray has been developed.[4] The glucome microarray comprises 153 probes from different gluco-oligosaccharides, prepared by partial depolymerization of glucan polysaccharides isolated from plants, fungi and bacteria and multiple chromatographic methods, or are synthesized chemically. The glucan probes contain different chain lengths (from DP2 up to DP-13 or DP-16) and diverse sequences: linear sequences with 'homo' linkages of 1–2-, 1–3-, 1–4- or 1–6- with α- or β-configurations; and 'hetero' linkages of 1–3-, 1–4- or 1–6-; also branched oligosaccharide sequences with 1–3- and 1–6-linkages. The sequence and linkage were pre-determined by high-sensitivity negative-ion ESI-CID-MS/MS.[4] The combined use of microarray binding analysis with MS sequencing enabled the high sensitivity detection and unambiguous assignment of specificity of glucan recognition by CBMs.

Examples of the binding patterns obtained for bacterial CBMs are given in Figure 18.5, highlighting different specificities for glucan chain length and sequence requirements. Importantly, the chain length requirements observed in the microarray analysis of these CBMs were correlated with the different modes of β1–3-glucan interaction in solution using STD-NMR and with their assigned functional types as follows: type B *Tm*CBM4-2,[84] for which the binding site is a cleft that accommodates four residues, and interacts most strongly with internal residues of β1–3-linked gluco-oligosaccharides

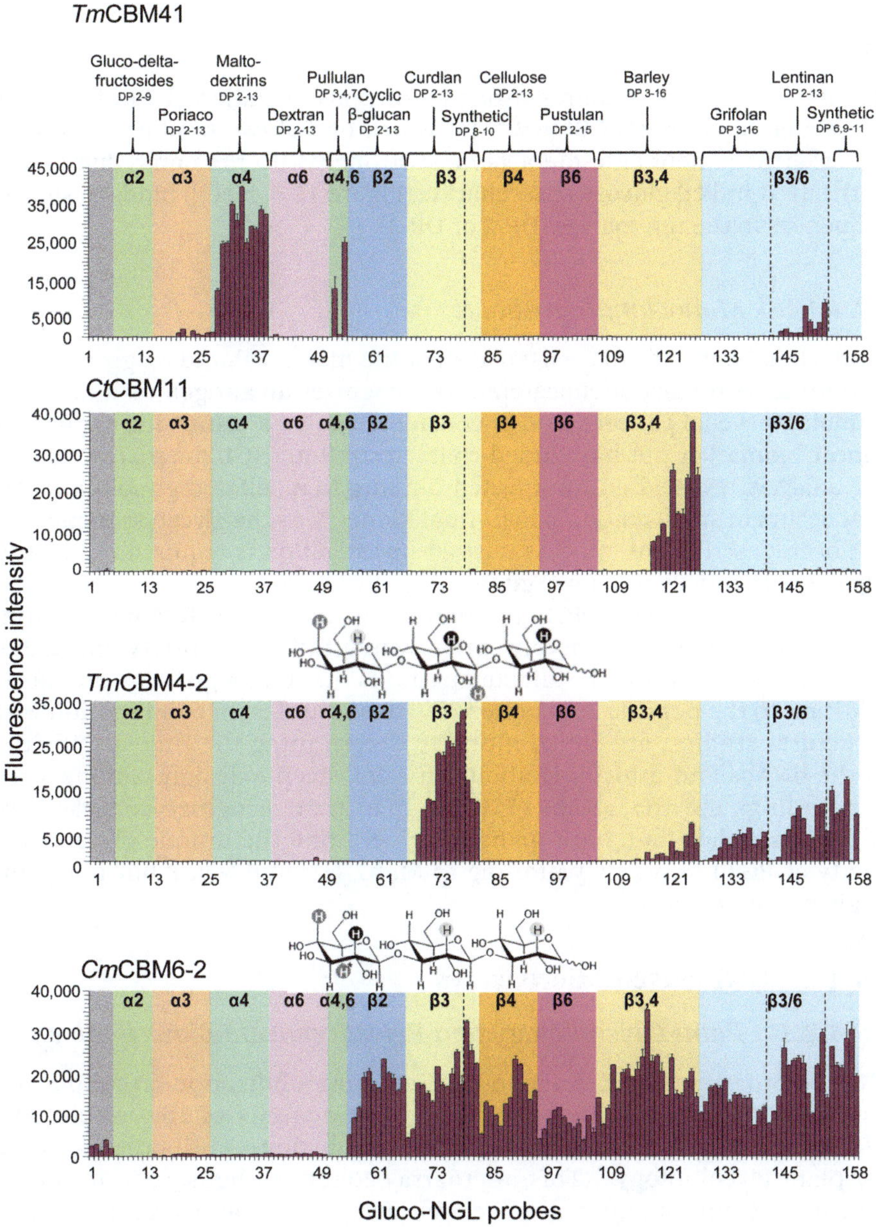

Figure 18.5 Differing specificities and chain length requirements obtained in the microarray analysis of glucan-binding CBMs from different CAZy families: family 41 and family 4 CBMs from the marine hyperthermophile *Thermotoga maritima* (*Tm*CBM41 and *Tm*CBM4-2, respectively); family 11 CBM from *Clostridium thermocellum* (*Ct*CBM11); and family 6 from the aerobic soil bacterium *Cladanthus mixtus* (*Cm*CBM6-2); the inset shows the binding epitopes as determined by STD NMR of β1–3-linked D-glucose trisaccharide in the presence of *Tm*CBM4-2 and *Cm*CBM6-2 (dark, medium and light grey circles indicate strong, medium, and weak STD effects, respectively). Depiction of the sequence diversity in the microarray is on the top of the panel. DP: degree of polymerization. Reproduced from ref. 27 with permission from The Royal Society of Chemistry.

(Figure 18.5, inset), requires a degree of polymerization (DP) of four for detectable binding; *Cm*CBM6-2,[85] for which one of two binding sites has a type C arrangement and interacts most strongly with the non-reducing end of the β1–3-linked glucose trisaccharide (Figure 18.5, inset), binds to shorter sequences in the microarray (DP-2 or DP-3).

18.4.1.5 Monoclonal Antibodies

Monoclonal antibody AE3 raised against the major sialomucin glycoprotein of murine mammary adenocarcinoma recognizes an antigen on a variety of human epithelial cancers, and is considered to be a promising serological cancer biomarker but has eluded characterization. NGL microarray screening analysis revealed an unexpected binding to a sulfated glycolipid SM1a that occurs in small amounts in normal kidney.[86] As the glycan sequence has not been described on mucins, further investigation is required to identifying the authentic ligand on *O*-glycans.

A focused *N*-glycan NGL microarray is among different platforms being used in studies of potent, broadly neutralizing anti-HIV antibodies (bNAbs), cloned from HIV-infected patients that target gp120 *N*-glycans in addition to the peptide backbone.[87–89] These studies, in combination with structural studies, are contributing to determining the fine specificities of the bNAbs and informing about rational vaccine design. Among striking findings are the bNAbs (PGT121-like) that recognize complex-type *N*-glycans rather than high-mannose type,[88] and the unique glycan specificity of PGT135, namely binding to Man7(D1) and Man8 but not Man9 *N*-glycan sequences.[89]

18.4.2 NGP-based Microarrays

18.4.2.1 Plant Glycobiology and Polysaccharide Biotechnology

NGP microarray based on non-covalent immobilization principle was established primarily for high-throughput analysis of the specificities of CBMs and monoclonal anti-plant carbohydrate antibodies as tools for plant glycobiology.[19] This microarray contains oligosaccharides from glucans, xylans, mannans, galactans, xyloglucans and arabinans after enzymatic or chemical hydrolysis or generated by chemical synthesis. These NGPs are arrayed non-covalently together with plant polysaccharides onto nitrocellulose-coated glass slides. This microarray was used to probe unknown CBMs of *Ruminococcus flavefaciens* cellulosome, revealing six previously unidentified CBM families targeting β-glucans, β-mannans and pectic homogalacturonan.[90] This study was important to gain knowledge on the complexity of the *R. flavefaciens* cellulosome and its extended repertoire of CBMs for efficient plant cell-wall degradation in absence of CBMs that target cellulose. More recently, the sequence-diversity in these microarrays was expanded to contain linear, branched and phosphorylated α1–4-linked maltooligosaccharides.[91] These were applied to characterize the starch-binding domain CBM20 of *Aspergillus niger* for its use as a tool in high-throughput screening of starch structures during development and germination.

18.4.2.2 Cancer Diagnosis and Vaccine Development

The covalent NGP microarray printed on epoxide-functionalized glass slides has been applied for the development of cancer diagnostics, vaccines, and therapeutic agents. This research frequently requires evaluating large numbers of glycan probes and samples (*e.g.* patient serum) and/or conditions (*e.g.* series of dilutions). The microarray containing 54 NGP probes was initially validated with 24 lectins,[18] and demonstrated for investigation of antibody specificity towards the Tn antigen (GalNAcα1-Ser/Thr). The Tn antigen is a carbohydrate antigen expressed in most carcinomas, during embryogenesis, on pathogenic parasites, and on HIV. It has been evaluated extensively as a potential diagnostic marker. However, there have been some discrepancies in the literature regarding Tn expression and the use of anti-Tn antibodies and lectins for its detection. Three frequently used Tn-binding proteins (HBTn1, Bric111, and VVL-B4) were evaluated using the NGP microarray and some unexpected cross-reactivity was identified.[92] In addition to the Tn antigen, VVL-B4 also bound to GalNAcα1–6Gal and GalNAcα1–3Gal, while Bric111 bound additionally to blood group A, GalNAcα1–6Gal and GalNAcα1–3Gal. HBTn1 showed better selectivity, but still displayed moderate binding to blood group A epitope.

Advances in whole-cell cancer vaccines have been limited by an incomplete understanding of what constitutes a favorable immune response, what factors lead to such a response, and why responses vary among patients. Although whole-cell vaccines display a diverse assortment of glycans to the immune system, little is known about responses to carbohydrates and their contributions. Glycan microarray using NGP as glyco-antigens was used to profile IgG and IgM responses induced by GVAX Pancreas, a whole-cell cancer vaccine in late-stage clinical trials for pancreas cancer.[93] The microarray results revealed more diverse responses to carbohydrate antigens than previously reported for whole-cell cancer vaccines. Both IgG and IgM antibody responses were induced to a variety of tumor-associated carbohydrate antigens, blood group antigens, and other glycans (Figure 18.6A–C). Large responses to two non-human antigens, bovine fetuin and α-Gal, were also observed. Responses to these glycans were found to be derived from fetal bovine serum used in the growth medium for the cells. Responses to α-Gal antigens correlated inversely with survival and appear to compete with productive responses to the vaccine. Large IgG responses to α-Gal were also detected in patients treated with an H1299 lysate vaccine but not in patients

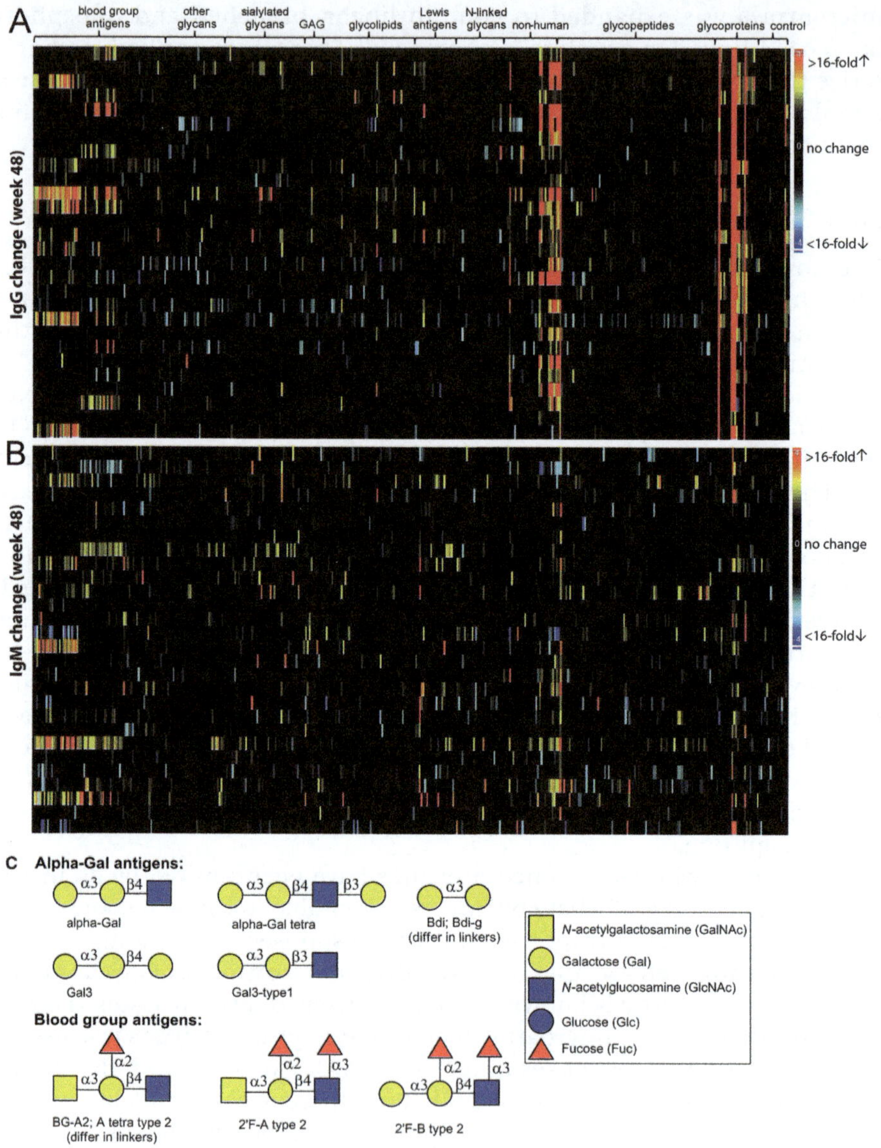

Figure 18.6 Glycan microarray analysis of antibody changes in serum of pancreatic cancer patients at week 48 after 3 GVAX Pancreas vaccine doses. (A) IgG changes. (B) IgM changes. Each row contains data from the same patient and each column contains data from the same antigen. Antigens are grouped by families. The magnitudes of antibody changes are shown as heatmaps denoted by different colors: black (no or low change), yellow-red (signal increase) and cyan-blue (signal decrease). (C) Structures of a few targeted antigens. GAG, glycosaminoglycan. Reproduced from ref. 93 with permission from Elsevier, Copyright 2016.

treated with a K562-GM vaccine. The former is produced with fetal bovine serum and the latter is produced in serum-free medium.[93] Therefore, the microarray results highlight the usefulness of glyco-antigen arrays for evaluating immune responses to vaccines and suggest that removal/reduction of FBS could potentially be a simple strategy to improve vaccine efficacy.

18.5 Conclusions

Glycan microarrays have become essential in glycan-recognition research in the post-genomic era. The selected examples of glycan microarray methods overviewed in this chapter highlight the importance of semi-synthetic neoglycoconjugate approaches to present in a multivalent format the structural diversity of glycans from different sources (mammalian, plant and microbial). The successful application of neoglycoconjugate microarrays has advanced our knowledge on glycan structures and their functions across different biological fields. Illustrated in this chapter are the roles of these microarrays in detection of glycan interactions in endogenous glycan-recognition systems, immune system and infection, virus–host and parasite–host interactions, polysaccharide recognition by bacterial systems, plant biology and in cancer diagnosis and vaccine development. The increase of sequence-defined glycan probe libraries in number and diversity, assembled from multiple sources and by different methods, is critical to increase the scope of these microarray approaches. Technological advances in mass spectrometry and other sequencing methods, chemical/enzymatic synthesis and isolation of selected glycan species from natural sources and bioinformatics, are expected to contribute to more comprehensive or specifically designed glycome-based microarrays for wider application in biology, biomedicine and biotechnology.

Acknowledgements

For financial support, the authors acknowledge Wellcome Trust Resource grant (108430/Z/15/Z), March of Dimes Prematurity Research Center grant, and the Portuguese Science and Technology Foundation (FCT-MEC) through grants PTDC/QUI-QUI/112537/2009, PTDC/BBB-BEP/0869/2014; Unidade de Ciências Biomoleculares Aplicadas-UCIBIO UID/Multi/04378/2013; POCI-01-0145-FEDER-007728.

References

1. K. Drickamer and M. E. Taylor, in *Recognition of Carbohydrates in Biological Systems, Part A: General Procedures*, Academic Press, 2003, vol. 362, pp. 560–567.
2. M. E. Taylor and K. Drickamer, *Curr. Opin. Struct. Biol.*, 2014, **28**, 14.
3. A. S. Palma, T. Feizi, Y. Zhang, M. S. Stoll, A. M. Lawson, E. Díaz-Rodríguez, M. A. Campanero-Rhodes, J. Costa, S. Gordon, G. D. Brown and W. Chai, *J. Biol. Chem.*, 2006, **281**, 5771.

4. A. S. Palma, Y. Liu, H. Zhang, Y. Zhang, B. V. McCleary, G. Yu, Q. Huang, L. S. Guidolin, A. E. Ciocchini, A. Torosantucci, D. Wang, A. Luísa Carvalho, C. M. G. A. Fontes, B. Mulloy, R. A. Childs, T. Feizi and W. Chai, *Mol. Cell. Proteomics*, 2015, **14**, 974.
5. M. Maccarana, B. Casu and U. Lindahl, *J. Biol. Chem.*, 1993, **268**, 23898.
6. W. Chai, J. G. Beeson and A. M. Lawson, *J. Biol. Chem.*, 2002, **277**, 22438.
7. R. Roy, *Curr. Opin. Struct. Biol.*, 1996, **6**, 692.
8. E. J. Toone, *Curr. Opin. Struct. Biol.*, 1994, **4**, 719.
9. Y. C. Lee and R. T. Lee, *Acc. Chem. Res.*, 1995, **28**, 321.
10. M. Mammen, S.-K. Choi and G. M. Whitesides, *Angew. Chem., Int. Ed.*, 1998, **37**, 2754.
11. R. T. Lee and Y. C. Lee, in *Neoglycoconjugates Preparation and Applications*, ed. Y. C. Lee and R. T. Lee, Academic Press, 1994, ch. 2, pp. 23–50.
12. W. B. Turnbull and J. F. Stoddart, *Rev. Mol. Biotechnol.*, 2002, **90**, 231.
13. R. Roy, T. C. Shiao and K. Rittenhouse-Olson, *Braz. J. Pharm. Sci.*, 2013, **49**, 85.
14. J. M. de la Fuente and S. Penadés, *Biochim. Biophys. Acta, Gen. Subj.*, 2006, **1760**, 636.
15. C. P. Stowell and Y. C. Lee, in *Advances in Carbohydrate Chemistry and Biochemistry*, ed. R. S. Tipson and D. Horton, Academic Press, 1980, vol. 37, pp. 225–281.
16. T. Feizi, M. S. Stoll, C.-T. Yuen, W. Chai and A. M. Lawson, *Methods Enzymol.*, 1994, **230**, 484.
17. S. Fukui, T. Feizi, C. Galustian, A. M. Lawson and W. Chai, *Nat. Biotechnol.*, 2002, **20**, 1011.
18. J. C. Manimala, T. A. Roach, Z. Li and J. C. Gildersleeve, *Angew. Chem., Int. Ed.*, 2006, **45**, 3607.
19. H. L. Pedersen, J. U. Fangel, B. McCleary, C. Ruzanski, M. G. Rydahl, M. C. Ralet, V. Farkas, L. Von Schantz, S. E. Marcus, M. C. F. Andersen, R. Field, M. Ohlin, J. P. Knox, M. H. Clausen and W. G. T. Willats, *J. Biol. Chem.*, 2012, **287**, 39429.
20. J. L. Magnani, D. F. Smith and V. Ginsburg, *Anal. Biochem.*, 1980, **109**, 399.
21. P. W. Tang, H. C. Gool, M. Hardy, Y. C. Lee and T. Felzi, *Biochem. Biophys. Res. Commun.*, 1985, **132**, 474.
22. W. Chai, M. S. Stoll, C. Galustian, A. M. Lawson and T. Feizi, *Methods Enzymol.*, 2003, **362**, 160.
23. O. Blixt, S. Head, T. Mondala, C. Scanlan, M. E. Huflejt, R. Alvarez, M. C. Bryan, F. Fazio, D. Calarese, J. Stevens, N. Razi, D. J. Stevens, J. J. Skehel, I. van Die, D. R. Burton, I. A. Wilson, R. Cummings, N. Bovin, C.-H. Wong and J. C. Paulson, *Proc. Natl. Acad. Sci. U. S. A.*, 2004, **101**, 17033.
24. V. Padler-Karavani, X. Song, H. Yu, N. Hurtado-Ziola, S. Huang, S. Muthana, H. A. Chokhawala, J. Cheng, A. Verhagen, M. A. Langereis, R. Kleene, M. Schachner, R. J. De Groot, Y. Lasanajak, H. Matsuda, R. Schwab, X. Chen, D. F. Smith, R. D. Cummings and A. Varki, *J. Biol. Chem.*, 2012, **287**, 22593.

25. Y. Liu, T. Feizi, M. A. Campanero-Rhodes, R. A. Childs, Y. Zhang, B. Mulloy, P. G. Evans, H. M. I. Osborn, D. Otto, P. R. Crocker and W. Chai, *Chem. Biol.*, 2007, **14**, 847.
26. A. Varki, R. D. Cummings, M. Aebi, N. H. Packer, P. H. Seeberger, J. D. Esko, P. Stanley, G. Hart, A. Darvill, T. Kinoshita, J. J. Prestegard, R. L. Schnaar, H. H. Freeze, J. D. Marth, C. R. Bertozzi, M. E. Etzler, M. Frank, J. F. G. Vliegenthart, T. Lütteke, S. Perez, E. Bolton, P. Rudd, J. Paulson, M. Kanehisa, P. Toukach, K. F. Aoki-Kinoshita, A. Dell, H. Narimatsu, W. York, N. Taniguchi and S. Kornfeld, *Glycobiology*, 2015, **25**, 1323.
27. D. O. Ribeiro, B. A. Pinheiro, A. L. Carvalho and A. S. Palma, in *Carbohydrate Chemistry: Chemical and Biological Approaches Volume 43*, The Royal Society of Chemistry, 2018, vol. 43, pp. 159–176.
28. D. Wang, S. Liu, B. J. Trummer, C. Deng and A. Wang, *Nat. Biotechnol.*, 2002, **20**, 275.
29. O. Kosík, R. P. Auburn, S. Russell, E. Stratilová, S. Garajová, M. Hrmova and V. Farkas, *Glycoconjugate J.*, 2010, **27**, 79.
30. W. G. T. Willats, S. E. Rasmussen, T. Kristensen, J. D. Mikkelsen and J. P. Knox, *Proteomics*, 2002, **2**, 1666.
31. K. S. Ko, F. A. Jaipuri and N. L. Pohl, *J. Am. Chem. Soc.*, 2005, **127**, 13162.
32. Y. Guo, H. Feinberg, E. Conroy, D. A. Mitchell, R. Alvarez, O. Blixt, M. E. Taylor, W. I. Weis and K. Drickamer, *Nat. Struct. Mol. Biol.*, 2004, **11**, 591.
33. E. L. Shipp and L. C. Hsieh-Wilson, *Chem. Biol.*, 2007, **14**, 195.
34. T. Feizi and W. Chai, *Nat. Rev. Mol. Cell Biol.*, 2004, **5**, 582.
35. P. H. Seeberger and D. B. Werz, *Nature*, 2007, **446**, 1046.
36. B. Xia, Z. S. Kawar, T. Ju, R. A. Alvarez, G. P. Sachdev and R. D. Cummings, *Nat. Methods*, 2005, **2**, 845.
37. H. H. Wandall, O. Blixt, M. A. Tarp, J. W. Pedersen, E. P. Bennett, U. Mandel, G. Ragupathi, P. O. Livingston, M. A. Hollingsworth, J. Taylor-Papadimitriou, J. Burchell and H. Clausen, *Cancer Res.*, 2010, **70**, 1306.
38. J. Gao, D. Liu and Z. Wang, *Anal. Chem.*, 2008, **80**, 8822.
39. X. Song, B. Xia, Y. Lasanajak, D. F. Smith and R. D. Cummings, *Glycoconjugate J.*, 2008, **25**, 15.
40. A. R. De Boer, C. H. Hokke, A. M. Deelder and M. Wuhrer, *Anal. Chem.*, 2007, **79**, 8107.
41. S. Park and I. Shin, *Angew. Chem., Int. Ed.*, 2002, **41**, 3180.
42. M. Shipp, R. Nadella, H. Gao, V. Farkas, H. Sigrist and A. Faik, *Glycoconjugate J.*, 2008, **25**, 49.
43. O. Srinivas, N. Mitra, A. Surolia and N. Jayaraman, *J. Am. Chem. Soc.*, 2002, **124**, 2124.
44. C. Galustian, R. A. Childs, M. Stoll, H. Ishida, M. Kiso and T. Feizi, *Immunology*, 2002, **105**, 350.
45. M. S. Stoll, E. F. Hounsell, A. M. Lawson, W. Chai and T. Feizi, *Eur. J. Biochem.*, 1990, **189**, 499.
46. W. Chai, G. C. Cashmore, M. S. Stoll, S. J. Gaskell, R. S. Orkiszewski and A. M. Lawson, *Biol. Mass Spectrom.*, 1991, **20**, 313.

47. T. Mizuochi, R. W. Loveless, A. M. Lawson, W. Chai, P. J. Lachmann, R. A. Childs, S. Thiel and T. Feizi, *J. Biol. Chem.*, 1989, **264**, 13834.
48. D. Solis, T. Feizi, C. T. Yuen, A. M. Lawson, R. A. Harrison and R. W. Loveless, *J. Biol. Chem.*, 1994, **269**, 11555.
49. C. T. Yuen, A. M. Lawson, W. Chai, M. Larkin, M. S. Stoll, A. C. Stuart, F. X. Sullivan, T. J. Ahern and T. Feizi, *Biochemistry*, 1992, **31**, 9126.
50. C. Gao, Y. Liu, H. Zhang, Y. Zhang, M. N. Fukuda, A. S. Palma, R. P. Kozak, R. A. Childs, M. Nonaka, Z. Li, D. L. Siegel, P. Hanfland, D. M. Peehl, W. Chai, M. I. Greene and T. Feizi, *J. Biol. Chem.*, 2014, **289**, 16462.
51. M. S. Stoll, T. Feizi, R. W. Loveless, W. Chai, A. M. Lawson and C. T. Yuen, *Eur. J. Biochem.*, 2000, **267**, 1795.
52. R. N. Iyer and D. M. Carlson, *Arch. Biochem. Biophys.*, 1971, **142**, 101.
53. W. Chai, M. S. Stoll, G. C. Cashmore and A. M. Lawson, *Carbohydr. Res.*, 1993, **239**, 107.
54. W. Chai, T. Feizi, C. T. Yuen and A. M. Lawson, *Glycobiology*, 1997, **7**, 861.
55. G. Yu, Y. Zhang, Z. Zhang, L. Song, P. Wang and W. Chai, *Anal. Chem.*, 2010, **82**, 9534.
56. Y. T. Li, C. W. Chou, S. C. Li, U. Kobayashi, Y. H. Ishibashi and M. Ito, *Glycoconjugate J.*, 2009, **26**, 929.
57. R. Ghidoni, S. Sonnino, M. Masserini, P. Orlando and G. Tettamanti, *J. Lipid Res.*, 1981, **22**, 1286.
58. W. Chai, J. Luo, C. K. Lim and A. M. Lawson, *Anal. Chem.*, 1998, **70**, 2060.
59. Y. Liu, R. A. Childs, A. S. Palma, M. A. Campanero-Rhodes, M. S. Stoll, W. Chai and T. Feizi, in *Carbohydrate Microarrays. Methods in Molecular Biology Methods and Protocols*, ed. Y. Chevolot, Humana Press, 2012, vol. 808, pp. 117–136.
60. M. A. Campanero-Rhodes, A. Smith, W. Chai, S. Sonnino, L. Mauri, R. A. Childs, Y. Zhang, H. Ewers, A. Helenius, A. Imberty and T. Feizi, *J. Virol.*, 2007, **81**, 12846.
61. Y. Zhang and J. C. Gildersleeve, in *Carbohydrate Microarrays. Methods in Molecular Biology Methods and Protocols*, ed. Y. Chevolot, Humana Press, 2012, vol. 808, pp. 155–165.
62. X. Wu, C.-C. Ling and D. R. Bundle, *Org. Lett.*, 2004, **6**, 4407.
63. J. Wang, S. M. Schiller and P. G. Schultz, *Angew. Chem., Int. Ed.*, 2007, **46**, 6849.
64. S. I. Van Kasteren, H. B. Kramer, H. H. Jensen, S. J. Campbell, J. Kirkpatrick, N. J. Oldham, D. C. Anthony and B. G. Davis, *Nature*, 2007, **446**, 1105.
65. A. A. Karelin, Y. E. Tsvetkov, L. Paulovičová, S. Bystrický, E. Paulovičová and N. E. Nifantiev, *Carbohydr. Res.*, 2009, **344**, 29.
66. Y. Liu, A. S. Palma, T. Feizi and W. Chai, in *Chemical Glycobiology: Monitoring Glycans and Their Interactions*, ed. B. Imperiali, Academic Press, 1st edn, 2018, vol. 598, pp. 139–167.
67. M. Crusat, J. Liu, A. S. Palma, R. A. Childs, Y. Liu, S. A. Wharton, Y. P. Lin, P. J. Coombs, S. R. Martin, M. Matrosovich, Z. Chen, D. J. Stevens, V. M. Hien, T. T. Thanh, L. N. T. Nhu, L. A. Nguyet, D. Q. Ha, H. R. van Doorn,

T. T. Hien, H. S. Conradt, M. Kiso, S. J. Gamblin, W. Chai, J. J. Skehel, A. J. Hay, J. Farrar, M. D. De Jong and T. Feizi, *Virology*, 2013, **447**, 326.

68. R. A. Childs, A. S. Palma, S. Wharton, T. Matrosovich, Y. Liu, W. Chai, M. A. Campanero-Rhodes, Y. Zhang, M. Eickmann, M. Kiso, A. Hay, M. Matrosovich and T. Feizi, *Nat. Biotechnol.*, 2009, **27**, 797.
69. A. S. Palma, T. Feizi, R. A. Childs, W. Chai and Y. Liu, *Curr. Opin. Chem. Biol.*, 2014, **18**, 87.
70. T. Feizi, *Ann. N. Y. Acad. Sci.*, 2013, **1292**, 33.
71. T. Schallus, C. Jaeckh, K. Fehér, A. S. Palma, Y. Liu, J. C. Simpson, M. Mackeen, G. Stier, T. J. Gibson, T. Feizi, T. Pieler and C. Muhle-Goll, *Mol. Biol. Cell*, 2008, **19**, 3404.
72. Y. Liu, A. S. Palma and T. Feizi, *Biol. Chem.*, 2009, **390**, 647.
73. J. Stevens, O. Blixt, T. M. Tumpey, J. K. Taubenberger, J. C. Paulson and I. A. Wilson, *Science*, 2006, **312**, 404.
74. C.-C. Wang, J.-R. Chen, Y.-C. Tseng, C.-H. Hsu, Y.-F. Hung, S.-W. Chen, C.-M. Chen, K.-H. Khoo, T.-J. Cheng, Y.-S. E. Cheng, J.-T. Jan, C.-Y. Wu, C. Ma and C.-H. Wong, *Proc. Natl. Acad. Sci. U. S. A.*, 2009, **106**, 18137.
75. Y. Liu, R. A. Childs, T. Matrosovich, S. Wharton, A. S. Palma, W. Chai, R. Daniels, V. Gregory, J. Uhlendorff, M. Kiso, H.-D. Klenk, A. Hay, T. Feizi and M. Matrosovich, *J. Virol.*, 2010, **84**, 12069.
76. U. Neu, M. S. Maginnis, A. S. Palma, L. J. Ströh, C. D. S. Nelson, T. Feizi, W. J. Atwood and T. Stehle, *Cell Host Microbe*, 2010, **8**, 309.
77. K. Reiss, J. E. Stencel, Y. Liu, B. S. Blaum, D. M. Reiter, T. Feizi, T. S. Dermody and T. Stehle, *PLoS Pathog.*, 2012, **8**, e1003078.
78. U. Neu, S.-a. A. Allen, B. S. Blaum, Y. Liu, M. Frank, A. S. Palma, L. J. Ströh, T. Feizi, T. Peters, W. J. Atwood and T. Stehle, *PLoS Pathog.*, 2013, **9**, e1003688.
79. U. Neu, Z. M. Khan, B. Schuch, A. S. Palma, Y. Liu, M. Pawlita, T. Feizi and T. Stehle, *PLoS Pathog.*, 2013, **9**, e1003714.
80. T. M. A. Blumenschein, N. Friedrich, R. A. Childs, S. Saouros, E. P. Carpenter, M. A. Campanero-Rhodes, P. Simpson, W. Chai, T. Koutroukides, M. J. Blackman, T. Feizi, D. Soldati-Favre and S. Matthews, *EMBO J.*, 2007, **26**, 2808.
81. N. Friedrich, J. M. Santos, Y. Liu, A. S. Palma, E. Leon, S. Saouros, M. Kiso, M. J. Blackman, S. Matthews, T. Feizi and D. Soldati-Favre, *J. Biol. Chem.*, 2010, **285**, 2064.
82. L. Lai, J. Bumstead, Y. Liu, J. Garnett, M. A. Campanero-Rhodes, D. P. Blake, A. S. Palma, W. Chai, D. J. P. Ferguson, P. Simpson, T. Feizi, F. M. Tomley and S. Matthews, *PLoS Pathog.*, 2011, **7**, e1002296.
83. J. Marchant, B. Cowper, Y. Liu, L. Lai, C. Pinzan, J. B. Marq, N. Friedrich, K. Sawmynaden, L. Liew, W. Chai, R. A. Childs, S. Saouros, P. Simpson, M. C. R. Barreira, T. Feizi, D. Soldati-Favre and S. Matthews, *J. Biol. Chem.*, 2012, **287**, 16720.
84. A. B. Boraston, D. Nurizzo, V. Notenboom, V. Ducros, D. R. Rose, D. G. Kilburn and G. J. Davies, *J. Mol. Biol.*, 2002, **319**, 1143.

85. J. L. Henshaw, D. N. Bolam, M. R. Pires, M. Czjzek, B. Henrissat, L. M. A. Ferreira, C. M. G. A. Fontes and H. J. Gilbert, *J. Biol. Chem.*, 2004, **279**, 21552.
86. A. S. Palma, Y. Liu, R. A. Childs, C. Herbert, D. Wang, W. Chai and T. Feizi, *Biochem. Biophys. Res. Commun.*, 2011, **408**, 548.
87. R. Pejchal, K. J. Doores, L. M. Walker, R. Khayat, P.-S. Huang, S.-K. Wang, R. L. Stanfield, J.-P. Julien, A. Ramos, M. Crispin, R. Depetris, U. Katpally, A. Marozsan, A. Cupo, S. Maloveste, Y. Liu, R. McBride, Y. Ito, R. W. Sanders, C. Ogohara, J. C. Paulson, T. Feizi, C. N. Scanlan, C.-H. Wong, J. P. Moore, W. C. Olson, A. B. Ward, P. Poignard, W. R. Schief, D. R. Burton and I. A. Wilson, *Science*, 2011, **334**, 1097.
88. F. Klein, C. Gaebler, H. Mouquet, D. N. Sather, C. Lehmann, J. F. Scheid, Z. Kraft, Y. Liu, J. Pietzsch, A. Hurley, P. Poignard, T. Feizi, L. Morris, B. D. Walker, G. Fätkenheuer, M. S. Seaman, L. Stamatatos and M. C. Nussenzweig, *J. Exp. Med.*, 2012, **209**, 1469.
89. L. Kong, J. H. Lee, K. J. Doores, C. D. Murin, J. P. Julien, R. McBride, Y. Liu, A. Marozsan, A. Cupo, P. J. Klasse, S. Hoffenberg, M. Caulfield, C. R. King, Y. Hua, K. M. Le, R. Khayat, M. C. Deller, T. Clayton, H. Tien, T. Feizi, R. W. Sanders, J. C. Paulson, J. P. Moore, R. L. Stanfield, D. R. Burton, A. B. Ward and I. A. Wilson, *Nat. Struct. Mol. Biol.*, 2013, **20**, 796.
90. I. Venditto, A. S. Luis, M. Rydahl, J. Schückel, V. O. Fernandes, S. Vidal-Melgosa, P. Bule, A. Goyal, V. M. R. Pires, C. G. Dourado, L. M. A. Ferreira, P. M. Coutinho, B. Henrissat, J. P. Knox, A. Baslé, S. Najmudin, H. J. Gilbert, W. G. T. Willats and C. M. G. A. Fontes, *Proc. Natl. Acad. Sci. U. S. A.*, 2016, **113**, 7136.
91. V. Tanackovic, M. G. Rydahl, H. L. Pedersen, M. S. Motawia, S. S. Shaik, M. D. Mikkelsen, S. L. Krunic, J. U. Fangel, W. G. T. Willats and A. Blennow, *Sci. Rep.*, 2016, **6**, 1.
92. J. C. Manimala, Z. Li, A. Jain, S. VedBrat and J. C. Gildersleeve, *ChemBioChem*, 2005, **6**, 2229.
93. L. Xia, D. S. Schrump and J. C. Gildersleeve, *Cell Chem. Biol.*, 2016, **23**, 1515.

CHAPTER 19

Current Stage of Commercial Glycan Microarrays

LEI LI*[a], WANYI GUAN[b], ZHIGANG WU[c] AND SHUAISHUAI WANG[a]

[a]Department of Chemistry, Georgia State University, Atlanta, GA 30303, USA; [b]College of Life Science, Hebei Normal University, Shijiazhuang, Hebei 050024, China; [c]College of Bioscience and Bioengineering, Hebei University of Science and Technology, Shijiazhuang, Hebei 050018, China
*E-mail: lli22@gsu.edu

19.1 Introduction

Along with nucleic acids, proteins and lipids, carbohydrates (or named glycans, saccharides) are one of the four major classes of macromolecules, which are present on the surface of literally every living organism. The other three classes often conjugate with glycans, *e.g.*, nucleic acids themselves contain monosaccharides ribose or deoxyribose, whereas proteins and lipids often require attached glycans for proper activity. Such glycans or glycoconjugates contain information that is essential for a wide variety of biological processes, including inflammation, immune responses, cell-trafficking and adhesion, receptor-mediated signaling, tumorigenesis, pathogenic infections, *etc.*[1] Glycans mediate these processes through glycan-binding proteins (GBPs) that recognize them as ligands.[2–5] For example, cell-surface glycans mediate cell trafficking,

Chemical Biology No. 11
Synthetic Glycomes
Edited by Wanyi Guan, Lei Li and Peng George Wang

Published by the Royal Society of Chemistry, www.rsc.org

adhesion and signaling by association with GBPs. A well characterized example is the selectin family adhesion, which recognizes glycan ligands and mediates critical interactions between blood cells and vascular cells in a variety of normal and pathological situations.[6] Furthermore, particular glycan structures on the surface of pathogens are recognized by various receptors of the immune system, which leads to immune responses to bacteria or viruses.[3] On the other hand, pathogen GBPs (*e.g.*, toxins, viral hemagglutinins) could recognize specific human cell surface glycan structures and initiate the invasion process.[4,7]

As a consequence, functional and specificity studies of GBPs and their related glycan ligands are critical to understanding underlying glycan–protein recognition events, and have the potential to provide deep insight into various biological processes, as well as designing diagnostic tools and even therapeutics. Traditional approaches to analyzing glycan–protein interactions, including surface plasmon resonance (SPR), isothermal calorimetry (ITC), enzyme-linked lectin assays (ELLAs), and ELISA assay, require large amount of glycans, and usually focus on interactions between a specific GBP and one glycan structure. The microarray technology, on the other hand, provides a high-throughput strategy to solve the problem, requiring only miniscule amounts of glycans.

The microarray technology takes advantage of the high precision of robotic arraying, which spots large numbers of materials (while using only miniscule amounts of each material) onto a solid support in a spatially defined arrangement, and the high-resolution imaging, which accurately transforms binding results into readable formats. As a natural extension of the successful development of DNA microarrays in 1995 and protein microarrays a few years later,[8,9] glycan microarrays were introduced in 2002 by several independent groups.[10–16] Since then, many research groups and organizations have developed their own glycan microarrays *via* various immobilization strategies, which have been well reviewed previously.[17–22] For example, the most famous Consortium for Functional Glycomics (CFG) mammalian glycan microarray v5.2 contains 609 structurally varied glycans with varied chemical linkers, and the microarray produced by the Gildersleeve group contains 371 bovine serum albumin (BSA) conjugated glycan structures.[21] These microarrays were used for many applications, including rapid analysis of glycan–protein interactions, GBPs specificity profiling, detection of pathogens and whole cells, determination of IC_{50}, enzyme substrate specificity profiling, biomarker discovery, profiling of immune response, *etc.*[18]

19.2 Production of Glycan Microarray and Binding Assay

The typical procedure for the production and binding assay of a glycan microarray is illustrated in Figure 19.1. Firstly, glycans with a linker and an active group (*e.g.* amine group) were diluted into an appropriate printing

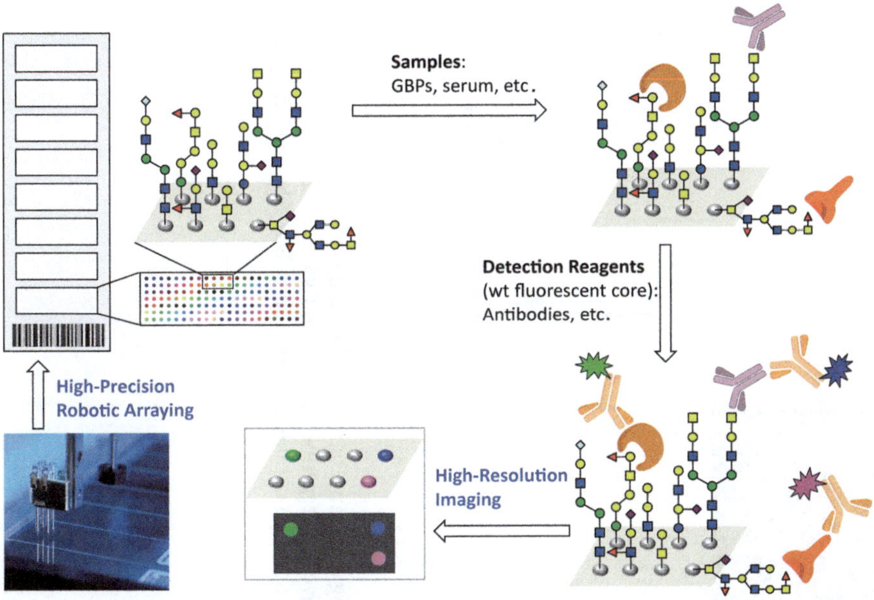

Figure 19.1 Production and binding assay of a glycan microarray. The microarray is produced by high-precision robotic arraying of numerous glycoforms onto a solid support (usually an activated glass slide). In the binding assay, the slide is first blocked to deactivate reactive functional groups on surface. After blocking, it is incubated with the sample of interest, and then the captured GBPs are detected with fluorophore-labeled secondary reagents. Binding is quantitated by a high-resolution fluorescent scanner.

buffer in appropriate concentrations (usually 100 μM), followed by high-precision robotic arraying (*e.g.*, a Scienion sciFLEXARRAYER S3 spotter) onto a pre-activated solid support (*e.g.* NHS-activated glass slides). The amine group on glycans and the NHS group on glass slides will then crosslink to form a permanent covalent bond. After dehumidification and blocking to cover un-reacted NHS groups, the microarray can be dried and stored desiccated at −20 °C until use. The binding assay is similar to typical ELISA experiments, by first applying target samples, followed with repeated washing, and then applying a secondary reagent that is conjugated with a fluorescent core. After washing, the array can then be visualized *via* high-resolution imaging (*e.g.*, a 10 μm GenePix microarray scanner), and finally transform into precise readouts for data analysis.[23]

An example of the microarray image and analyzed data is shown in Figure 19.2. In the example, we prepared 98 glycoforms, majorly *N*-glycans, with a few gangliosides. All of these glycans were derivatized to attach an amine group, and printed on NHS-activated glass slides, in replicates of six for each glycan. The specificity of a commonly used lectin, wheat germ agglutinin (WGA, 10 μg mL^{-1}) was then evaluated against the array, which was

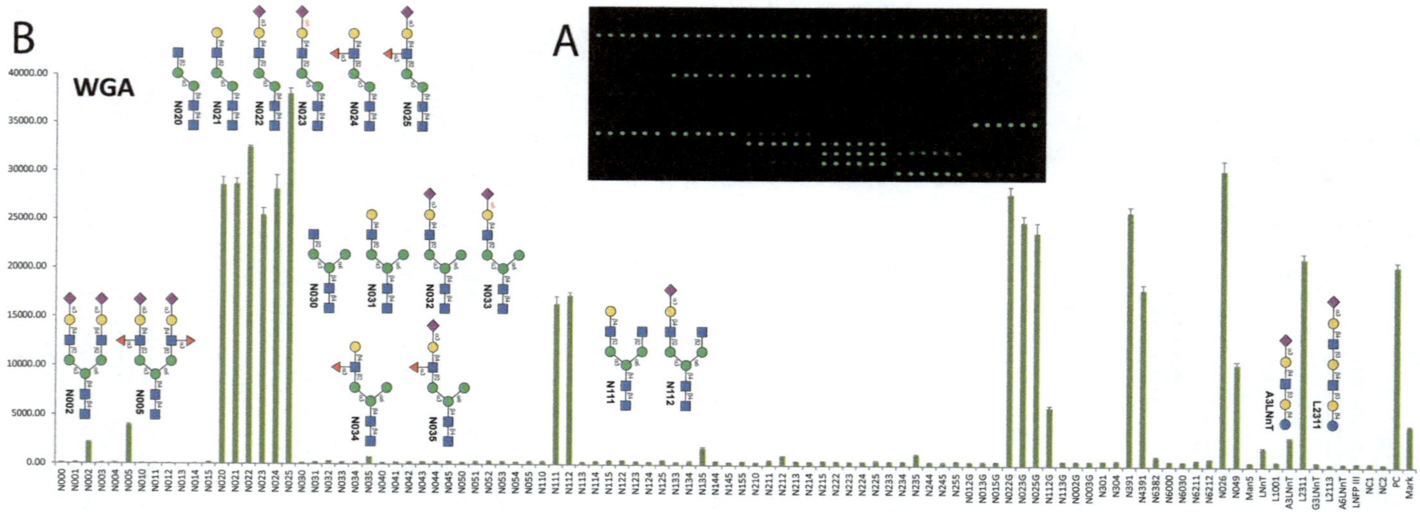

Figure 19.2 Specificity assay of wheat germ agglutinin (WGA, 10 μg mL^{-1}) using a glycan microarray. (A) Image of results; (B) analyzed results. The *x* axis shows the glycans, and the *y* axis shows the relative fluorescence, readout by Cy5-streptavidin. NC = printing buffer, PC1 = biotin-PEG2-amine, PC2 = rabbit IgG, M = human IgG–Cy3 conjugate and human IgG–Alexa647 conjugate.

visualized using a 10 μm-resolution microarray scanner (Figure 19.2A). Fluorescent spots on the image were then transformed into relative intensities to reflect binding profiles (Figure 19.2B). WGA was reported to bind *N*-acetylglucosamine (GlcNAc) and sialic acid (Sia) residues,[24] whereas surprisingly, the result showed that WGA strongly binds to various *N*-glycans and gangliosides, which may or may not contain terminal GlcNAc or Sia. Interestingly, WGA showed strong bindings to glycans N02x, but no binding to N03x that has an α1–6-Man for branching, which suggests that the branch may either result in local structural change or block binding directly. In addition, it was found that WGA showed distinct binding towards two linear glycans (A3LNnT and L2311) that both contain terminal Siaα2–3Galβ1–4GlcNAc trisaccharide but differ in overall glycan length, with over five times higher binding towards the longer glycan L2311 than that towards the short A3LNnT. Such fine details are invaluable information for deeper understanding of glycan–protein interactions.

19.3 Current Stage of Commercial Glycan Microarrays

The glycan microarray has become a major tool for studying glycan–protein interactions, especially specificity of GBPs. However, regardless of the tremendous efforts made by a number of research groups, there is an increasing gap between available glycan microarrays and fast-growing demands in various research areas and applications.[17,18] Furthermore, the majority of researchers are not able to access either various glycans or unaffordable high-precision robotic spotters to produce glycan microarrays, which makes such technologies only available to a small group of specialists. Commercial sources represent a practical alternative that may fill the gap. While numerous nucleic acid, peptide and protein microarrays are commercially available nowadays, there have been few glycan and glycoconjugate microarrays from commercial sources until recently. One major reason is the inaccessibility of large numbers of structurally well-defined glycoforms, which is due to inherited complexity, diversity and most frequently low abundance of such molecules. Only as recently as the past five years has rapid access to large numbers of glycoforms become possible, thanks to tremendous advances in chemoenzymatic synthesis.[25–30] As a consequence, commercial glycan microarrays and services have emerged (Table 19.1), even though the number of such products is still quite limited.

As listed in Table 19.1, a total of 15 glycan microarrays are commercially available to date, the majority of which focus on mammalian glycans. To better describe the arrays, a unique number (*e.g.*, GM01, GM as in Glycan Microarray) has been assigned to each product (Table 19.1). Even with very limited glycan structures, MoBiTec's GlycanArray I and II are among the earliest launched glycan microarray products. GlycanArray I (GM01) immobilized various types of glycans, with a total number of 28 (Figure 19.3A),

Table 19.1 A summary of currently commercially available glycan microarrays. All information is from online product instructions.

Entry	Glycan Microarray	Number of Glycans	Type of Glycans	Concentration of Glycans	Solid Support	Immobilization Approach
MoBiTec (https://www.mobitec.com/cms/products/bio/08_glycobio/analysis.html)						
GM01	GlycanArray I	28	*N*-glycans (7), *O*-glycans (2), GAG (5), Lewis series (4), Lac series (7), ABO series (3)	GAG (0.5, 0.1 mg mL^{-1}) *N*-glycans (0.2, 0.04 mM) Others (1.0, 0.2 mM)	Plastic slide	
GM02	GlycanArray II	24	Ganglioside (15), Globo-series (9)	0.25, 0.1 mM	Plastic slide	
Dextra Laboratories Ltd (https://www.dextrauk.com/products/glycans/product/924-glycan-array-kit)						
GM03	GAK001	104	*N*-glycans (8), GAG (21), blood group antigens (12), others (63)		Epoxy-coated glass slide	Amine-epoxy
Semiotik-Array (http://semiotik-array.com/en/product/)						
GM04	Microarray	591	Mammalian glycans (372), microbial polysaccharides (219), peptide/proteins			

RayBiotech LLC (https://www.raybiotech.com/products/arrays/glycobiology-arrays/glycan-arrays/); Chemily Glycoscience (http://www.chemilyglycoscience.com/glycan-arrays/)						
GM05	Glycan array 100	100	Mammalian glycans (95), aminoglycoside (5)	0.1 mM	NHS-coated glass slide	Amine-NHS
GM06	Glycan array 300	300	Mammalian glycans (95), aminoglycoside (5), *N*-glycans (80), HMO (50), gangliosides (70)	0.1 mM	NHS-coated glass slide	Amine-NHS
Z Biotech LLC (http://www.zbiotech.com/products.html); Chemily Glycoscience (http://www.chemilyglycoscience.com/glycan-arrays/)						
GM07	100 Glycan array	100	Monosaccharides, mammalian glycans		Hydrazide-coated glass slide	Aldehyde-hydrazide
GM08	*N*-glycan array	100	*N*-glycans	0.1 mM	Hydrazide-coated glass slide	Aldehyde-hydrazide
GM09	HS glycan array	24	Heparan sulfates	0.1 mM	Hydrazide-coated glass slide	Aldehyde-hydrazide
GM10	GAG glycan array	34	Glycosaminoglycans	0.1 mM	Hydrazide-coated glass slide	Aldehyde-hydrazide
GM11	Mucin glycan array	12	*O*-GalNAc glycans, glycopeptides	0.1 mM	NHS-coated glass slide	Amine-NHS
GM12	Neu5Ac/Neu5Gc array	71	Sialylation *N*-glycans	0.1 mM	NHS-coated glass slide	Amine-NHS
GM13	HMO array	46	Human milk oligosaccharides	0.1 mM	NHS-coated glass slide	Amine-NHS
GM14	Glycosphingolipid array	58	Gangliosides, globo series	0.1 mM	NHS-coated glass slide	Amine-NHS
GM15	Bacterial glycan array	22	Bacterial polysaccharides, oligosacharides			

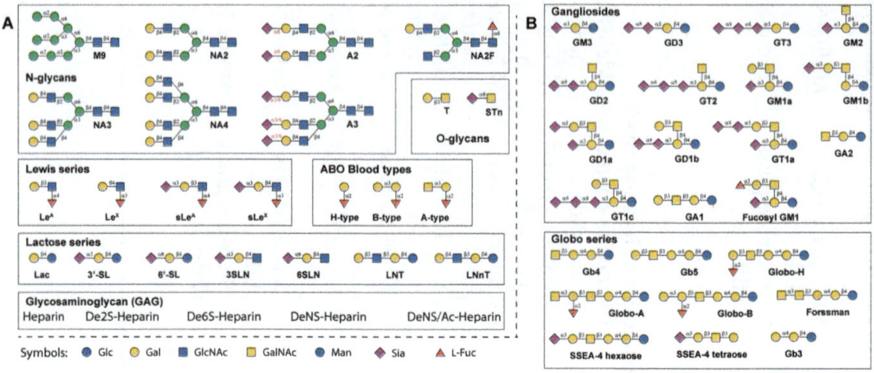

Figure 19.3 Glycan structures of MoBiTec's Glycan Microarray I (A) and II (B).

which is suitable for screening primary specificities of GBPs. On the other hand, GlycanArray II (GM02) focuses on glycosphingolipids (GSLs), with a total number of 24 (Figure 19.3B). GSLs are a subclass of glycolipids found on the cell membranes of organisms from bacteria to humans, and are the major glycolipids of animals. GSLs are closely involved in nervous system communication and bacterial pathogenesis.[31] Both arrays were validated and utilized by many researchers, yielding encouraging results.[32–36]

Over the past few years, several broad-spectrum glycan microarrays that contain various glycoforms were launched successively. For example, in 2016, Dextra Laboratories and the Institute for Glycomics at Griffith University partnered to launch a glycan microarray (GM03) that contains 104 glycoforms, including tumor antigens (8), oligomannose (6), *N*-glycans (8), glycosaminoglycans (GAG, 21), common oligosaccharide antigens (51), and other oligosaccharides. RayBiotech, on the other hand, in collaboration with Chemily Glycoscience (a start-up company that focuses on the production of glycans and glyco-related enzymes), launched two glycan microarrays. The glycan array 100 (GM05) contains oligosaccharide antigens, and the glycan array 300 (GM06) contains 200 extra complex glycoforms, including 80 *N*-glycans (isomers and extended ones), 50 human milk oligosaccharides (HMOs), and 70 gangliosides. In addition, Z Biotech released its own 100 glycan array (GM07). Detailed structural information is available in online product instructions (Table 19.1). All of these four glycan microarrays contain large numbers of terminal glycan epitopes (or glycan determinants)[37] that are normally found in mammalian glycomes, providing valuable tools for discovering GBPs and screening their primary specificity.

So far, the largest commercial glycan microarray was launched by SemiotikArray, a Moscow based biotechnology R&D Company. The array (GM04) contains 372 mammalian glycans, mostly monosaccharides to octosaccharides that cover a large portion of mammalian glycan determinants.[37] In addition, 219 bacterial polysaccharides were included, representing the first and largest commercial bacterial polysaccharide microarray. The majority of these bacterial polysaccharides are lipopolysaccharides from pathogenic bacteria, including 69 *Escherichia coli* O-serotypes, 23 *Salmonella enterica* O-serotypes, 18 *Pseudomonas aeruginosa* O-serotypes, 36 *Proteus* strains, and 41 *Shigella* strains. Such a mammalian/bacterial glycan microarray serves not only as a powerful probe for investigating GBPs, but also as valuable tools for diagnosing pathogen infections and understanding host–microbe interactions.[38]

According to the nature of the linkage to aglycone (protein or lipid), glycans are grouped into various classes, *e.g.*, *N*-glycans, *O*-glycans, proteoglycans (hyaluronan, heparan sulfate, *etc.*), glycosphingolipids, *etc.*[39] Different classes of glycans may be involved in different biological processes, and play distinct roles. As a result, focused glycan microarrays (*e.g.*, MoBiTec's GlycanArray II) that contain a relatively small number (but comprehensive structures) of a particular class of glycans are valuable for in-depth understanding of related biological processes. Z Biotech LLC (http://www.zbiotech.com/) and Chemily Glycoscience LLC (http://www.chemilyglycoscience.com/) now hold the most diversified glycan microarray product line as listed in Table 19.1. For example, the *N*-glycan array (GM08) solely presents *N*-glycans, including high-mannose type, complex type, and hybrid types. The arrays also feature glycan isomers and closely-related structures (Figure 19.4A),[25] and serve as a unique tool for mining fine specificities of GBPs. The HMO array (GM13), on the other hand, including 46 linear and branched glycans and derivatives identified in human milk (Figure 19.4B), is the sole HMO focused microarray on the market.[40] The Neu5Ac/Neu5Gc array (GM12) contains terminal *N*-acetylneuraminic acid (Neu5Ac) or *N*-glycolylneuraminic acid (Neu5Gc) residues that are attached to the same underlying substructures (Figure 19.4C), and the glycosphingolipid array (GM14) contains gangliosides and globo-series glycans with various sialic acid molecules and derivatives (*e.g.*, Neu5Ac, Neu5Gc, 8-methyl-Neu5Ac, deaminated Neu5Ac, *etc.*) (Figure 19.4D). Such focused arrays are ideal tools to identify specific GBPs, and probe detailed functions of various Sia molecules. For example, by using the Neu5Ac/Neu5Gc array (GM12), a mutated toxin, SubB2M, was demonstrated to be specific for Neu5Gc-containing glycans, but not Neu5Ac-containing ones.[41] The HS array (GM09) presents 24 heparan sulfate tetrasaccharides to decasaccharides (Figure 19.4E), and the GAG array (GM10) has 34 chemoenzymatically prepared glycosaminoglycans (Figure 19.4F). A comparable microarray with up to 52 such structures (only for microarray assay services) is also available in Glycan Therapeutics (https://www.glycantherapeutics.com/). Even though the current mucin glycan array (GM11) presents only 12 basic *O*-GalNAc glycans and glycopeptides (*e.g.*, T antigen, Tn antigen, and sTn antigen) (Figure 19.4G), a comprehensive array with complex mucin-type *O*-glycans will be co-launched soon by Z Biotech/Chemily Glycoscience (personal communication).

Other than these glycan microarray products, glycan microarray assay services are also provided by a few biotechnology startups. For example, Glycan Therapeutics offers glycan microarray analysis covering up to 52 structurally defined heparan sulfate oligosaccharides with a size range

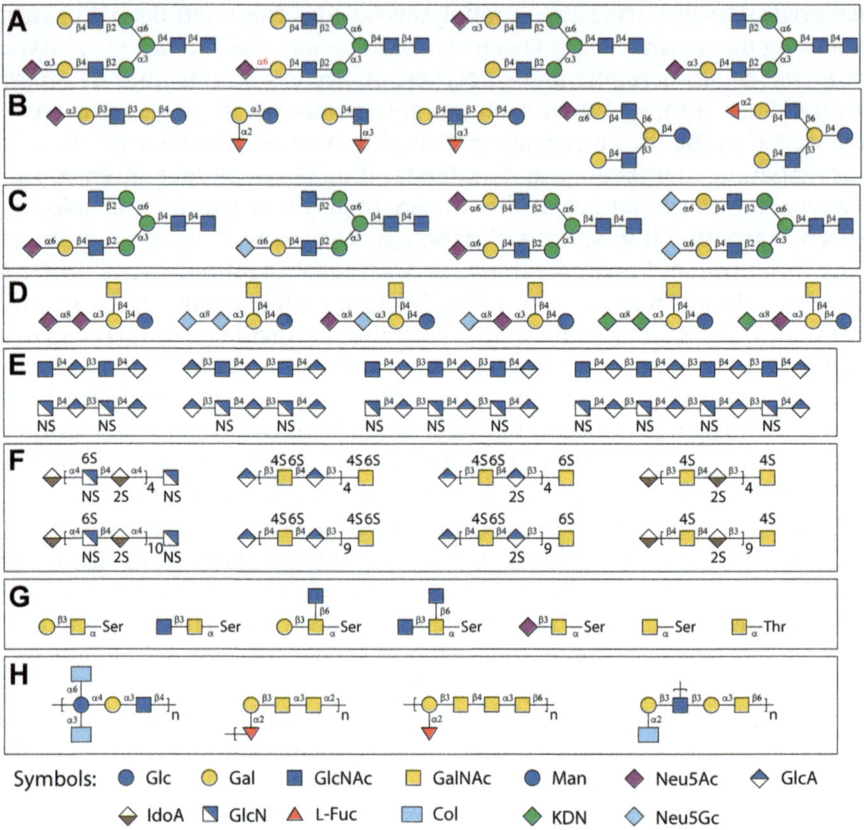

Figure 19.4 Signature glycan structures on commercial focused glycan microarrays, (A) *N*-glycan array; (B) HMO array; (C) Neu5Ac/Neu5Gc glycan array; (D) Glycosphingolipid array; (E) HS microarray; (F) GAG glycan array; (G) mucin glycan array; (H) bacterial glycan array.

from tetrasaccharides (4-mer) to dodecasaccharides (12-mer). And Creative Proteomics currently provides glycan-related microarray assays towards mammalian glycans, microbial glycans, and glycopeptides (structure unknown).

19.4 Prospective

Over the past 15 years, the glycan microarray has become a major tool to investigate glycan–protein interactions. It has been used to discover and evaluate hundreds of GBPs, and been successfully employed in many research and application areas.[18,20] Commercialization of such tools, on the other hand, is in a preliminary stage, and much less effort has been put into it compared with DNA and peptide/protein microarrays. Challenges in accessing large numbers of structurally well-defined complex glycoforms have been the major roadblock. Glycans in nature possess inherited complexity, diversity and low abundance (for most glycoforms), which is mainly due to the variable and multiple connectivity of glycan building blocks and the processes by which they are assembled in biosystems (not template-driven like nucleic acids and proteins). Such properties make isolating glycans from natural sources nearly impossible, and preparation of such molecules usually requires special expertise and tedious steps. Until recently, several carbohydrate chemistry groups developed various chemical or chemoenzymatic strategies to rapidly generate glycan libraries,[25–30,40,42,43] named synthetic glycomes in this book. In addition, recent advances in chemoenzymatic synthesis and polymer support development enabled solid phase generation of glycans, shedding light on machine-driven automatic carbohydrate synthesis (unpublished data and personal communications). On the other hand, facile strategies developed recently makes isolation of complex natural glycoforms from natural sources more practical. For example, the cellular "reporter/amplification" strategy enabled large scale production and isolation of *O*-glycans, in which a peracetylated benzyl-α-*N*-acetylgalactosamine was added to living cells, yielding a large variety of secreted *O*-glycan derivatives.[44] In another example, bleach was used as an efficient reagent to release glycans from kilograms of natural materials, enabling bulk isolation of various classes of natural glycans.[45] All these advances in synthetic glycomes and natural production isolation will greatly expand the glycan library accessible to academies and industries in the next decade, prompting the commercialization of glycan microarrays. Furthermore, researchers have put efforts into systematic analysis of glycan microarray data.[46–49] It is thus predictable that many more glycan microarray products and assay services will be commercially available in the next few years, accessible to people from various research areas, which will greatly facilitate the study of glycan–protein interactions and underlying biological processes.

References

1. A. Varki and P. Gagneux, in *Essentials of Glycobiology*, ed. A. Varki, R. D. Cummings, J. D. Esko, P. Stanley, G. W. Hart, M. Aebi, A. G. Darvill, T. Kinoshita, N. H. Packer, J. H. Prestegard, R. L. Schnaar and P. H. Seeberger, Cold Spring Harbor Laboratory Press, Cold Spring Harbor, NY, 3rd edn, 2015, pp. 77–88.
2. A. A. Kulkarni, A. A. Weiss and S. S. Iyer, *Med. Res. Rev.*, 2010, **30**, 327.
3. Y. van Kooyk and G. A. Rabinovich, *Nat. Immunol.*, 2008, **9**, 593.
4. A. Imberty and A. Varrot, *Curr. Opin. Struct. Biol.*, 2008, **18**, 567.
5. M. E. Taylor and K. Drickamer, *Curr. Opin. Cell Biol.*, 2007, **19**, 572.
6. R. D. Cummings and R. P. McEver, in *Essentials of Glycobiology*, ed. A. Varki, R. D. Cummings, J. D. Esko, P. Stanley, G. W. Hart, M. Aebi, A. G. Darvill, T. Kinoshita, N. H. Packer, J. H. Prestegard, R. L. Schnaar and P. H. Seeberger, Cold Spring Harbor Laboratory Press, Cold Spring Harbor, NY, 3rd edn, 2015, pp. 435–452.

7. R. Raman, K. Tharakaraman, V. Sasisekharan and R. Sasisekharan, *Curr. Opin. Struct. Biol.*, 2016, **40**, 153.
8. M. Schena, D. Shalon, R. W. Davis and P. O. Brown, *Science*, 1995, **270**, 467.
9. G. MacBeath and S. L. Schreiber, *Science*, 2000, **289**, 1760.
10. S. Fukui, T. Feizi, C. Galustian, A. M. Lawson and W. Chai, *Nat. Biotechnol.*, 2002, **20**, 1011.
11. D. Wang, S. Liu, B. J. Trummer, C. Deng and A. Wang, *Nat. Biotechnol.*, 2002, **20**, 275.
12. M. C. Bryan, O. Plettenburg, P. Sears, D. Rabuka, S. Wacowich-Sgarbi and C. H. Wong, *Chem. Biol.*, 2002, **9**, 713.
13. F. Fazio, M. C. Bryan, O. Blixt, J. C. Paulson and C. H. Wong, *J. Am. Chem. Soc.*, 2002, **124**, 14397.
14. B. T. Houseman and M. Mrksich, *Chem. Biol.*, 2002, **9**, 443.
15. S. Park and I. Shin, *Angew. Chem., Int. Ed.*, 2002, **41**, 3180.
16. W. G. Willats, S. E. Rasmussen, T. Kristensen, J. D. Mikkelsen and J. P. Knox, *Proteomics*, 2002, **2**, 1666.
17. C. D. Rillahan and J. C. Paulson, *Annu. Rev. Biochem.*, 2011, **80**, 797.
18. S. Park, J. C. Gildersleeve, O. Blixt and I. Shin, *Chem. Soc. Rev.*, 2013, **42**, 4310.
19. T. M. Puvirajesinghe and J. E. Turnbull, *Microarrays*, 2016, **5**, 3.
20. J. Y. Hyun, J. Pai and I. Shin, *Acc. Chem. Res.*, 2017, **50**, 1069.
21. L. Wang, R. D. Cummings, D. F. Smith, M. Huflejt, C. T. Campbell, J. C. Gildersleeve, J. Q. Gerlach, M. Kilcoyne, L. Joshi, S. Serna, N. C. Reichardt, N. Parera Pera, R. J. Pieters, W. Eng and L. K. Mahal, *Glycobiology*, 2014, **24**, 507.
22. V. Padler-Karavani, X. Song, H. Yu, N. Hurtado-Ziola, S. Huang, S. Muthana, H. A. Chokhawala, J. Cheng, A. Verhagen, M. A. Langereis, R. Kleene, M. Schachner, R. J. de Groot, Y. Lasanajak, H. Matsuda, R. Schwab, X. Chen, D. F. Smith, R. D. Cummings and A. Varki, *J. Biol. Chem.*, 2012, **287**, 22593.
23. J. Heimburg-Molinaro, X. Song, D. F. Smith and R. D. Cummings, *Curr. Protoc. Protein Sci.*, 2011, 12.10.1–12.10.29, ch. 12.
24. M. Monsigny, A. C. Roche, C. Sene, R. Maget-Dana and F. Delmotte, *Eur. J. Biochem.*, 1980, **104**, 147.
25. L. Li, Y. Liu, C. Ma, J. Qu, A. D. Calderon, B. Wu, N. Wei, X. Wang, Y. Guo, Z. Xiao, J. Song, G. Sugiarto, Y. Li, H. Yu, X. Chen and P. G. Wang, *Chem. Sci.*, 2015, **6**, 5652.
26. I. A. Gagarinov, T. Li, J. S. Torano, T. Caval, A. D. Srivastava, J. A. Kruijtzer, A. J. Heck and G. J. Boons, *J. Am. Chem. Soc.*, 2017, **139**, 1011.
27. Z. Wang, Z. S. Chinoy, S. G. Ambre, W. Peng, R. McBride, R. P. de Vries, J. Glushka, J. C. Paulson and G. J. Boons, *Science*, 2013, **341**, 379.
28. S. S. Shivatare, S. H. Chang, T. I. Tsai, S. Y. Tseng, V. S. Shivatare, Y. S. Lin, Y. Y. Cheng, C. T. Ren, C. C. Lee, S. Pawar, C. S. Tsai, H. W. Shih, Y. F. Zeng, C. H. Liang, P. D. Kwong, D. R. Burton, C. Y. Wu and C. H. Wong, *Nat. Chem.*, 2016, **8**, 338.

29. S. Wang, Q. Zhang, C. Chen, Y. Guo, M. R. Gadi, J. Yu, U. Westerlind, Y. Liu, X. Cao, P. G. Wang and L. Li, *Angew. Chem., Int. Ed.*, 2018, **57**, 9268.
30. H. Cao, C. Meng, A. Sasmal, Y. Zhang, T. Gao, C. C. Liu, N. Khan, A. Varki and F. Wang, *Angew. Chem., Int. Ed.*, 2018, **57**, 9003.
31. R. L. Schnaar and T. Kinoshita, in *Essentials of Glycobiology*, ed. A. Varki, R. D. Cummings, J. D. Esko, P. Stanley, G. W. Hart, M. Aebi, A. G. Darvill, T. Kinoshita, N. H. Packer, J. H. Prestegard, R. L. Schnaar and P. H. Seeberger, Cold Spring Harbor Laboratory Press, Cold Spring Harbor, NY, 3rd edn, 2015, pp. 125–135.
32. Y. Sugawara, M. Iwamori, T. Matsumura, M. Yutani, S. Amatsu and Y. Fujinaga, *FEBS J.*, 2015, **282**, 3334.
33. K. Kato, K. Yahata, B. Gopal Dhoubhadel, Y. Fujii and H. Tachibana, *Sci. Rep.*, 2015, **5**, 13901.
34. T. Kasai, T. Hamaguchi and M. Miyata, *J. Bacteriol.*, 2015, **197**, 2952.
35. M. Oda, A. Fujita, K. Okui, K. Miyamoto, M. Shibutani, T. Takagishi and M. Nagahama, *Biochem. Biophys. Res. Commun.*, 2013, **431**, 164.
36. M. Oda, M. Kabura, T. Takagishi, A. Suzue, K. Tominaga, S. Urano, M. Nagahama, K. Kobayashi, K. Furukawa, K. Furukawa and J. Sakurai, *J. Biol. Chem.*, 2012, **287**, 33070.
37. R. D. Cummings, *Mol. BioSyst.*, 2009, **5**, 1087.
38. S. R. Stowell, C. M. Arthur, R. McBride, O. Berger, N. Razi, J. Heimburg-Molinaro, L. C. Rodrigues, J. P. Gourdine, A. J. Noll, S. von Gunten, D. F. Smith, Y. A. Knirel, J. C. Paulson and R. D. Cummings, *Nat. Chem. Biol.*, 2014, **10**, 470.
39. A. Varki and S. Kornfeld, in *Essentials of Glycobiology*, ed. A. Varki, R. D. Cummings, J. D. Esko, P. Stanley, G. W. Hart, M. Aebi, A. G. Darvill, T. Kinoshita, N. H. Packer, J. H. Prestegard, R. L. Schnaar and P. H. Seeberger, Cold Spring Harbor Laboratory Press, Cold Spring Harbor, NY, 3rd edn, 2015, pp. 1–18.
40. Z. Xiao, Y. Guo, Y. Liu, L. Li, Q. Zhang, L. Wen, X. Wang, S. M. Kondengaden, Z. Wu, J. Zhou, X. Cao, X. Li, C. Ma and P. G. Wang, *J. Org. Chem.*, 2016, **81**, 5851.
41. J. Wang, L. K. Shewell, A. W. Paton, J. C. Paton, C. J. Day and M. P. Jennings, *Biochem. Biophys. Res. Commun.*, 2018, **500**, 765.
42. A. R. Prudden, L. Liu, C. J. Capicciotti, M. A. Wolfert, S. Wang, Z. Gao, L. Meng, K. W. Moremen and G. J. Boons, *Proc. Natl. Acad. Sci. U. S. A.*, 2017, **114**, 6954.
43. Y. Xu, S. Masuko, M. Takieddin, H. Xu, R. Liu, J. Jing, S. A. Mousa, R. J. Linhardt and J. Liu, *Science*, 2011, **334**, 498.
44. M. R. Kudelka, A. Antonopoulos, Y. Wang, D. M. Duong, X. Song, N. T. Seyfried, A. Dell, S. M. Haslam, R. D. Cummings and T. Ju, *Nat. Methods*, 2016, **13**, 81.
45. X. Song, H. Ju, Y. Lasanajak, M. R. Kudelka, D. F. Smith and R. D. Cummings, *Nat. Methods*, 2016, **13**, 528.

46. Z. Klamer, B. Staal, A. R. Prudden, L. Liu, D. F. Smith, G. J. Boons and B. Haab, *Anal. Chem.*, 2017, **89**, 12342.
47. D. Kletter, S. Singh, M. Bern and B. B. Haab, *Mol. Cell. Proteomics*, 2013, **12**, 1026.
48. P. Xuan, Y. Zhang, T. R. Tzeng, X. F. Wan and F. Luo, *Glycobiology*, 2012, **22**, 552.
49. N. Zhao, B. E. Martin, C. K. Yang, F. Luo and X. F. Wan, *Sci. Rep.*, 2015, **5**, 15778.

Subject Index